NATURAL SCIENCE IMAGING AND PHOTOGRAPHY

This book provides an in-depth exploration of scientific photography. Highlighting the best practices needed to make, distribute, and preserve scientific visual information using digital photographic methods and technologies, it offers solutions to some of the biggest challenges facing photographers.

Written by a team of international, award-winning image makers with over 300 years of cumulative experience, this comprehensive resource explains the foundations used, the tools required, and the steps needed for creating the optimal photograph in a range of environments and circumstances. Topics covered include:

- ethical practices
- aerial photography
- close-up and macro photography
- computational photography
- field photography
- geological photography
- imaging with invisible spectrums
- photographing small animals in captivity
- time-based imaging
- image processing in science

Showcasing modern methods, this book equips readers with the skills needed to capture and process the best image possible.

Designed for basic and intermediate photographers, *Natural Science Imaging and Photography* exists as an essential contemporary handbook.

Michael R. Peres is an award-winning photo-educator and biomedical photographer. A professor at Rochester Institute of Technology, Peres has been a leader in the field for more than 40 years. He has lectured and led workshops in more than 15 countries and authored more than ten books, including the *Focal Encyclopedia of Photography, Fourth Edition* and *Laboratory Imaging and Photography*.

"Michael R. Peres has produced yet another outstanding technical text, *Natural Science Imaging and Photography* which addresses the various techniques and methods for capturing images from science and nature. Along with seven expert contributors, Peres has created a go-to source for all image makers attempting to record high-quality images of both visual and invisible subjects. Starting with a historical overview of photography's origins, up through emerging digital technologies, this volume provides concise data which provides how to successfully navigate the imaging chain with its the various parameters, considerations, preparations, practical applications, and best practices.

Natural Science Imaging and Photography addresses a variety of imaging techniques and treatments, including microscopic, macroscopic, small animal, time based, and even aerial images captured using aircraft & drones. Nineteen chapters are offered filled with B&W and color photographs, illustrations, equations, and formulas to help simplify and explain the principles of imaging natural science subjects. Careful selection of imaging equipment, including cameras, lenses, filters, lab tools, etc. are presented in sufficient depth to guide one through the stages of image making. Plus digital imaging processing is presented in great detail, including extensive helpful step-by-step protocols. In addition to the technical imaging data, emphasis is stressed regarding creating scientific images while being careful to work safely, cause no harm, and obey professional standards and ethical guidelines.

This outstanding technical book which Peres and his co-authors have created, each contributing their respective specialty, is hallmarked by one overriding theme: *passion for their craft*. Whether in the controlled lab or out in a remote field location their passion is palpable as they all clearly approach their work with the goal of attaining perfection. That passion is clearly reflected in their willingness to share their knowledge via this text. Anyone interested in learning about detailed techniques on capturing natural science images, from student to seasoned professional alike, should add this outstanding volume to their personal library, and keep it handy for frequent reference. Natural Science Imaging and Photography should be required-reading for all natural science image makers as it is truly an absolute 'gold mine' of topical information!"

Robert C. Turner, RBP, MBA, FBPA, Director (retired), BioMedical Graphics Department, The Scripps Research Institute, La Jolla, California

"Professor Michael R. Peres' book *Natural Science Imaging and Photography* is a complete tome on the techniques and practices required for the photographic documentation of natural science. His decades of experience as the head of RIT's Biomedical Photography program prepared him well for this book. Peres invited some of the world's specialists in their fields to be contributing authors covering aerial photography, ethical practices, field photography, geological photography, imaging processing, and best practices. These authors together with Peres have written a wonderful and comprehensive manual for natural science applications.

Photographing subjects that are invisible is explored, as well as photomicrography, close up techniques, and photographing small animals in captivity. There are also chapters devoted to time-based imaging and ethics.

Everything from the foundations of photography, optics, light measurement, electromagnetic radiation, digital technology and lighting is included. For a scientist who needs to photograph their research or the photographer interested in science, you will find this book to be an all-encompassing resource for the techniques and practices needed in natural science as well as scientific photography. It is written using clear and everyday language, so you don't have to be a scientist to enjoy the book."

Constance J. Johansen, RBP, BFA, FBCA, Retired from the National Geospatial-Intelligence Agency

ABOUT THE SERIES
APPLICATIONS IN SCIENTIFIC PHOTOGRAPHY

Like all fields, there are many subspecialties within the broad field of scientific photography. While each seemingly specialized, there are areas within the individualized fields of science where photography is practiced in generalized ways as well. The **Applications in Scientific Photography** series was produced to become an industry resource, in homage to the seventeen books created for the Time Life Photography series in the 1970s, which outlines various aspects of photography for the working professional and student, including technology, methods, applications, tools, and important photographers.

The second book in the **Applications in Scientific Photography** series is, *Fundamentals in Forensic Photography: Practical Techniques for Evidence Documentation on Location and in the Laboratory*. This book offers practical techniques for the most common situations encountered in forensic documentation. The topics include proper equipment selection, effective settings and scientific strategies for forensic documentation. Emphasis is placed on techniques used in both laboratory and field settings to present the factual, true and accurate imagery that is vital in all aspects of forensic science and crime scene investigation. The text concludes with methods for archiving and retrieval of imagery, including scene, injury, autopsy, and evidence documentation; and HDR, high speed, and time-lapse imaging.

Each subsequent title in the series has been designed to share the best practices from each discipline using applied and practical strategies. Each book in the series will deconstruct and emphasize how images are considered scientific data and the end facts of an experiment or photographic documentation/survey.

Titles in the Series

Laboratory Imaging and Photography: Best Practices for Photomicrography & More
Michael R. Peres, Rochester Institute of Technology

Fundamentals of Forensic Photography: Practical Techniques for Evidence Documentation on Location and in the Laboratory
Keith Mancini and John Sidoriak

Digital Ultraviolet and Infrared Photography
Adrian Davies

Ophthalmic Imaging: Posterior Segment Imaging, Anterior Eye Photography, and Slit Lamp Biomicrography
Christye P. Sisson

Natural Science Imaging and Photography
Michael R. Peres, Rochester Institute of Technology

NATURAL SCIENCE IMAGING AND PHOTOGRAPHY

Edited by Michael R. Peres
with contributions by
Norman Barker, Adrian Davies,
James Hayden,
Charles Hedgcock,
Ted Kinsman, Staffan Larsson
and Keith Mancini

Routledge
Taylor & Francis Group
NEW YORK AND LONDON

First published 2021
by Routledge
52 Vanderbilt Avenue, New York, NY 10017

and by Routledge
2 Park Square, Milton Park, Abingdon, Oxon OX14 4RN

Routledge is an imprint of the Taylor & Francis Group, an informa business

Library of Congress Cataloging-in-Publication Data
A catalog record for this title has been requested

ISBN: 978-0-367-34742-0 (hbk)
ISBN: 978-0-367-34740-6 (pbk)
ISBN: 978-0-429-32761-2 (ebk)

Typeset in Photina MT and Conduit
by Newgen Publishing UK

This book is dedicated to the curious and passionate science photographers, who for more than 180 years, inspired and communicated the world of science using photographs in new and innovative ways.

Contents

Acknowledgements

I am deeply grateful to so many people who played a role in the creation of this important book. Writing a book like this, or any book for that matter, is no small task. It takes a supportive family and a village of friends and colleagues who provide unqualified support to you as you get lost in the process. For me, a loving and supportive family has been an important ingredient during nearly 40 years of professional adventures. I want to share a special debt of gratitude to my wife Laurie Peres, who has been a constant source of inspiration and encouragement since 1981.

My Publisher

Thank you to my publisher Routledge who granted me the opportunity to write this, my fourth Focal Press Book. This book's journey began with a phone call with then acquisitions editor Simon Jacobs in the fall of 2018. I am appreciative of his encouragement to create this volume built heavily upon content from other books in the Applications of Scientific Photography series. I also appreciate the support and assistance from Emma Tyce, Katherine Kadian, John Makowski, and Alyssa Turner in helping guide me through the publishing processes and my work along the way. I also wish to acknowledge the expertise and contributions of the copy-editors and designers at Routledge.

A special thank you goes to you Bob Turner and Connie Johansen for providing the book's first review. Your generosity is greatly appreciated.

Below is a list sharing the names of people and organizations who played a big role in the creation of this book. Some shared encouragement, some shared knowledge, and many shared writing, expertise, and photographs. There is not enough space, or words to adequately thank everyone. Please know your contributions big or small made a difference to me and will contribute to this book's success.

Contributing Authors

I feel a huge debt of gratitude to my contributing authors Norman Barker, Adrian Davies, James Hayden, Charles Hedgcock, Ted Kinsman, Staffan Larsson, and Keith Mancini. You contributed a lifetime of knowledge and images to this book. It is no secret publishing is in a dynamic state of change and your passion and sweat equity helped me make this extraordinary book possible. Thank you.

People and Organizations Who Encouraged My Work

Thank you Rochester Institute of Technology and the School of Photographic Arts and Sciences for the opportunity to come to campus every day for the last 34 years and try new things. My time at RIT has taught me more than I could have ever imagined. I would also be remiss not to thank the hundreds of students who have made me better every day. Your joy of learning has been a welcome source of daily inspiration. Thank you also to colleagues and institutions from near and far for the important opportunities to collaborate and practice my skills at different times in my life. A special thank you to Therese Mulligan, Ph.D., Frank Cost, Willie Osterman, and Christye Sisson, colleagues at RIT, for your friendship, sharing of your knowledge, and your encouragement for a very long time.

Artists, photographers and organizations who generously provided illustrations, reproductions, photographs, permissions, and subjects:

Adobe Inc., San Jose, California
Anna Atkins
John James Audubon
Audubon Society, USA
Norman Barker
The Biomedical Photographic Communications Program
Louis Boutan
Canon, USA
Professor Dr. Fabrizio Carbone, École Polytechnique Fédérale de Lausanne, Switzerland
Amon Carter Museum of American Art, Ft. Worth, Texas
Cognisys, Traverse City, Michigan
Library of Congress, Prints and Photographs Division, Washington, D.C.
Ashley Crichton
Carl Curman
Torgeir Dahlen
Adrian Davies
William Henry Fox Talbot
Laura Gilpin
Gretag/Macbeth
The Hagströmer Medico-Historical Library, Karolinska Institutet, Stockholm, Sweden
Adam Hartley
James Hayden
Charles Hedgcock
History of Medicine Division, National Library of Medicine, Washington, D.C.
Robert Hooke
Institutes of Health, Washington, D.C.
William Henry Jackson

Cecilia James
Keyence, Osaka, Japan
Ted Kinsman
Eastman Kodak Company, Rochester, New York
Caroli Von Linne
Officer Sean C. Lonergan, Westchester County Police, Westchester, New York
Joseph Mancini
Keith Mancini
Metropolitan Museum of Art Collections, New York
Torey Miller
Eadweard James Muybridge
National Geographic Magazine
National Park Service, Department of Interior, Washington, D.C.
Michael R. Peres
Nikon, USA
Profoto, USA
RIT Big Shot, School of Photographic Arts and Sciences, Rochester Institute of Technology, Rochester, New York
Nanette Salvaggio
Service Regional d'Identité Judiciaire, Préfecture de Police, Paris, France
Josh Shagam
ShutterStock Images
Spencer Collection, The New York Public Library, New York
John Syme
Swedish Museum of Natural History, Stockholm, Sweden
Leonardo da Vinci
The Wellcome Collection, London, UK
Westchester County Police, Westchester, New York
Westcott Inc., Maumee, Ohio
Steven H. Westin
Joshua White
White House Archive, Washington, D.C.
Yellowstone National Park Archive, Department of the Interior, Washington, D.C.
Brigham Young Archive, Provo, Utah
Carl Zeiss Microscopy, USA
Teri Zgoda

About the Authors

The Editor

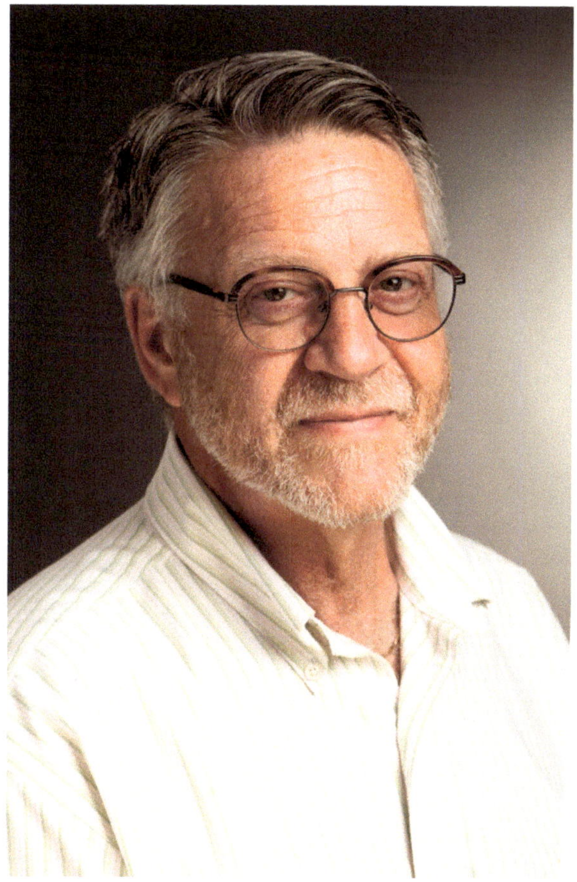

Michael R. Peres, Professor of Biomedical Photographic Communications, Program Director, Media Arts and Technology, MS; Rochester Institute of Technology, Rochester, New York. http://mrppph.cad.rit.edu

Michael R. Peres is an award-winning photo educator, author, artist and science photographer. Peres joined the teaching faculty of Rochester Institute of Technology's School of Photographic Arts & Sciences in 1986 where he is Professor of Biomedical Photographic Communications. He teaches courses in photomicrography, biomedical photography, and other applications of photography used in science. He served as the chair of the biomedical photographic communications department for nearly 25 years. He has lectured and led workshops world-wide including Europe, Australia, Africa, Asia, and North America.

Peres has enjoyed a varied photographic career that began in 1973. He has been publishing most of his career and in 2007, served as the editor-in-chief of the *Focal Encyclopedia of Photography*, Fourth Edition. In 2016, he wrote *Laboratory Imaging and Photography: Best Practices for Photomicrography and More* (Focal Press). Some of his professional activities include co-coordinating the RIT Images from Science project and RIT Big Shot.

Peres has received numerous awards including RIT's Eisenhart outstanding teaching award and the Schmidt medal from the BioCommunications Association for lifetime achievement in the field of bio-communications. He holds a master's degree in instructional technology and bachelor's degrees in biology and biomedical photographic communications. He is also a registered biological photographer.

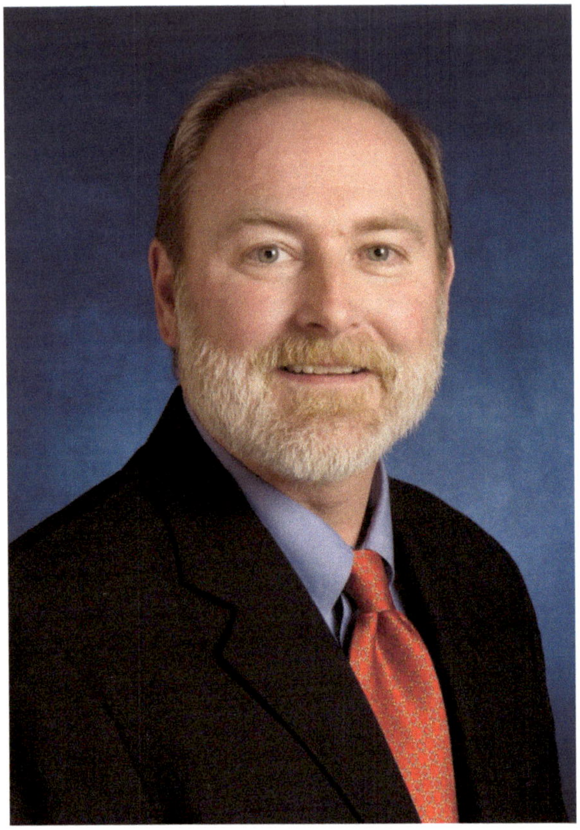

The Contributing Authors

Norman Barker, Professor of Pathology and Art as Applied to Medicine and Director, Pathology, Photography And Graphic Arts, The Johns Hopkins School of Medicine, Baltimore, Maryland. medicalart.johnshopkins.edu/norman-j-barker-ms-professor/

Norman Barker is Professor of Pathology and Art as Applied to Medicine at the Johns Hopkins University School of Medicine. He specializes in photomicrography, macro and natural science photography. Barker is director of the Pathology, Photography, and Graphic Arts Laboratory. He received his undergraduate degree from the Maryland Institute College of Art. He earned his M.S. in education from Johns Hopkins and an M.A. in publications design from the University of Baltimore. He joined the Johns Hopkins faculty in 1983.

He is an award-winning photographer, writer and designer. His work appears in textbooks, journals and museums worldwide. His photographs are in the permanent collections of more than 40 museums including The Smithsonian Museum of Natural History, The George Eastman Museum, The American Museum of Natural History, The Nelson-Atkins Museum, and The Science Museum in London. He has published seven books and numerous scientific articles. His latest book collaboration, *Hidden Beauty: Exploring The Aesthetics of Medical Science* shows the beauty of medicine and the human body and was on a museum tour from 2014 to 2020.

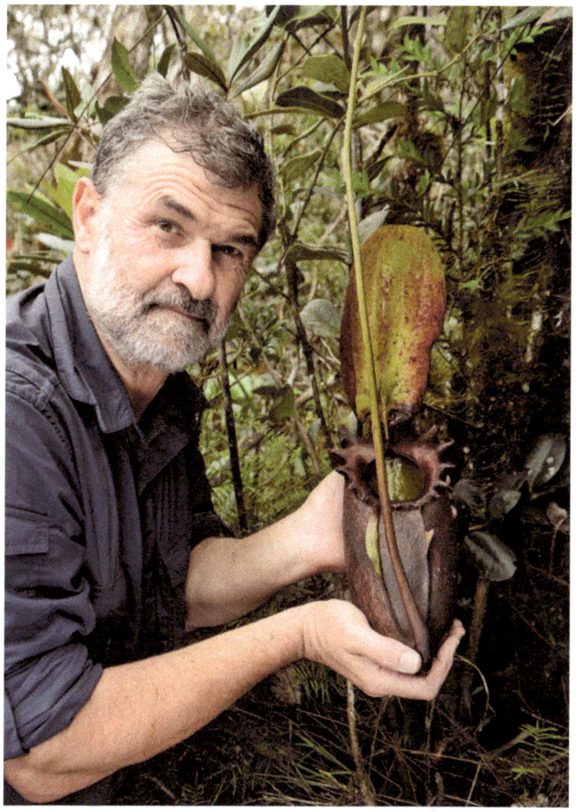

Adrian Davies, Natural Science Photographer and Author, Cambridge, United Kingdom. www.adriandaviesimaging.com; www.imagingtheinvisible.com

Adrian Davies is a wildlife and natural history photographer, author and highly experienced lecturer. He was one of the first people in the UK to use a digital camera over 20 years ago (the mighty Kodak DCS100) and has been an advocate of digital imaging ever since. He has run workshops for organizations such as Nikon, the British Institute of Professional Photographers, the Royal Photographic Society, the Field Studies Council, the Royal Horticultural Society and numerous others. He co-wrote, with Phil Fennessy, one of the first textbooks on digital imaging (*Electronic Imaging for Photographers*, Focal Press, 1994) a title which was printed in four editions. He has since written several other books and dozens of articles for various magazines.

Adrian has been photographing all aspects of wildlife and natural history for over 30 years, both in the UK and abroad. Increasingly, he is covering environmental issues such as the effects of global warming, and the effect of alien species in the British countryside. He is also currently looking at how plants and other natural history subjects appear in ultraviolet radiation and has published two books on the subject *Digital Ultraviolet and Infrared Photography* (Focal Press) and *Photographing the Unseen World* (Crowood Press) His pictures have appeared in hundreds of books and magazines.

Adrian is represented by two picture agencies, Nature Picture Library and Alamy. His images can also be purchased directly from him.

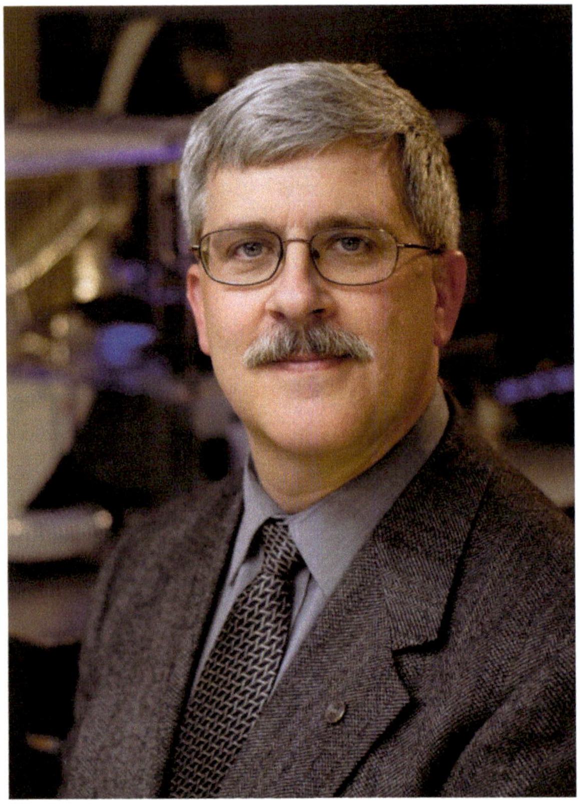

James Hayden, RBP, FBCA, Managing Director, Imaging Shared Resource, The Wistar Institute, Philadelphia, Pennsylvania

James Hayden is managing director of the Imaging Core facility at The Wistar Institute in Philadelphia, Pennsylvania. Hayden earned a BA in biology and biophotography and was employed by the University of Pennsylvania School of Veterinary Medicine in 1983. He also ran a freelance photography business called Bio-Graphics and served as an adjunct professor at the University of the Sciences in Philadelphia. He was hired by The Wistar Institute in 2002. Although specializing in photomicrography, he photographs everything from wildlife to large groups of people. Hayden's responsibilities include customized imaging, consultation on experimental design, advanced imaging training, and quantitative image analysis. Hayden has lectured extensively and led workshops in both domestic and international venues. He has won numerous awards, as well as appearing in books and on the covers of such journals as *Nature*, *Science*, *Proceedings of the National Academy of Sciences of the United States of America*, and *Cell*. Hayden's images have been featured in the Nikon International Small World photomicrography competition where he has also served as judge. His written publications range from technical papers to contributions in scientific research journals. He is a Registered Biological Photographer, Fellow, and Louis Schmidt Laureate of the BioCommunications Association.

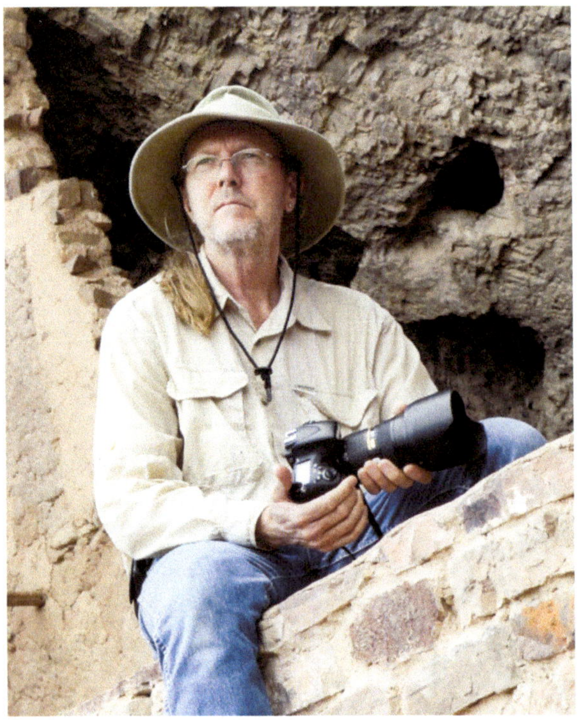

Charles Hedgcock, Photographer, Biological, Scientific, Natural History and Fine Art, Tucson, Arizona. www.CharlesHedgcock.com

Charles Hedgcock is a Registered Biological Photographer with more than 27 years of experience working with scientists both in the laboratory and the field. He has worked as a photographer in the medical field, the life sciences, and in the fine arts to create a unique vision of the natural world. Brooks Jensen, editor of *Lens Work* Journal, stated that Charles "is to bugs what [Edward] Weston was to peppers!" Hedgcock participates in frequent biological expeditions as the primary photographer for such groups as The Sky Island Alliance and GreaterGood.org where he is responsible for documenting the biodiversity of unique, seldom explored mountain ranges in the Southwestern U.S. and Northern Mexico. He exhibits his photographic work nationally and has received numerous awards and grants. Most recently, Hedgcock was invited to work as Artist in Residence for Biosphere 2, outside of Tucson, Arizona, and provided a unique and artistic contribution to the scientific interpretation of plant diversity in a world class research facility.

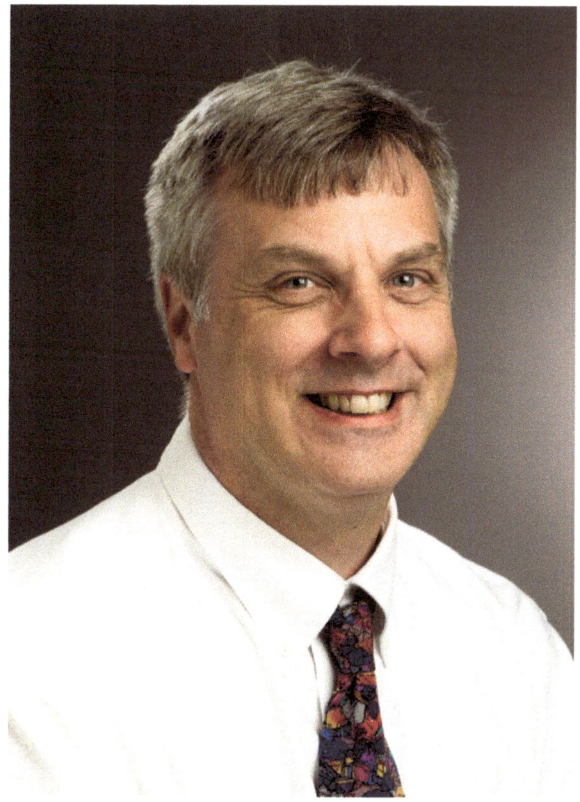

Ted Kinsman, Assistant Professor Photographic Science, Rochester Institute of Technology, Rochester, New York. http://www.tedkinsman.com

Ted Kinsman has an AS in Optics, a BS in Physics, and an MS in Science Education. He worked as an optical engineer, a physicist, and a physics instructor before coming to RIT to teach the technical side of imaging. Kinsman's work explores highly specialized images needed for books and magazines. His work has appeared on *CSI: Crime Scene Investigation*, *The X-Files*, *South Park*, *The Tyra Banks Show*, The Discovery Channel, ABC, NBC, PBS, CBS and the British Broadcasting Corporation (BBC). Recently, he worked on the BBC series *The Frozen Planet*, and James Cameron's movie *Avatar*.

In 2015, Kinsman won the National Science Foundation's Imaging Science contest with an X-ray image of a turtle with eggs. Kinsman's work predominantly focuses on using images to teach science. Kinsman's latest book is *Cannabis: Marijuana Under the Microscope* (Schiffer Publishing, 2018). His fresh and visually stunning survey celebrates the extraordinary beauty and diversity of the world's most controversial plant, *Cannabis sativa*.

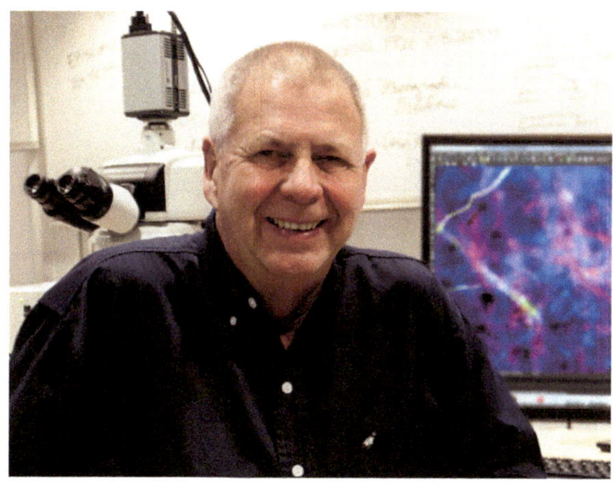

Staffan Larsson, Imaging Consultant and Scientific Imaging Expert, KTH, Stockholm Sweden

Staffan Larsson lives in Stockholm, Sweden and earned an AS degree in photography in 1972 and a national certificate in digital image workflow delivery systems. In 2018, he earned his National Swedish Masters Diploma in Photography.

Larsson has over 45 years of experience in medical and scientific photography. He has worked at Karolinska Institute Medical University and Stockholm County Council Healthcare. In 2009, he joined the Royal Institute of Technology's School of Technology and Health faculty and Clinical Science Intervention and Technology (CLINTEC) at Karolinska Institute. Larsson has also worked in the arts, photojournalism and in nature. As a photographer his work has illustrated countless science and medicine articles featuring images created using new and interesting ways. He has led workshops and courses for students all across Sweden and the U.S. Larsson was pivotal in the formation of the Lennart Nilsson Award in 1997 and developed initiatives supporting its mission for more than 15 years. He was an author in *The Focal Encyclopedia of Photography*, Fourth Edition, and wrote a chapter for *Laboratory Imaging and Photography* (Focal Press, 2016). In 2017, he retired as a research engineer in medical and scientific photography at the School of Technology and Health, Royal Institute of Technology where he taught and assisted scientist-working groups as a consultant for communications and public relations. He is currently a consultant to the Image Center, located at The Jonasson Centre for Medical Imaging/3D Visualization Lab, KTH Royal Institute of Technology, Stockholm, Sweden.

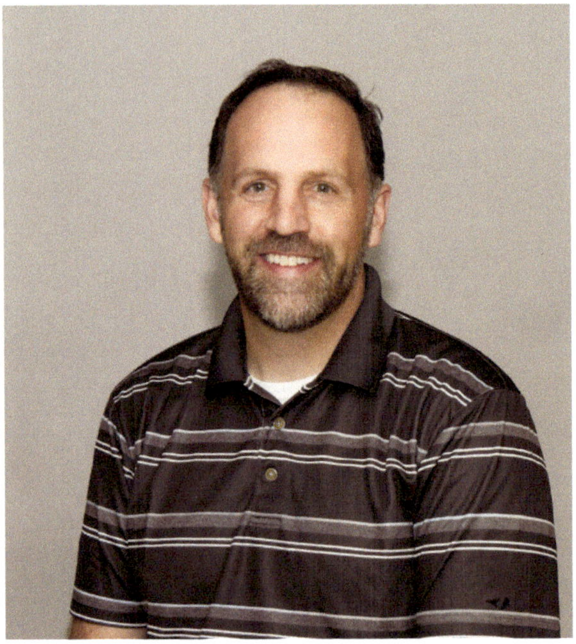

Keith Mancini, Director of Forensic Photography, Westchester County Crime Lab, Westchester, New York

Keith Mancini is a Forensic Photographer in the Westchester County, NY Forensic Lab. He has served in that role for over 20 years. Keith is a certified Forensic Photographer and Forensic Video Examiner with the International Association for Identification where he also serves on the Forensic Video Certification Board. Keith is also a member of the National Institute of Justice's Organization of Scientific Area Committees (OSAC) Video and Image Analysis Subcommittee, and a member of the Scientific Working Group for Digital Evidence (SWGDE) where he serves on the Photography task groups.

Preface: Why I Wrote this Book

During my career, I have learned that I really enjoy making books. It's a challenging job and is full of opportunities to do things in new ways. Writing a book is a process. While working on a book, you can never exactly know what is going to happen until it happens. The unknown creates a certain excitement and a lot of anxiety. Being successful requires knowledge, access to resources, passion, and good fortune. The process of writing a book also requires a long commitment to a journey where many unexpected situations will arise. I hope you find this BIG book useful and filled with interesting and practical information exploring this unique genre of photography. I have tried to make it interesting, modern, and not preachy.

One of the greatest attributes of a science image is its ability to communicate visual information and tell stories to both science and lay audiences. For the image to be successful, it must be good. The image making process can be fraught with unseen and unexpected challenges. Early on in my teaching career, I learned that it takes just as much effort to make bad pictures as it does good ones. I share this because even today, unfortunately, many mediocre pictures are shared in various forums. I believe this happens as a result of lack of knowledge, interest, and/or time. It does not have to be that way.

I wrote this book to be a resource. I wanted to help scientists adopt informed and better practices when photographing in natural science. The book consists of four Parts. Chapters 1 to 5 focus on important considerations investigating imaging foundations and best practices. I believe having a context as to why something is important can be very helpful and lead to more success. Chapters 6 to 9 investigate and describe the theoretical aspects of tools and imaging practices. I would consider these chapters to be a sort of "nuts and bolts" of imaging tools and how images are formed. Chapters 10 to 18 cover a wide variety of applications and Chapter 19 is an excellent guide for image processing approaches used in science. Because of space constraints, we did not include content about video tools and production.

Natural science is a large and diverse field. The last book written specifically exploring this subject was *Natural Science Photography: Laboratory and Field Procedures* published in 1972. It was written by H. Ross Jackson. Other books that have explored the broader subjects of science and photography include Sidney E. Ray's *Scientific Photography and Applied Imaging* (Focal Press, 1999), Graham Saxby's *The Science of Imaging* (CRC Press, 2002), Charles S. Johnson's *Science for the Curious Photographer: An Introduction to the Science of Photography* (Focal Press, 2017), Enrico Savazzi's, *Digital Photography for Science* (2010) and Felice C. Frankel's *Envisioning Science* (2004). Examining their contents, I learned they each were very different than this book. Some of these books were written before digital had been adopted and they emphasized film photography. Others seemed to include more theoretical knowledge or were more applied with a little technical thrown in for good measure. I believe this BIG book shares a very good balance of "why to" and applied theories about "how to" make great images in a wide spectrum of disciplines.

When Jackson wrote *Natural Science Photography: Laboratory and Field Procedures*, the making of pictures required film. Film had to be developed using any number of chemicals. There were many challenges and variables for photographers. Making highly technical pictures of challenging subjects was then, and remains now, difficult. Science photographers used many specialized films sold by Kodak and other companies during that time to do their magic. Special emulsions were needed to photograph experiments or make photographs of nearly invisible things. There were significant obstacles on the way to success. There was a constant need to manage contrast and control the image during film and print processing. Choosing to use a different film was done because each film had various spectral and sensitivity advantages. While the world of film photography in science is long gone and nearly everything about photographic technology has changed, many of the current approaches were informed by the experiences and failures used for more than a century.

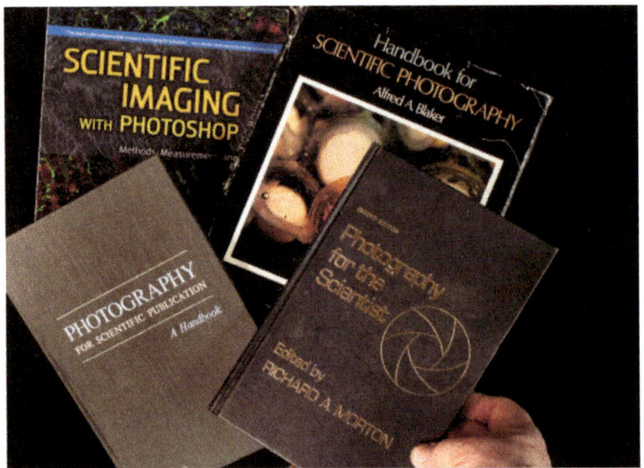

Several important scientific photography books published during the last 50-plus years including *Photography for the Scientist* (R. Morton, 1984), *Photography for Scientific Publication* (A. Blaker, 1965), *Scientific Imaging Using Photoshop* (J. Sedgwick, 2008) and *Handbook for Scientific Photography* (A. Blaker, 1977).

Natural Science Imaging and Photography was written to share the best practices needed to make, distribute, and preserve scientific visual information using digital photographic methods and technologies. This book also represents the last book in the Applications in Scientific Photography series. While writing *Laboratory Imaging and Photography*, the idea for the series was suggested by the publisher. This book, like the others in the series, is rich with contemporary information required for various applications. I was – and remain – honored to have been invited to be the series editor. While bittersweet to consider that this book is the last in the series, I am proud of the work that was produced by Keith Mancini, John Sidoriak, Christye Sisson, and Adrian Davies during the last 6 years.

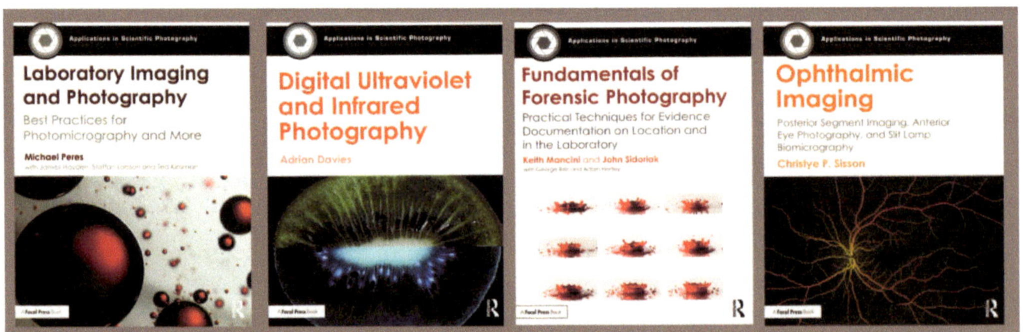

Books in the Applications in Scientific Photography series.

This book has been designed to be a contemporary handbook for individuals who might not have been formally trained in photography or imaging but now perform imaging as a significant part of their work. The book was written to co-exist with the Internet, where the sharing of information about the newest equipment and software is accomplished in ways traditional publishing cannot.

Today, there is no shortage of websites that feature photographic practices but only a few are dedicated to the technical subjects featured in this book. While the Internet is an extraordinary and vital resource, Routledge, the authors, and I still believe there is a place for a concise and modern book that contains in one volume, the best practices required for making natural science images.

Because the subject of natural science is vast, I invited noted experts who are friends to collaborate with me. They are award-winning, smart, and curious image makers. Most importantly, they were willing to share their knowledge in this book for the opportunity to be involved. These authors could easily be on a list of who's who in imaging. I am appreciative and grateful for their contributions. I think you would be impressed if you performed a web search and read about the many accomplishments of Norman Barker, Adrian Davies, James Hayden, Charles Hedgcock, Ted Kinsman, Staffan Larsson, and Keith Mancini. In these authors, there is more than 300 years of cumulative experience.

Because of the diversity of the authors and their backgrounds, you may find some minor redundancy in some of the book's chapters. Each author included what they felt was most important to include. Once I received their manuscripts, I conformed some variability while not changing the meaning or information that was shared.

Traditions and Practices

I am sure in the future, when historians chronicle the science from this time, it will be described as the "Age of the Image" because every type of scientific investigation includes imaging. Images accomplish what words cannot and allow the barely visible to be recorded and studied over time. When practiced at a high level, photography is a powerful communication tool and an important language required for effective and truthful story-telling of complex ideas, subjects, and behaviors.

There is nothing new about using images in science. Anna Atkins, a British botanist and photographer, is considered to be the first to publish a book dedicated exclusively to using pictures in the mid-1800s. Using photographs to convey science rather than drawings was not common at that time. Photographs made at the time were singular pieces because there was no simple way to make multiple copies. Each image was an original and if other copies were needed, new originals had to be created. Most of all, it was very difficult to make photographs at all. Producing a book of images meant each copy of the book was unique and had its own set of images.

Today, we share work electronically to worldwide audiences nearly synchronously from the moment of an image's creation using platforms such as Instagram®, LinkedIn® or

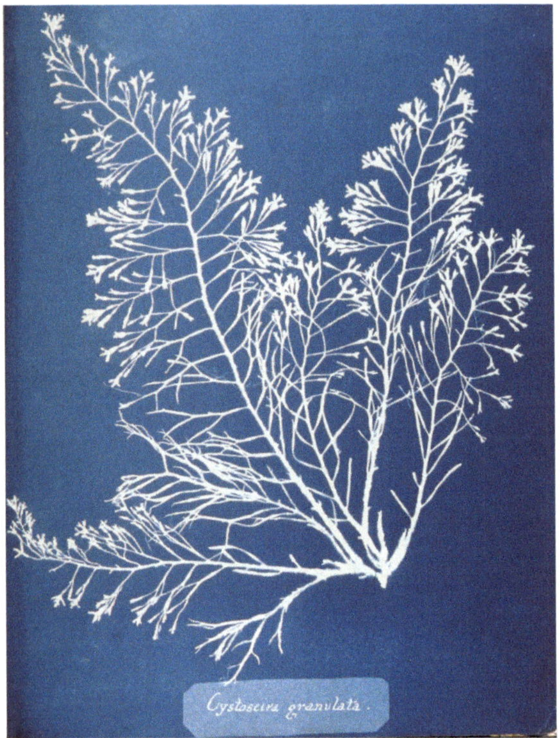

This photograph, *Cystoseira granulata*, cyanotype photogram is from *Photographs of British Algae: Cyanotype Impressions*, by Anna Atkins (1799–1871). The image was digitized from Sir John Herschel's copy of the book from the Spencer Collection/The New York Public Library and is in the public domain.

Twitter®. Self- and open-source publishing has also become the new normal for many contemporary scientists and it is being used more frequently than print editions of many journals. That being shared, the motivation as to why people make and share images as well as their value in communicating science has not changed since photography's invention, and only the tools and the ease of making photographs has.

Today, every publication and area of research is heavily invested in images, imaging practices, and imaging technology. Science first adopted photography as soon as the technology and materials needed to form permanent images was discovered. What is absolutely new today is how people can create, edit, display, and share images. The advancement of technology remains on a trajectory of unparalleled transformation and the release of new products and tools does not seem to be slowing. Continual changes and applications coupled with better tools will result in new and groundbreaking informational images, affect the way scientists work, and reveal things that have never been seen. We hope this book will prove to be an excellent resource to build new skills for years to come.

It is important to accept the simple fact that there will be continued changes for tomorrow's imaging and applications. Technology will never stop improving and the need to develop new knowledge and secure new tools should remain an expectation for anyone who works in this field. Embrace the changes, it will be less frustrating.

Defining Contemporary Photography

While ubiquitous in society, photography is a fascinating and sometimes complex subject. It requires skills, knowledge, equipment, and is often described as an artform. Photography however in science is not described as an artform by its practitioners. It is an important tool used to provide highly accurate photographic representations of events or objects. Its images are needed for the collection of factual visual information or data about a subject.

A frequent discussion amongst photographic colleagues is what the word photography means in today's contemporary world. Like all things that are more than 175 years old, nothing stays the same. While today's photography is completely different than the photography that was practiced more than century ago, photography remains both a technology and a visual language. Also the word photography has come to mean different things to different people in a constantly changing landscape of users and disciplines.

The term imaging is also a word that has come to mean photography amongst other things. These two words have become interchangeable in many contemporary publications. Pictures, snapshots, images, photographs, or digital assets are also terms used to describe the outcomes of the photographic process. For this book, the authors used the term photography to describe single shot applications and the term imaging to describe outcomes that include the use of many images, computing and software, or are the product of image processing. It is important to note that there is truly no absolute definition of photography that is acceptable to all and in most instances the words are nearly synonymous. In the end, the subtle – if any – differences are truly not that important.

Make no mistake, achieving effective pictures that clearly communicate science and preserve dynamic events remains challenging. To achieve good work requires a desire to work through the numerous challenges and use industry accepted best practices. I have found that learning about a subject before starting to photograph and paying careful attention to the smallest of details are vital ingredients to successful imaging outcomes. While not suggesting that they are photographers, scientist photographers must have the interest to make good photographs or it will not happen. A real commitment to the process of making images is at the foundation of successful imaging outcomes. Without the interest to work through the obstacles, the making of good photographs will be a lucky accident. There are no short-cuts for creating excellent results day in and day out.

I believe you will benefit from the practical information shared in this contemporary resource written by some of the best practitioners in the world. I hope you enjoy using the information in this book as much as I did creating it.

Michael R. Peres
Professor, Biomedical Photographic Communications
Rochester Institute of Technology

References

Photographing the Unseen World, by Adrian Davies, Crowood Press, 2020.
Cannabis: Marijuana Under the Microscope, by Ted Kinsman, Schiffer Publishing, 2018.
Digital Ultraviolet and Infrared Photography, by Adrian Davies, Focal Press, 2017.
Focal Encyclopedia of Photography 4th ed., by Michael R. Peres, Focal Press, 2017.
Science for the Curious Photographer: An Introduction to the Science of Photography, by Charles S. Johnson, Focal Press, 2017.
Laboratory Imaging and Photography: Best Practices for Photomicrography and More by Michael R. Peres, Focal Press, 2016.
Hidden Beauty: Exploring The Aesthetics of Medical Science, by Norman Barker, Schiffer Publishing, 2014.
Digital Photography for Science, by Enrico Savazzi, Lulu.com, 2010.
Scientific Imaging Using Photoshop, by J. Sedgwick, New Riders, 2008.
Envisioning Science, by Felice C. Frankel, The MIT Press, 2004.
The Science of Imaging, by Graham Saxby, CRC Press, 2002.
Scientific Photography and Applied Imaging, by Sidney E. Ray, Focal Press, 1999.
Electronic Imaging for Photographers, by Adrian Davies, Focal Press, 1994.
Photography for the Scientist, by R. Morton, Academic Press, 1984.
Handbook for Scientific Photography, by A. Blaker, W.H.Freeman & Co, 1978.
Natural Science Photography: Laboratory and Field Procedures, by H. Ross Jackson, 1972.
Photography for Scientific Publication, by A. Blaker, W.H.Freeman & Co, 1965.
Photographs of British Algae: Cyanotype Impressions, by Anna Atkins, 1853.

Part I

Industry Expectations and Attitudes

Chapter 1 Why Best Practices as an Approach

Introduction

Trying to do things in the very best ways and seeking improvement should always be a personal goal regardless of the discipline. This book was written to demystify imaging and its tools and suggest a professional framework to operate in. Sometimes there are things you can do to improve that are not related to technology. Adding small tricks or thinking about things in different ways can be extremely valuable for improvement. The first four chapters of the book were written to establish the importance of practices that are not grounded in tools or technology. These chapters will explore the image in science and the importance of Best Practices.

Best Practices is a term used to describe approaches that are repeatable and lead to the very best results. Sample selection, proper use of lighting tools, developing new knowledge and skills, as well a personal motivation, play an important role in the successful creation and distribution of scientific images. Techniques are the framework of what Best Practices are built upon.

Defining Best Practices and an Imaging Workflow

Best Practices should be considered your steps (when optimized) that are used when making photographs. Workflow is also a term used to describe the process of making digital photographs. Workflow is a part of Best Practices and was a term primarily used in the graphic arts industry many decades ago. Best Practices and workflows are similar and maybe synonymous in some ways. They also overlap one another. But they are not exactly the same. When people talk about workflow, the primary focus describes the digital pipeline. In my mind, workflow begins in choosing the camera settings and image capture preferences. It includes image processing, distribution, and culminates in archiving. Best Practices includes both image formation activities as well as the digital workflow. It also includes usage considerations.

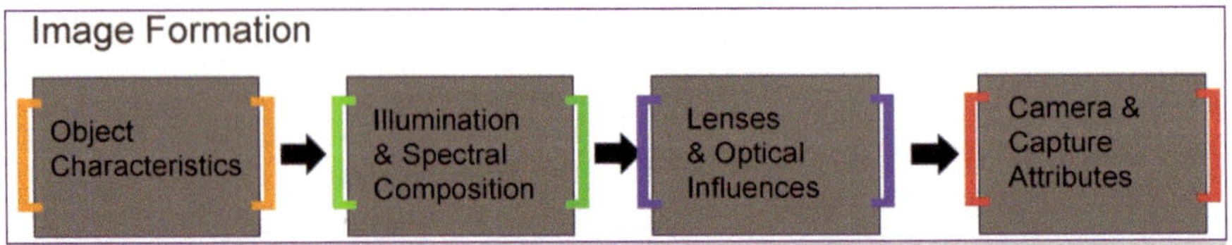

Figure 1.1 Making photographs requires using equipment and managing variables and negative influences. The subject, illumination, the lens, and camera capture settings all play a role in the attributes that an image will contain.

A best practice describes an approach that is accepted as professional and conforms to the standards and expectations of a specific industry. Best Practices describes approaches that are optimized and managed from the beginning to the conclusion of the work. Best Practices must use repeatable approaches that eliminate variables or manage them to the maximum extent possible. This includes everything, no matter how trivial. What's the expression, the devil is in the detail. How samples are handled, how devices are selected and used, what lenses and apertures are selected, or what file is chosen for saving are each a part of a systemized goal. Another important facet of Best Practices is the absolute expectation that over time, your best practices will evolve. Best Practices are affected by many things and should continually be evaluated for currency, technology, and relevance.

Individuals who have not been formally trained in photography and imaging might not "think like an image maker" in the earliest stages of their new role. In fact, today's cameras are so easy to operate that sometimes their capabilities are taken for granted and we get lulled into complacency about what is really being accomplished by the devices we use daily. The ease of operation can frequently make us forget the challenges we are addressing as image makers. To be successful, you must adequately plan for photographing including inevitabilities and the unexpected. Image makes need to be organized and create consistency. There will be constant problem solving and a need for critical thinking in new work and subjects.

I consider the making of photographs as having two distinct activities. The first activity is forming the best image and this includes everything prior to pushing the shutter button on the camera capturing the photograph. The second activity is the digital workflow that includes: capturing the file using optimized camera preferences, ingestion, and image processing to improve visibility, culminating in display. Each activity within the continuum of steps must use standardized approaches which are defined by the tools and many other factors. No element is too minor to pay attention to.

Creating a Workspace

It seems obvious to say, but it might not be. Making photographs begins in a place. This could be where you work every day, or possibly you will go out to a new location. Photographs can be made nearly anywhere and sometimes in challenging places that are less than ideal. To achieve the best result, you may have to establish an optimized space to work in. **Best Practice: Create a GOOD space to work in.** Each situation will be different and require adaptation and maintaining a general awareness about how to use the space to your advantage.

Often there will not be enough wall outlets for all the various devices that require electricity. Computers, fiber-optic lighting, microscopes, or other devices all take up space, generate heat, create vibrations, and can produce electrical feedback to an imaging system. All pieces of the equipment will have a power cord and an outlet. The tangle of cords can look like spaghetti falling off of a plate around your work table. I have seen someone's foot catch a cord and – well you can finish the story. To make good work, you need to create an uncluttered work space. Staying organized and having access to various tools and accessories while performing an imaging experiment is very important to successful outcomes. I almost always tape down my electric and other cords.

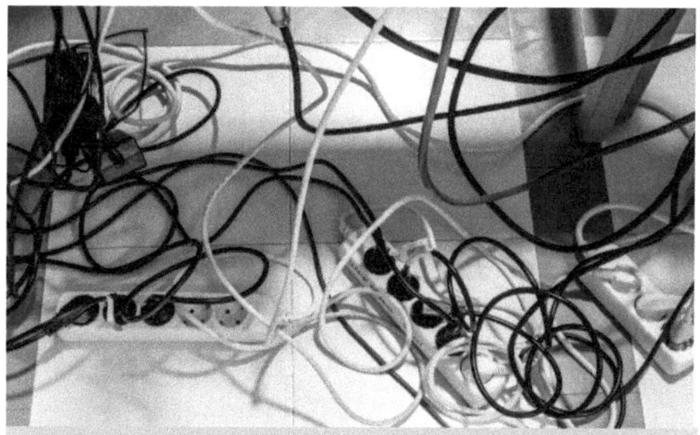

Figure 1.2 Staying organized and safe is an important requirement for making photographs. Getting things properly set-up right from the beginning of a session is a best practice. It can be very helpful to cable tie loose cords and keep plug bars easily accessible.

Since a person or "standard viewer" determines how successful a photograph will be, creating an environment that helps you "see better" will lead to better results. I know you have tried to look at your phone outdoors and have experienced how the ambient brightness competes with screen visibility. That being shared, pay careful attention to other light and light sources in your work area. **Best Practice: Be aware and manage ambient or room lighting (windows) during imaging.** This includes desk and ceiling lighting as well as windows, doors, and monitors.

Using Effective Camera Supports

Using sturdy tables, tripods, and camera stands can also be an important element required for successful outcomes. Always select the best tool for the job. It may be the only tool you have and so it will be the best tool. Always buy the best tools you can afford. All camera stands and tables can lead to image shake or blurry pictures.

Not all tables will be good for photographing. Some are simply not stable and contribute vibration to the camera. A sturdy imaging table is vital to success. Heavy tables are much more desirable than light and portable tables, but sometimes portability is required. Tables with wheels can be an advantage or disadvantage. Where a table is located can also negatively influence results of highly sensitive systems such as a photomicroscope. It might be helpful to have a dense, heavy material top added to a table to increase mass and possibly absorb, but not eliminate, table shake. This need for structure and stability applies to the need for an excellent and sturdy tripod or camera stand as well. **Best Practice: Use a professional quality table, tripod, or camera stand to control and/or isolate vibration from the system.**

Cleanliness of the Imaging Area

Of great importance is having a clean space to work in. One of the most commonly photographed subjects in science is dirt and particulate matter. Dirt will be a chronic problem and one that needs constant attention. **Best Practice: Pay attention to the presence of**

dirt and where it is coming from. Frequent dusting of surfaces using lint free cloths can be helpful to keep dirt from a sample or finding its way into imaging equipment, cameras, lenses or other locations. Wash floors as needed as part of a best practice. Dirt can be introduced to an imaging room from shoes, from clothing (fibers), and sometimes the HVAC which brings in outside dust and debris unless you work in a cleanroom. Whether dirt is big problem or a minor problem, being aware of its potential presence will lead to better results. I sometimes use a piece of copy paper as a surface to lay and prepare my samples on even when I have cleaned my tables.

Carpets are never good in imaging rooms. Carpets are made of fibers, which over time will break down. As they age, fibers will become airborne and can find their way onto samples and pretty much anywhere. In my lab, we have added additional air filters to the inflowing HVAC vents.

Selecting and Using the Right Imaging Tools

Chapters 6, 7, and 8 explore cameras, lenses, lights, and other image forming considerations. Each tool and optical component has a particular reason why it should be selected and what to expect from the image attributes it produces. Image contrast, camera sensitivity, image noise, and color fidelity are a few of the choices that influence camera selection and operation. Additionally, a lens will impart image shape and structural information at capture. **Best Practice: Select the right tools for the job as required by the sample and subsequent image use.** Forming the best image is an expectation. Image defects can be introduced by using incorrect tools or from equipment that is not of the proper caliber for the job.

Best Practice: Use the camera in its optimal mode of operation. There are significant advantages to operating a camera using manual settings when possible. More is shared about this in Chapter 6. While a camera's AUTOMATIC features offer some advantages and efficiencies, automatic operations will lead to variability in file attributes. These variations may require additional post processing to equalize the differences. Inconsistencies that come from using auto-white balance settings will be evident as image color variability in the various files. Using manual mode is more time consuming but will create more precision and uniformity across the images.

Always choose capture settings that create the optimal and largest image data set. This is achieved through choosing a camera's highest bit depth and largest color space. Chapters 6 and 8 share a great deal about this. Many devices default to an sRGB color space. The sRGB color space is the smallest possible choice and produces the most compressed color. Colors are squeezed together when recorded using sRGB and the color gamut is less than ideal. For DSLR cameras, Adobe RGB 1998 is a better choice than sRGB and creates more colors. When advanced image processing is required, the creation of the "cleanest" digital file is an ideal to strive for.

Optimizing Camera Settings

It was previously shared that Best Practices include using the same camera, settings, and other related tools. When starting a new imaging session, and the tools have been selected, it is imperative to confirm the capture device's settings are optimized. Haste leads to mistakes, and assuming everything is still as it was will open the door for poor outcomes. **Best Practice: Review all possible capture settings prior to starting any NEW work.** This includes checking ISO or sensitivity, white balance, the selection of the file type, color space, sharpening filters, and other camera image enhancement settings including noise reduction, etc. Changes can occur to capture settings by choice or accident. This happens frequently when cameras are used by multiple users.

It is also recommended to clean the sensor on a DSLR camera and reformat the media card(s) frequently. These small steps that can absolutely contribute to the formation of images that are less prone to corruption and need less post processing.

Use a Clean Lens for Imaging

Lenses and other optical elements often require surface cleaning – using an appropriate brush or lens tissue – before taking pictures. It is rather remarkable how fast airborne particles and fingerprints can find their way onto the front surface of a lens. You should always check the lens before starting to use it. Particulate matter or fingerprints will diminish image contrast and lower visibility. If fingerprints or other oily marks are located on the surface of the lens, use an approved lens cleaner or isopropyl alcohol applied on a cotton-tipped wooden stick. When cleaning, never apply pressure but rather use a light circular motion with a very light touch. Pretend you are polishing the lens surface softly. Glass and other optical elements are vulnerable to scratching and care should be taken when cleaning and removing oily marks. I actually use an air blower before surface cleaning. **Best Practice: Clean optical surfaces frequently.**

Make a Number of Exposures and Not Just a Single Frame

Once everything is prepared, make several exposures of the photograph and do not simply expose one frame. Surprisingly, even when everything stays the same, individual frames may experience small differences in image quality or variances in exposure. It's a Murphy's Law sort of thing. **Best Practice: When everything is ready, make multiple exposures of the photograph and not just one.**

Monitors and Video Displays

Digital images cannot be seen until they are displayed. When they are on a hard drive they are invisible. They remain invisible as digital data until they are opened in an image editor. The digital image once captured, is stored and requires software for display to make the file visible. Display devices exist on the back of cameras, on smartphones, on monitors from laptops, tablets, and other flat-screen devices. Since the digital image can only be seen when displayed – and if the monitor is not properly managed for its brightness and color – characteristics of the digital image's display attributes will be incorrect [for reproduction] because of a lack of calibration. Additional content sharing how to calibrate a monitor's display characteristics is covered in Chapter 19. **Best Practice: Calibrate and monitor a computer's display and settings.**

LCD technology (Liquid Crystal Display) has moved CRT (cathode ray technology) to the closet. A significant advantage of LCD displays is they provide more brightness and color constancy over time and use less power. Drift of brightness and color still happens to LCD monitors but much more slowly and in very small increments. Because images are processed using the display characteristics of the monitor, which include brightness, color, and gamma, it is necessary to calibrate a display or screen using appropriate tools. In a perfect world, image processing would be accomplished using pixel brightness information and describe image attributes as numbers. In reality, the majority of image processing is done based on the image as viewed on a monitor. We assume, sometimes incorrectly, that a monitor is correct. While significantly better than CRT technology, LCD screens are vulnerable to burn-in because pixels can permanently lose their ability to return to their relaxed state after a continued static use profile – so don't trust a monitor display completely until it has been calibrated.

Since the image is displayed on a monitor and decisions about image processing are based largely on the display, errors can occur. It can be frustrating when image processing is performed on one device that is not calibrated and then evaluated on another. When this happens, the image will look different on the two devices. Because of the variables of the human visual system and environmental influences on seeing, calibrating a monitor using a "digital tool" can be useful and lead to precision and control. Calibrating a monitor – or at least, being aware of the importance of calibration and what is required to calibrate – can be very helpful in the production of more effective results displayed across various devices.

Color Management

Since digital images are recorded by one device, displayed on another device, and possibly printed using yet another device, variances of a file's color appearance can occur in each situation. The manner in which an image's color and color space is managed on each device and across the various devices can be a challenge for you and the devices. The process used by various devices and their conversion of the digital color language used to describe an image is called color management. The appearance of color is the outcome of mathematical modeling using algorithms. Because the computer code is coordinated, color profile standards have been created to work in managing color across all the platforms. To help users who work with images on various computers that use different operating systems, color profiles have been created by the International Color Consortium or ICC. Assigning a profile to an image allows the attributes to remain the same regardless of the device (in theory). **Best Practice: Requires the use of color management and use of profiles where practical.**

When considering color gamut, color management facilitates the translation of the colors across platforms working to preserve them. Chapter 19 shares a great deal about gamut. The translation of color data across devices is managed by a process called gamut mapping. Since some colors are difficult to describe (mathematically), there are ICC specifications used allowing all devices and users to manage the losses or gains of color accuracy in the best way.

Color language uses four types of rendering when the color gamut is outside the capability of a device. These include absolute colorimetric, relative colorimetric, perceptual, and saturation intents. Rendering is a protocol that helps to describe an image's appearance when the device cannot handle all of the data. Each description provides the user with advantages and environments where choosing one over the others makes sense. Personal preferences often guide the final selection of one rendering tool over another. **Best Practice: Use a calibrated color target for optimizing color reproduction and color management evaluation.** See Chapter 8.

To achieve the tightest color accuracy, include a color checker in the one frame when making a photograph. Relative to a subject's size, a traditional color checker might be perfect or too large to include in the frame with the subject. If that were the case, almost anything can become a standard if it contains specific characteristics. A sample that contains a neutral white and black patch can be used if nothing else exists.

Software, Upgrades, and Optimizing the Computer

There are times when new software products are released too frequently for comfort. Often you will feel pressured to acquire new products or related software that are advertised to make "better" work. Almost everyone I talk to about upgrades and new products has experienced the angst about what to upgrade, when to upgrade, and why to upgrade. These considerations are intertwined with the costs of both imaging devices and software.

Upgrading may or may not be needed for Best Practices. Digital imaging is now at an interesting period in its evolution. Being capable of performing complex imaging activities with relative ease, nothing seems impossible. New features, lower costs, and simpler operations are driving innovation, growth, and adoption in every industry. In the beginning of digital photography's adoption, cameras, software, and devices were constantly changing, which was expensive and frustrating. The need to acquire new tools occurred over short periods of time relative to today's new product releases. Rapid changes and lack of standards made it complicated to use cameras and related software for more than a few years. The hardware had its own set of challenges needed to communicate across the devices and platforms. Since their first release, camera sensors have gone from 1MP to 64MP in just over a decade. Prices went from tens of thousands of dollars to less than a thousand dollars for a baseline DSLR camera. Digital technology is no longer in a period of explosive change. Consider as well, cameras in some smartphones make better files than some low-end DSLR cameras. Big changes to devices and capabilities occur with much less frequency and the cost benefit to upgrade might not be worth the price. There are so many considerations. **Best Practice: Be aware of why and when it is best to upgrade your software and hardware.**

Having a reliable imaging workflow requires the right tools and software that is both grounded in practical approaches and uses compatible tools. It will not be possible to stay current on everything, all of the time. If a system is working but performs slower than a brand new system, that alone may not be a reason to upgrade.

Many universities upgrade software packages annually because of site licenses. This makes sense for an organization. This approach minimizes variables and controls usage for a larger number of clients. Not all organizations though have a unified policy for managing software with an institutional point of view. Individual users will have different challenges that are grounded in the costs associated with securing upgrades. Sometimes changes offered in an upgrade might be small and new features are not matched for the type of work being produced. There is no one answer for all.

Operating systems, like all software, can be upgraded as well. An upgrade to an operating system may have consequences for hardware devices and other software, that can affect everything including the learning curve needed to operate it. Drivers, plugins, transfer protocols and a host of other factors come into play when evaluating an operating system upgrade. Each year countless dollars of products and services become obsolete because devices can no longer talk to each other. This occurs when companies stop supporting the drivers for scanners (for example) and these devices may not work with new operating systems. Going backwards and trying to open files made on previous editions of a software package once an upgrade has been made can lead to other unexpected issues and frustrations. It is vital to back-up important data before proceeding with software changes. This all can be maddening.

Reviewing Images

During capture, it was recommended that multiple frames be exposed. This was suggested because individual frames may reveal small differences within the set of exposures. Now it is important to carefully look for small variances in exposure, color information, or the presence of artifacts. Most important, identify any variances in image sharpness and identify files without vibration or image shake. **Best Practice: Carefully review and select the best frame from a series of exposures for optimizing.**

The Image Pipeline, Folders and Naming Files

Staying organized is a requirement for any digital photographer. **Best Practice: Create a workflow that uses the same procedures time in and time out.** Having an effective file naming convention and a folder management system will pay measurable dividends over time. Not using a structured file management strategy creates challenges when trying to find pictures taken months or years earlier. Many imaging experiments are conducted over time and creating a practical and easy-to-use library or cataloguing system is invaluable.

It seems obvious but for the sake of sharing the obvious, each experiment (event) should have its own unique folder. These folders should be located in a master experiment folder. In this experiment folder, other folders can be added as needed. My foldering/filing system uses a main folder for each session. It might be named Snowflakes_January_12_2020, for example. As soon as I make this folder, I then make three new subfolders within that folder. One of the folders is used to hold the native files or RAW files, another folder is used for the image-processed TIFF or PSD files, and the third folder is used for resized JPEG files used for social media applications. As soon as the photography session is completed, I ingest the camera native files to the RAW folder where they will live forever. I NEVER rename them. Once a RAW file is opened and the image processed, it is saved as a TIFF or JPEG using the command SAVE AS. It is recommended to SAVE AS into another folder named "processed" or something useful to your system. Overwriting a native file may have terrible consequences, if at later date, that file was required for a number of reasons. You can never go backwards once you destroy or delete data. RAW is an excellent file format for science applications as long as the original file is preserved. Once processed, a RAW file can always be reset to its native capture characteristics by discarding the sidecar or XMP file. This action will revert the data to the unprocessed state.

It is also crucial to keep the native camera generated file number in the file's name for the newly processed image files. For example, February_14_2019_**DSC3190**.tif is how I name processed files. By keeping the camera raw DSC number in the file name, it is much easier to locate the native file – should a file or folder become lost. It is relatively easy to search for a file with a very specific name such as a number. It makes me smile to think back to when I first started making digital pictures in the mid 1990s. Twenty years ago, there were no norms for naming. Sometimes I would call my file "best" or in the case of family pictures, Jon_1 (my son's first name). After only 1 year, I had countless Jon_1 files. It was a perfect plan when launched, but quickly was not.

There are a number of programs that can help a photographer stay organized such as Adobe Lightroom®, Adobe Bridge® and other image database products. One valuable consideration for selection of any product is its ability to talk to the larger IT environment or the Cloud if used for back-up. Unfortunately there is no one answer or strategy for all situations. Numerous variables affect a system and personal preferences play a role in developing a Best Practices solution for a particular environment where images are archived. Sometimes in an ideal circumstance, there is an IT group who oversees database access and management. Always archive using file types that are backwards compatible. TIFF, JPEG, and DNG are considered to be versatile and should be selected over others. **Best Practice: Design and manage your digital asset system in a timely and consistent way over time.**

Adding metadata is an expectation to any file management approach. Adding metadata takes time and can be boring work. There is a tedium to this task but it is required for science and will be time well spent when that data is needed. The more data that can be added to the metadata, the more mileage it will provide. I have found that when creating a new article, the information added as metadata has been invaluable. **Best Practice: Add metadata fields at the time of ingestion as part of your digital asset management approach.**

Archiving, Data Redundancy, and Backing Up

When storing physical documents such as film or photographic prints, they could be touched, retrieved, and were managed using visible processes. A material's pH could be measured if warranted, environmental temperatures could be maintained, humidity monitored, and the brightness of the environmental lighting controlled. When these factors were properly managed, images could last for what was believed to be forever. Archivists developed strategic methods for filing, storing, retrieving, and handling documents gained from decades of experiences. Degradation was visible and systems could be physically checked. Physical degradation came from water, humidity, the environment, or from fire, and was the primary contributor to image loss. Importantly, photographers knew without too much effort where an image – whether a print or film – was located.

Digital technology and digital file storage is very different and requires an operator to consider how to create and manage files and adopt appropriate strategies. Because data is electronic, it is vulnerable to different problems that are not physical. I am not a "bit head" but staying aware of trends, products, and monitoring a product's functionality remains an important element for success. The most important element of this topic is the absolute requirement that you MUST preserve digital assets as a personal PRIORITY.

Digital data is stored on hard drives, which are electronic devices. Hard drive technology – like all technology – is evolving. There are two major types of drives, spinning disk and solid state. Hard disks with one or more spinning "platters" that are used to store and retrieve digital data as needed are an older technology. Data is "written" to the drive using a specific code on a particular region of the disk that is determined by the operating system. Spinning disk technology is being replaced by newer, more expensive, and more stable hard drive technology that uses solid-state drives. Solid-state drives (SSD) are less prone to failures and are faster but are a bit more expensive. SSD costs have dropped significantly in the last few years and the jury is still out as to when spinning disks will no longer be used in computers.

Images can be stored on internal hard drives or external drives, sometimes called portable hard drives. Both types of drives have a similar need for monitoring performance. Interestingly there are now careers in digital conservation and image archiving. Medical and scientific stock photography companies, university and research organizations, and many other institutions have literally billions of images to manage, store, preserve and access. Data has both financial and emotional importance to organizations and the creators. **Best Practice: Develop and USE a digital image preservation strategy.**

Archiving software and technology is changing continually and the ability to maintain access to digital data is something to pay attention to. Older products and devices are frequently replaced by newer, faster and cheaper devices. Scholars refer to this as technological obsolescence because while the devices physically operate, they do not operate effectively in the environment where they are used. Often as they fill, hard drives will become slower. Once a new product gains a place in the market, it may not always be backwards compatible with older devices or file types. USB 2 devices, for example, do not plug into USB C ports. Files made with a proprietary device may also create problems for users. Obsolescence can happen to storage media as well.

When considering hard drives, certain criteria are thematic including capacity, physical size, portability, and how they communicate with the CPU. Both internal and external drives continue to grow in capacity. When storage media was first marketed, the average cost of 1MB in 1999 was approximately $1. Today a 1TB hard drive may cost as little as $100. Storage and NOT the size of files should be the only concern in this contemporary era. Bigger drives are simply going to be needed when making bigger IMAGES. Bigger image files will be the norm for the foreseeable future and there are currently DSLR cameras that create 70MB TIFF files right out of the camera. There are no universally accepted strategies for the handling of electronic media, however, at one time or another, you will be involved with your own digital conservation.

Hard disks, and all computer hardware for that matter, are vulnerable to electronic fatigue and ultimately failure. It is not a question of if a device will fail but rather when. Planning for failure and managing data preservation is an incredibly important responsibility. Data should exist in multiple places and be redundant. Redundancy and archiving data takes time, commitment, and may seem like time not well spent until there is a problem. Because data is invisible, the notion of archiving DATA is not important until you need it to use it. I liken it to having insurance. For the record, jump or thumb drives and memory sticks should NEVER be considered for use as long-term storage devices. These devices, while incredibly useful, are more vulnerable to problems. **Best Practice: Frequently back-up data and assess drive "health".**

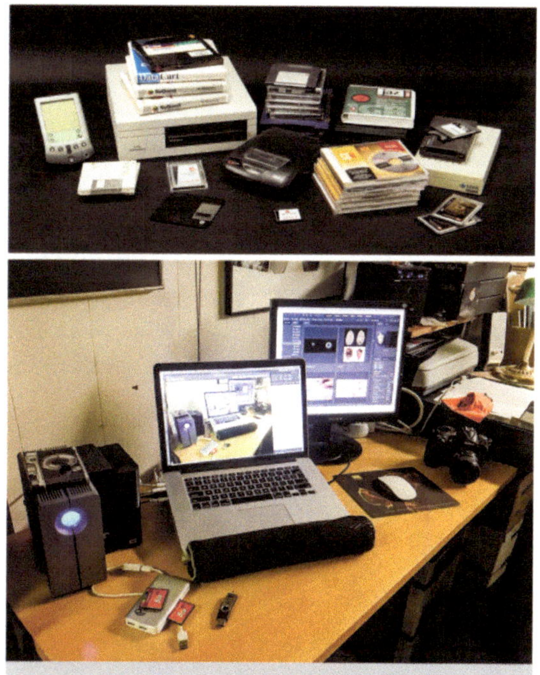

Figure 1.3 Top: examples of 25 years of removable media and drives including Syquest®, ZIP® and JAZ® drives, Kodak Photo CD®, and a very small hard drive (20MB). Bottom: contemporary devices needed to produce and archive digital files including multiple external hard drives.

Specific challenges and solutions are unique to each person and environment. An important strategy for storage lies in the questions as to whether multiple users need access to the data physically on-site or remotely. Answers to this question or others will influence decisions about how to create an archive. A valuable type of hard drive used to minimize data loss is a RAID drive. RAID is short for "redundant array of independent disks". In this system, when data is copied to the RAID, it is written to separate partitions on two drives. In this manner, should one drive fail, the other drive will have the same data. Each drive is large and matched. When necessary, I manually back-up data from my computer's drive up once a month or as needed to another drive. Being concerned with data loss, I also maintain other hard drives kept both at home and at the University. I use the same foldering and naming convention on all my drives. This all takes time to manage. I have lost data only once and it felt similar to losing a relative. I also learned the hard way that data can rarely be recovered. There are companies that provide recovery services but the cost and lack of certainty about prospects for success looms as a real obstacle. Lost data cannot be replaced. And so experiments, images, articles, and the work of a lifetime requires a certain responsibility to preserve and manage looking forward, not backwards. Various products exist for backing up drives directly to a computer's drive. In the Mac environment, Time Machine® is an excellent choice for AUTO backing up when the external Time Machine drive is physically attached to the CPU. There are similar products for a Windows platform.

Removable and writeable media can also degrade. The material that removable media is made from will not last forever. It is not impervious to use, damage, or corruption. In some cases, the degradation will occur more quickly than might be imagined. A DVD or other removable media can become unusable if it has been scratched or delaminated. A scratch results in the disk being unreadable and the data being lost. This happens more often than you would imagine and the loss of media holding countless digitized images has a variety of costs. Hard drives are much better for storing data long term than DVDs. Most modern laptop computers no longer come with drives and DVDs are no longer a standard for storage.

As an endnote to this topic, one of the most important strategies required for digital conservation is the continual migration of data. It will always be critical to migrate data from older technologies to new technologies before the ability to do so no longer exists. **Best Practice: Never wait until the bitter end of a technology's life to move files to a newer product**. It can be helpful to have a specific time of year to migrate older data and important files forward annually. This activity could include consolidating older hard drives and other related media. I know organizations that keep older machines working for the sole purpose of transferring and reading files. It is a lot of work but can have value.

Digital Housekeeping

As evidenced by all of the suggestions in this chapter, being successful requires time and attention to detail that might not be measurable or important in the moment. Paying attention to them will lead to better outcomes over time. When talking about these concepts I sometimes compare an imaging session to cooking. There is preparation time, there is cooking time, followed by eating time, and then the dreaded clean-up activities. While components of the process are more rewarding, they all play a role in success.

Making images can make digital "messes" within a computer and camera media. When working on large files, the messes can affect almost every facet of your device's performance. Digital bits and bytes, files and folders from the process can end up all over a drive and in many folders. There can also be invisible "residual digital artifacts" that are in left in RAM, in the CPU, or in the caches in the software. I frequently use my desktop and other locations in my CPU while working on big projects. While somewhat organized, my digital workspace can become a dynamic place with various files and folders here and there. While writing this book, it was not uncommon for me to have countless files open and saved onto the desktop before moving them to my more permanent filing system. Sometimes my desktop held so many folders and files, I simply could not find anything without a concerted effort. This unfortunately happens to most of us at one time or another. I know photographers who keep all folders and images in the operating system's picture folder no matter what. In this fashion, every image file can be copied to other drives without concern for location or leaving things behind. I also know other scientist photographers who make shortcuts from their desktop to their image folders facilitating a simple workflow.

Any digital clutter over time is not good for a machine's performance. It is good to tidy things up as often as necessary. Many folders and files will seem to just show up and busy science photographers can end up with a mess on the desktop or in the main directory. Having an imaging session can be dynamic and it is easy to not clean up after a session because if you are like me, you move onto the next crisis. Over time and when not routinely paying attention to keeping things organized, the digital environment can become similar to the aftermath of a digital hurricane. It is not an overly large chore to keep things tidy when there is only a small volume of data to manage but it takes considerably more time and interest to do so when there is more. This can be especially true when there are a great number of folders and files haphazardly organized. **Best Practice: Develop good computer work habits that ensure a device is tuned up and files are saved into folders located in proper directories**. There is no right or wrong method, simply one that achieves the goals of the organization/photographer. Image directories could be considered a sort of personal Digital Dewey Decimal System.

Keeping Your Computer System Tuned Up

Apple and Windows operating systems are not the same in how they accomplish certain tasks but both systems create random digital artifacts during use. These outcomes in turn affect performance. It is nearly impossible to prevent files from creating this outcome no matter how careful an operator is. Every time a computer is used, this happens to one degree or another. The residual pieces of files remaining over time will slow the processor and clock speed of a computer down.

Best Practice: Pay attention and make a concerted effort to monitor a system's performance. It is important to run a disk utility as needed. There are core things to do, which include, as already mentioned, keeping files and folders where you want them, and removing the digital data artifacts and corrupted time codes and hierarchies when possible using the appropriate software tools.

Something that is also helpful is to keep directories optimized and fixed when corruptions occur. Macintosh computers use a program called Disk Utility®, which looks for problems across its directories. It fixes dates and times so that all applications and files are correct and do not cause conflicts within the computer's operating system and software. When the operating directories are corrected, the computer will perform operations more efficiently. Windows operating systems use a product called Disk Cleanup®. This software helps users free up space on a hard disk by finding files that can be safely deleted. Disk Cleanup performs the following: removes temporary Internet files; removes downloaded program files; empties the Recycle Bin; removes Windows temporary files; removes optional Windows components no longer in use and removes installed programs no longer in use. While each of these activities individually is minor, together they can degrade a computer's overall performance.

It is important to run anti-malware or anti-virus programs frequently as well. There are many companies that sell these products including McAfee® Total Protection for Windows. These products can block the installation of viruses, malware, ransomware, spyware, and unwanted programs. Once a computer has been infected, it may run like a turtle if at all. Also disable any start-up programs that are not required. VirusBarrier® is an excellent product for the Mac environment.

A Summary of Best Practices

- Bring an attitude that allows for success no matter what the challenges.
- Be patient. Successful imaging takes time.

- Prepare for the imaging experiment by assembling all the necessary supplies and equipment before starting. Anticipate what will happen and what will be needed to finish. Remain organized.
- Establish a good space to work and use a sturdy table, tripod, or camera stand.
- Evaluate the sample and select the best choice possible from the inventory. Always consider the ethical aspects for sample selection.
- Develop an appropriate imaging treatment plan.
- Manage the environmental conditions such as the presence of ambient light.
- Select the correct background for the sample based on the intended use of the image.
- Determine the ideal light and organize lighting modifiers if needed.
- Once established, never change the lamp brightness during the photography session.
- Select the best camera for the job.
- Review all the capture settings on the camera prior to starting.
- Select RAW file capture as a priority if possible.
- Select the largest color space possible such as Adobe RGB 1998.
- Select the highest bit depth.
- Select the proper lens for the imaging requirements of the size and scope of the sample.
- Establish the best aperture setting for the sample and imaging objective.
- Closing an aperture (smaller opening) will create more DOF but images will not be as sharp.
- Clean your lens.
- Measure the available light and select the best choice of shutter speed and aperture combination for the sample and imaging requirements. There are a number of combinations of time and brightness that can be used to create reciprocal exposures.
- Never use autofocus.
- Carefully place focus in the best location and align the subject to the sensor.
- Establish the proper white balance setting. Never use auto-white.
- Trigger the camera using methods that do not create or minimize vibration.
- Make multiple exposures of the same event. One will be better than the others.
- Calibrate the monitor prior to image processing.
- Create an appropriate image processing workflow that creates accepted outcomes.
- Carefully review the exposures and determine the best one for image processing.
- Perform image processing to impart minimal personal bias and embellishment.
- Always archive native files.
- Always save image-processed files using SAVE AS.
- Stay organized.
- Practice digital conservation.

Conclusions

This chapter has tried to summarize all the various non-photographic components required to create the best results. Little things will make BIG differences. Making effective scientific images requires attention to details and the desire to do so. Creating successful outcomes takes time and effort. It is easy to lose track of time when making photographs. Time will simply evaporate right before your eyes. It is not overly difficult to develop a Best Practices approach and with experience the approach will pay dividends in improvements to imaging outcomes. Being told to do something will never produce the expected results. The desire to improve must be personal and come from the desire to do so.

Suggested Reading

Color Management in Digital Photography: Ten Easy Steps to True Colors in Photoshop, by Brad Hinkel. Rocky Nook Press, 2006.

Digital Asset Management, by David Austerberry. Focal Press, 2007.

Digital Photography Best Practices and Workflow Handbook: A Guide to Staying Ahead of the Workflow Curve 1st ed., by Patti Russotti and Richard Anderson. Focal Press, 2010.

Chapter 2 Optimizing Imaging Outcomes

The Product is the Sum of Parts

Photographing natural science subjects creates challenges and responsibilities different than the approaches used to make photographs for art or in advertising. The advertising photographer's role is to make "things" look better than they do. Embellishment and amplification are understood to be part of the process. Viewers expect that significant retouching has been done to advertising photographs. One could conclude the role of an advertising photograph is to entice a buyer to become more interested in an item based on emotional reactions to the subject and not the subject's reality. In the fine arts, there are no rules about how to photograph, image process, or display images. Creating successful work is challenging for different and non-technical reasons.

In science, the embellishment of a subject's characteristics or making photographs that include false truths is not ethical. See Chapter 4. Science photographs must be absolute reproductions of an object requiring the formation of highly precise images that use industry accepted methods. Photographs made for science require an absolute verisimilitude to the extent that the technology can achieve. Using a neutral point of view is mandated. It is imperative that images have no personal or photographic bias imparted into an image. This is expected and needed for careful analysis and the preservation of scientific data.

The process of making photographs for science requires the use of appropriate tools. Because of the continual release of new products – with ever-increasing capabilities – there will always be the challenge of staying current and having work evolve with the tools and software. However, the objective in scientific photography has remained the same since photography's invention, and that is to create effective images using the right tools and accepted practices.

Making permanent records of science is sometimes accomplished using specialized lights or selecting appropriate optical methods such as interference needed to make things visible. Making things visible is part of the continuum of linked events that a science photographer must complete. Once an image has been formed, it can be recorded. Recording of an image is a step or an activity within this process of linked events. Samples themselves also play a role in imaging outcomes.

The Imaging Chain

It might be useful to consider that there are many individual events (or steps) and pieces of equipment that are required to form, see, record, display, and store images. Without deconstructing the entire process or repeating what has been covered in Chapter 1, the sequence of events that leads to the formation of an image is called the imaging chain. The imaging outcome is the result of the various decisions made by an operator, their choice of tools, and their ability to form "optimized" images during the process that uses Best Practices. Scientist photographers must select methods that maximize what is possible to see and needed to create visibility. Science photographers must also operate within the accepted practices of their field. These considerations and strategies are called Industry Best Practices and should inform what is adopted in a workflow. Chapter 1 was dedicated to Best Practices.

Making photographs in a natural science environment should be characterized by a series of independent and linked events or steps. These events are grounded in the reality that each

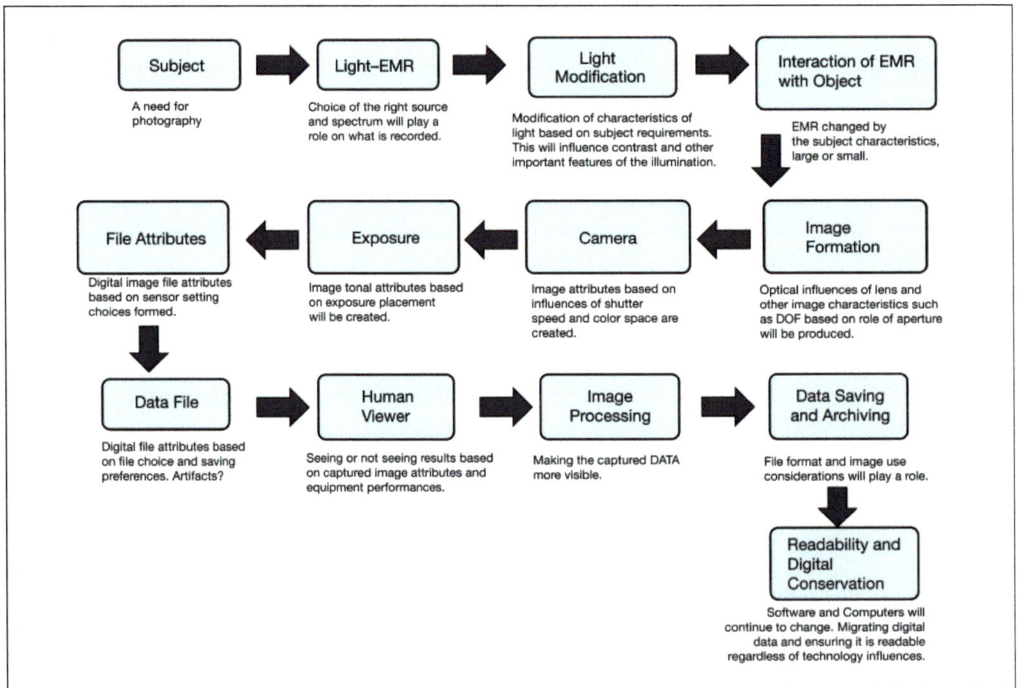

Figure 2.1 There are many contributing factors that play a role in the formation of a high quality permanent photographic image of a subject. Composed of independent steps or locations in a continuum, each step contributes to enhancing or degrading image characteristics. Image attributes such as sharpness, optical resolution, or contrast are altered by various choices or the process itself. While each location in the continuum may introduce only minor changes to image quality, when compounded over the entire sequence, degradation or unwanted enhancement to the image will become evident.

situation is unique. When taking a holistic view of photography that is applied to science, the figure below summarizes why – at each location – the image of an object is affected by the process, by the sample, by the operator, or by other factors.

Pre-Photographic Considerations

It can be helpful to previsualize the result you seek before beginning an imaging session. Establishing the essential expectations for an image by pre-visualizing goals will influence equipment choices. Using effective lighting or choosing camera equipment will define image attributes. Every decision leads to influences on the image's characteristics and might be called a treatment. Depending on the particulars of a sample, a treatment could be pretty much anything.

All samples are unique, and each presents different management and imaging challenges. For example, samples themselves can present difficulties by changing appearance during an imaging session. Samples can be opaque, transparent or both. Samples can be monochromatic or exhibit birefringent properties. Birefringence is the display of multiple refractive indices within a sample and is characterized by a display of rainbow colors. A sample – in the right conditions – might exhibit emissions or behaviors (a sort of fingerprint) that are unique to it. Samples can be motile or without contrast. These and other factors must be considered prior to starting to photograph. Creating sample visibility will directly lead to making good photographs. Identifying what image attributes need to be revealed is a priority. The selection of tools must be carefully considered against the sample's needs prior to starting because once an imaging session has begun, revising something in-progress creates real challenges.

Preparing for Imaging

Successful imaging, like all things, requires effective preparation and planning. Selecting the right equipment is step number one. Collecting the proper accessories and having tools ready and organized before starting will provide a huge advantage when making new work. Photographically the list of "getting ready" steps might include collecting media cards and formatting them, charging batteries, and identifying any auxiliary lighting that might be needed. It is helpful to have a collection of accessories as well. These could include clean glass slides, hand-held dust blowers, lens cleaning supplies, containers, tape, small brushes, and pieces of black and white paper. Your tools could include razor blades, needles, scissors, scalpels, pins, toothpicks, or dark modeling clay. These might be part of the cadre of things that could be helpful. Other useful accessories are Sharpie® markers, aluminum foil, rulers, glass dishes, silicon caulk, magnets, bibulous paper, syringes, and a ready supply of distilled water if working with live materials. Lens tissue is an absolute must. Having the right supplies and accessories will contribute small but important advantages when working on imaging initiatives.

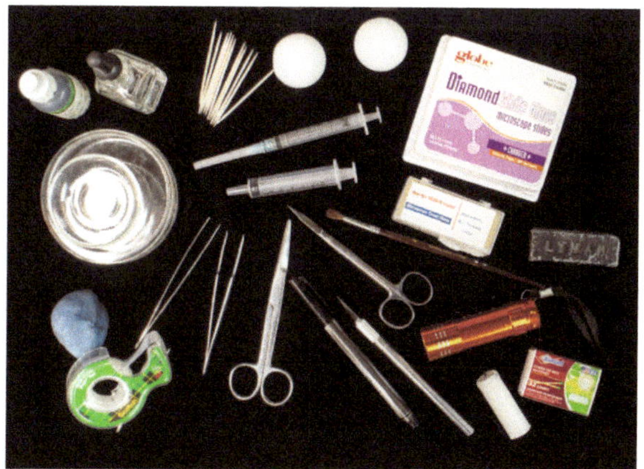

Figure 2.2 Having a good collection of accessories in your imaging "tool box" can provide advantages when making highly precise and challenging images of delicate or rare objects. Image courtesy of Michael R. Peres.

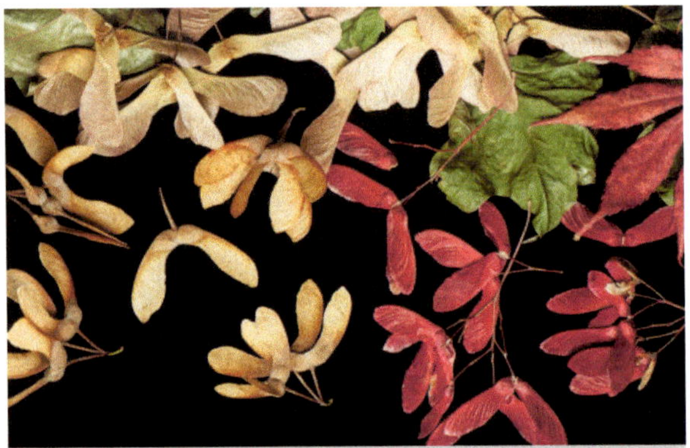

Figure 2.3 A variety of maple seeds. It is easy to see a wide range of attributes for each of the various species. Care should be taken to select a good sample for the type of illustration being produced. Image courtesy of Michael R. Peres.

Selecting a Sample

A sample is not just a sample. A sample serves as the foundation of the permanent image. By previsualizing the goal for an image, you can choose the best sample/subject with great care. The selection of the best sample is one of the MOST important decisions you can make. Selecting the ideal sample based on appearance might not be something that is routinely considered. Because there is variability in nature, samples can exhibit a wide range of presentations. Selecting the right sample that does not have anomalies or damage takes additional time, and can be tedious, but it will contribute greatly to an improved result. Care should be taken in the selection of a sample that is not misleading either. Identify a subject that would be considered typical and not an exception. See Chapters 4 and 5.

Bad samples will lead to inferior images. Sometimes only one sample is available but when there are choices and flexibility, it is important to carefully evaluate all of the samples. All the imaging skills in the world cannot improve a lousy sample. Careful evaluation and selection of a good sample can lead to noticeably better photographs without doing too much else.

Sample Preparation and Artifacts

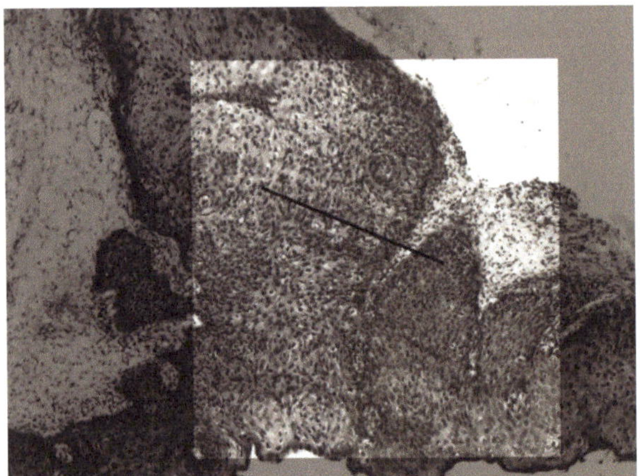

Figure 2.4 When composing a picture, care should be taken to not to include regions that contain artifacts. The white mark is a knife mark and is the result of the sectioning equipment. The blade had a nick and left this divot. In this figure, the white line (thin region) moves from the top left to the bottom right corner. I have added a black line to help draw attention to the artifact. Image courtesy of Michael R. Peres.

When samples are poorly selected or prepared, it can lead to images that are inadequate. This is most evident when an image is larger than the subject, such as in microscopy. When images are magnified, the characteristics of the sample become the features of an image and not the object itself. Sometimes the object itself is no longer recognizable. Bad preparation of less than ideal samples may lead viewers to see the imperfections as the subject. This in turn can lead to false conclusions. This occurs because the defects become more pronounced when magnified. Sometimes during sample preparation, damage occurs to a sample. This type of damage is called an artifact, which is visible as a tear, fold, or surface damage. It might be the result of a tool mark left when preparing the sample. Whatever the case, any evidence of the dissection or preparation of the sample should be minimized or better yet, avoided. It is for this reason and this reason alone, that great care should be taken to evaluate, prepare and select samples more carefully for imaging than for visual inspection. If only one sample is available and it has artifacts, try to exclude them in the framing of the picture. Because visual analysis is a dynamic process, sometimes imperfections in the sample are overlooked or seen as irrelevant because you can simply look past them. How many times when looking for something, do you simply not see it when you start? Images are permanent records and will be carefully studied. Pick your sample wisely.

Getting Started

It does not meet professional expectations to simply take a picture with no regard to methods unless there is no other way forward. Once a sample has been selected, the imaging session begins by using optimal methods to create the previsualized image. The role of the scientific photographer is to isolate the subject and improve the visibility of the science it contains.

A photographer can take a picture, or make a photograph. This refers to the notion that it takes effort to make something good. In digital photography, there is a common philosophy that artifacts or other unwanted components can be easily removed during image processing. This approach will compromise image integrity, should be avoided, and should not be part of anyone's Best Practices. Sloppy work habits will contribute to the presence of artifacts in images. It is best to spend a few minutes before photographing to manage or eliminate unwanted elements from a composition. Some imaging experiments create hundreds of files and if each image requires image editing, it will lead to days of work. Spending a few moments in the beginning of an imaging session to manage samples will create files that need less editing.

Controlling Dirt and Extraneous Materials

In Chapter 1, I mentioned how frequently dirt is photographed. Equipment should be routinely cleaned, and the vessels used to hold liquid samples cleaned using air blowers or lint free cloths. When using black velvet or other material for backgrounds, it is useful to use tape and remove surface dust prior to placing a sample on the background. It can also be practical to consider what type of clothing is worn when working in confined spaces. Fibers

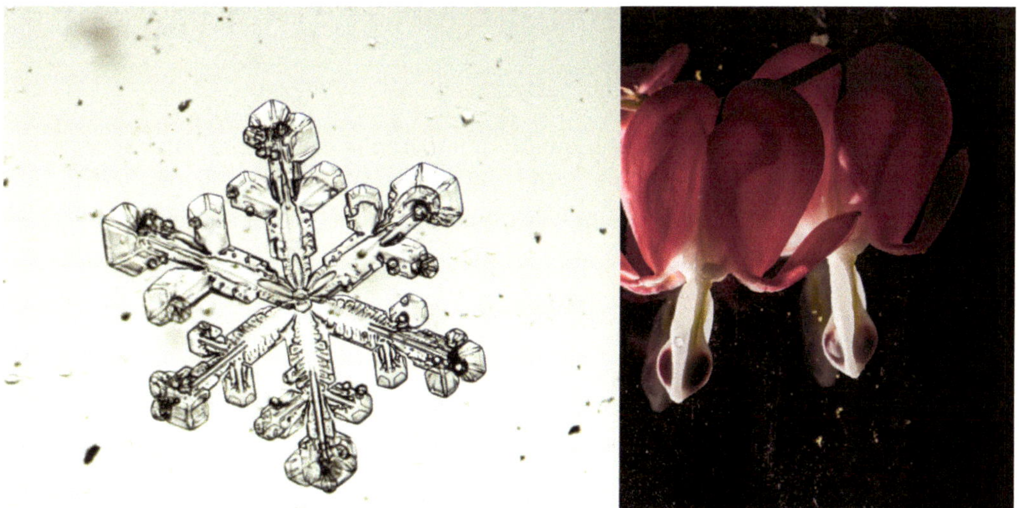

Figure 2.5 Dirt and particulate matter can easily find its way into an image. In this photograph of the bleeding heart flower or snowflake, particulate matter is evident throughout. Image courtesy of Michael R. Peres.

from clothing such as sweaters routinely find their way into imaging environments. Lab coats can be a practical solution to minimize airborne fibers ending up in images.

Handling Samples

Objects should be handled with great care when they are to be photographed. While many subjects are robust and hardy, many are delicate and easily damaged. In the forensic sciences, there is a theory about material transfer. In this theory, it is suggested that when two objects come into contact with one another, content from both is transferred to each subject. The more forceful the interaction, the greater the transfer. This knowledge is very useful when used to solve crimes by the identification of the foreign material. Having foreign material on a subject is not desirable for natural science photography. Care should be taken when touching and moving objects. Use a method that does not apply pressure or damage a sample's surface. Creating ways to move samples from here to there is really important.

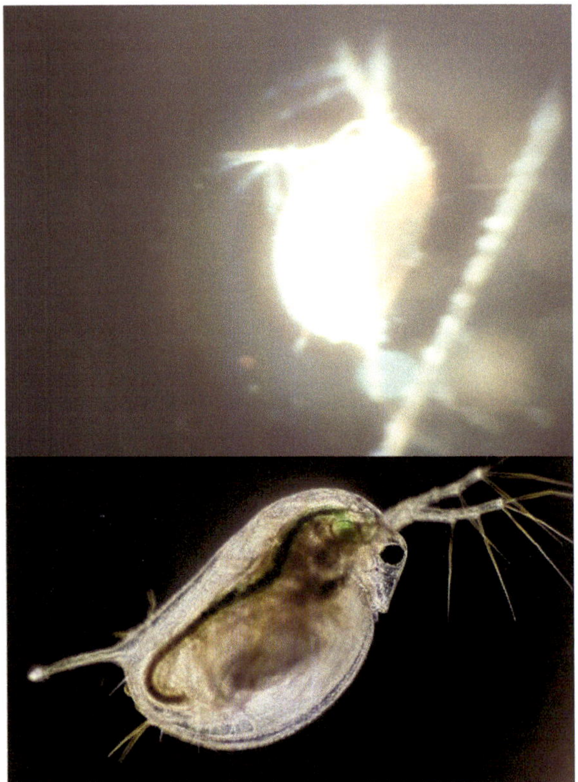

Figure 2.6 These photographs clearly illustrate why operator skills and optimizing an imaging system matters. Both photographs feature a live *Daphnia magna*. Exposure, sharpness, composition, treatment, and the accurate recording of the organism's appearance are criteria of successful photomicrographs. The upper photograph is quite bad while the bottom photograph exhibits proper focus, exposure, isolation and magnification. Images courtesy of Michael R. Peres.

Sometimes a sample will need to be dissected or taken apart. Care should be taken before engaging in the cleaning or preparing of a sample. You should seriously consider the risks and advantages of dissection. It is practical to photograph the object as a whole prior to starting any dissection. It can be relatively easy to damage a fragile sample. It can be a waste of time to photograph damaged samples unless the damage is the focus of the picture. Damage typically will be evident on the surface of the sample, but not always. Move or touch things only once to avoid cross contamination. Once damaged, a sample might be considered useless. In the end, getting better results revolves around the controlling of many little things.

Bad Pictures May NOT Always be Bad Pictures

Making high quality photographs can be challenging for many reasons. That being said, "making an image is better than making no image". To be successful, science photographers must be able to innovate and control a multitude of variables that affect the photographic outcome.

The Role of the Background

Once a subject has been identified, selecting an appropriate background is next. In science photography and certainly in other applications, the background plays an important role for isolating a sample in a photograph. Backgrounds need to be simple, neutral, uncluttered and suited for reproduction or display.

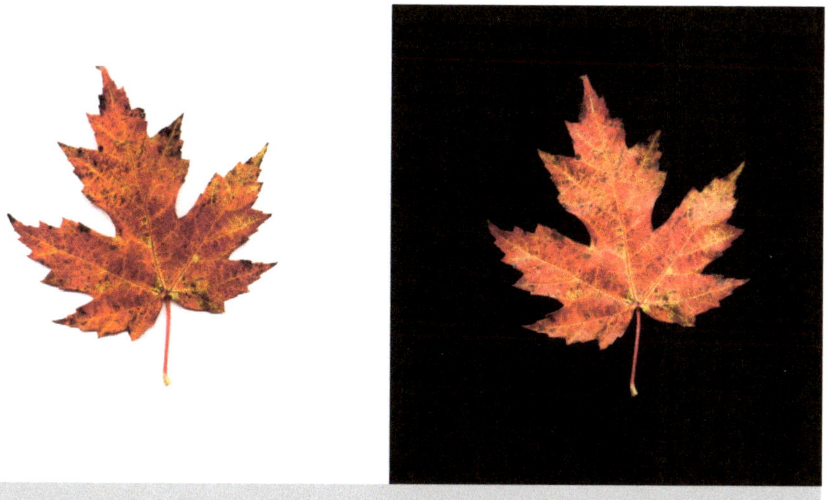

Figure 2.7 This illustration reveals how differently one sample can appear when photographed in front of both a white and black background. The silver maple leaf might look perceptually brighter against black and might look perceptually darker when situated in front of white. Images courtesy of Michael R. Peres.

Backgrounds should never be an afterthought. Poor background choices and management of them will lower image effectiveness and create poor visibility in the photograph. If an ineffective background is selected it will introduce competing visual elements into a photograph.

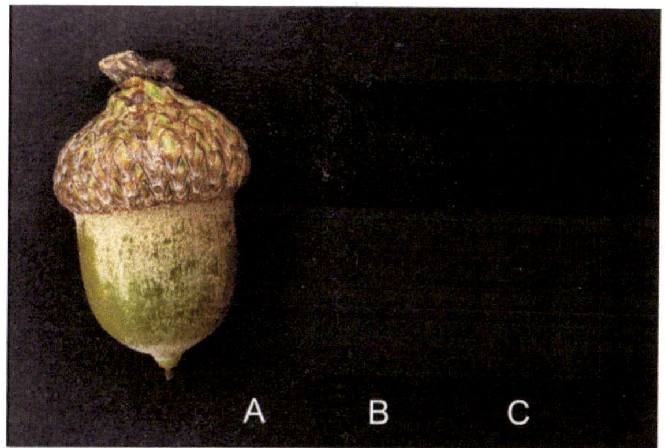

Figure 2.8 All black materials do not behave in the same way when used as a background. In this picture an acorn was photographed on black paper (left). Image B was made using black felt, and image C used black velvet. The velvet created the darkest and best outcome. The black tones is changeable because of surface texture and "blackness" from the background material itself. Images courtesy of Michael R. Peres.

Choosing a Neutral Background

White backgrounds are logical choices for photographing in science. They are useful when there is a need to isolate an object against a bright tone to communicate the shape and delineate the edges of an object.

It can be easy or difficult to make multiple pictures of samples that all contain the same white tone in the background. More is shared in Chapter 19 about image processing and color management. The creation of images with the same tone and color when using a white background has never been easier if the scientist/photographer pays careful attention to optimizing the capture settings and subsequent digital image processing strategies. When preparing for image capture, all settings should be accomplished using the camera's manual mode.

Black can also be an effective choice of background. Unfortunately, dark objects have a tendency to blend into the darkness of a black background when there is no edge lighting. Black can be very dramatic and is much easier to manage than white in some applications. For starters, there is should be no variability of color in black backgrounds. White can be

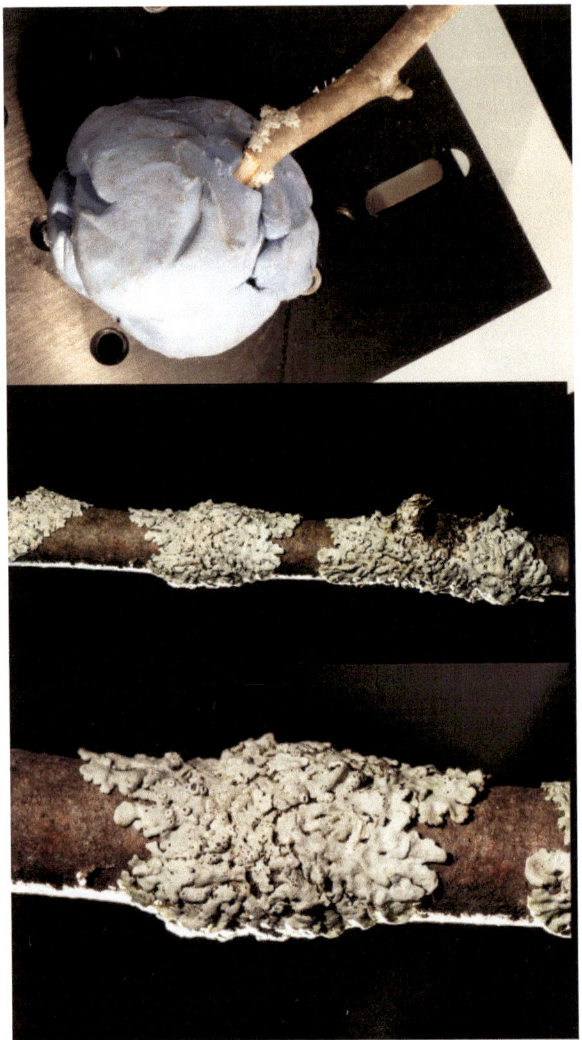

Figure 2.9 In this composite, lichens are visible growing on this branch. The branch was placed in clay as a way to allow for improved backlighting and alignment. Good alignment between the sample and camera leads to optimizing image quality. Simple approaches and effective use of tools can make a significant difference in imaging outcomes. Images courtesy of Michael R. Peres.

vulnerable to secondary influences whereas black is inert and does not reveal minor changes as easily.

Achieving black can be as simple as placing a piece of black material at some distance behind the sample. It is ideal not to shine light onto the background or worse yet cast shadows onto the background itself. A shadow box is a device that does not allow any light to fall on the background. Shadow boxes can be useful in controlling scattered light and the creation of the darkest shadow. For light toned subjects, a black background can be very useful to demonstrate the structures contained within the sample.

Gray backgrounds can be practical but are challenging to use consistently when making numerous photographs. It presents similar challenges to white in its need for consistency. Minor variances in exposure and small lighting challenges are compounded with subtle color changes, which can make gray hard to work with. When multiple photos are to be reproduced as a group, small variances in color and darkness will become emphasized in the composite. It is far easier to manage black and white for this reason.

Create Simple Solutions

Every sample will be unique and require different strategies for display in front of the camera. Sometimes elevating a sample for backlighting is required. One of my favorite accessories is modeling clay. By using toothpicks, pins, or floral wires, a sample can be situated in a variety of ways as needed by locating the sample into the clay. When using clay, keep in mind it may transfer onto the sample when using it. For this reason, care should be taken when and how you use it for support. One of its greatest features of clay is how quickly samples can be repositioned without much work. I keep different colored clay in small plastic bags when not in use.

Chapter 3 The Image in Science

It is widely accepted that images are extremely valuable for communicating and preserving scientific facts. Beginning with the earliest images that featured science, there is compelling evidence to suggest that the chronicling and preserving of observations was valued.

It is also accepted that communicating science using only words is challenging. This becomes evident when considering the multitude of obstacles needed to describe visual characteristics of objects using words. For example, consider describing the color blue with only words. It is near impossible. However, if a color is described mathematically or described using a highly precise photographic reproduction, this problem of accurate representation is mostly mitigated.

The earliest pictures created by people can be found on rocks, in caves around the world, in museums, or in private cultural heritage collections. The images share information about the makers and things that they or societies were interested in at the time of their creation. There are countless examples of early scientific discoveries that advanced knowledge and sometimes were accompanied by illustrations. Creating drawings then or making of photographic images now, both require the use of tools and materials. Good practitioners require skills, experience, and knowledge about how to use materials and tools.

Beginning with the discovery of how to make permanent silver-halide photographic images in the early nineteenth century, photography has not slowed in its transformation into today's digital platforms. What has not changed during this time is the motivation as to why people take pictures in the first place. That motivation remains the same and is to create a precise record of a subject or time preserving visual information. It does not matter if the picture was for personal or professional reasons. Beginning with the very first Anna Atkins or Fox Talbot photographs, the role of the scientific image has been to preserve the scientific record. It requires no additional explanation why mastering tools is a requirement for making effective scientific photographs.

This chapter was written to highlight a small portion of scientific photography's history and explore the beginnings of scientific photography applications and inherent challenges,

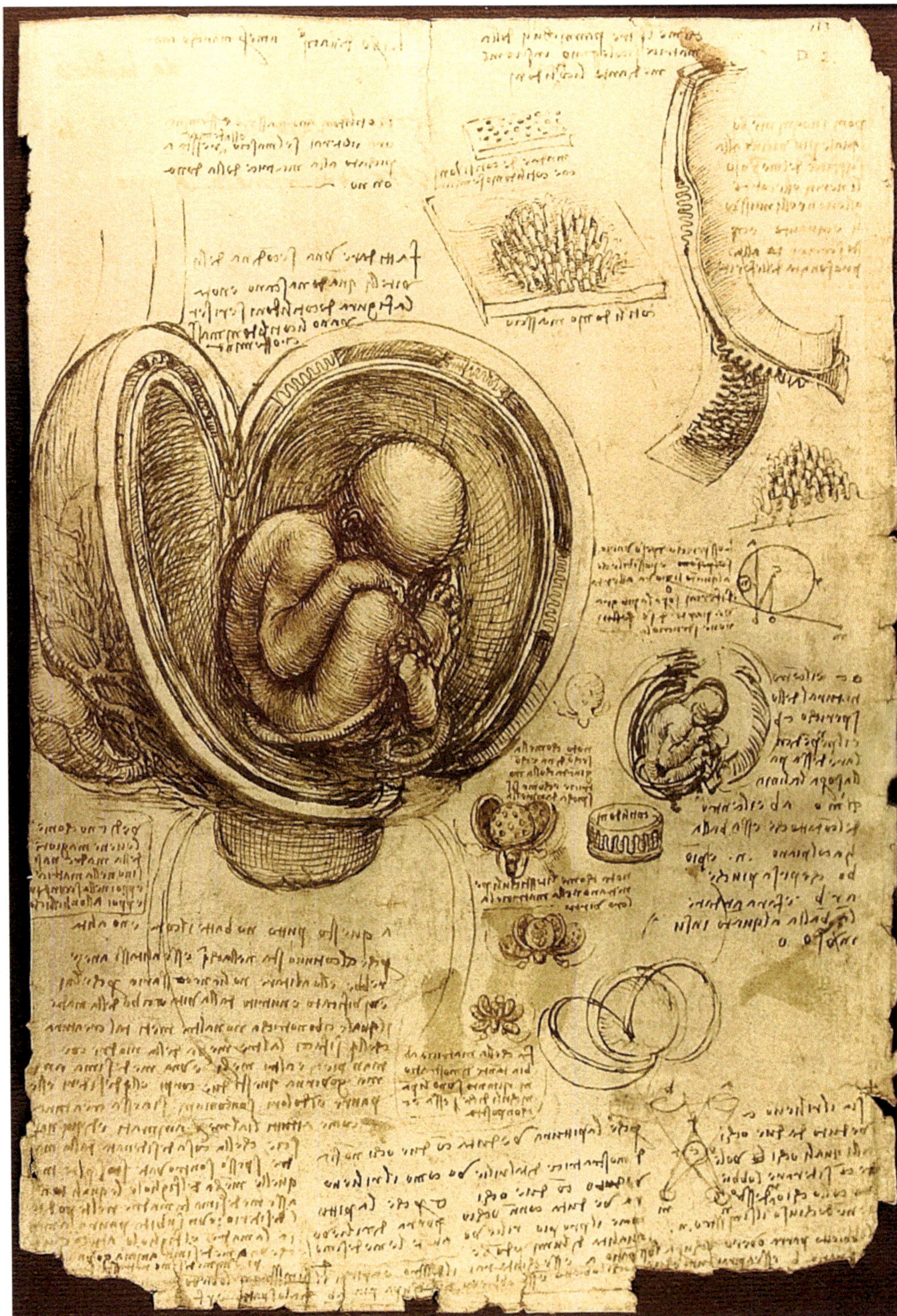

Figure 3.1 This drawing was part of the *Studies of Embryos* by Leonardo da Vinci. It was created using ink on top of red chalk between 1510–1513. The image was downloaded from Wiki Commons and is in the public domain.

as well as the advancement of tools. Sprinkled throughout the chapter will be references to current approaches, key inventors, important equipment, and fundamental practices spanning more than 200 years.

Making a photograph of an object or event imparts a certain importance to that object or event. It is interesting that many of today's Best Practices have origins tied to the time of photography's invention and before. Pictures can record time or produce a pictorial representation of an object. Pictures can be called photographs or images. In science, they are visual information or data and showcase experimental outcomes. As knowledge and technology have advanced, photographic practices have become even more integral to exploration, discovery, and publishing. Images can record what is not visible to the human visual system and they make permanent records of transient events that can serve as images for analysis. Pictures chronicle things where words would mostly be inadequate and images "remember" what the mind will forget. These imaging outcomes are important criteria used to define the science image.

Tenets of the Science Image

Identifying scientific images from other types of photographic images requires the use of criteria. Shared below are three imperatives required for a science image.

- Science images must use industry-accepted practices.
- Sciences images must – at the least – feature science as the subject.
- Science image-makers must use scientific methods.

While there are many subjective practices in photography, it is possible to find many objective elements as well. Many of today's practices were initially proposed for adoption in natural science in the seventeenth century. These include:

- Perform a systematic observation of a subject using the same methods or tools over time.
- Perform a careful analysis of all characteristics of a subject that are visual.
- Develop a protocol (imaging) or experimental approach that includes the formulation, analysis, and possible modification of a method used to create repeatable outcomes (photographs) over time.

Standardized Approaches and Repeatability

When making scientific images, the use of scientific methods is an absolute requirement. Using standardized practices leads to the production of high quality results that will be consistent over time. Much was shared about these ideas in Chapters 1 and 2. Defining scientific methods for some disciplines might be easier than in photography because of photography's subjective practices.

While standardization is the goal, it goes without saying that people will always bring individual approaches to problem solving and thinking. Innovation and problem solving approaches infer that solutions will be unique and lead to departures from standard protocols. Human thought is full of personal bias, distortions and is frequently based on incorrect assumptions. This is what differentiates people from computers. It is assumed the quality of work produced by a person who works in science will be heavily influenced by their experience and biases. Having good analytical skills is crucial for success and is required for systematic methods. This is foundational for the formation of successful and repeatable imaging practices that minimize adding personal biases to methods and outcomes.

Managing Variables

Photographing in science requires the elimination or minimizing of variables required for the creation of highly repeatable steps. When photographing, this would include steps such as:

- Use the same camera (sensor) and create NO changes to camera settings or object(s) over time.
- Use the same lens and settings.
- Use the same lights and lighting.
- Use the same aperture and its depth of field.
- Implement a standardized treatment of the object, creating a neutral perspective including where focus is placed, etc.
- Use the same background.
- Use the same point of view and magnification.

The only thing that should change between photographs is time. Photographing with a neutral point of view and treatment is core to scientific methods. This approach will introduce minimal distortions or embellishments to the pictorial result. In scientific photography, the

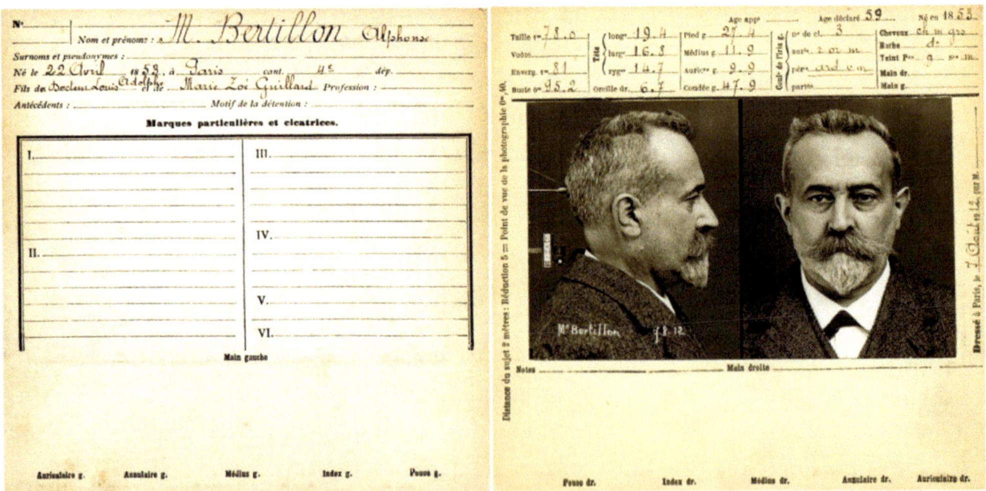

Figure 3.2 An anthropometric data sheet (both sides) from Alphonse Bertillon (1853–1914). Bertillion was a pioneer of the Scientific Police and the first head of the Forensic Identification Service of the Prefecture de Police in Paris (1893). Image courtesy of Jebulon, from the Service Regional d'Identité Judiciaire, Préfecture de Police, Paris and is in the public domain.

goal of the process is to create a precise facsimile of the object. The image should not be negatively affected by the equipment or operator. In science, images must be science facts, not science fiction.

Alphonse Bertillon (1853–1914), a French police officer, was an early adopter of standardized photographic practices. His work led to the creation of the "mug shot" system as a part of an identification system based on the physical attributes and measurements of suspects. The approach was influenced by the field of anthropometry. It became the first scientific system used by police to identify criminals, but over time was replaced by the more accurate fingerprint photographic processes. In 1888, Bertillon standardized his photographic process and it remains as valuable today as the day it was proposed.

Science Images Require the Inclusion of a Scale

Images operate on various levels and in many different environments. Human perception plays an important role in what is seen, observed, or what is believed to have been seen. Photomicrographs can operate as science images or sometimes as art images. This duality is based on the point of view of the observer and how an image is displayed or used. Astronomical photographs operate similarly. Scientific photography might be described differently depending on who is asked for the definition. These definitions might include reference to the equipment that was used such as in photomicrography or it might be described by the application such as in radiography. It can be complicated to categorize images as being only of one type.

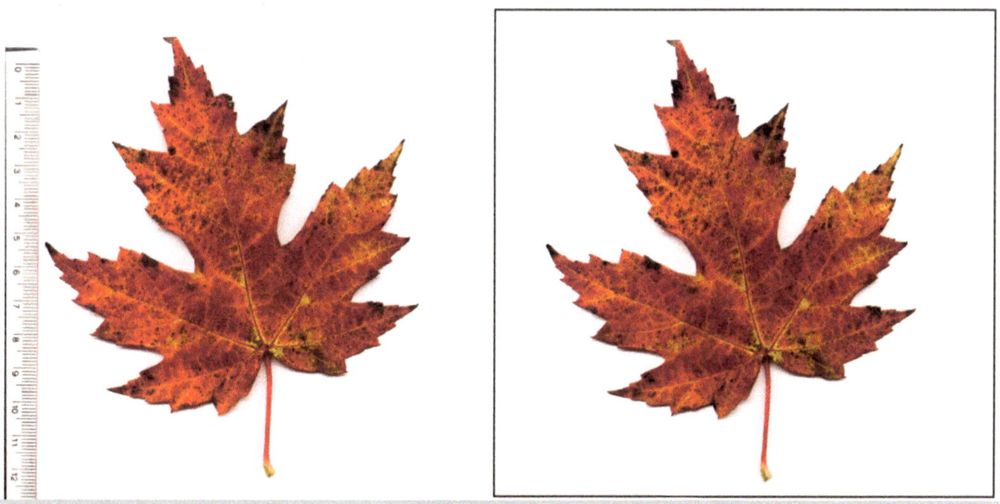

Figure 3.3 These photographs feature a silver maple leaf that recently had morphed into its fall colors. It has been photographed with a scale (left) and without a scale (right). The inclusion of a scale clearly implies a science image. Without the scale, a viewer might see the image's characteristics of shape, color, and natural design before assigning the image, as a science image (or not). Image courtesy of Michael R. Peres.

One absolute requirement for the identification of a science photograph as opposed to other types of photography is the need for the inclusion of – and proper use of – a ruler scale or time stamp. Whether a ruler is used for size, scale, or as a time stamp, the inclusion of scale clearly delineates a science image from any other type of photography.

Photographer's Intent and Subject Matter

For many, the intent of the photographer might also be used to discriminate scientific photography from other types. Sometimes subtle and not prominent elements play a role in this process. Sometimes the need to distinguish one image from another is important while at other times it is not. It is actually quite easy to photograph science subjects in non-scientific ways and it is also possible to evaluate science images – after creation – using a lay point-of-view. The images in Fig 3.4 were produced for very different reasons and at different times by two different photographers. In both images, the subject might be considered to be the same, a butterfly. A quick assessment of the images might lead a viewer to conclude they are both science photographs because of the subject. While the content in both images is similar, the intent of the photographers could not be more different and the treatment of the subject similarly different. In conclusion, a subject by itself cannot be used as the sole criteria for the categorization of an image. It is difficult if not impossible to measure the intent of an image-maker.

A Picture is Truly Worth a Thousand Words

The need for – and the importance of – scientific illustrations has a well-documented history. Carl Von Linneaus was an extraordinary artist scientist who chronicled – to the best of his abilities and the capabilities of his materials – the exploration of biology and botany. Initially created as scientific illustrations, these extraordinary drawings and many others over time

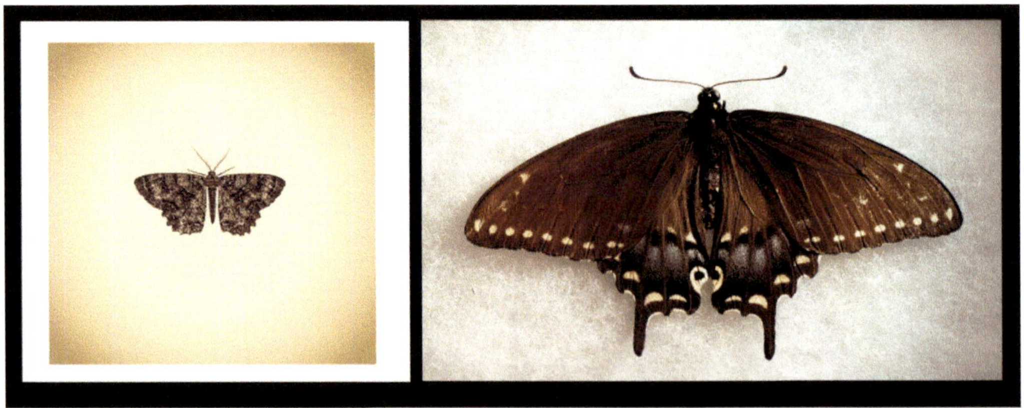

Figure 3.4 Sometimes it can be challenging to discriminate a science photograph from a fine art photograph. Often the boundaries of distinction are not clearly delineated. The subjects in both photographs above are members of the family Lepidoptera. The photograph on the left was made by Joshua White, a fine artist and photo-educator. The image was reproduced courtesy of Joshua White. The image on the right was made by and courtesy of Michael R. Peres.

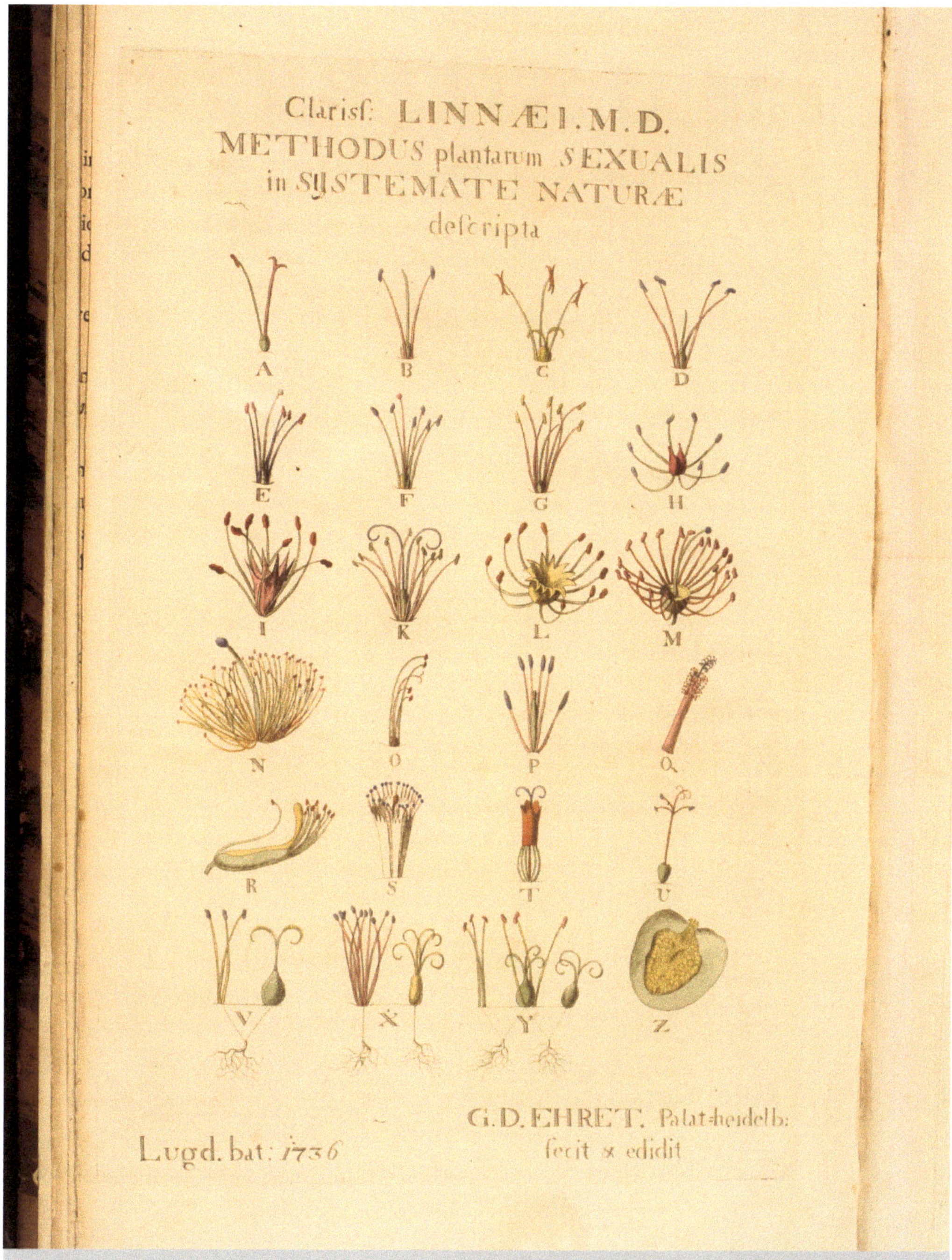

Figure 3.5 A page from *System Naturae* by Caroli Von Linne from 1768. It is a notable example demonstrating the newly "forming" importance of scientific methods used for illustrating science in the pre-photographic period. The publication is part of The Hagströmer Medico-Historical Library, Karolinska Institutet Stockholm, Sweden and is in the public domain.

have become valuable pieces of art and collectible. At the time of their making, they were produced to chronicle science.

A Quick History of Photography and Scientific Photography

The earliest known examples of photographic images using light-sensitive materials were made at the turn of the nineteenth century. Thomas Wedgwood and Humphry Davy made pictures using an emulsion of silver nitrate. In their experiments, leaves were placed onto a light sensitive material. The material was then exposed to sunlight and got dark over time. Unfortunately the photographs could not be made permanent. In the end, they did not develop a process that would make their images permanent but their discovery served as the foundation of silver halide photography that was to come.

Permanent silver halide images were first achieved in 1826 or 1827 and attributed to Joseph Nicéphore Niépce. His picture entitled *View from the Window at Le Gras* was the outcome of years of experimentation. Louise-Jacques-Mandé Daguerre was also working on a process later named the Daguerrotype. Numerous daguerrotypes of scientific work have survived time. William Fox Talbot was also a pioneer of a photographic positive/negative process and he developed the calotype process. Talbot published much of his early work in a book, *Pencil of Nature* during 1844–1846.

Talbot's earliest pictures were photographic traces of botanical specimens that were laid on sheets of chemically sensitized paper. As a serious botanist, Talbot envisioned the accurate recording and ease of sharing of natural specimens to be one of the most important functions of his invention. This image is from a print in the collection of the Metropolitan Museum of Art and is in the public domain.

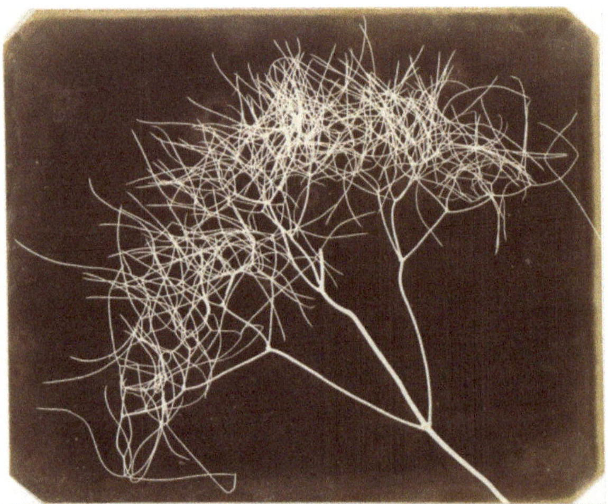

Figure 3.6 This photograph of wild fennel was made in 1841–1842 by William Henry Fox Talbot. It represents one of the earliest permanent photographs of a scientific image.

One of photography's greatest attributes was that it depicted subjects with a precision that unskilled artists or their drawings could not. Photography was considered real and because it was relatively permanent, it preserved facts. Soon after photography's invention, scientists, researchers, and physicians began incorporating photography into their work. From the earliest beginnings, science used more and more photography. The earliest scientist photographers believed that a photomechanical process would be viewed as more credible than drawings which had communicated ideas to scholars. Today the use of digital technologies and the ability to manipulate images using software creates credibility issues about whether an image is truthful or contains false information.

Anna Atkins's book *Photographs of British Algae: Cyanotype Impressions* produced in October 1843 is considered the first book to include solely scientific photographs. Working with Sir John Herschel, they invented the cyanotype process. Blueprints, as they were later called, are still produced for artistic reasons but the blue print process used for architectural drawings has been mostly retired. One of Akins's illustrations of *Cystoseira granulata* is included in the Preface.

As photographic knowledge grew and materials improved, the adoption of photography in science also grew. Photographs became a new tool for science and allowed for a careful inspection of things difficult to observe in real time. Permanent accurate images provided an opportunity for critical analysis of an event long after it was concluded. In many cases the photography could be accomplished in a controlled environment. While many events are transitory, the photographs provided a new way for the collection of scientific facts.

Carl Curman (1833–1913) a Swedish professor of medicine was also a science photographer who worked at Karolinska Institutet in Stockholm, Sweden in the mid 1800s. He was an early adopter of photography and used it to collect and preserve scientific facts about many of his medical subjects and his other interests in science. He, like many, realized photographs could preserve scientific facts in an objective manner and record even the tiniest of details if careful attention was paid to the work. Making photographs was certainly much faster than trying to draw the object with the same levels of detail. In Figure 3.7, Curman's photographs shows objects against dark and light backgrounds. We can assume he observed how different backgrounds changed the visual tonality of his subjects and tried different ones as needed.

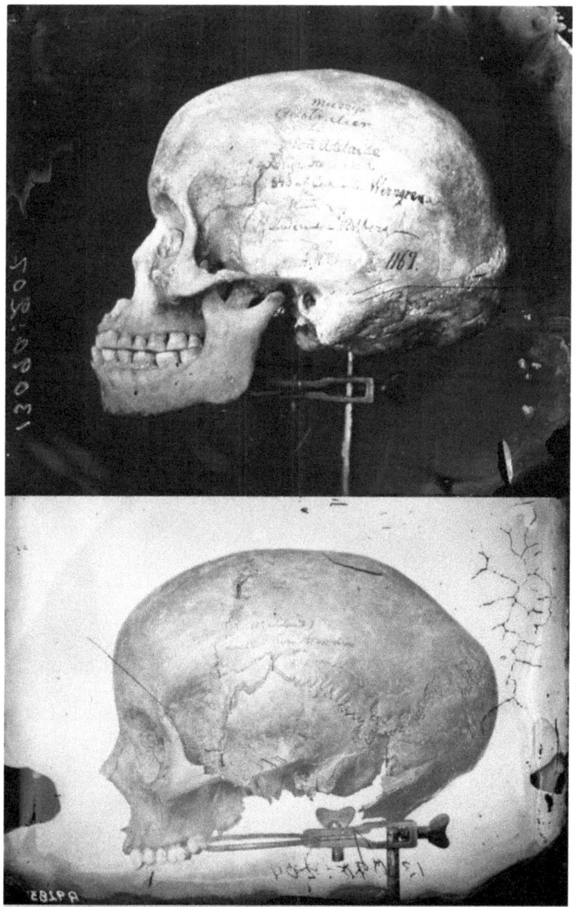

Figure 3.7 These pictures are early examples of laboratory photography made by Swedish physician and scientific photographer Carl Curman at the Karolinska Institutet, circa 1865. The images are in the collection of The Hagströmer Medico-Historical Library, Karolinska Institutet, Stockholm, Sweden and are in the public domain.

Making the Invisible Visible

Scales, subject matter, and intent all play a role in discriminating scientific photography from other types of photography but there are also other factors. Creating images that contain empirical facts (data) can define a science image (at the time of their making) as well. An important early example of this can be found in Eadweard James Muybridge's photograph of a galloping horse. Muybridge was hired in 1872 by Leland Stanford to make photographs of a galloping horse in the hopes that the photograph might show all four feet of the animal off the ground at the same time. Muybridge successfully captured this outcome in 1877 by assembling twelve cameras that created consecutive exposures at regular and pre-programmed intervals. He published eleven volumes of his motion analysis work in 1887.

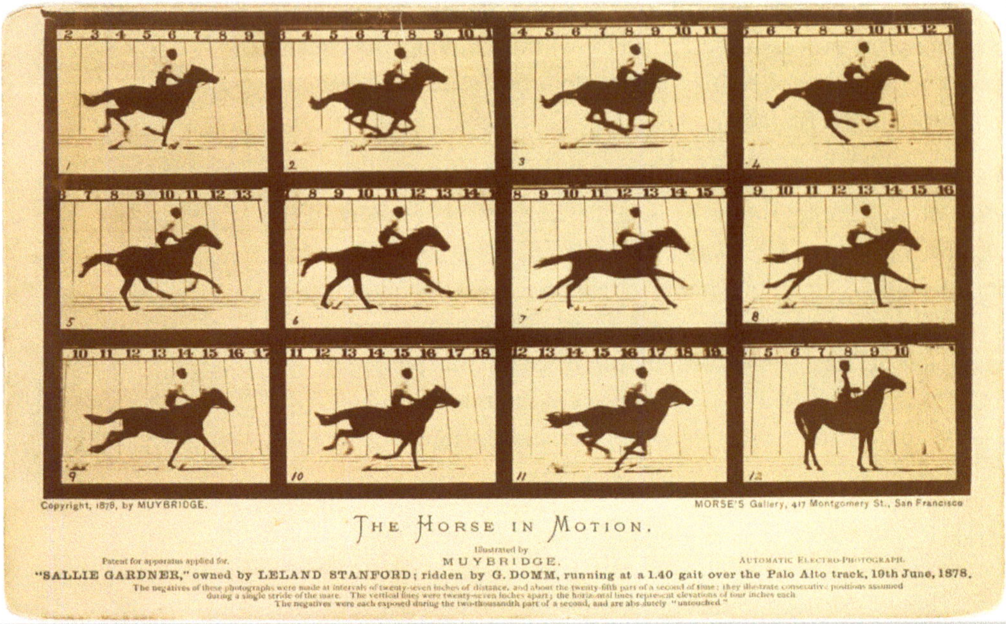

Figure 3.8 This composite photograph titled *Galloping Horse* was made in 1877 by Eadweard James Muybridge. The image is courtesy of Library of Congress, Washington D.C. and is in the public domain.

Instrumentation

It can be argued that art, because of its interpretative nature, creates work that explores ideas or emotions. Scientific photography, in contrast, requires veracity and telling whole truths. Image integrity is an absolute expectation in science. There are, however, elements of the human visual system including perception that will influence how an imaging system operates. The human motivation to see and learn more has been fundamental to discovery. Whether trying to increase or extend the sensitivity of a sensor, or enabling an instrument with special lenses to create higher resolution, there are ultimately limitations to what is possible and this will influence outcomes. The need for extending vision has been a constant challenge for photographing what is invisible since the invention of photography itself.

Two early and vitally important optical instruments that added to scientific exploration and discovery were the telescope and microscope, both invented in the early 1600s. These instruments predated photography by almost 200 years. While predating photography by centuries, these instruments contributed significantly to the creation of detailed drawings of observations and new knowledge. Robert Hooke was an early pioneer of applied microscopy and is credited with suggesting the term "cells" used to describe the things he observed under the microscope. He referred to the structures as cells because they reminded him of monk's cells.

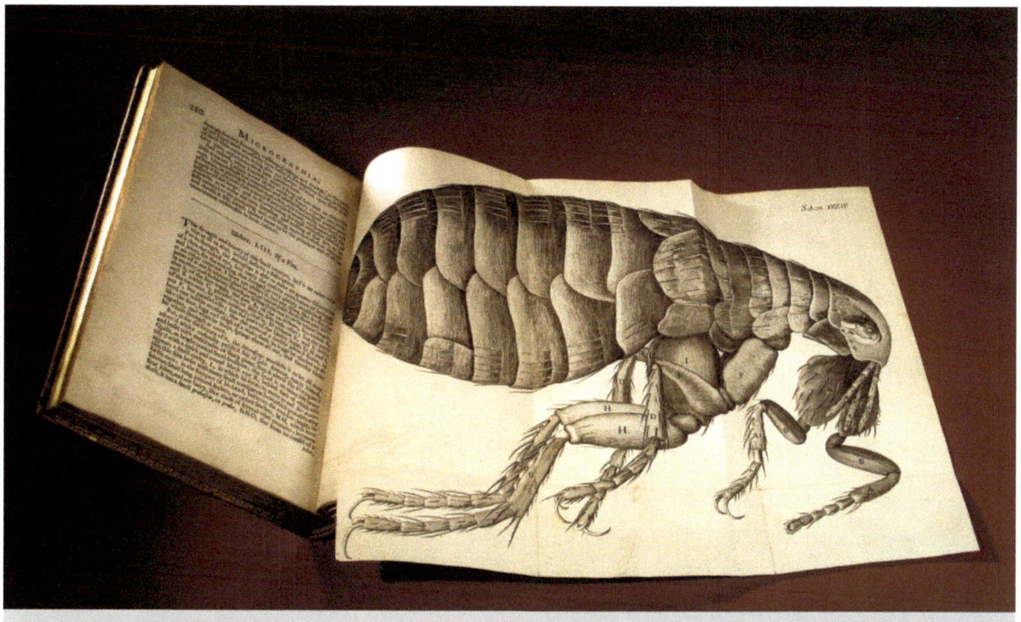

Figure 3.9 This is a drawing of a flea from Robert Hooke's book *Micrographia*, 1636. The image is from the Wellcome Collection and shared using Attribution 4.0 International (CC BY 4.0) and is in the public domain.

Scientific Instruments and Optical Improvements

Contemporary microscopy and other optical instruments have remained powerful tools for research, discovery, and imaging. Carl Zeiss Microscopy has remained a leader in optics for more than 180 years. Optical Instruments was the name of the first company founded by Carl Zeiss (1816–1888) which opened as a lens manufacturer in 1846. Zeiss became notable because he produced high quality lenses that formed very large and bright images. Zeiss's business made contributions to lens manufacturing that aided the production of improved, modern, and specialized optics used in the production of microscopes. In 1872, physicist Ernst Abbe joined Zeiss and invented ways to correct a lens's optical performance that formed images which had noticeably improved sharpness and contrast. Abbe made numerous significant contributions to the field of optics, most notable of which was his groundbreaking work on diffraction theory and image resolution that still remains relevant even today. In 1886, Abbe with Otto Schott produced a new type of glass that transformed lens making for microscopes. This discovery in improving glass made possible the production of a new type of microscope objective, the apochromatic lens. The APO lens remains the gold standard for optical performance.

Invisible Spectrum Photography

Infrared radiation (IR) was discovered in 1800 by astronomer William Herschel. Herschel became aware of this radiation because of the affect heat had on his thermometer. The discovery of ultraviolet (UV) radiation was also made through observation 1 year later. In 1801, German physicist Johann Wilhelm Ritter discovered UV while watching silver salts darken when exposed to sunlight. Ritter's curiosity arose when darkened silver chloride-soaked paper happened more quickly when in the sunlight than when using violet light by itself. Ritter called the result oxidizing rays.

Photographing using the invisible spectrum was nearly impossible at that time because emulsions were incapable of recording near UV or IR radiation. The first infrared photographs were made by Robert W. Wood and published in 1910. Wood's photographs were taken using an experimental film using very long exposures. The IR photography was accomplished using plates that contained special emulsions created by Charles Edward Kenneth Mees who worked at the photographic company Wratten & Wainwright. The Wratten & Wainwright Company would be purchased and became a part of Eastman Kodak in 1912.

The discovery of X-ray radiation and its ability to see within an object was also a significant tool for science and imaging. Wilhelm Conrad Röntgen (1845–1923), a German scientist, produced and detected electromagnetic radiation of short wavelengths and high energy

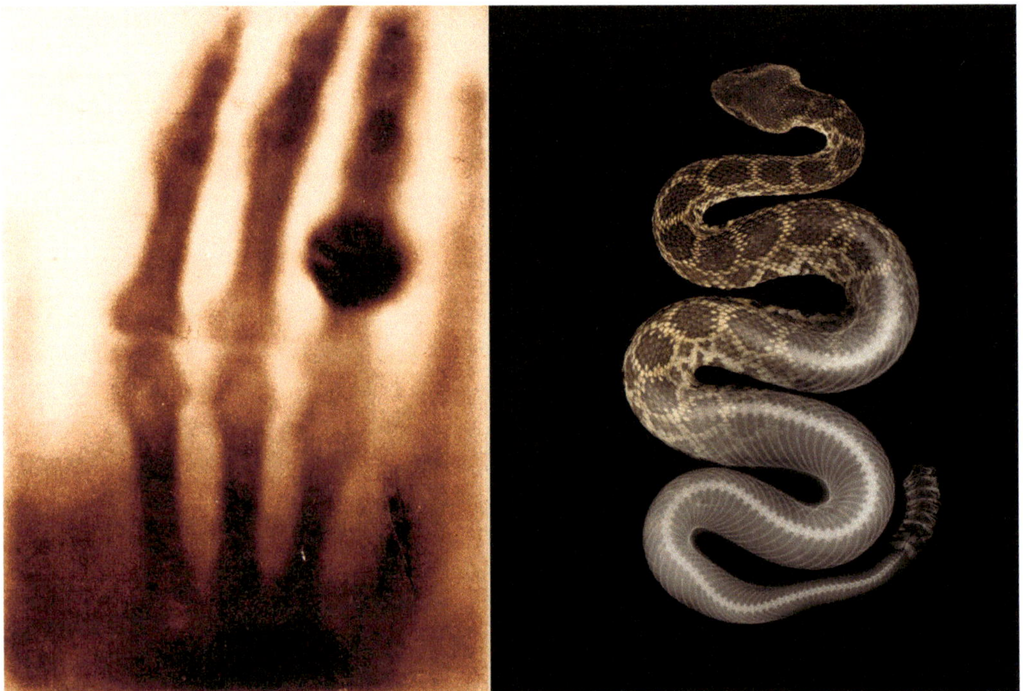

Figure 3.10 The radiograph on the left features the hand of Mrs. Wilhelm Roentgen and was captured in the first X-ray image made in 1895. This image was originally reproduced in Otto Glasser's *Wilhelm Conrad Röntgen and the Early History of the Roentgen Rays*, London, 1933. The image is courtesy of The History of Medicine Division, National Library of Medicine of the National Institutes of Health and is in the public domain. The photograph on the right was made by author and notable natural science photographer Ted Kinsman whose work explores the boundaries of visible and invisible spectrums. By combining X-ray images with visible light photographs, Kinsman's image of a rattle snake provides a better understanding of the structural and external features of the snake. Image courtesy of Ted Kinsman.

today known as X-rays. At the time of their discovery in 1895, they were called Röntgen rays. Röntgen referred to the radiation as "X" because it was an unknown type of radiation. Röntgen earned the first Nobel Prize in Physics in 1901 for this discovery. Radiography was immediately embraced as a powerful tool for seeing the invisible and inside of solid materials and medical images were being produced 1 year after its discovery.

Advancements in Photographic Technology

Cameras, lenses, and gaining a better understanding of radiated energy all played important roles for photography in the late1800s. The capabilities of the materials and the equipment available at the time ultimately determined what photographs could be made. Eastman Kodak Company made the most photographic products used in science. There were others companies including Agfa, Ilford, FujiFilm, 3M, and Polaroid who also produced important products for science applications for decades but not on the scale of Kodak.

Kodak officially opened in 1888 after 7 years as the Eastman Dry Plate Company. George Eastman pioneered the making of "dry" film. Prior to this invention of dry plate technology, photography was accomplished primarily using the wet collodion process, which used wet emulsions on glass plates. It was not a standardized process. By 1885, Kodak had invented

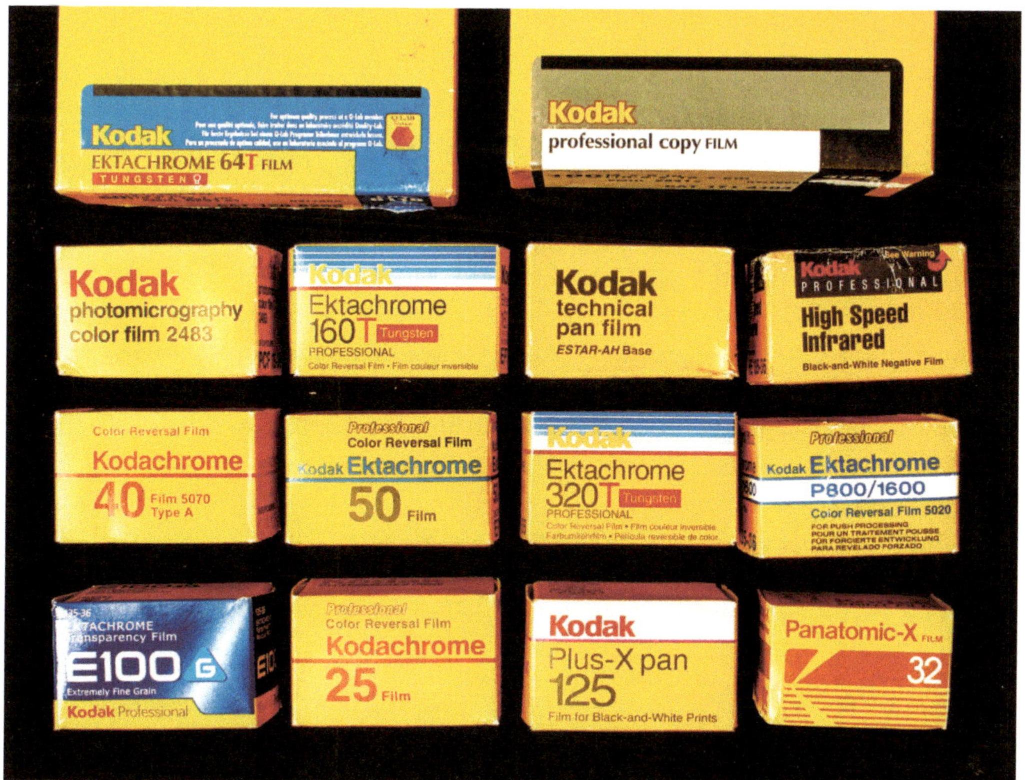

Figure 3.11 Various boxes of important Kodak scientific films. Kodak film technology pushed the boundaries of what was possible to record using sensitized silver halide products. Product images reproduced courtesy of Eastman Kodak Company.

flexible roll film. During most of its 140-plus year history, Kodak influenced what was possible to photograph using film. Their ever increasing interest in making new and special films and related photographic processes played a profound role in scientific imaging.

As photography became mainstream at the turn of the twentieth century, the invention of new products and services from Kodak exploded and was critical for discovery. New Kodak discoveries allowed film to be used in more demanding environments in science and astronomy. Kodak significantly improved the film base materials and gelatin required to suspend the light sensitive salts. Compounded with chemical development, products featured improvement to contrast, resolution, and sensitivity (ASA, light speeds, or now called ISO). Image stability was also greatly improved and would continue for the decades that would follow.

Eastman Kodak first produced infrared sensitive emulsions in the 1920s. In 1935, the Kodachrome slide film was invented, and in 1963, the Lunar BIMAT emulsion was created for NASA and John Glenn's orbit around the earth. Ektachrome, a color slide film, was initially invented in the early 1940s. In the late 1970s, E-6 processing was invented to support that emulsion for easier development.

Electron microscopy emulsions were released in the 1960s and UV emulsions found their way to market in the late 1960s. In 1971, photomicrography color film was invented and was followed shortly by an extremely high resolution emulsion, Technical Pan film which was able to differentiate 200 lines/mm. In 1982, Kodak created tabular grain silver halide films that resulted in the production of a whole new generation of silver halide products at the height of film photography's usage.

While Kodak was predominantly in the U.S. market, Agfa and Ilford were large players in the European market. Agfa originated from a German company that produced chemicals and dyestuffs in the early 1900s. In the early 1920s, the manufacturing of photographic products became dominant. Agfa developed a grain screen process in 1916 that enabled color transparencies to be produced. Agfa patented a triple-layer reversal color film in 1935 that went into production as Agfacolor. The new reversal film came in 35mm and 8mm sizes. A few years later, Agfa introduced a subtractive color negative film. Agfa manufactured and distributed a sizable amount of its photographic films, processing equipment, paper, and chemicals to the minilab/retail, professional, wholesale, and consumer markets. Some of the notable products from Agfa included: Agfa Contour Agfa DIA COPEX R, Radiography films, Agfa Otho 25, Diazo films, PMT processes and Scientia 10 E.

Short Duration Light, Electric Flash and Stroboscopes

One of the challenges in science photography has been the need to create short duration lighting. In the late 1800s, sparks were used for the sole purpose of stopping the action of an event. In 1917 Etienne Oehmichen invented an electrical stroboscope that was used for examining engines while they were running. By 1935, the Oehmichen system was capturing pictures at a rate of 1100 frames per second. Following the invention of this technology,

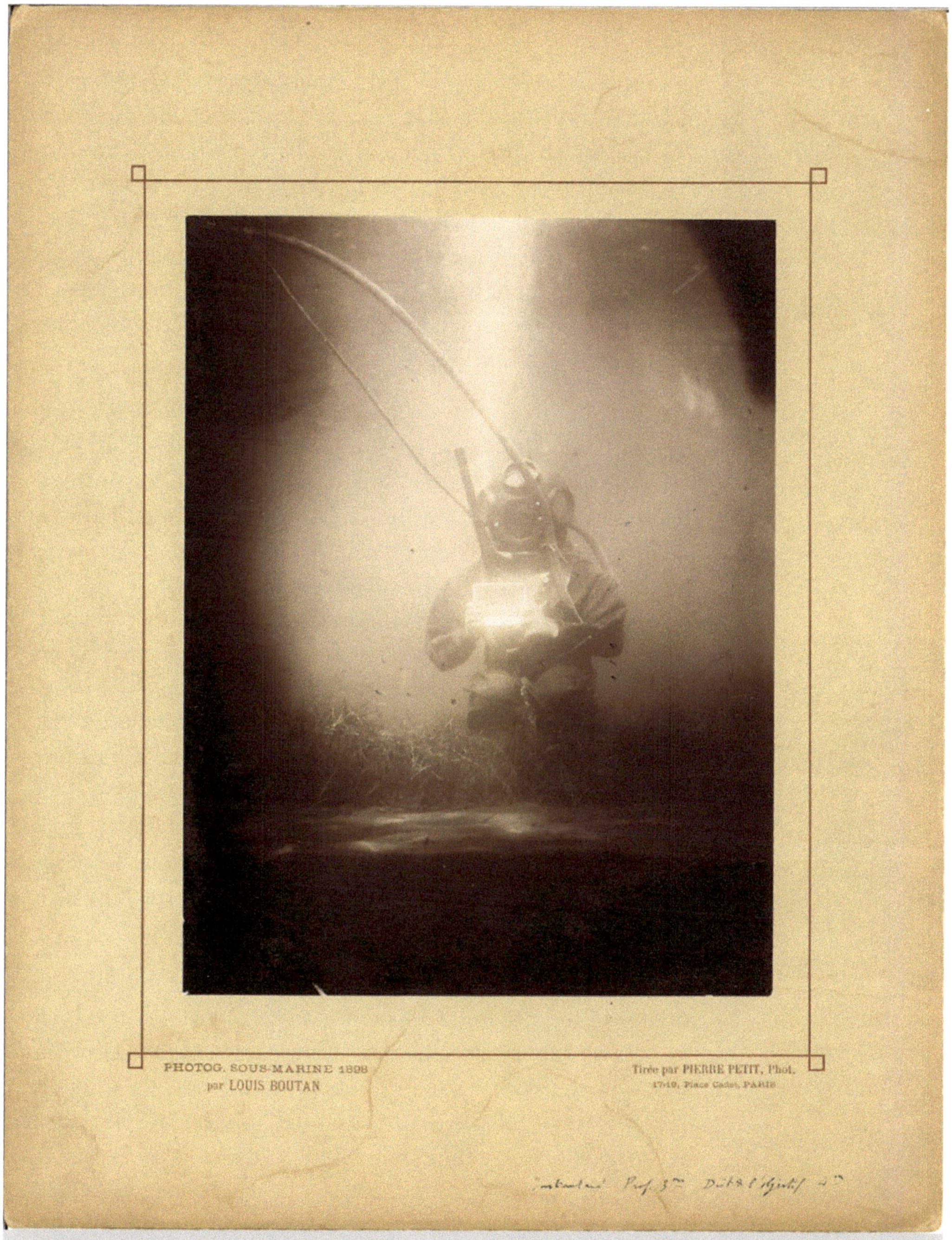

Figure 3.12 While this image of Louis Boutan is not the first image to be taken underwater, it is one of the earliest. Boutan worked with his brother, Auguste, to build this special underwater camera which allowed them to make adjustments to the equipment without flooding it completely. Because there was no high-speed film, many of their exposures lasted for 30 minutes. Eventually, they also figured out a way to discharge a flash underwater using oxygen, magnesium, and an electric current. This photograph was made by Louis Boutan, ca. 1893 and is shared using Attribution 4.0 International (CC BY 4.0) and is in the public domain.

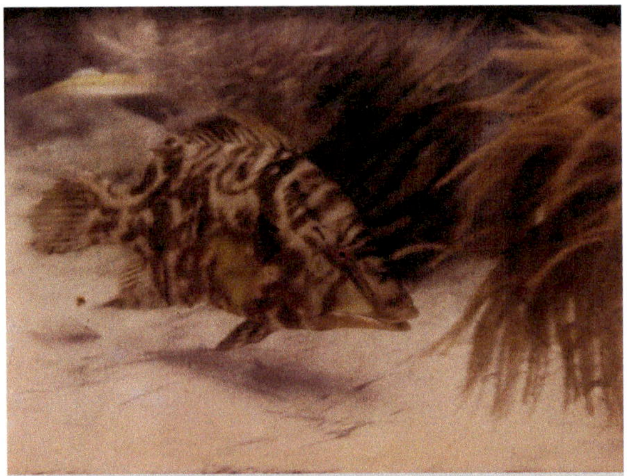

Figure 3.13 Innovation was required to make the first underwater color photograph in 1926. Can you imagine setting off an explosion to make a photograph? The photograph features a hogfish. Image courtesy of National Geographic.

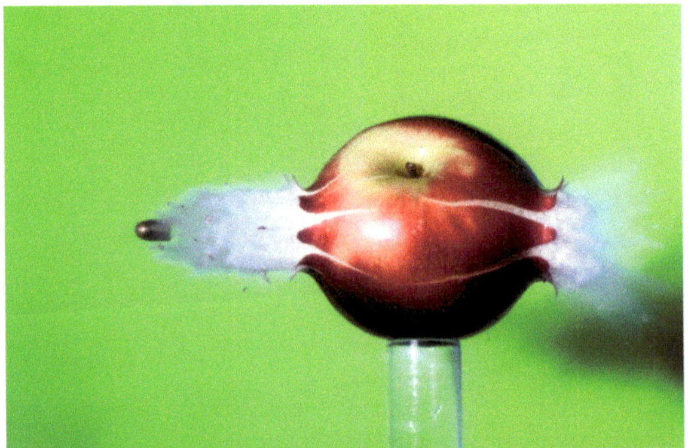

Figure 3.14 This high speed photograph was made by Professor Ted Kinsman. It uses the same techniques pioneered by Harold Edgerton's in his seminal work. The image features a .22 caliber bullet. The flash duration was 1/100,000sec. Image courtesy of Ted Kinsman.

Augustin and Laurent Seguin discovered that by accumulating electricity using a capacitor, they could create a spark with more energy than could be provided from a battery by itself. Until this point, all stroboscopes used a single electrical circuit that ionized the gas, which gave off light when electricity passed through it. Various switches were needed to deliver the electrical energy to the gas-filled tube and the equipment was bulky and hazardous. Some early underwater photography used explosions for artificial lighting since the light available at certain depths and the speed of the photographic film did not allowed for a usable photographic exposure.

The first underwater color photograph was made in 1926 and featured a hogfish. See Figure 3.13. It was photographed off the Florida Keys by Dr. William Longley and National Geographic staff photographer Charles Martin. The camera was encased in waterproof housing and pounds of highly explosive magnesium flash powder required for underwater illumination was set off when ready. The flash powder was located on a frame that was placed above the area where the camera was to be located. The explosion of light from the magnesium flash powder allowed Longley and Martin to photograph underwater life for the first time. When ready to photograph, the photographers activated the camera's shutter and triggered the magnesium powder explosion, which illuminated down to 15 feet (4.6 meters). Unfortunately during this expedition, Dr. Longley was burned by the explosion and spent 6 days in the hospital recuperating following the making of this photograph.

Harold Edgerton is considered to be one of the most important figures in the history of modern electronic flash lighting. While at MIT, he quickly elevated the research and capabilities of short duration lighting technology. Short duration flash equipment had proven useful to scientists and the industry by allowing the photographing of events that could not be seen any other way. During the 1930s, Kodak improved its emulsion sensitivities to respond to short duration energy and the development of high-speed color films allowing this important work to find new applications. Edgerton's work led to the creation of a flash system that had the same intensity as 40,000 standard 50W bulbs. This was incredibly powerful at that time.

Most of Edgerton's work was accomplished in a darkened room, which was used to record the effects of the stroboscope using an open shutter. Edgerton is also well known for the development of effective methods for synchronizing short duration events required for timing with flash discharges. His high speed techniques have remained mainly unchanged for nearly 80 years later.

Modern Technologies – Digital and Electronic Photography

In 1975, the first digital camera was invented by Kodak imaging scientist Steve Sasson. His invention used a charged couple device (CCD) as the sensor. The non-commercial camera was very heavy and cumbersome. It wrote data files to cassette tapes. The sensor had 10,000 pixels, created a grayscale image, and took 23 seconds to be recorded following the exposure. The first digital camera that created images directly as a digital file was the prototype Fuji DS-1P camera. In 1988, this camera recorded directly to a 16MB Toshiba internal memory card that required a battery to preserve the data in its memory. Today's modern cameras remain influenced by these achievements.

Electronic pictures were actually pioneered by NASA 10 years prior to Sasson and the Kodak invention. A frame grabber was used to capture one field of video in those early systems. Mariner 4 flew past Mars in July 1965 and used an imaging system that utilized a video camera equipped with a tube. The image was moved from the tube and was sent to a digitizing board. The camera was designed and built by imaging scientists at the NASA Jet Propulsion Lab. The digital image was written to a 4-track tape and transmitted to earth using a slow but effective radio communication technology. The spacecraft took 8 months to get to Mars. The imaging system produced 22 photographs and it required 4 days to transmit the files back to Houston. Today images can be made using a smartphone and images containing millions of pixels and can be transmitted worldwide in milliseconds.

The Duality of Science Images

Images that feature science can often create conflictions for scientific viewers because scientists are taught that science should be factual and not be emotive. Truthfully I find that many science images inspire, amaze, even astound while sharing important scientific information. Some science students are taught to look at science in a certain way and to remain objective. This can lead to confusion because at its core, photography is an applied art and should be interesting to look at.

Corey Keller in his essay "Sight Unseen: Picturing the Invisible", published in *Brought to Light, Photography and the Invisible 1840–1900*, shared that "nineteenth-century popularizers of science, did not see the lay response of wonderment and pleasure as inappropriate but rather they counted on this reaction to encourage scientific interest from the general public." His research suggests that in the early nineteenth century in Britain, the use of pictures in science education was a hot topic of debate.

Anne Secord in her article "Botany on a Plate: Pleasure and the Power of Pictures in Promoting Early Nineteenth-Century Scientific Knowledge"[1], evaluated this phenomenon when she looked at how botanists used illustrations in the late 1880s. These botanists recognized the importance of allowing viewers to experience pleasure while imparting scientific knowledge. Secord's research brought attention to "the role of pleasure in intellectual pursuits."

The enjoyment of science as art continues in modern days. It is evidenced in the numerous Science as Art activities such as the RIT and Johns Hopkins University Images from Science 3 exhibition or the Nikon Small World Photomicrography Exhibition now celebrating its 45th year.

Note

1 "Botany on a plate: Pleasure and the power of pictures in promoting early nineteenth-century scientific knowledge", by Anne Secord. *Isis* 93: 28–57, 2002. Cited in the article 'Do botanical illustrations promote scientific knowledge?' by Tania Marien, 2010. Online: http://artplantaetoday.com/2010/04/01/botanical-illustrations-promote-scientific-knowledge/

Suggested Reading

Ancient Microworlds, by Giraud Foster and Norman Barker. Custom & Ltd Editions, 2000.

Beauty of Another Order: Photography in Science, by Ann Thomas. Yale University Press, 1997.

A Child is Born 4th ed., by Lennart Nilsson. Delta, 2004.

Brought to Light: Photography and the Invisible, 1840–1900, by Corey Keller. Yale University Press and the San Francisco Museum of Modern Art, 2009.

Die fantastische Welt des Unsichtbaren: Entdeckungen im Mikrokosmos, by Peter-Matthias Gaede, Oliver Meckes, and Nicole Ottawa. Gruner + Jahr Geo, 2002.

Focal Encyclopedia of Photography, Digital Imaging, Theory and Applications, History, and Science 4th ed., by David Malin and Michael R. Peres. Focal Press, 2007.

Heaven and Earth: Unseen by the Naked Eye, by David Malin. Phaidon Press, 2004.

Hidden Beauty: Exploring the Aesthetics of Medical Science, by Norman Barker and Christine Iacobuzio-Donahue. Schiffer Publishing, 2013.

Images from Science 3, by School of Photographic Arts and Sciences. RIT Press, 2019.

Micronavigating Between Science and Art, by Martin Oeggerli. Blurb, 2014.

The Story of Kodak, by Douglas Collins. Harry N Abrams, 1990.

Chapter 4 Ethical Practices

James Hayden

Introduction

The majority of this book is dedicated to specialized applications and technical considerations needed for the production of high-quality images of the natural world. In this chapter, I will examine the relationship that exists between the photographer, the subject matter, and the person viewing the images that are created in this process. There is no over-arching written code providing specific guidance for adoption of ethical practices in scientific photography, but it is a very important consideration. As with all scientific pursuits, the documentation and analysis of truth relies on an implied understanding that the work being presented is untainted in any way. Scientific veracity depends on an objective viewpoint and an unflinching analysis of the results. Too often, however, the scientist/photographer uses their skills to influence the process, thereby altering the apparent message being presented to fit a subjective viewpoint and sometimes creating harm in the process. The implementation of an ethical framework is a way to reduce imparting bias and protect the subject matter providing more accurate and humane results.

Ethics, in general, is a social and moral construct which guides us in how we should or should not act and is a reflection of the social mores of the day. Ethics can be defined in broad strokes: a correct way of doing things and an incorrect way of doing things; a right and a wrong. However, there are many subtle intermediate ways in which scientific images can misrepresent their conclusions if the photographer is not careful and does not fully understand their methodologies or interactions with their subject matter. Applying ethics to natural science photography begins with understanding and employing the best imaging practices as outlined in this book and continues with careful consideration of how those practices affect and influence the subjects at hand. To complicate matters, these concepts are in dynamic flux, constantly changing as we learn more about the subjects and continue to develop new technologies for recording them. It is imperative, therefore, that a photographer be willing to adjust practices when necessary to respond to concerns that may alter the underlying safety and objectivity of the process.

There is a significant creative component to photography and so there is argument that there is no correct way to make a photograph, use a camera, or an imaging device. With scientific photography, however, the moral imperative is to use the equipment to create the most accurate reproduction of the subject at hand. Any alteration of that interpretation represents a failing standard. This is not to say that there is no role to play for creativity, but the primary responsibility of a scientific document is to provide an unvarnished view of the natural world. Employing ethical practices in scientific photography will help to ensure this happens.

Do No Harm

The idea of ethics applied to photography is a continually shifting notion that takes us beyond the question of "can it be done?", which is purely a technical question, and into the realm of "should it be done?", which explores the relative values of the day. People forget that practices that were perfectly acceptable 50 or more years ago are currently disparaged, now that we have a better understanding of the consequences of those actions. In the lab, acceptable practices can even change based on the results of the last experiment, when final analysis of that experiment indicates negative effects relating to the imaging process. When recording a dynamic object, especially a living one, any effects that photography imparts on the morphology or behavior of the subject can contribute to a misrepresentation of the natural condition. This outcome may invalidate the points of the photographic work itself. It may even harm the subject. Taken together, these issues can be looked at in an historical context to better understand why the rules keep changing.

Subjects represented in this book go beyond the lab and include wildlife, plant life, and even samples and physical processes of the natural world. Before the advent of practical photography, visual records of these subjects were quite different than they are today. Scientific illustrations were based on collecting specimens and using these as the raw material for detailed drawings. The master works of John James Audubon began at the point of his gun and his final drawings were sometimes contorted into somewhat unnatural poses to conform to the dimensions of his drawing materials (Figure 4.1). In his time, the purpose for such illustrations was to create better environmental awareness for these creatures that most people would never the opportunity to see. Today, however, the idea of shooting subjects with guns instead of cameras to display in a new publication would be horrifying and be in opposition to the very purpose for which the images were created. The technology of Audubon's time, however, limited his options. What was considered standard practice and acceptable for scientific documentation in the early 1800s would be ethically and morally irresponsible today.

As technology has advanced, photographing the natural world became more achievable, but certainly not easier. Images taken by William Henry Jackson in 1871/72 during the Hayden expeditions to the Yellowstone area, were used to help convince Congress to vote for the formation of Yellowstone National Park in 1872. It took seven men with mules to carry and attend to the photographic equipment (Figure 4.2). Climbing on the fragile hot springs or the grounds around Old Faithful was an unrecognized hazard when those original explorers were documenting the area in 1871 (Figure 4.3) but would be unthinkable

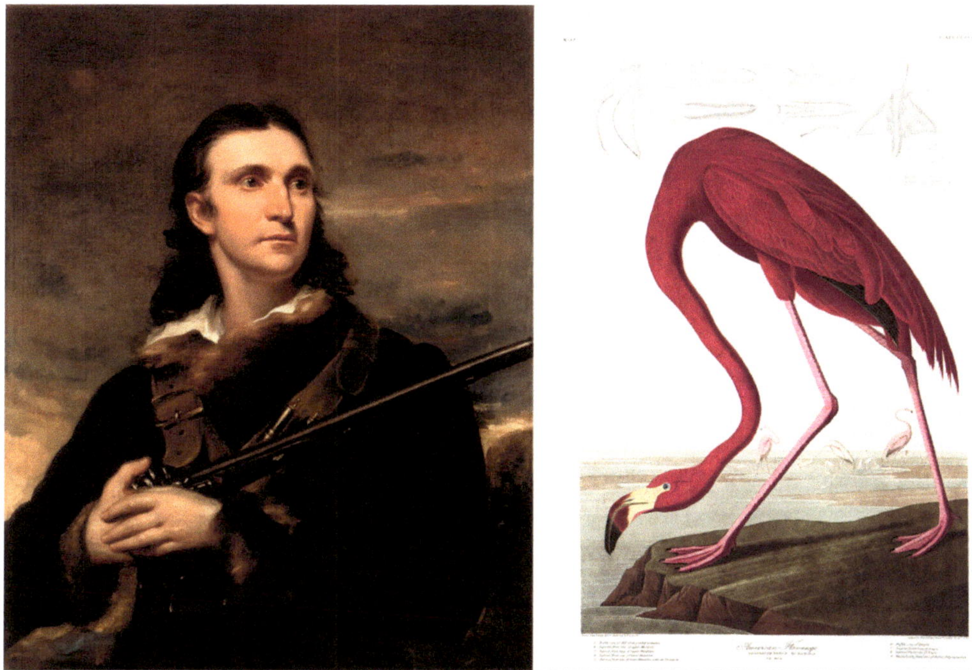

Figure 4.1 Left: John James Audubon, naturalist and wildlife artist, in a painting using oil on canvas by John Syme in 1826. Image courtesy of White House Archive, Washington, DC. Right: "American Flamingo" illustration from the *Elephant Folio*, John James Audubon, 1832. Image courtesy of Audobon Society and is in the public domain.

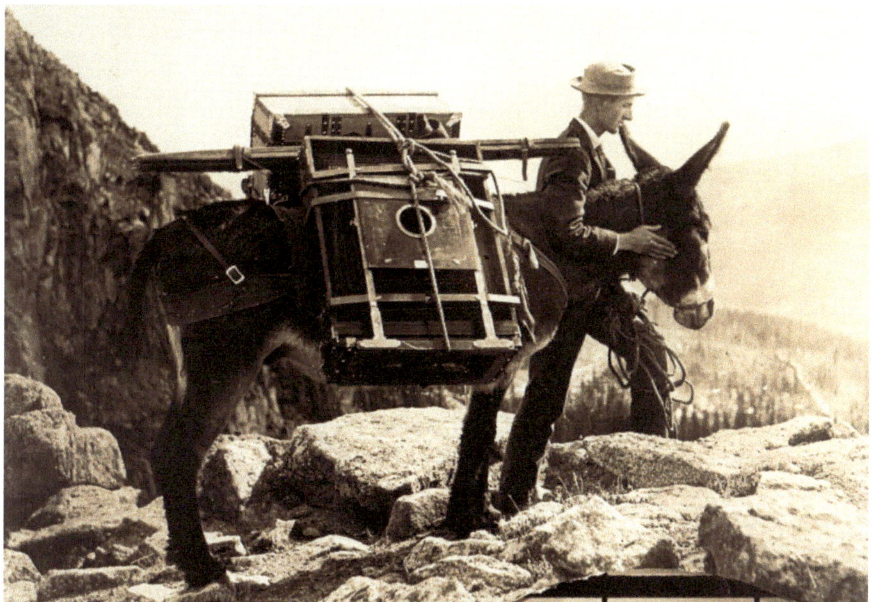

Figure 4.2 Jackson's assistant was photographed leading a mule loaded with photo equipment into Yosemite Valley; Photograph by William H. Jackson in 1873; Image courtesy of Brigham Young University Digital Archive and is in the public domain.

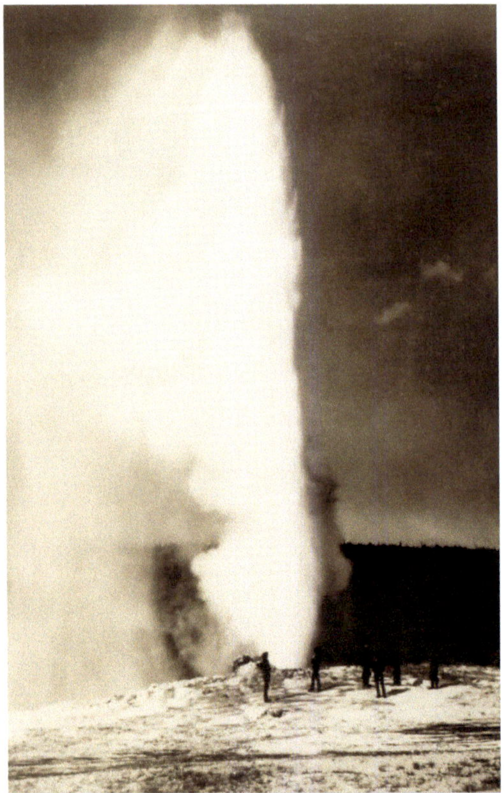

Figure 4.3 The first picture of an Old Faithful eruption. Photograph by William H Jackson in 1872; Image Courtesy Yellowstone National Park, U.S. Dept of the Interior and is in the public domain.

today because more than 4 million people visit each year. That number of visitors, trampling the same paths followed by Jackson, would create a cumulative effect of destruction, negating the accomplishments of those early visionaries.

For a more recent example, fast forward to the 1950s and 60s and the beautiful images of birds taken by Eliot Porter. Using the best large format films, new electronic strobes and remote triggering technology, he was able to photograph bird behavior at the nest (Figure 4.4). However, it required considerable effort to set up the equipment and images could still only be taken one at a time. His published technical descriptions included methods to make sure that the birds were acclimatized to both the equipment and his presence so as not to scare them from their nests and abandon their eggs or young. Even with these constraints, however, Porter used techniques that would be frowned upon today, going so far as to cut the branches that included the nests and lowering them into a better position for photography.

Catastrophic effects on specific bird populations can also be partially attributed to photographic activities. The great Indian bustard, for example, is a very large bird that lives in the grassy plains of India and is almost extinct partly due to overzealous photography. Well-meaning people who want to capture images of the bird have continually encroached upon their territory to get photographs. Unfortunately, the birds, ever vigilant, can't tell the difference between a poacher and a photographer, so they treat both as a threat, using massive amounts of energy to protect themselves and disrupting their reproductive success. Published studies have shown a 90 percent population decline in only 50 years. As of 2018, there were less than 150 individuals still in existence. It has gotten so bad that the Indian government has now banned photography of the great Indian bustard during breeding season.

Doing physical harm to a photographic subject extends to the laboratory, where living cells in petri dishes are used in critical experiments surrounding basic cancer research and similar areas. Modern microscopy equipment uses intense light and high-powered lasers to create fluorescence excitation, which has the ability to kill or mutate the cells as they are being studied or change the way they react under different conditions. When comparing results of controlled experiments, where certain conditions should be affected by a drug treatment and others should not, the results could easily be skewed if such phototoxic effects are not taken into account and mitigated.

Harm can also befall the scientific process when biased imaging practices result in erroneous interpretation of research results. In this age where "fake news" has become a buzzword, public perception of all scientific information can be affected by even a few well-publicized mistakes, or worse – intentional misrepresentations. This type of harm is easily countered, however, by application of strict acquisition protocols and standardized processing steps. Imaging in the lab is no longer merely a documentary activity. Images generated by digital technology now provide the ability to generate quantitative data directly from the recording devices, thereby making the photographic equipment the same as any other scientific instrument in the lab. Only by applying the strictest, unbiased, reproducible methods, will the scientific record accurately reflect real results. For this to happen, the operators of the imaging systems need to understand the nuances of their instruments and how their choices can influence those final results. Controls become extremely important to verify claims based on such platforms. Utilizing controls demonstrates that nothing changes between comparative images other than the topic being studied. The control image, when employed in a proper experimental design, shows that no other harm befalls the subject.

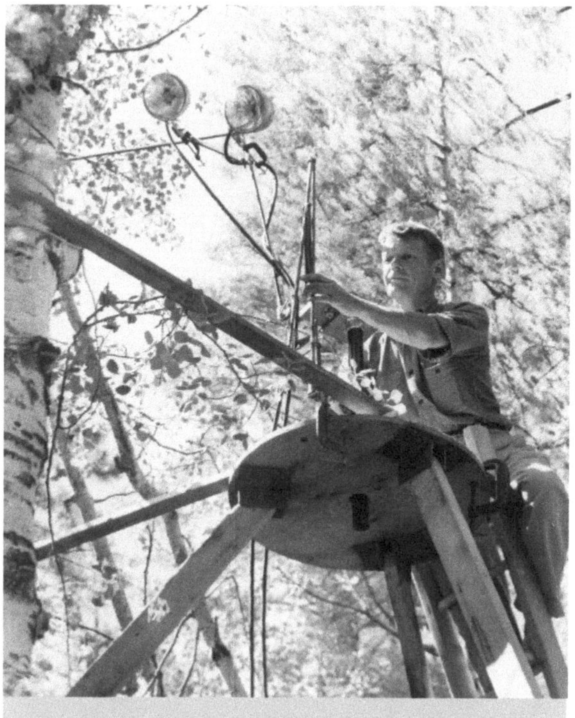

Figure 4.4 Eliot Porter setting up equipment for photographing a Lewis's woodpecker in New Mexico. Photographed by Laura Gilpin in 1952. Image Courtesy Amon Carter Museum of Amercian Art.

In the field, controls can be equally important. Images taken out of context can often be used to push a biased opinion or claim. For example, a single image of a receding glacier used in a treatise on climate change could also be used for its opposite claim – that the glacier is advancing. This is a classic example of the glass being half full or half empty; a single image is not enough to tell the story. Only by using a comparative image (a control), taken under the same circumstances (time of year, same physical location, with identifiable features) and with the same photographic conditions (focal length, aperture, perspective, etc.), can the change over time be visualized and properly compared. Of course, in this example, the "before" image is usually an historical document and the "after" image is a modern photograph of the same view. To provide a valid match, a bit of sleuthing must be done to determine the conditions of the earlier photograph so as to match them with modern equipment.

Finally, public trust is harmed when scientific photographers use their skills with specialized techniques to influence scientific interpretation. Incorrect application of lighting techniques, poor choice of acquisition parameters, biased selection of image areas, and overenthusiastic image processing can all help to propagate scientific falsehoods, whether by accident or by design. Regardless of the variety of ways in which they can occur, false visual representation must be avoided at all costs. If the result of the photographic activity alters the subject in any measurable way or misrepresents its inherent properties, the documentation of the experiment is, at best, questionable. A second image, taken under the same conditions, with the same hardware configurations, would not necessarily provide the same information. In fact, it would not be repeatable if during the first imaging session, the subject was damaged or its behavior changed somehow. These mistakes can be avoided by understanding the effects of conscious acquisition and processing decisions and applying an ethical framework to avoid harming subjects, the science, and the truth.

Equipment and Exposure Standardization

Photography requires making choices and balancing options to get the ideal exposure. Ethics can be applied whenever there are choices to be made. Those decisions can ensure the image is captured as accurately as possible, or they can be chosen to consciously and knowingly provide sub-par results. The latter approach will alter the scientific interpretation. Keep in mind that a decision to ignore this simple concept is a decision itself. The photographer must first decide what equipment will be used for the project based on both their knowledge of the subject and their knowledge of how the equipment will render the information. These concerns apply to both controlled situations, such as those in the lab, as well as the singular situations generally encountered in the field. Care must always be used to choose the equipment best suited for the subject being studied.

Photographic equipment comes in all shapes and sizes, with all sorts of capabilities, from extremely basic to very advanced. Useful images can be captured with a pocket point-and-shoot camera, a digital SLR, or high-end laboratory equipment such as a confocal microscope. Regardless of the hardware, there are some generalities in common to all of these imaging devices. They all use lenses and have other optical elements. They all have a sensor and they all have acquisition controls. Each of these elements has features that must be considered to be effectively employed in accurate and reproducible imaging. If the features are not adjustable then it is imperative that the photographer understand the limitations and what effect they will have on an image. If there are options, then all of the choices must be considered so that the one best suited to document the subject is selected (Figure 4.5).

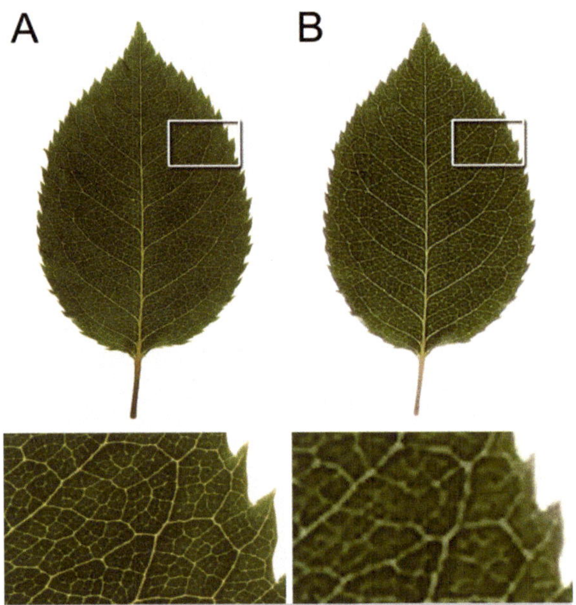

Figure 4.5 Comparison of rose leaf photographed with (A) a Nikon D500 DSLR using a 105 MicroNikkor lens and in (B) an Apple iPhone 7 photographed at the same distance. Both instruments provide adequate quality for characterization and identification, but only the DSLR resolves finer details in the venation (insets). You should always choose the right tool for the job. Image courtesy of James Hayden.

One of the key factors in any experimental design is standardization. Images are often recorded to compare conditions and that can only be done accurately if the equipment configurations remain the same. When choosing lenses or objectives, the properties of those optics must be considered. Focal length, aperture, working distance, and magnification will all have an effect on perspective and distortion. When there is a choice of a camera or sensor, the overall sensitivity, bit depth, physical size, number of pixels and size of those pixels can also factor into the quality and accuracy of the captured image. Acquisition decisions can be as simple as choosing to use manual or automatic. The key is to make sure the settings are repeatable. Knowing how equipment works is step one to form a well-defined and reproducible scientific image.

In scientific imaging, as with all forms of photography, one of the most important elements of the hardware is the optics. When considering lenses, not all optics are created equally. Camera lenses may be fixed focal length or zoom and anything from wide angle to telephoto. They may be macro lenses, designed for close focusing. They may also be highly efficient optics, transmitting a lot of light and allowing for more options in exposure settings. Because of this variability, evaluating images taken with different lenses will not give comparable results (see Figure 7.12). Focal lengths should be chosen to maintain a useful distance from the subject both to protect the subject as well as avoid distortions that would alter the shape of the specimen in the image. For comparative images in particular, the focal length should stay the same. If not, the magnification, perspective and apparent depth of field will vary between the images. An aperture should be selected which provides adequate depth of field for the specimen. With macro lenses, especially those mounted on a bellows, the magnification should be set and focusing should only be done by adjusting the working distance, either by moving the specimen, or moving the entire camera setup. If the lens barrel is rotated like a traditional lens to "focus" the subject, it actually changes the magnification. Cropping the images after acquisition to correct for this difference results in uneven image quality. Cameras with lower quality lenses, such as point-and-shoot cameras with "macro" settings often create noticable distortion at close focusing distances. This type of lens will not create images that have straight lines and will introduce bending to objects located near to the edges of the frame.

Microscope objectives provide variability that is similar to camera lenses. They come in different magnifications, which will provide different fields of view. Each objective also has different light transmission efficiencies, meaning the exposure time needed for one will not be the same as another. Each objective also has a specific numerical aperture (N.A.) which is a measure of resolution. The higher the N.A., the higher the resolution (more detail). See Chapter 16 for more. Making a photograph using a low magnification, the objective wil have a lower N.A. When comparing it to an image taken with a higher magnification, a higher N.A. lens will not provide the same image or information within the image. Higher magnification objectives will also concentrate fluorescent light onto a very small area, often creating problems with specimen bleaching, so choosing a lower magnification objective with a lower N.A. may give less resolution but save the specimen from damage. Best Practices dictate that images taken for comparative purposes must be taken with the same objectives and it could be that solution creates the least amount of damage to a specimen.

Scientific images are used for both quantitative, as well as qualitative purposes. The amount of information captured is dependent on the specifications of the camera's sensor. When considering qualitative issues, sensor resolution and sensitivity are important for high quality results, but when capturing quantitative information, those options become critical. Quantitation is a numerical operation, requiring the most information available. This requires choosing the highest bit depth for acquisition. Choosing a lower bit depth will result in less accurate intensity measurements and can invalidate some results. When using microscope cameras, or other specialized imaging equipment, selecting the highest bit options (12-, 14- or 16-bit) is requried to capture the full numerical data at each pixel. With traditional DSLR cameras, it means shooting in RAW format. The quality of the sensor will also factor into sensitivity options (gain or ISO) for the individual recording device. Generally, the higher the gain, the noisier the image, so, if detail is important, higher gains should be avoided (see Figure 4.6). However, each sensor has a range of gain settings that would be considered acceptable and that range should be determined ahead of time by careful experimenting and comparing the results. Most photographers understand that there is an upper limit to acceptable noise, but what constitutes "acceptable" will vary. Finally, spatial

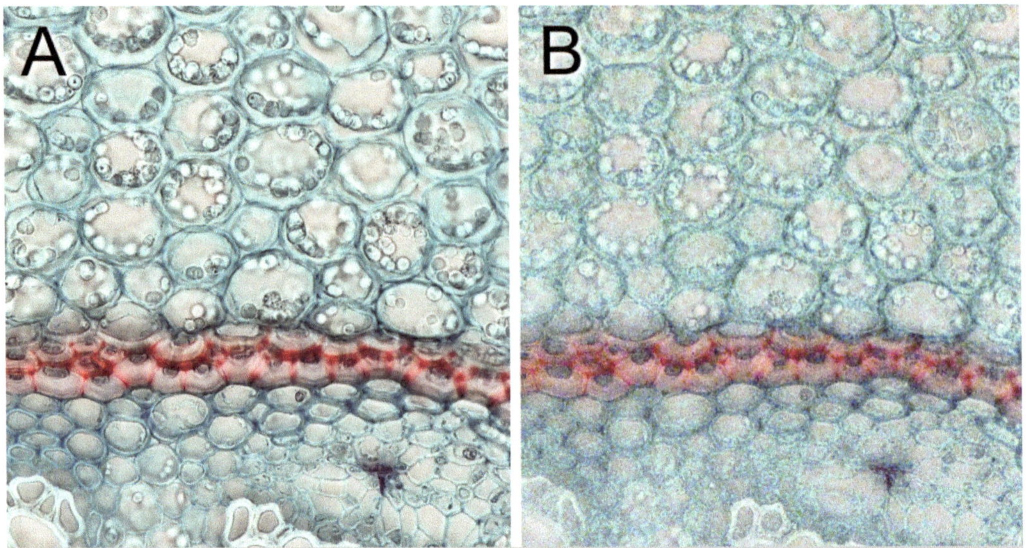

Figure 4.6 Light micrographs of a *Convallaria* cross section taken with (A) a gain of 1 (low) and (B) a gain of 46 (high). The noise effects of high gain make this image unacceptable. Knowing this in advance, the photographer can increase light levels to ensure the use of low gain, thereby providing better results. Image courtesy of James Hayden.

resolution (number of pixels in the image) and physical pixel size are also important hardware considerations that affect the resolution and clarity of an image but are not adjustable features. Decisions about sensors are generally made when choosing the camera to be used. Smaller pixels will provide better resolution, but can also be less sensitive, leading to more noise in the images. Larger sensors will provide an increased fields of view and more options for binning or cropping. Choosing the best camera for a situation will always be a compromise based on the needs of the subject and the capabilities of the equipment.

When standardizing acquisition and exposure settings, AUTO is a four-letter (bad) word! Choosing to work with auto exposure means putting those choices into the hands of the computer inside the camera or in the acquisition software. Light measurement generally relies on evaluating the light intensity that falls onto the sensor, normalizing the measurements and providing a middle gray reading. Unfortunately, as the imaging conditions change, the light intensity also varies, thereby changing the auto exposure settings. This is very apparent in situations with low light or dark backgrounds such as animals in an open field or against dark foliage, in fluorescence or brightfield micrographs, or when placing specimens on solid black or white backgrounds. To avoid the variability created with auto-exposure, the best choice for exposure reproducibility would be to use manual acquisition settings instead. In this way, the known values will not change between images and the results will be fully comparable. In the lab, there is usually plenty of time to set up and prepare the specimens for photography and the conditions can be easily controlled. In the field, however, events can happen in a split second, and with little advance warning, so being prepared in advance ensures transient opportunities are not missed. For example, understanding the basic requirements for incident vs. reflected light measurement or simple photographic rules, such as “Sunny 16”, will provide better results than most basic Auto settings.

Incident light is a measure of the light falling onto a subject and does not change unless there is a light intensity change, while reflected light measures the light reflecting from an object and is influenced by the tones and colors of the subject. A DSLR camera uses a reflected light, through-the-lens meter to provide a starting point for exposure information and can create inconsistent results. For example, when tracking a flying bird, it will often fly through a variety of backgrounds (bright sky, dark woods, open field, etc.) where the exposure should stay the same as long as the light falling on the bird stays the same. However, the meter will take into account the intensity of the changing backgrounds and constantly change the exposure values with it. These changes can be frustrating and create unusable images. In such situations, a fixed manual exposure based on the light falling onto the subject (ambient), rather than the light reflected from the subject, will result in more reproducible results.

A Rule of Thumb

"Sunny 16" is a photographic approach grounded in film photography, but still applies in digital technology. It states that the correct exposure in full sun at midday is *f*/16 at a shutter speed equivalent to 1/ISO (sec.). The rule comes with corollaries as well, such as −1 stop for light clouds, −2 stops for open shade or full clouds, −2 or −4 stops as the sun goes down, and −4 stops for full shade.

Why do hardware and acquisition decisions fall into the realm of ethics? When choices are made that directly affect the ability of the imaging system to capture accurate and reproducible images, there is a chance of inserting personal bias into the process. It is the role of the scientist/photographer to understand basic equipment issues and apply Best Practices to the situation. Pointing a camera at a subject requires understanding the equipment at hand and knowing what will happen when the shutter is depressed. Willful ignorance of equipment function is a poor excuse for not obtaining data that provides the proper scientific result.

Photographic Techniques and their Effects

Perhaps the first choice made when creating an image is to decide exactly what is to be photographed. At first, this sounds trite, but a message can be altered by consciously choosing a representative area that is not actually representative at all and pushes a point of view. If the purpose of the image is to demonstrate a quantitative value, like a population density, or number of cell colonies in a culture dish, then choosing an area with few subjects vs. selecting an area that shows a high concentration, changes how the information in that image is perceived (Figure 4.7). This biased selection can be overcome by taking many images, quantifying the data, and averaging the results. Unfortunately, there is generally not enough space to show all of this raw data, so finding a single image to represent the derived values depends on being as objective as possible and choosing a single area that exemplifies the calculated average. It is still a subjective choice, but one that is based on hard data.

Photography is a term derived from the Greek which is roughly translated as "drawing with light." Without light and shadow, there would be no image. This applies to traditional visible

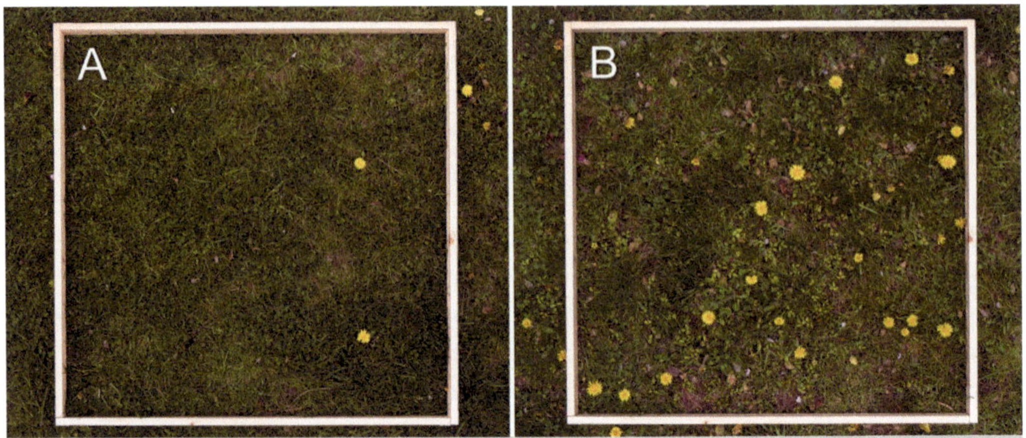

Figure 4.7 The importance of proper sampling is well understood. When quantifying the number of dandelion flowers per square meter of lawn, selecting an area with (A) 2 blooms vs. (B) an area with 23 blooms, results in a level of magnitude difference in the extrapolated total. Photographing many areas and averaging the results provides the most accurate measurements. Image courtesy of James Hayden.

wavelengths as well as images created from other parts of the electromagnetic spectrum. As long as a sensor is sensitive to the wavelengths in use, and regardless if your eyes can see it, it will be recorded by the pixel as information in the same way. Much of this book deals with the idea of learning how to use light in controlled ways and understanding the effect that light has on interpretation of the final image. The more a photographer understands how the light can affect the visual representation of the subject, the more the light can be controlled to show the desired effect. Therein lies the potential for problematic usage and biased interpretation of the scientific information. Lighting which is appropriate for one type of subject is not necessarily appropriate to show something else. The unethical photographer can make use of this knowledge to push an agenda, rather than strictly documenting a result. In effect, they could use the photography to lie.

A common example can be found when using lighting incorrectly when making a comparison of raking vs. broad lighting. To best demonstrate dimensionality and texture in a subject, rake lighting uses a low-angle source to create strong shadows. These shadows exaggerate the texture and provide images with the three-dimensional information. Unfortunately, the strong shadows and high dynamic range of such an image effectively mask the overall tonal information. If accurate color and tones are important, then flat, broad lighting should be used instead. This technique is designed to eliminate as much shadow as possible, leaving just the tonality to create any contrast in the image. However, the lack of shadows in flat lighting effectively eliminates the texture information (Figure 4.8). These techniques are used frequently for more nefarious purposes in questionable advertisements for "wrinkle creams" where the "before" image is shot with high contrast rake lighting and the "after" image is captured with broad, flat lighting. At a basic level, this is supposed to show the result of an experiment where the "before" image is presented as the control and the "after" image shows the effects of the cream. However, even though both images are technically photographs of their respective points in time, the rake lighting accentuates the wrinkles and the flat light hides the wrinkles due to the lack of shadows. In this example, the photographer used their photographic knowledge of lighting to apparently reduce the amount of wrinkling, thereby changing the visual result of the experiment. To maintain integrity of any comparison, all conditions should be kept the same for both images, especially the lighting. If there is an interest in both color and texture information, then both types of lighting should be used at each time point and presented as a group.

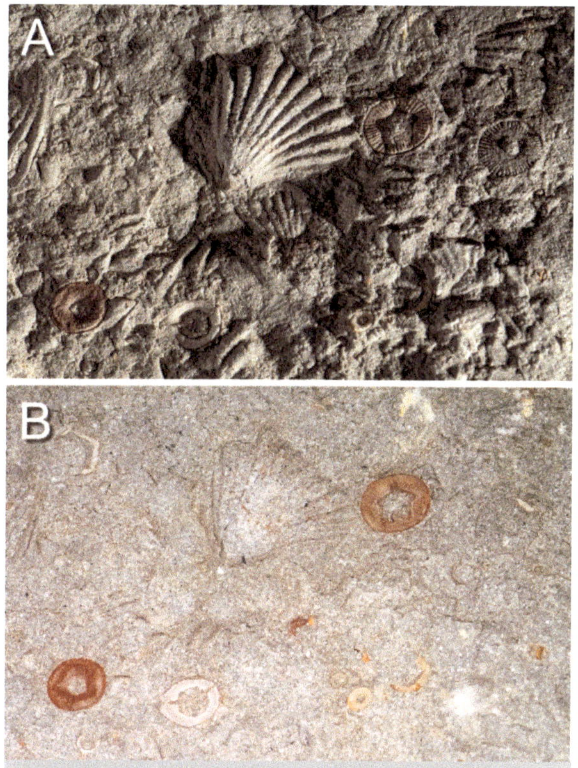

Figure 4.8 A comparison of (A) rake (texture) lighting vs. (B) broad (flat) lighting in fossil shale from the Allegheny Mountains in Pennsylvania showcases how an image can be affected by the choice of lighting. Each technique highlights the mutually exclusive and very different details in dimensionality and color. Image courtesy of James Hayden.

Other ways that lighting can be used to create a false interpretation are by taking advantage of the different qualities of light at different times of day or using different light sources when creating a visual comparison of subjects. The harsh light of noon does not look the same as the soft golden light in the hour before sunset. Likewise, natural light has a very different look than direct flash. Tungsten light has a different color temperature than LED lights. Mixing and matching different lighting conditions when making comparative images makes it difficult to accurately compare color information. Standardizing the light sources, while also standardizing the acquisition settings, makes a valid comparison possible.

Time-based imaging is another photographic technique that has the potential to be abused and create a biased interpretation of facts. With time-lapse studies, there are two time intervals to consider: the total length of time from the beginning to the end of the experiment and the interval time between each of the images in the sequence – see Chapter 18. These types of experiments generally compare growth or movement over time between two or more samples. There is an implied assumption with the presentation of sequential images that equivalent time points are selected for comparison but this may not always be the case. If comparative images are chosen from different time points in the sequences, then they will not show equivalent timing and any comparison of the visible record would be invalid. The issues in this example go beyond the actual acquisition parameters (which should be completely standardized) and concentrate more on the responsibility of proper presentation of the collected images.

Luckily, there is a simple way to check the timing information in a time-lapse sequence and that is by checking the original metadata captured with the images. Metadata should be considered as a part of all captured images and must be maintained as part of the scientific record. It includes all of the original acquisition parameters and generally cannot be altered. Metadata can, however, be lost by saving files improperly or in unsupported file formats, so it is imperative that the original images be saved and stored untouched while additional work should only be done with copies. This applies to all scientific images, not just time-lapse. All images should first be saved in the native format of the recording software to retain the pertinent information about hardware, acquisition settings and timings. Selected images can then be saved separately in a standard, non-compressed format such as .tif and used for presentation and analytical purposes. It is up to the scientist/photographer, then, to choose the appropriate images for valid comparisons.

A Pixel is Worth a Thousand Words

Perhaps the most obvious place to create ethical problems with scientific images is with post-production during image processing and digital manipulation. The old adage "A picture is worth a thousand words" can certainly be truthful, but how many of those words are accurate? Advances in digital technology have changed the way images are made and people are now predisposed to question everything that they see and hear. So, if they don't believe the picture, they don't believe the science that is being represented in the picture. It is up to the scientist/photographer, therefore, to determine what is acceptable in scientific imaging and what is not. When information is miscommunicated through images, the perception of the underlying science suffers and the field enters into morally ambiguous territory. Even small adjustments to a digital image will change the underlying data and must be avoided altogether or at least disclosed to foster reproducibility. It is important to consider the effects of both acquisition choices and post-processing at the pixel level. A close analysis of the pixel data reveals numerical intensity information overlaid with additional color data (Figure 4.9). At this level, it makes sense that changing any of the numbers would change the information for that specific pixel. Change enough pixels and you change the information embedded in the image.

The concept of "pixel-as-data" lies at the heart of image analysis and, by extension, image manipulation. At its basic level, it means that the scientific information in an image can be derived objectively from the pixel values and if the pixels are changed, the data is changed. Starting with this knowledge, you must consider all of the post-processing operations that are being considered before making final decisions. Many adjustments are considered acceptable. These adjustments are contrast, brightness, and overall color balance which seem to be innocent functions. These adjustments are made to an image with hardly a thought, even though these practices change the individual pixel values in the process. It is not always the problem whether the pixels are modified, but how they are modified and in what context is the problem.

Applying stringent rules for acceptable manipulation to ensure scientific clarification can be difficult, as what may be appropriate for one situation may not be for another –see Chapter 19. Valid adjustments can also be taken too far, losing data by manipulating to an extreme level. A general starting point for image processing is to use corrections that may have once been done in darkrooms. These adjustments, like brightness and contrast, generally work by applying a uniform modification to all pixels in the image so that the intensity relationships between pixels remains the same. Another rule of thumb states that, while global adjustments to an image are generally fine, manipulations to selective areas are not. Selective functions include such processes as cloning, dodging and burning, painting and erasing; basically anything that affects adjacent pixels differently. One last generality is when images are combined, such as stitching images in two dimensions, or panoramas,

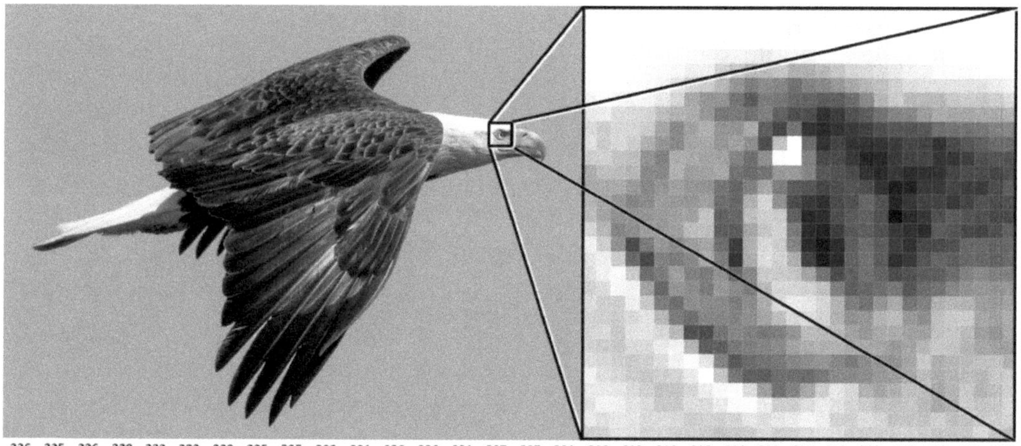

226	225	226	229	232	232	229	225	225	222	221	226	226	231	227	227	221	219	219	216	213	214	220	223	218	224	227	225	223	216
226	223	224	227	229	229	226	225	224	226	221	220	229	225	210	215	217	215	215	216	218	220	221	215	217	220	221	219	219	214
224	223	222	223	222	221	221	224	221	223	225	220	218	221	219	225	230	226	220	218	220	226	225	220	221	226	228	229	226	221
219	220	219	216	213	211	210	210	204	195	200	194	192	203	205	203	216	215	212	211	210	216	221	220	211	220	229	229	225	220
219	221	220	216	210	205	200	193	193	174	164	153	156	160	148	147	156	164	175	179	182	189	198	199	209	209	209	213	221	221
216	219	217	211	205	198	185	174	169	152	127	107	115	112	95	112	116	128	142	152	158	169	180	183	183	184	183	184	190	193
200	209	198	198	195	187	175	141	125	126	100	77	106	102	88	85	79	78	94	117	155	169	155	156	155	165	169	165	168	178
188	188	175	176	174	158	150	147	129	104	105	99	108	82	68	69	50	55	70	78	99	113	116	132	147	141	143	144	144	150
177	167	174	174	167	140	115	136	148	128	127	147	78	60	101	53	46	53	56	60	69	73	72	77	95	100	101	103	110	113
182	168	175	169	148	127	118	140	143	129	138	125	68	163	244	95	51	63	53	50	54	56	64	61	49	65	75	80	90	92
171	171	162	158	124	111	147	144	139	140	115	97	125	222	227	113	65	68	56	40	46	56	64	64	53	54	60	70	71	68
145	158	158	151	115	116	153	133	138	147	100	93	100	129	131	117	78	73	69	51	47	56	50	51	51	64	69	66	63	65
143	142	145	131	116	128	138	128	127	104	88	113	144	108	78	75	65	70	77	70	54	42	43	59	58	65	59	55	66	79
153	140	126	115	120	134	130	129	138	96	88	151	179	125	68	58	54	68	73	90	78	46	56	63	65	58	64	73	68	69
158	139	122	127	128	130	137	136	165	85	87	160	172	169	83	46	50	69	79	85	73	30	50	56	72	83	77	72	65	64
173	151	113	122	136	129	137	159	161	82	79	142	165	176	106	47	42	59	77	83	72	58	51	50	80	77	68	64	69	79
195	168	122	65	105	129	117	132	140	98	56	150	180	170	115	46	33	47	82	98	82	60	51	68	66	72	90	93	86	87
187	195	171	90	65	113	107	118	147	111	50	150	183	173	149	60	39	46	85	101	80	50	56	75	87	82	90	90	85	89
180	194	205	167	94	107	134	111	107	129	93	115	170	168	166	141	102	68	84	99	87	69	72	82	82	96	112	122	110	108
172	179	201	198	138	86	127	121	108	132	121	108	154	178	177	173	162	111	82	87	98	106	78	95	100	115	125	129	126	129
180	170	169	193	167	107	97	124	121	121	108	120	146	183	196	189	180	153	110	98	128	147	104	113	152	157	153	134	130	137
174	170	166	175	187	156	107	97	115	116	115	124	141	158	180	183	175	150	125	129	155	159	130	124	161	169	170	143	137	145
173	168	164	184	185	177	157	69	68	103	118	112	103	108	126	158	157	158	139	131	147	147	140	138	162	167	158	137	128	142
172	167	163	177	176	189	198	153	78	82	103	110	114	117	138	146	149	149	145	145	142	131	140	155	167	162	153	144	137	134
180	175	169	164	169	181	198	213	189	113	76	93	99	103	114	121	127	113	104	119	127	125	142	168	147	152	156	160	161	147
195	180	173	166	163	168	174	189	230	165	101	79	78	93	94	93	102	115	116	119	128	130	143	158	146	152	158	153	165	168
188	187	184	175	167	159	162	173	193	205	212	181	134	95	111	140	169	188	176	147	139	147	159	159	140	150	166	151	159	168
182	196	192	181	175	165	165	179	176	190	210	214	193	171	173	182	190	183	174	169	160	161	166	155	142	148	169	163	164	173
188	181	184	183	176	178	175	166	170	169	172	185	179	186	190	199	190	172	170	180	173	160	162	160	169	158	160	167	167	171
167	169	183	180	172	173	168	168	165	160	162	167	176	170	171	176	183	181	169	169	168	169	172	167	163	161	163	177	162	153

Figure 4.9 Images are data. Every pixel in an image is represented by a specific numerical intensity value. Changing a pixel changes the data. Image courtesy of James Hayden.

or merged in three dimensions like extended depth of field or HDR, this image processing should be noted in any caption.

While these guidelines are useful recommendations as a starting point, there are always exceptions. There are fine lines between manipulating for clarity versus misrepresentation. Photographs are ofter the result of artistic license and intepretation often falls upon a subjective consderation about the guidelines when deciding what is actually acceptable. The question at the heart of any ethical decision is really about what the image modification is for. Is the manipulation designed to present a better-looking image or is it to hide some important bit of data embedded in the pixel information? An image that has been adjusted to be more visible and possibly more visually appealing can often be used as a general representation of a project, but images used to derive data should not be adjusted at all. For publication, a compromise may be necessary. A modified file with an appropriate caption can be used to visualize the experiment, while the derived information abstracted from the unadjusted file should be presented in a graphical manner as well. In the end, just about any adjustment can be justified, as long as:

- the manipulation is disclosed,
- the original, raw, and unadjusted image (with metadata) is retained for review if needed,
- the image processing (manipulation) is reproducible.

Decisions about image manipulations can best be made by understanding exactly what effect the adjustment has on the image, specifically at the pixel-level information. For example, while adjusting contrast or brightness is generally considered appropriate, understanding how it works will help define the limits of how change is acceptable. Analyzing the image histogram will provide some insight about how far an adjustment can go before it becomes unacceptable. A histogram is a visual representation of the intensity values of all pixels that make up an image. The standard histogram shown in image editing software, is a horizontal graph with a value of 0 (pure black) to the left and the full value of the pixel (pure white) to the right. In an 8-bit image, there would be 255 tones. An average image has a curve that approximates a bell curve, with only a few very dark or very light pixels and the rest distributed throughout the tonal range, often with a peak in the middle region. By making an image lighter, either by changing acquisition parameters or manipulating the file in post-production, the entire curve is essentially shifted to the right. In this way, the brightest pixels in the curve are represented in the histogram very close to the farthest region that can display information. A problem will occur when the brightness is adjusted too much and the whole curve slides past the end of the 255 level in the graph. In such a case, all of the pixels represented by the part of the curve that went beyond the boundary are cut out of the image and replaced with a solid white level of 255. Any fine detail in those lighter areas will be lost. Likewise, an image that was darkened would be shifted to the left and could be shifted too far and down past a 0 level. In this case, the pixel data is replaced by black. Analyzing the shifting curve allows you to objectively see if any information will be lost. These results are demonstrated in Figure 4.10 where lost details in the highlights and shadows are represented by the vertical lines at either extreme of the histogram. Note that the non-continuous curves in the light and dark files indicate that the file has been manipulated. Original files (middle) will show a solid curve. This is a mechanism by which files can be investigated to discover manipulations, whether acceptable or not.

Shifting the curve as a whole is considered a linear adjustment. The actual values of the pixels change, but the relationship between the pixels does not. Linear adjustments are predictable, quantifiable and generally acceptable in science. Non-linear adjustments, on the other hand, are ones that apply a sliding modification to the image, depending on where the specific pixel falls in the scene. The classic example for this is gamma. Adjusting the gamma in an image adjusts the mid-tone values for each pixel, but applies the algorithm differently to each one, depending on the original numerical value for the pixels individually. Adjusting the gamma brings out better visual information in an image, but it does so by changing the relationships between the pixels. This makes the result of the modification non-quantifiable and renders the image unacceptable for deriving numerical data in the future. For this reason, many scientific journals do not allow non-linear adjustments without full disclosure.

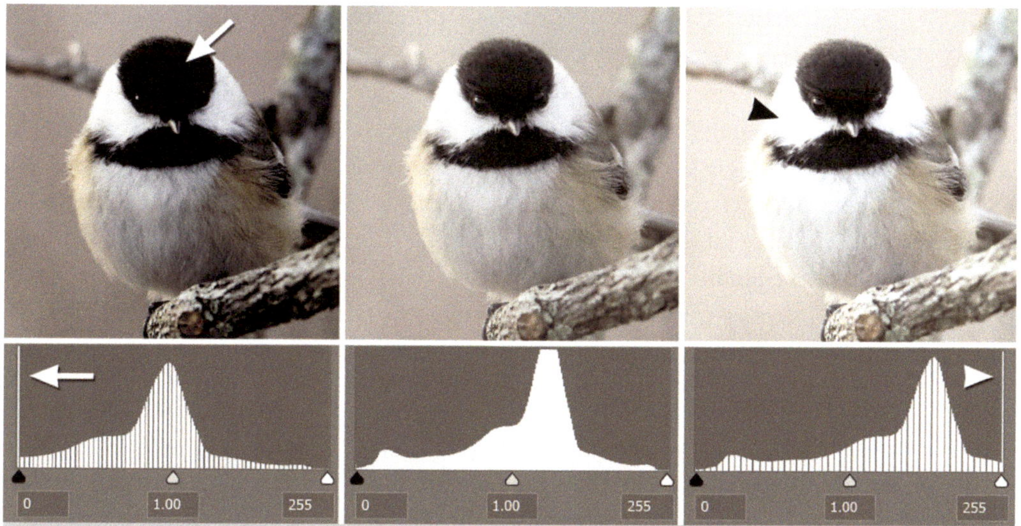

Figure 4.10 Visualizing out-of-range pixels in an image histogram can be important. The original file (middle) shows a proper exposure, with a fully distributed curve. An over-adjusted bright file (right) shows a peak at 255 (arrowhead), while a dark version (left) shows a peak at 0 (arrow). Image courtesy of James Hayden.

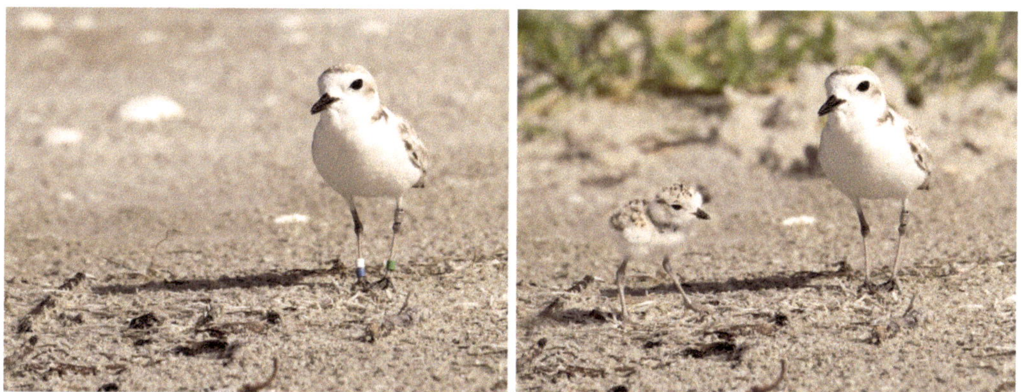

Figure 4.11 Digital manipulation. This composite image features an adult snowy plover and its chick (right). It was created by adding the chick and background to the original image (left). The colored leg bands on the adult were also digitally removed. Image courtesy of James Hayden.

Taking non-linear manipulation to the extreme happens when the software is used to falsify an image, or fabricate new parts by cloning, erasing, painting, copy/pasting or using other destructive tools (Figure 4.11). Scientific images should stand on their own as direct representations of a specimen and never be subjected to such overt manipulation. By adding, removing, or changing selectively different parts of an image, the photographer introduces their biases and perspectives into the record which will alter the perception of what is seen in the image. In Figure 4.11, although both images were taken minutes apart on the same stretch of beach, the chick never came that close to its mother. Also, by removing the identifying leg band, important information about the subject was removed from the image. If the purpose of a scientific image is to document the truth, then it stands to reason that changing anything in that image would alter that "truth". The image reproduced here never actually happened. The possibility of people creating this type of outcome should be of critical concern. The scientist/photographer stands in as a representative for science in general. Their use of objectivity, skills and even some creativity while creating a visual for those who cannot see the subject directly comes with great responsibility. When it is discovered that an image has been faked, either accidentally or on purpose, the credibility of all of the associated work will come into question. Even if the transgression seems relatively minor, the overall effect can have a major impact on the belief and trust of the general public.

Questionable manipulation practices are often associated with subjects that are difficult to photograph. Whether an animal isn't behaving in just the way you want, or a thunderstorm isn't dramatic enough, or even an image already taken doesn't meet certain requirements, it is required to carefully create interpretations of those situations that are honest. It is tempting to create a scene when real life doesn't deliver. In 1982, *National Geographic* provided one of the first examples of digital image manipulation when it altered the positioning of the Pyramids at Giza in a cover image so that it better fit the vertical format of the magazine. The global outcry was extensive and resulted in an institutional commitment to reject all photomanipulation from that point on. The credibility of *National Geographic*, one of the most well-known and respected science magazines, took a big hit, but it shows what can happen when deadlines drive behaviors. The magazine needed the image quickly, and rather than use proper and accepted scientific methods or re-shoot, they took a shortcut instead.

Ethics in the Field

Natural science photography might be accomplished in the controlled environment of a lab as well as in the field. Scientific imaging in the field presents many unique photographic challenges for the scientist/photographer – see Chapter 13. Among other issues, photographic opportunities can often be fleeting and the lighting conditions can change rapidly. This makes it difficult to create reproducible images with acquisition parameters. Standardization in the field can be difficult, but important aspects of exposure can still be controlled. Metadata can still capture all of the specifics required to make informed observations. In this way, the mechanics of the photographic process can still be standardized and scrutinized at a basic level. Aperture, shutter speed and ISO can be chosen to best illustrate the subject and techniques can be modified to account for the specific issues raised by the situation and subject. For example, when the sun gets low and the shutter speed is necessarily lengthened, a good tripod makes an important field companion! White balance can still be properly determined regardless of the time of day and decisions about focal length, magnification and perspective will not change that much.

Ethics in the field actually have more to do with the treatment of the subject and the environment than with the photographic process, although all of the details previously mentioned need to be considered. Photographing living subjects in their natural environment requires a gentle touch and a healthy dose of non-interference by the photographer. Intrusive actions by even the most well-meaning researcher will affect the behavior of the subject and result in images that don't quite capture the natural conditions. When dealing with living beings, this can be problematic. Additionally, unique natural objects can be damaged in a headlong rush to capture the perfect photograph if the photographer is not vigilant and aware of the surroundings at all times. Stomping on an endangered plant species while trying to capture the definitive shot of a rare bird defeats the purpose of non-interference. Which species is more important? Who decides that? Ethical considerations can sometimes take a back seat to the thrill of the chase. Establishing some personal guidelines is a good way to protect animals, the environment, and the accountability of the photographic process.

To encourage the safety of living subjects and maintain the integrity of the natural environment, the U.S. National Park service embraces the concept of "Leave No Trace" and this provides a helpful guide for anyone working or playing in the wild (Figure 4.12). This approach was developed in cooperation with U.S. Forest Service, National Park Service, and Bureau of Land Management in the mid-1980s. These agencies established the Leave No Trace Center for Outdoor Ethics.

The full guidelines can be found on the National Park website.[1] The following selected points stress the issues that commonly apply to photographing in the field.

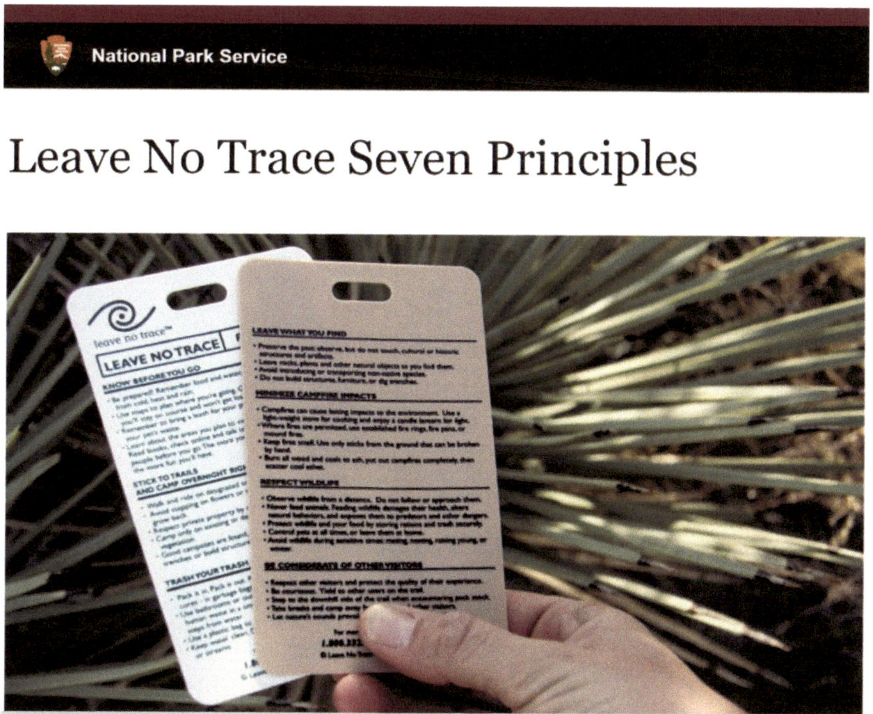

Figure 4.12 Helpful outdoor guidelines from the National Park Service. Image Courtesy of Department of the Interior, Washington, D.C.

- *Plan Ahead and Prepare*: Know the regulations and special concerns for the area you'll visit.
- *Travel and Camp on Durable Surfaces*: These include maintained trails and designated campsites, rock, gravel, sand, dry grasses or snow.
    - In popular areas, concentrate activity on existing trails and campsites. Walk single file in the middle of the trail, even when wet or muddy.
    - In pristine areas, disperse use to prevent the creation of campsites and trails. Avoid places where impacts are just beginning.
- *Dispose of Waste Properly*: Bring it in, take it out.
- *Leave What You Find.*
    - Preserve the past: Examine, photograph, but do not touch cultural or historic structures and artifacts.
    - Leave rocks, plants and other natural objects as you find them.
    - Avoid introducing or transporting non-native species.
- *Respect Wildlife.*
    - Observe wildlife from a distance. Do not follow or approach them.
    - Never feed animals. Feeding wildlife damages their health, alters natural behaviors, [habituates them to humans], and exposes them to predators and other dangers.
    - Avoid wildlife during sensitive times: Mating, nesting, raising young, or winter.

These guidelines provide instructions to make sure that that nothing is affected by a person being at a particular site. They are good rules for ensuring that photographers have minimal impacts on their subjects. "Take only pictures, leave only footprints" is another old adage that has stood the test of time. These recommendations provide a way to maintain the status quo, allowing for wildlife behaviors and the natural environment to be recorded honestly and nothing introduced that would bring current or future harm to the subjects. As with all guidelines, these points are meant to provide awareness and changes in attitude. Often, image requirements are at odds with these rules, so it is important to understand the spirit in which they were created and intended. It comes down to respect for the subject and your responsibility for your role in the natural world.

Scientific curiosity begins with the observation of natural processes and animal behaviors and documenting those events is the role of the natural science photographer. Some occurrences are repetitious, like specific courtship behaviors in birds, while others are transient and fleeting, like the sudden strike of a predator. Some events are known in advance, like a solar eclipse, while others happen with little warning, like a falling meteor or a lightning flash. It is no wonder, then, that understanding your subject becomes one of the most important elements in the photographer's arsenal. Comprehensive research about the subject will allow a photographer to know where to be, as well as when to be there and increase the chances of photographing a specific situation. With known behaviors, the photographer can prepare in advance, make sure they do not interfere with the subject and arrive with the best combination of equipment and choice of position. Setting up a blind along a well-used trail, for example, can yield the kind of close-up images of animals that the casual photographer dreams about, while maintaining the non-interference needed for the welfare of the animals (Figure 4.13). Serendipity places a role as well, as even the best laid plans can be foiled by bad weather or uncooperative subjects.

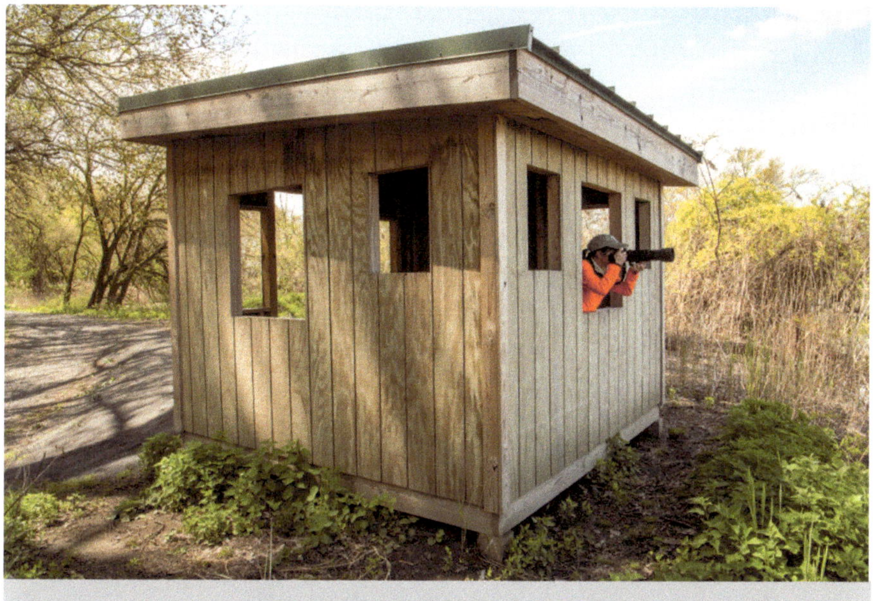

Figure 4.13 Using a blind, especially in areas frequented by a lot of people, provides a way to get unobstructed views of the subject while maintaining healthy non-interference. Image courtesy of James Hayden.

It is helpful to remember the purpose for capturing images of the natural world. All scientific images have an obligation to document reality. Images of the world stand in proxy for the actual views that they capture and it is the role of the conscientious photographer to make sure that the information is displayed as accurately as possible – warts and all. An experienced science photographer can take their considerable expertise to bear and overcome some of the challenges encountered in the field, but this goal is sometimes at odds with getting the "ideal" shot. In a perfect world, the photographer should be no more than a fly on the wall and only an observer. Unfortunately, this approach conflicts with deadlines and the need to photograph a specific behavior. It is because of these pressures that staging, baiting, and other questionable practices are chosen.

All photographers that work in the field must answer the question of what to do when animals don't cooperate with a pre-visualized idea. It is tempting to use external means to attract those animals, or to re-create a behavior for the camera, but this is where the photograph changes from an unbiased photograph to biased and subjective storytelling. The idea of recreating a scenario for the benefit of the camera is not new. There have been many examples and accusations of setting up shots in game farms that rent out animals to be used in wildlife films or television programs such as *Wild America*, or *Mutual of Omaha's Wild Kingdom*, but these programs also stoked the public imagination and paved the way for programs with higher ideals, such as BBC's *Nature* or *Planet Earth* series. In some ways, it can be argued that the questionable ethical practices of the earlier films were necessary, as the technology of the time and the understanding of the subject matter was not as fully realized as it is today. These programs raised awareness of the natural world and instilled a sense of wonderment that influences people to this day. Perhaps, in another 25 years, the work created now will be seen as equally barbaric by the mores of that future time.

Arguably, one of the most egregious examples of lapsed ethical behavior by wildlife cinematographers was in the 1958 Academy Award-winning documentary *White Wilderness* by Walt Disney. The film shows lemmings "jumping" from a cliff into the ocean and the dramatic voice-over intones "This is the last chance to turn back ... yet over they go ... casting themselves bodily out into space." The implication in these clips is that migrating lemmings throw themselves into the ocean to drown. This idea has become so ingrained in modern culture, that it is barely questioned. An entire generation believes that there is an animal that uses mass suicide as a form of population control. Unfortunately, this story was fabricated and presented as a form of entertainment in the guise of scientific "facts". The film crews could not find such behaviors naturally occurring in the wild, so they re-created what was thought to be the actual scenario. The lemmings were corralled and pushed from the cliff to make the dramatic point. Whether any animals actually died in the process is not even the point, although it should be stated for the record, no animals should die for a photograph. In this story, an artificial situation was established and presented to audiences as scientific fact. The cinematographers used their skills, their reputation, and took advantage of viewers' natural willingness to believe what they see, to create "facts" rather than merely showing what truly was. This was not based on any scientific method and illustrates the responsibility of the scientific photographer to uphold the basic principles of science and appropriate practices.

Staging a scene can occur in different ways and not always as overtly as with the lemmings. In some cases, restrictions go beyond merely guidelines and fall under laws. One way to get an animal to engage in activity is to provide bait. The advantage to the photographer is that the animal comes to you, rather than you trying to track it down. Unfortunately, baiting wild animals creates a dependency on humans, changes their natural behaviors, reduces their ability to compete effectively with others of their own species, and reduces their overall chances of survival. This is why baiting can be illegal. Beyond the general guidelines of Leave No Trace, federally controlled areas have rules about baiting. One marine biologist discovered this the hard way in 2014 when they were fined $12,000, had their boat impounded and faced 3 years of probation for baiting killer whales, trying to invite them come closer to the boat for cinematography in the Monterey Bay National Marine Sanctuary.

Feeding wildlife is generally considered taboo, but why then are backyard bird feeders (see Figure 4.14) not painted with the same negativity? Bird seed attracts a great variety of birds that come to depend on the food source for their survival and surely makes them dependent on humans. What about audio recordings to call birds or other animals and lure them out of hiding to get a better image? Such recordings might be anything from alarm calls to courtship songs. Animals that expend energy to lure away danger or find and confront an interloper are being drawn away from the important work of living their natural lives. Consideration of these separate dilemmas illustrates a classic problem in the application of ethical factors on a broad scale. One size does not fit all and the answer to whether such activities are acceptable is invariably "it depends". In the case of the bird feeders, the difference between backyard feeding and baiting is that bird feeders are regularly filled and are provided as a steady supply of food – not just a one and done situation. In the case of the audio recordings, there is much debate about their acceptability and many wildlife refuges specifically ban the use of recordings within their boundaries, while other locations may not even mention the practice. Ultimately, it will fall to the individual to decide what is right by weighing the effects on the wildlife and balancing that with the images that are captured.

NANPA (North American Nature Photography Association) is an organization that has carefully considered many of these situations and offers additional guidance to photographers in the field. As with the National Park Service, many of their suggestions deal with maintaining an awareness of the surroundings and making sure that the welfare of the animals and the environment comes first. The following is an extract from "Principles of Ethical Field Practices", available on the NANPA website[2]:

- Learn patterns of animal behavior so as not to interfere with animal life cycles.
- Do not distress wildlife or their habitat.
- Respect the routine needs of animals.
- Use appropriate lenses to photograph wild animals. If an animal shows stress, move back and use a longer lens.
- Acquaint yourself with the fragility of the ecosystem. Stay on trails that are intended to lessen impact.
- Do not participate in or endorse the use of a live mammal as bait for photographic purposes.
- Research the subject beforehand and avoid risking harm to capture an image.
- Help minimize cumulative impacts and maintain safety.
- Learn the rules and laws of the location. If minimum distances exist for approaching wildlife, follow them.
- Be a good role model, both as a photographer and a citizen. Educate others by your actions; enhance their understanding.

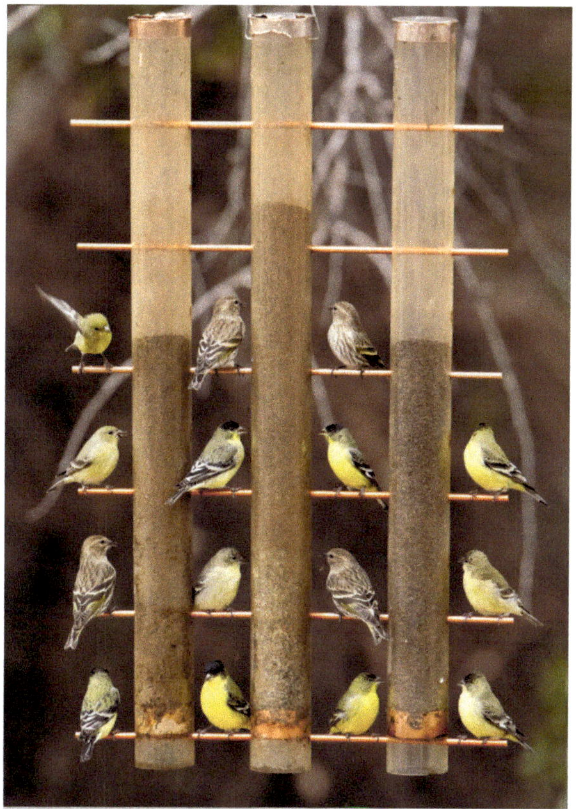

Figure 4.14 Attracting wildlife is not the same as baiting wildlife. These lesser goldfinches and pine siskins depend upon a steady source of food from the Santa Rita Lodge in Arizona. Image courtesy of James Hayden.

These guidelines from NANPA and the National Park Service offer specific ways that photographers can mitigate their personal impact in the field. By following these suggestions, the welfare of the subjects and the integrity of the environment are maintained and this can help to ensure that both are still here for future generations to admire and study.

Image Usage

Ethical considerations in scientific imaging fall along a continuum of practices. This range begins with destroying an animal or it's habitat to get a photograph which falls squarely in the "No" column, and others, like adjusting the brightness of an image in post processing, which would be considered "mostly acceptable". This leaves a wide range of situations and practices somewhere in-between with no specific instructions for yes or no answers. The guidelines presented in this chapter provide a useful place to start, but application of those guidelines is mostly a personal interpretation with little enforceable oversight. However, ethics is about an awareness of what should be right for a subject, for its environment, for science and for public trust, and should not focus on just what you can get away with. All scientific images should be subjected to full disclosure at any time if they are to have any value. That includes information about how they were taken and how they were processed. Metadata is a useful tool and part of every image, but it only provides part of the story. Images can be factually correct, yet still be scientifically inaccurate. The intent of the image and the post-processing must also play a part in the presentation of the information.

Captions are an important part of this process. While much of what is visible can be obvious, there are many details that are best left to a written explanation. Captions can include specific information that identifies the subject in an image as well as an explanation of a behavior or an experimental result. They can also include information about how and where the image was taken, as well as any processing that was done after the fact. NANPA has also considered the importance of captions with wildlife images and offers additional guidance to provide a better understanding of how the images were captured and modified. While a beautiful image of a wild animal might stand on its own in some situations, questions can arise about the circumstances in which it was acquired, and a caption can help provide some of those answers. The following terminology from NANPA is offered as a suggestion for image clarity[3]:

- "Wild": applies to any animal having the freedom to go anywhere and to disregard artificially set boundaries, with the exception of preserves established to protect the animal for its own sake, and where it lives in a natural state.
- "Captive": applies to any animal that is living under human care and control for an extended period of time. This includes, but is not limited to, zoos, game farms, falconry birds, rescue facilities, and research facilities. Including the name of the facility in the caption may also help convey the captive nature of the animal. Many educational and documentary photo users require this information.
- "Controlled": applies to an animal temporarily under human control but otherwise wild. Examples include animals in a cage, net, trap, temporary aquarium, or in refrigerated, drugged or tethered conditions.
- "Baited": refers to the use of live, dead, or fake food to attract a predatory bird/mammal/reptile.
- "Attracted": applies to using food, water, scent, audio, decoys, or any other introduced means to attract a non-predatory animal. It also applies to using anything other than food (e.g., decoys) to attract a predatory bird or other animal.
- In some situations, providing additional information about how the image was taken can help educate viewers about how to photograph their subject more responsibly. For example:
  - Captured under permit, or with appropriate permissions.
  - With guidance and supervision.
  - Taken with remote camera.

Beyond the specifics of the original acquisition, information should also be included that points to any additional image processing:

- "As Photographed": Nothing has been added to the original image and nothing removed. Standard digital adjustments may have been employed, such as global or selective adjustments to contrast, brightness, and saturation, as well as cropping, sharpening, dust spot removal, and noise reduction.
- "Cleanup": Minor elements, such as a stick, have been removed. No new elements have been added to the original image.
- "Manipulated": Major elements have been removed and/or new elements have been added, while maintaining the basic natural history of the scene at the time of capture. Examples of added elements include adding canvas, replacing eyes, adding wing tips.
- "Composite": Visual elements from separate single images were combined to create a new scene. Examples include replacing an entire background or adding a subject.
- "Multiple Exposures": The image is comprised of multiple single images. These may have been captured at the same location at the same time, captured in the same location at different times, or captured in different locations at different times. They are merged in-camera or during post-processing. Examples include HDR, stitched panoramas, focus stacking, and multiple in-camera exposures. Photographers may caption such images with the technique used, such as, simply, "HDR" or "stitched panorama."
- "Effects": An artistic filter or texture has been used to dramatically alter the appearance of all or part of the image beyond anything that could have occurred in nature.

While the suggestions from NANPA and the National Park Service are useful guidelines, they still lack much authority to force photographers to follow them. In the realm of scientific imaging, however, there are entities that hold sway over the presentation of these images and they are the journals that the images are published in.

Scientific journals generally have specific statements that govern the use of images in their pages and authors are required to follow those rules. Over the years, there have been several high-profile cases in the news where the questionable use of images or manipulations significantly affected the scientific record and forced major retractions, loss of standing and loss of funding for those involved. In 2004, a seminal paper was published in the *Journal of Cell Biology* by Rossner and Yamada entitled "What's in a picture? The temptation of image manipulation."[4] This was arguably the first major paper to address the issue of image manipulation in scientific images and helped to raise awareness for the issue in the scientific press. Since that time, most journals have adopted guidelines that lay out what is acceptable and what is not. The "Instructions to Authors" generally contain this information which must be adhered to by every author if they hope to be published. The following is taken from the "Image integrity and standards" statement for all of the *Nature* research journals.[5] Beyond their specific instructions, they offer these general points to consider:

- Images submitted with a manuscript for review should be minimally processed. Authors should retain their unprocessed data and metadata files, as editors may request them to aid in manuscript evaluation. If unprocessed data are unavailable, manuscript evaluation may be stalled until the issue is resolved.
- A certain degree of image processing is acceptable for publication (and for some experiments, fields and techniques is unavoidable), but the final image must correctly represent the original data and conform to community standards. Authors must also

take care to exercise prudence during data acquisition, where misrepresentation must equally be avoided.

- When submitting revised final figures upon conditional acceptance, authors may be asked to submit original, unprocessed images.

While scientific journals can enforce specific image requirements, a vast majority of nature-related images do not always have the same tight constraints. Higher-impact magazines like *National Geographic* have rules that prohibit image manipulation, but not all magazines do. Images that appear in the press, on the web, or in calendars or books for public consumption do not always provide disclosure about how their images were captured or processed and must be taken at face value. In many ways, where an image is to be used can dictate the level to which ethical standards will be upheld. This is not to say that all natural science images are suspect and not to be trusted, but it would be safe to say that the casual viewer should be aware that images may not always be what they seem if there is a lack of regulations to follow.

There is one popular destination for wildlife imagery that does look more closely at the way in which the images were obtained and processed and that is with wildlife photography contests. These competitions attract thousands of photographers and images every year, with the winners gaining world-wide renown while highlighting the wonders of nature to the general public. While there are many of these competitions available, there are a few that garner a lot of attention every year and have occasionally uncovered some rather questionable images.

The Natural History Museum in London sponsors the annual competition "Wildlife Photographer of the Year." This contest was created in 1965 and enjoys an impeccable reputation based on public trust in the integrity of the institution. To protect that standing and ensure that selected images are above reproach, the Museum publishes an extensive list of regulations that all images must adhere to. These rules are meant to protect the subjects in the images and maintain the scientific truth that they represent. Along with specific warnings for image manipulation, they also include a section on ethics, which includes, in part[6]:

- Entries must not deceive the viewer or attempt to disguise and/or misrepresent the reality of nature.
- Caption information supplied must be complete, true and accurate.
- Entrants must not do anything to injure or distress an animal or damage its habitat in an attempt to secure an image. An animal's welfare must come first.
- Live baiting is not permitted. Any other means of attraction, including bird seed or scent, must be declared in the caption for the Jury and Owners to review.
- If the Owners suspect that an entry has been achieved through cruel or unethical practices, the entry will be disqualified and the Owners reserve the right to report the entrant to the applicable authorities.

Even with this highly specific set of rules, there have been transgressions. In 2010, an image of a rare Iberian wolf jumping over a fence at night won the competition. It was later determined that the image, while technically impressive and photographically untarnished, was actually a staged image taken of a tame animal known by the name of "Ossian". The winner was stripped of his prize and banned from the competition. His actions were a gross breach of the rules and misrepresented the reality of the natural environment since the wolf was not wild. In 2018, a similar issue arose, but this time it was because an anteater, the main subject of an award-winning image, was determined to be a taxidermy specimen. The picture was made at the entrance to a park in Brazil, where the otherwise honest image was captured. It is interesting to note that both of these digital images were disqualified for reasons that had nothing to do with the digital technology or photographic technique. The underlying truthfulness of the images was the problem in each case, and that is an issue squarely in the realm of ethics.

Conclusion

No one works in a vacuum. Natural science images are meant to be seen, analyzed, and enjoyed. They are reproduced in a wide range of publications ranging from serious academic treatises to supermarket tabloids. They saturate the Internet and are accessible to millions. It is not surprising, then, that mistakes and irregularities are quickly recognized, reported and commented on, with results that can resonate around the world. No one is immune from the watchful eyes of Internet trolls or serious scientific agencies. This is why it is amazing that photographers still try to take shortcuts and attempt to fabricate truth or hide reality with their imaging skills. Every year, there are more accounts of misrepresentation revolving around images from individuals and institutions that strive to provide only the unvarnished facts. Despite their stringent rules against photomanipulation, *National Geographic* still published an article in 2019 with images of the Milky Way galaxy that had been extensively cloned. In that same year, over 30 papers in several high-impact scientific journals spanning 17 years were also retracted or brought into question, and a human clinical trial was halted after it was found that an important scientific research study on cardiac stem cell regeneration was founded on fabricated image data. These examples show how important it is to remain vigilant and demand accountability from anyone wielding a camera in the name of science.

The scientist/photographer has an important part to play in the documentation and validation of scientific information via the image data they create. It is only by fully understanding the subject matter and applying acceptable photographic protocols that accurate and unbiased reporting of the scientific record can be produced. It is the responsibility and obligation of an imaging professional to ensure that their work is accurate, truthful and reproducible to the extent that it can be, while looking out for the health and welfare of the subject matter and the environment. There are only a few basic guidelines that need to be addressed to ensure that the integrity of this process is maintained:

- Do No Harm:
  - to a subject
  - to the environment
  - to the scientific integrity of the photograph.
- Do not misrepresent the factual details of the subject or the image.
- Always photograph in a raw format and maintain the original files.

- Only make adjustments to copies.
- Do not make spot manipulations to images.
- Do not fabricate images.
- Disclose any manipulations.
- Provide accurate captioning information.

The reality of digital imaging is that, in skilled hands, it is easy to manipulate photographs and practices in ways that hide or alter the truthful facts. It is the objective of the scientific photographer to make sure this does not happen. Not so long ago, we looked at compelling images and wondered how they were captured. Now, when we look at a beautiful image, we wonder if it is real.

Suggested Reading

Laboratory Imaging & Photography: Best Practices for Photomicrography & More (Applications in Scientific Photography series), by Michael R. Peres. Routledge, 2016.

Websites

1 National Park Service: "Leave No Trace Seven Principles". https://www.nps.gov/articles/leave-no-trace-seven-principles.htm (originally published 1999 by the Leave No Trace Center for Outdoor Ethics: www.LNT.org).

2 North American Nature Photography Association: "Principles of Ethical Field Practices". https://www.nanpa.org/wp-content/uploads/Ethical-Field-Practices-Revised-3-2018.pdf

3 North American Nature Photography Association: "Truth in Captioning Statement". https://www.nanpa.org/wp-content/uploads/Truth-in-Captioning-Statement-Revised-3-2018.pdf

4 Rossner, M. and Yamada, K.M. (2004) "What's in a picture? The temptation of image manipulation", *J Cell Biol* (2004) 166 (1): 11–15. https://rupress.org/jcb/article/166/1/11/34064/What-s-in-a-picture-The-temptation-of-image

5 Nature Research: "Image integrity and standards". https://www.nature.com/nature-research/editorial-policies/image-integrity.

6 Natural History Museum: "Wildlife Photographer of the Year – Rules". https://www.nhm.ac.uk/visit/wpy/competition/adult-competition/rules.html

Chapter 5 The Subject of the Photograph

Introduction

Making images rich with information in a highly precise and accurate manner is an expectation in scientific photo-documentation. Sometimes though, in a rush to produce images, it is easy to forget just how important the subject by itself is to the making of a good photograph. I first referenced this idea in Chapters 2 and 3. The subject will define an image and its potential for success even before making a photograph.

The Sample and Photographic Treatments

Effective photographs are the result of making and executing a good imaging plan that faithfully features sometimes less than cooperative samples, sample characteristics, or environmental situations. Successful and informative images are not lucky accidents. They are the outcome of a process where the sample itself plays a critical role in the creation of a successful image. Often scientist photographers are quick to photograph whatever ends up in front of their lenses. This is frequently a consequence of time constraints or not considering the role of the sample's appearance and its contribution to the image. A bad sample will always lead to an inferior photograph. A poor sample or damage within the sample may actually become the focal point of an image. This is not important unless the damage to the sample is the subject of the photograph.

Formulating your Plan

All samples are unique, and each presents different challenges in handling and photographing. Samples can change during an imaging session or they may stop exhibiting a specific behavior during imaging. Creating and maintaining the visibility of a sample's characteristics is fundamental to making good images for science. The sample could be damaged, destroyed, or changed in negative ways if changes are imposed during an imaging session. The objective of the scientist photographer is to make images of challenging objects without embellishment or without changing the specimen. Highlighting characteristics of a sample that otherwise might not be immediately evident or even visible, should be the goal of the photographic approach. Anticipation of future challenges can provide great advantages when developing the imaging plan.

Locating the Sample

Once a sample has been selected and a treatment identified, how to manage the imaging session should become the priority. In today's digital photography, there is an attitude amongst many practitioners that artifacts or other unwanted content can be edited or deleted during image processing. While this is possible, it can compromise a sample's integrity and should be avoided in science. Planning to retouch image artifacts out as a standard practice should never be part of a Best Practices approach. Sloppy work habits will contribute artifacts to images. It is best to spend time fixing things during the set-up phase rather than spending a lot of time on the backside of the imaging session to clean up errors. In some imaging experiments, hundreds of pictures will be produced and if each image requires time to fix a problem, this results in a significant commitment to a lot of post-production work that could have easily been avoided.

Isolating the Subject: Composition

One challenge a photographer has to resolve is sample isolation. Isolation can be achieved in many ways. One of most effective tools for isolation is the simple use of camera framing or composition of the photograph. Frequently, the framing of photographs is not adequately managed. "Less is more" when it comes to isolating a subject within the frame. The frame of a sensor is a rectangle and has an aspect ratio. Many full-size sensors in DSLR cameras have an aspect ratio of 3:2. This number is derived from a 36mm × 24mm imaging area. A cropped or smaller sensor camera may have a 4:3 aspect ratio. The first number refers to the width of the sensor and the second number the height of the sensor. A more panoramic sensor will feature an aspect ratio of 16:9. The dimensions of the sensor will play a role in what can be recorded in an image. Filling the frame is an important strategy for isolating. Making loose compositions or including extraneous content will diminish the effectiveness of photographs. In some situations it may be unavoidable. Changing magnification will be discussed later in the book, but using the frame to create an effective image magnification and the removal of competing elements is a very simple way to improve images. Anything in a composition will be considered by the viewer to be part of the photograph.

It is interesting to compare how artists create work and how photographers make pictures. Artists start out with a blank canvas and add content as desired, but a science photographer must constantly look for ways to subtract and isolate the subject from its background or environment. This is needed to create emphasis and isolation. A photographer should always locate important features within the frame that lead to a better understanding and not create confusion about the subject and its story. It is important not to position vital elements of the subject too close to the edge of the frame and be cautious about cropping out important details. There will always be compromises and limitations.

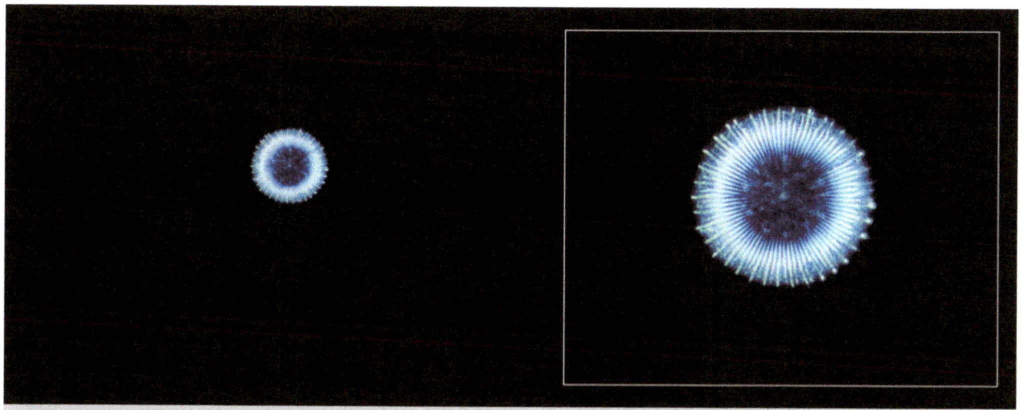

Figure 5.1 In this composite photomicrograph of a diatom, two different magnifications were used to make the pictures. Clearly in the image on the right, where the magnification is higher, there is more information to be seen. Images courtesy of Michael R. Peres.

Not surprisingly, there are numerous websites dedicated to framing and the composition of effective photographs. They all share the following key points:

- Simplify what is included in the frame.
- Consider using the longer dimension for a composition where useful and not always shooting photographs in the horizonal orientation where the sensor will define composition.
- Do NOT locate everything of interest in the middle of the frame. Sometimes it is possible to distribute the sample across the whole frame.
- Use the diagonal and locate important structures evenly throughout the frame if possible.
- Select and use an effective background.

Isolating the Subject: Magnification

Image magnification plays a role in isolation. Images that are made using high magnifications may result in the cropping of relevant information and images that do not have enough magnification will not adequately reveal as much data as they might. Hence an image's magnification will play a role in the formation of effective images. Depth of Field (DOF) or zone of focus changes with magnification. Higher magnifications have smaller zones of focus and lower magnifications form images with more DOF. Choosing the correct image magnification that contains the most information about the sample should always be a factor. Often this requires compromises required to achieve the best balance of both.

Wet Specimens and Managing Liquids

Samples that are wet or reflective can be challenging. A photographer has to decide whether to keep the sample wet or dry it. Drying can be accomplished in many ways. Sometimes the use of a hair dryer using a low heat setting will expedite drying. Sometimes objects simply need to be left out for a period of time and allow the moisture to leave on its own terms. Accelerating the drying of a sample may not be good and might cause shrinkage, or worse yet damage. It is not a great idea to use paper towels since invariably paper fibers will find their way onto the sample. A wet sample that becomes dry will frequently change in its appearance or morphology when no longer wet. It might get smaller, curl, exhibit color changes, or it might experience other changes that are not obvious until later. Wetting a subject can make it darker and enhance color saturation as well.

If the decision is made to keep the object wet, then there are other considerations that will come into play. For example, what type of container should be used for immersing the sample. Always use a vessel that is the right size. Never use a vessel that is too large. If the vessel is large relative to the sample's size, the object may float around and present other challenges when photographing. Cleanliness of the vessel is critical. I have found that using a weak acid wash such as white vinegar followed by a light rinse of distilled water can be helpful to manage mineral or other surface contaminants that might find their way into or onto the vessel and subsequently the photograph. Cleanliness of the vessel and ensuring no dust or particulate matter is present to the extent that is possible is important. This simple step of housekeeping can make a big difference to the amount of particulate matter that might have to be removed from the digital file if not removed from the scene.

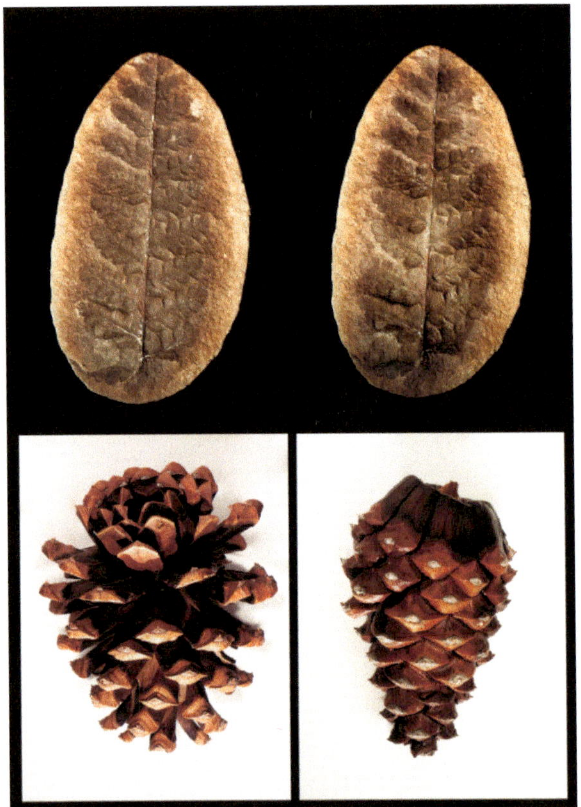

Figure 5.2 In this composite photograph of a fossil and pine cone, the subject was submerged in water. In each situation, the lighting was not changed. Because of the wetting, different features of the subject are revealed. It is interesting to see how the pine cone reacted to the addition of water. Images courtesy of Michael R. Peres.

One critical activity that requires attention is to minimize the contamination of the sample when adding liquids. Strategies for minimizing the presence of dirt and dust were shared previously. For wet samples, frequently it is useful to suspend samples in water or other liquids such as mineral oil or glycerin. When using water, the use of de-oxygenated, distilled, or de-ionized water is preferred. Tap water will often be full of dissolved air and minerals such as calcium. In a short time, the dissolved oxygen will form bubbles on the sample and walls of the container. Making an effective photograph when this happens will be tough. While sometimes interesting, the bubbles compete with the subject and might be seen as the subject themselves. Rarely is tap water a good idea to use, but if it is all that is available, allow the water to sit for a period of time before using. I have known photographers to run water through a non-woven coffee filter. On rare occasions, mineral oil or glycerin might be useful for suspensions if working with inert and not live samples. Another strategy is to locate the sample in clear gelatin such as Knox® gelatin, which can also be used to isolate and hold samples in suspension. You must allow it to set-up prior to use.

Many years ago, when attempting to photograph a developing root system, I planted seeds in a chamber filled with clear gelatin, similar to collagen. This material provided an ideal way to isolate the roots and micro-roots as they grew. This preparation provided an effective treatment for lighting and imaging the root system. Without this treatment the roots when removed from soil would collapse around themselves and have no separation. Additionally, photographing moist delicate structures in air would create morphological changes. The surface of the root would dry from the heat when shining lights on the root. This dehydration would result in color changes.

Making Chambers and Using Welled Slides

Sometimes when photographing wet samples, it is required that the sample remain in liquid. The objective of immersion is to keep a sample wet and to control reflections. When a light is shone onto a wet object, many reflections will be produced. These reflections will be random and may possibly obscure important details. For this reason, it remains practical and important to locate samples under water or other liquids. See Figure 9.11.

Often it is difficult to find the right vessel for this type of work. Petrie dishes or other laboratory glassware frequently include flaws in the glass that are formed when the glass is liquid and then cooled. This creates density differences in the glass. When illuminated from below, they will appear as varying dark bands in the background. These dark bands can also add false data or artifacts to a sample's appearance. For this reason, it might be desirable to fabricate a chamber using higher quality glass such as microscope slides, for example, or small pieces of picture frame glass based on the size requirements of the object.

Fabricating small glass chambers or "aquariums" is a relatively simple procedure but requires care to keep the pieces and parts clean and applying of the adhesives. The most useful glue for fabricating a glass chamber is clear silicon caulk used to seal bathtubs or for other waterproofing jobs. To make best use of the adhesive, it is useful to apply the adhesive using smaller tubes and lesser amounts. Less will be more. Too much caulk may produce flare when imaging. It might be helpful to assemble the sides first, allowing them to set-up and then adhere the sidewalls to the base. It is of course important to prepare the parts before starting the assembly. Chambers should be sized to the objects they are created for. If they are too small, they will compromise the work and if they are too large, the object might possibly float around. The area under the lens might be considered the "action field" and creating methods for keeping the sample under the action field is helpful. Chambers can easily be subdivided using additional pieces of glass cut to size. See Chapter 17 for more about making and using chambers.

Welled microscope slides are also useful tools for viewing live materials but they too have a few drawbacks for imaging applications. When photographing aquatic invertebrates, a welled microscope slide will keep the organisms in a controlled space but the depth of the well allows the organism to swim left and right as well as up and down. Using less water would be helpful and minimize the depth but the curvature of the well provides other challenges for lighting and imaging. The curvature will introduce flare and other non-image forming lighting outcomes in certain situations. Depending on the direction of the light or whether darkfield or other methods were selected, results may be less than acceptable and exhibit flare or other visual artifacts. It is possible to create a slightly different type of wet mount slide using broken coverslips to elevate a traditional coverslip. Break a number of coverslips and place the shards below a cover slip. The shards behave like legs of a small table and maintain a flat space for the organism and a water drop.

Staining to Reveal Features

It can be useful to add contrast to samples for the purpose of enhancing visibility where and when appropriate. Managing treatments where the efficacy of the sample is maintained should always be considered first before adding other materials. All objects are different and may have characteristics that need special attention. When a sample exists in a dry or wet state, it is not unusual for different features to be more or less visible. It might be practical to either dry or add water to a sample. Rehydrating a sample or wetting its surface can be accomplished in many ways. Simply locating the sample in a beaker of water maybe practical, and then soaking it as long as required. This process may also reveal other aspects of the subject. A pinecone when wetted will get darker and the cone protective seed leaves will retract when wet. Besides soaking, water can be applied using a small paintbrush or spray bottle selectively and across the whole object. See Figure 5.2.

Live aquatic organisms might benefit from using a vital stain, a chemical that is inert to biological samples. These stains are primarily water based and contain compounds that are inert and do not affect the morphology of a sample. To use vital stains, the sample to be stained should be prepared as a wet mount. This is a preparation where water has been added to the preparation and it typically uses a cover slip. Once the wet mount has been prepared, it is important to have a supply of bibulous (bib) paper nearby, along with the vital stains. Bib paper is a nearly fiber free paper and very absorbent. There are a variety of sources for bib paper and vital stains. Place one drop of stain next to the cover slip and hold the bib paper on the opposite side of the preparation, drawing the stain through the sample. Once the staining has occurred, it is important to then purge the excess stain out of the preparation. This is accomplished by adding a few drops of water where the stain was initially added and then placing another piece of bib paper across from this location and next to the coverslip. In this fashion, the excess stain will be drawn through the preparation leaving only the stained material and clear water around the subject. This can be very effective for making structures more visible for examination under magnification. Typically the vital stains come in small dropper bottoms and in various colors. Many fibers and aquatic invertebrates are subjects for this approach.

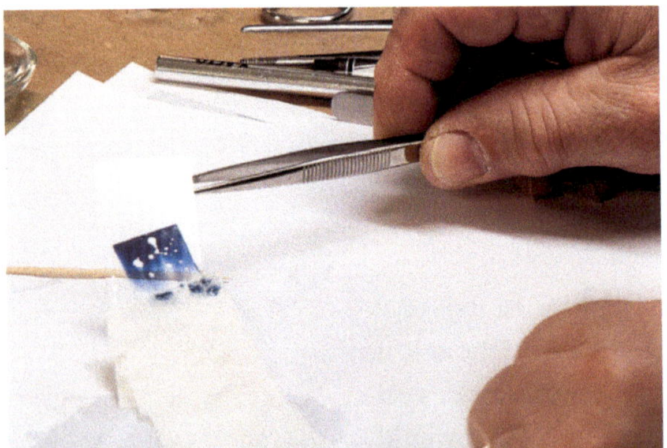

Figure 5.3 In this photograph, a vital stain has been added on the top of the coverslip and a piece of bib paper located on the bottom to draw the stain through the preparation. Image courtesy of Michael R. Peres.

Applying small amounts of liquids to preparations or removing small amounts of liquids can be challenging. Eyedroppers, pipettes and medical syringes can be helpful and capable of delivering small amounts of liquids precisely. It might be useful to keep a ready supply of various sized syringes for this task.

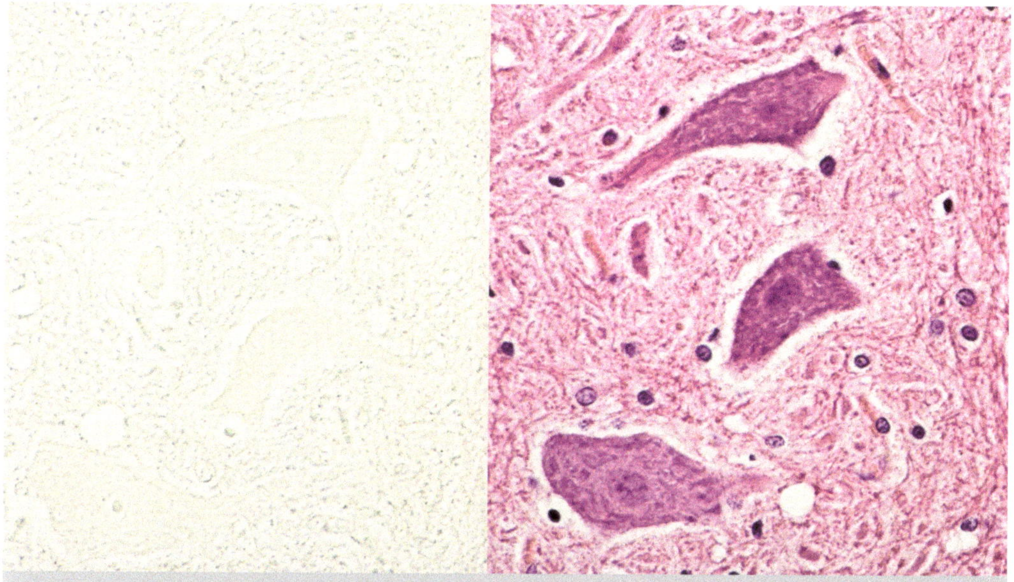

Figure 5.4 This illustration reveals how a section of brain tissue appears unstained (left) and stained (right) using hematoxylin and eosin. There is a significant change in sample and cell visibility. Staining is a complex but important process. Vital stains are useful for small live organisms. See Figure 5.3. Images courtesy of Michael R. Peres.

Specimen Platforms

Much of this chapter has focused on small objects but not all samples are small and can be opaque or semi-transparent and they can be large. Sometimes this type of sample will benefit from having light shone through it. For this reason, it is useful to fabricate a small transparent stage or specimen table to hold the sample. This is easily accomplished using wooden blocks and pieces of double strength window glass with polished edges. In the U.S., double pane window glass of ¼ inch thick is ideal. Polishing the edges is suggested for safety reasons. Depending on the size of the samples that will be photographed an 11 × 14-inch sheet of glass might be a practical size or a variety of sizes might be practical. The glass is placed on top of the wooden blocks and then light can be shined from below. Many lighting set-ups will be vulnerable to contamination from extraneous ambient light and reflections. To minimize lighting artifacts, it would be useful to paint the wood blocks black. A subject may have internal facets and surfaces that can refract and reflect light. Surrounding the area where photographing with black paper or material will reduce the potential for lighting artifacts to be introduced.

It should be noted that ceiling lights should always be turned off when photographing using glass surfaces. Many subjects are reflective and so is glass itself. Turning off ambient and ceiling lighting will minimize reflections of cameras and photographers. It is also sometimes practical to photograph through a black piece of cardboard that surrounds the lens. Black cardboard will be reflected in surface of the sample and will suppress any unwanted reflections in the sample's surface. Holding the black cardboard might be best accomplished by using a clamp or by hand. As experiences are gained, additional strategies, tools and

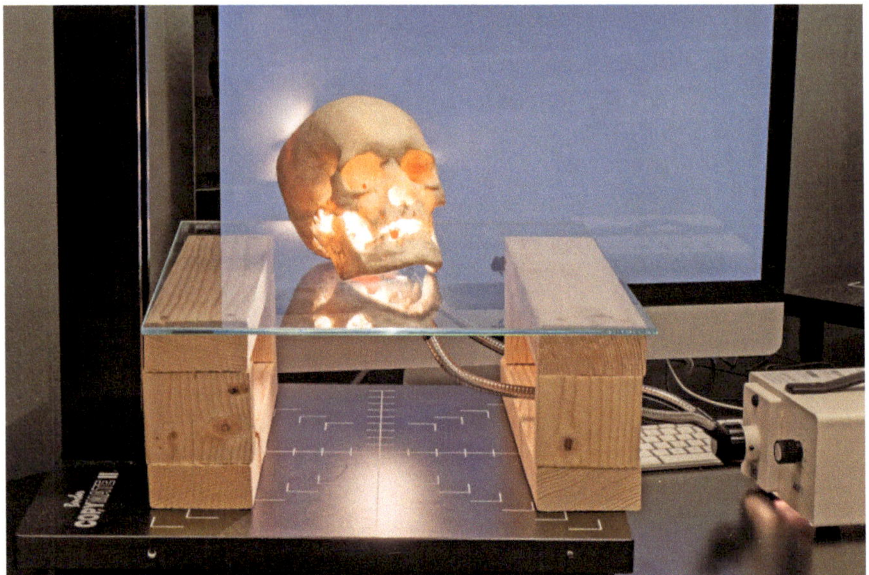

Figure 5.5 This photograph features a specimen table. It is possible to construct a base using framing lumber as seen in the picture and place a piece of sheet glass across the bridge. Painting the uprights black is recommended. Image courtesy of Michael R. Peres.

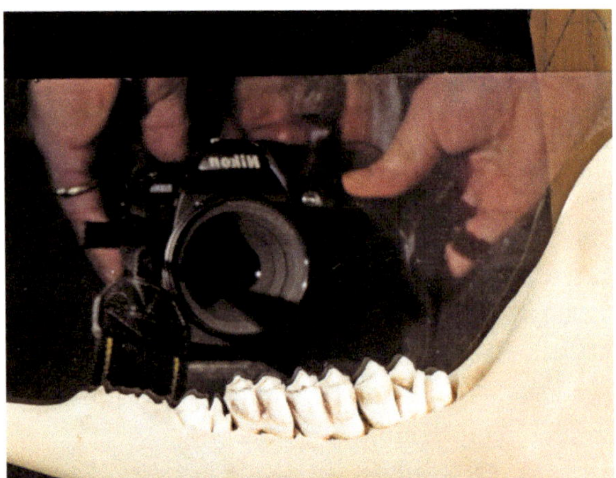

Figure 5.6 Glass is a highly reflective surface. When photographing into glass, be aware of reflections from the camera or other elements in the room. Pictured above is a reflection of me and my camera in the piece of glass used as a platform for the bone. Using a black piece of paper with a hole for the lens would eliminate this reflection. See Fig 11.3 Image courtesy of Michael R. Peres.

accessories will be collected that will greatly assist in solving problems associated with this work.

Mirrors

Sometimes the orientation of a sample is important. Choosing to photograph from the top or from the bottom of a sample can be important. Photographing a sample in its proper orientation can be challenging, especially if photographing the bottom or inside is required. When subjects are large, photographing the bottom of an object when there is an orientation concern can be accomplished by locating the subject on a glass platform (table) of an appropriate height and then photographing from below. Sometimes the characteristics of the sample are changed when not in their normal state. A useful example might be found if trying to photograph feet that are bearing weight. It is easy to photograph the bottom of the feet when not in a weight bearing position. A person could sit on a table and hold their feet outstretched, but this photograph would not be accurate since feet bear weight and would appear differently in this situation.

An effective solution is to use a front surface mirror. A subject could stand on a 1 inch piece of plexiglass located on an elevated platform. Under the platform, place a front surface mirror at a 45 degree angle to the camera. A front surface mirror has the reflective material coated onto the front of the glass rather than on the rear surface. Traditional mirrors have their reflective surfaces on the back of the mirror. Common mirrors produce two reflections, one dim reflection on the front of the glass and the primary and brighter reflection occurs from the silvering on the rear surface. Front surface mirrors have only one reflection because the reflective surface is on the front. By using a front surface mirror to reflect the bottom of weight bearing feet, a highly accurate photograph can be made. While you might not photograph feet, the idea of using a front surface mirror might be useful for various other photographic investigations including looking at a snail or snake moving. Front surface mirrors are used routinely in dental photography where access to the surfaces of the teeth would be impossible in any other fashion.

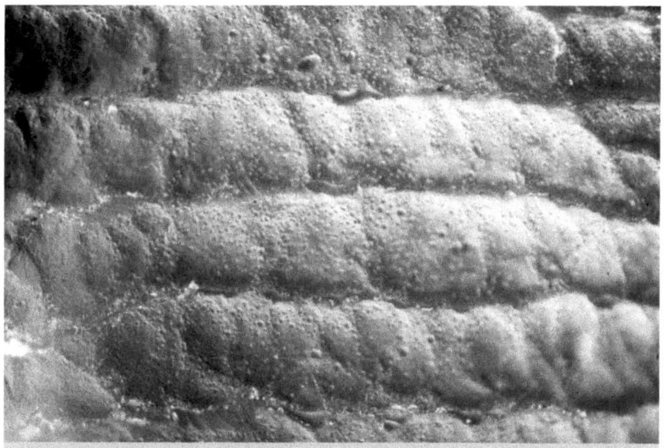

Figure 5.7 The surface of human skin. The photo itself is of a replica of the surface of the skin. Hot paraffin wax was used to create the surface replica featuring the surface of a human knuckle. Image courtesy of Michael R. Peres.

Surface Replicas

Photographing the surface of an object can be accomplished in many ways but when the magnification increases, it can present challenges. A simple strategy for seeing a surface of a subject is to create a replica of the surface. Different information about a sample will be revealed in a replica. Surface replicating can be accomplished using various materials. Two common materials that are used for surface replicating are NewSkin® and candle wax or paraffin. The later must be applied when in its liquid state and of course will be hot. For that reason, care should be taken when applying to surfaces.

NewSkin® can be applied by painting it across the surface of the sample or by laying an object into the Newskin applied to a clean microscope slide, for example. Allow the compound to dry for a short period of time, at a time to be determined before removing. Objects that might benefit from surface replicating could include hairs, fibers, leaves or other objects with surface structural information that otherwise would be difficult to see. Skins of various organisms can also be prepared for imaging this way.

Backgrounds

After a sample has been selected, the choice of what background should be used is no less important than anything else shared in this chapter. In science and certainly other applications, the background serves to isolate a sample in a photograph. Backgrounds need to be simple, neutral, uncluttered and suited for reproduction or display. Backgrounds for the uninitiated might be an afterthought but should always be a first consideration. Poor background choices often lead to lowering image contrast, visibility issues, or introducing competing elements into the photograph. Choosing the background and how to manage the background is an important consideration for making effective scientific illustrations. Wet objects will ooze liquids and leave marks on a background. There are many things to consider when selecting and using a background.

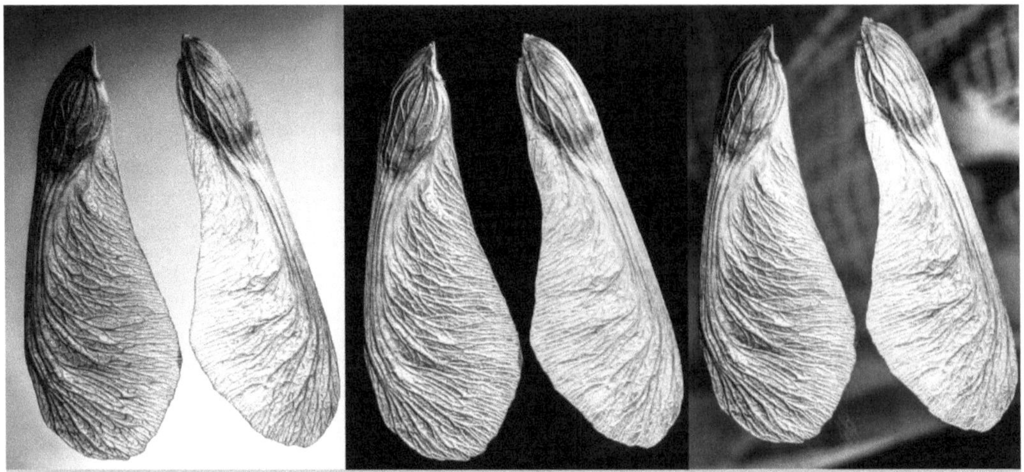

Figure 5.8 This shows how one sample can appear when photographed in front of various backgrounds. The seed leaves will look lighter against black, and darker in front of white. When located in front of a complex background (right), the various tones and structures in the background compete for attention with the subject. Images courtesy of Michael R. Peres.

The Use of Scales to Indicate Size

The inclusion of a scale in scientific images is an absolute must. Without a scale, the image might be considered unacceptable and not publishable without additional explanations of its size in the caption. The scale is part of the capture data. There are some considerations for the proper use and selection of a scale. The magnification or reduction of an object's size in an image will of course play a role. For close-up photography, it is practical to have several scales to choose from. There are scales that have white markings on black and other scales will have black markings on white. Metal rulers are not ideal because of their potential to introduce reflections, creating flare, and the difficulty in lighting them. It is also important to use scales that are calibrated using the metric system and not calibrated in inches. An inch ruler should not be used in science unless there are no other options. It can also be useful to have both opaque and semi-transparent rulers. I have found it very useful to also have an inventory of paper scales that have an adhesive back as well. There are a number of suppliers of these products. A quick web search can provide an inventory of vendors for these products. I have also found that forensic supply houses have the widest range of products.

Using scales requires a few considerations for the method to ensure the magnification of the scale is consistent with the magnification of the object. To accomplish this outcome, it is important to locate the ruler at the correct height or location in the photograph where the focus of the lens has been placed. This may require the use of a bracket or third arm to hold the scale at this location in space. The accuracy of a scale located at a different height to the camera will infer either more or less magnification. Objects placed more closely to a lens while possibly out of focus will be more magnified than objects placed further from the lens. In addition to locating the ruler scale at the location where the lens focus has been placed, it is also important to have the ruler parallel to the camera's sensor.

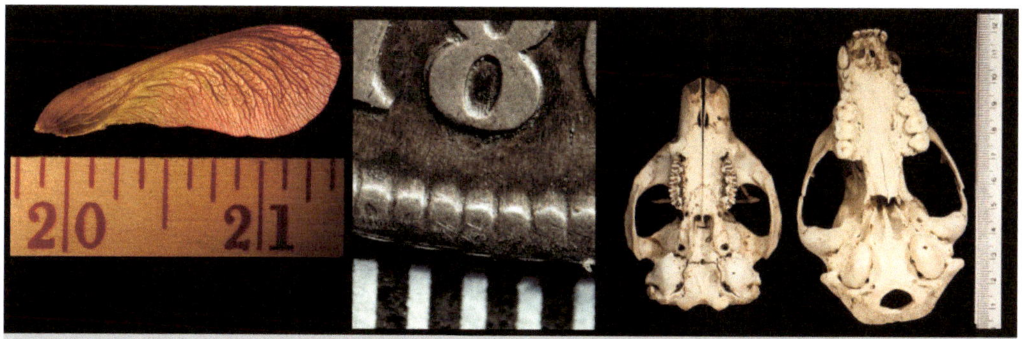

Figure 5.9 An *Acer rubrum* (red maple) seed and inch ruler, an 1804 coin with a mm scale, and two mammal scales using a mm ruler. Scales need to be placed at the correct height (plane of focus) and adjacent to the sample to be most effective. Images courtesy of Michael R. Peres.

Because the DOF of an image might be shallow, having the scale at the same height of the sample, as well as keeping the scale parallel to the sensor, will provide the most meaningful information. It is suggested to insert the scale into the field of view in a parallel orientation to the camera's sensor. Keeping all elements of the picture carefully composed and scales properly located will lead to images with more effectiveness. It takes just as much work to hastily locate a scale than to carefully locate the scale. Good photographers effectively manage all the details.

Scales for photomicrography are a bit more specialized and are called stage micrometers. They can be acquired for use with both reflected light or transmitted light microscopes. They can be calibrated into various units of microns either in .1mm distances or .01 unit distances. They might be available in 1 or 2mm distances. They can also be used for calibrating a microscope's internal measuring function or for calibrating an eyepiece micrometer. Many photomicroscopes will come equipped with software that can place bar scales into an image. This feature must be first calibrated before use. Once calibrated (only one time), the software will store the scale's information as long as the hard drive and operating system do not change. The instrument's software frequently is not programmed to determine specific distances but rather the number of pixels in a specific distance. To program the software, a reference distance using a scale of specific distance identifying the specific number of pixels in a specific distance such as .1mm for example. Each objective lens needs to be calibrated.

Suggested Reading

Envisioning Science: The Design and Craft of the Science Image, by Felice Frankel. MIT Press, 2002.

Photographic Composition: A Visual Guide, by Richard Zakia and David Page. Focal Press, 2010.

Photography for Scientific Publication: A Handbook, by Alfred Blaker. W.H. Freeman, 1965.

Part II

Tools and Technology Foundations

Chapter 6 Cameras and Photographic Practices

The evolution of cameras into powerful imaging devices represents an extraordinary technological transformation that began in 1827 when the first permanent photograph was made. That being said, the ultimate reason why people use cameras has remained basically the same since that time and that is to take a picture whether for personal or professional reasons. At its most basic level, a camera blocks all ambient light from entering the camera and keeps a light sensitive material in the dark awaiting exposure. Modern cameras are sophisticated and complicated devices, and yet their operation has become extraordinarily simplified.

Modern cameras come in a variety of sizes and shapes, and are equipped with many specialized features. Each type of camera has advantages, disadvantages, and reasons for using it. Smartphones (cameras), DSLR cameras, mirrorless cameras, and instrument or industrial cameras are a few of the varieties a scientific photographer can choose from. The basic components found in all cameras are sensors and shutters. Sensors when exposed to light convert the light into electrical signals of differing intensities that are displayed as an image. All sensors are not equal in their features and the image characteristics they create. Camera costs, supporting software, camera design, how images are transferred, as well as other operational features will vary greatly across brands and types.

Camera Components

All cameras – even the most basic ones – require physical or electronic controls to operate them. Switches or touch screen controls used for off and on are the most rudimentary controls needed to operate a camera. The sophistication and ease of operation for each device will vary model to model. A camera at its most basic function shields the sensor (film) from exposure to ambient light until the shutter is opened. Once opened, the shutter allows light to enter the camera and expose the sensor to different light values for a prescribed length of time. Typically, light enters a camera through a lens but pinholes or other optical elements play a role in image formation, transmission, and recording. A lens is used to improve image brightness and contrast as it focuses the image to a specific distance behind the lens.

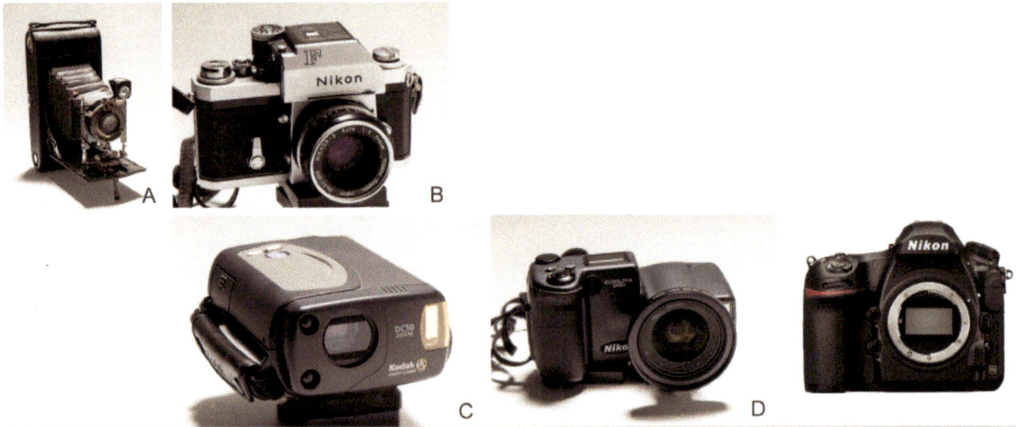

Figure 6.1 A variety of important cameras. (A) a Kodak No. 3A Autographic manufactured in 1914. (B) a Nikon F, Nikon's first SLR in 1959. (C) the Kodak DCS 50 that was marketed for use on a microscope in 1995. (D) a Nikon 990, a versatile 3.2MP camera sold in 2000 that was excellent for microscopy and other laboratory work. The image on the far right features a Nikon D850, a flagship DSLR that boasts a 45.7MP sensor released in 2017.

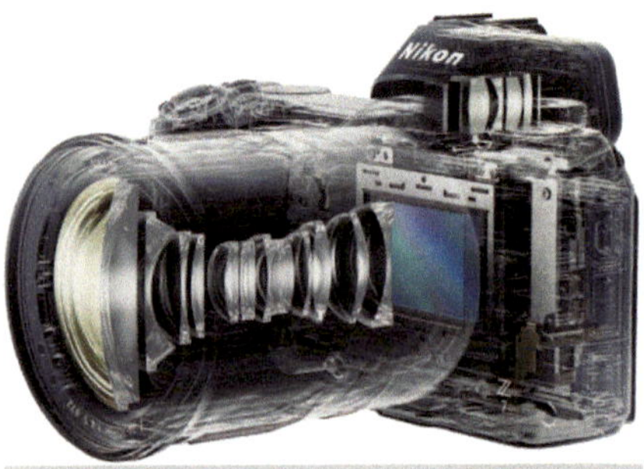

Figure 6.2 This digital illustration reveals the complex elements and design of a modern DSLR camera. Image courtesy of Nikon USA.

To make a properly exposed image, a camera must measure the available light and suggest the shutter speed and aperture. Light brightness will affect selection of a shutter speed required to form the proper exposure. There are other factors needed to produce a correct exposure. Most cameras will have a lens attached to the camera. Many lenses will have an adjustable opening called an aperture. Some lenses have fixed or non-adjustable apertures such as a smartphone. Apertures play a role in forming the exposure. The sensor itself plays a role in creating an exposure as a consequence of its sensitivity to light. Most cameras are equipped with ways to measure light as an "averaged" or integrated value. The light measurement will suggest what settings will lead to a "good" exposure under the conditions when the light was measured. Lighting and characteristics can change during an imaging session.

In addition to simple controls such as off/on, cameras will have a power supply and other controls. In addition to operating a camera in typical ways, some cameras can be operated remotely using a computer or an app on a smartphone. Power to the camera can be supplied using AC with an adapter or through the use of a battery, or both. A camera will also have a mechanism for storing/saving files to a removable media port, a USB or Bluetooth connection, and other ports as needed. Modern cameras have many ports that include video and audio input and output capabilities. The possibilities are endless and ultimately influence a camera's price. Camera technology is, and will continue to experience changes and upgrades to capabilities and features.

Modes of Operation

There are several modes used to operate a camera. These modes include manual or automatic. There are also several secondary modes that are available in modern DSLR cameras.

There are reasons why each mode should be chosen. Knowing your equipment's capabilities will greatly enhance its operations and results.

Manual Mode

Using a camera in manual mode provides total control over all elements of a photographic exposure, the camera's operation, and image attributes. There are three components to a photographic exposure. These components include the aperture setting on a lens, the duration or time increment (shutter speed) set, and the sensor's sensitivity (ISO) to light. When using the manual mode, careful attention must be paid to measuring the light available for photographing. Once the light has been measured, the photographer must interpret the information and use it to create a proper shutter speed selection and aperture (lens opening) choice needed for a proper exposure. A proper exposure creates images that have detail in the highlight (bright) tones and in the dark regions or shadows simultaneously. This outcome assumes a scene brightness ratio that is recordable and fits onto the sensor using a single exposure time. In Chapter 12, high dynamic range photography is discussed. HDR is a way to manage large dynamic ranges that are impossible to record using a single exposure time. Once the light has been measured, photographing in the manual mode provides the maximum control of variability caused from shutter or aperture changes, or variances in the samples itself, or other environmental changes. Using manual mode will produce significant advantages in the consistency of exposures created across many pictures if nothing changes during the imaging session. Files produced using the manual mode when conditions do not change are ideal for batch image processing. When working in challenging applications such as fluorescence microscopy, the use of the manual mode will facilitate tighter exposure control and create files with excellent management of important highlight information.

Automatic Mode

When operating a camera in automatic mode, the camera itself will determine some, if not all of the settings required to create a proper exposure. The camera will determine and recommend the exposure settings by choosing different combinations required to achieve the best result. The correct time needed to create a precise exposure; selection of the aperture; determining where to focus the lens; determining how and where to measure light; identifying the correct white balance; and how to adjust the camera's sensitivity are many of the automatic choices you will have to select. Some cameras have other modes of automatic operation as well.

Operating a camera in automatic mode can be accomplished using one of two methods: aperture priority and program priority. In the program mode, the camera calculates the shutter speed and aperture settings that are best to balance the brightness (aperture setting) and image stability (shutter speed) needs. Scientific photographers should not use program mode unless there is no alternative. There are situations where lack of control will produce image variances or possibly deficiencies that will take away from effectiveness. Aperture priority can be used more effectively in science because the image's range of focus (DOF) can be better managed. More about DOF is covered in Chapter 7. Being aware of how image attributes are affected when using long exposure times or how the aperture affects DOF will influence the decision to use shutter or aperture priority as a mode of operation.

Secondary Modes

Beyond these principle modes of operation, many cameras have other secondary settings that are useful for other operations. These might include burst or rapid-fire mode, where a number of photographs in quick succession are needed. There may also be a single shot mode that takes a single picture at a time when the shutter is tripped. Autofocus mode should be avoided in science since there is frequently no control as to where the lens's focus is placed. If warranted, the autofocus setting can be operated by either establishing a single location and then locking the lens to that distance, or by continuous active focusing. Using the single focus locking mode is especially useful for stationary subjects. When the focus is locked in, it will stay locked in if nothing changes. Continuous focusing mode is useful for moving subjects. Never use autofocus when photographing science unless there are compelling reasons to do so. Autofocus will often produce images that are in focus but create photographs where the focus might not be situated in the best location for the image.

Many cameras have various flash modes because many cameras are equipped with a flip-up flash that is built into the camera, or the camera can be adapted to use other electronic flash equipment. Using the camera in flash mode provides various flash settings such as fill flash or auto-flash. Red-eye reduction is another setting that discharges a pre-flash prior to the actual camera flash which causes a subject's pupils to become smaller and thus reduces red-eye. A pre-flash may also be used to trigger secondary flash units although these other units will not be synchronized to the camera's flash. The flashes will be discharged at slightly different times. If a subject is stationary this will be fine, however if there is subject movement, there is a chance there will be multiple images if the subject is moving quickly and the delay is pronounced. Operating the camera in the flash off mode will turn off the built-in flash.

The Photographic Exposure

Opening a camera's shutter and recording light values onto a light sensitive material is the definition of photography. This activity is sometimes called writing with light. As previously shared, there are three components that contribute to a photographic exposure. The length of time light that strikes the sensor (controlled by the shutter), the intensity of the light that strikes the sensor (controlled by the lens and/or aperture), and the sensitivity of the sensor itself to light. All of these contribute to the cumulative exposure. A photographic exposure is defined as a function of time and light brightness. It is explained using the following equation.

$$H = E \times t$$

In this equation, H represents the total exposure to the sensor, E is the brightness of the light (typically controlled by the lens aperture) and t is the time that light is allowed to fall onto the sensor. Any combination of time and intensity can be used to achieve the correct exposure when matched to a sensor's sensitivity. Making a good exposure can be accomplished using various combinations of aperture and shutter speeds. Bright light requires the use of short times and low light requires longer times.

Let's consider a drinking glass as an analogy to gain an understanding of exposure. Consider the glass to represent a pixel (sensor) and water (from a faucet) to represent the light energy required to make an exposure. If the faucet drips into the glass, the glass will fill, but it will take a long time. If the faucet is opened fully, it will create a large discharge and the glass will fill quickly. In either situation, the glass or pixels will be filled with water (or photons). When different combinations of t (time) and E (brightness) create the same H (exposure), it produces what are called reciprocal exposures.

As an example, the following times and aperture combinations will create the same cumulative exposure. However, the images produced when using these combinations will often exhibit different characteristics. These exposure combinations might produce image blur or variances in the range of focus (DOF).

1/1000 sec @ *f*/2.8
1/500 sec @ *f*/4
1/250 sec @ *f*/5.6
1/125 sec @*f*/8
1/60 sec @ *f*/11
1/30 sec @ *f*/16
1/15 sec @ *f*/22
1/8 sec @ *f*/32
¼ sec @ *f*/45

Light Measurement and Meters

Every modern camera has the capability to measure light. This is required for the creation of a properly exposed photograph. The component of a camera that measures light is called a light meter. It converts light energy into information needed to establish the correct time or aperture settings at a specific sensor sensitivity or ISO. Establishing the proper time (t) and aperture (E) setting is based on scene or object brightness. Meters can be built into cameras or they can be handheld devices. The use of handheld light meters has been in decline because digital cameras display images directly on the camera's LCD

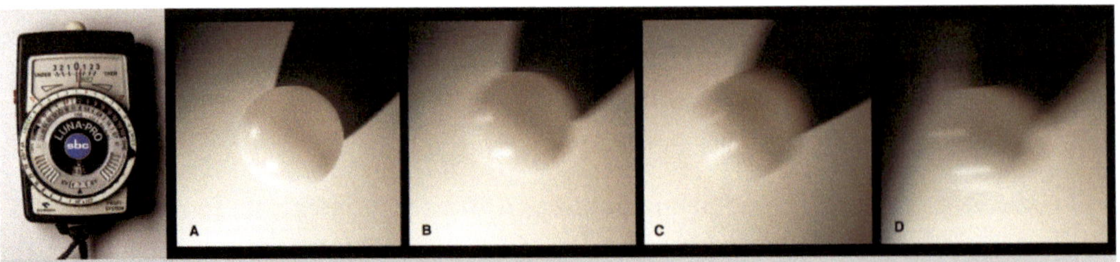

Figure 6.3 Reciprocal exposures produce the same exposure result using different combinations of time and intensity. This information is provided to a photographer as an exposure value or EV from a light meter. Many combinations of time and intensity create the same "volume" of light. Visible on the Gossen light meter are combinations of time and intensity creating an EV of 7. In this composite photograph are four photographs made using EV 7. The (A) photograph's camera settings were 1/250th sec at *f*/4, (B) 1/60th sec at *f*/8, (C) 1/15th sec at *f*/16, and (D) 1/4th sec at *f*/32. In each photograph, there are variances in image sharpness because the ping pong ball moved different distances during each exposure. Images courtesy of Michael R. Peres.

or onto a computer monitor at the time of creation. Photographers can take pictures of a subject to determine the correct exposure or by operating the camera in the automatic mode to determine the settings required for the correct exposure. There are also light meter apps for smartphones that can measure available light. Live View is another way to operate a camera which allows for accurate exposure information to be determined. Live View enables an image to be visualized leading to the selection of the correct camera settings. Live View is a very important tool but consumes a significant amount of battery life. Using Live View with an AC adapter and not using battery will be beneficial to an imaging session.

Exposure Values and their Application

Even though meters are in decline, they still play important roles for technical applications and providing very useful information prior to starting an imaging session. Light meters can measure light in two ways. The meter can measure light falling onto the subject or meter's photoelectric cell or it can measure light that is reflected back to the camera from the surface of the subject. In either application, the meter converts light brightness into an 18 percent gray value. This is important to keep in mind when comparing a subject's tonal range compared to an 18 percent gray tone. Reflective light metering is more difficult to use and is prone to making erroneous readings, but when operated correctly, provide the most precise information. The meter converts white objects to 18 percent gray tones and dark/black-like objects are also converted to an 18 percent gray value. To produce a corrected exposure when measuring light in the reflective mode, the information provided by the meter must be adjusted when compared to the subject. When photographing dark objects, it is necessary to reduce the exposure time by a factor of approximately –2 exposure values (EV) and when photographing bright objects, exposure times must be lengthened +2 exposure values. Changing an exposure by one EV in either direction, changes an exposure by a factor of 2, or 1/2. An EV unit shares the total energy available for an exposure using various combinations of time and aperture choice.

Meters can also provide exposure information by averaging all of the different brightness from the subject, or can measure only a small spot in the scene. Spot metering is very useful when photographing very challenging subjects such as fluorescence. Meters can also be operated using a center-weighted mode or measure the amount of light only in the periphery of the camera's field of view.

In the average mode, the meter integrates all of the brightness values from the entire field of view. In a center-weighted meter, the primary exposure information comes from the central 60–80 percent region of the field of view. Spot metering is the most precise method and can measure a very small region and a very specific region of a scene, sometimes down to a 1 degree area of the scene. Spot metering is very accurate and is not influenced by other areas in the frame. There are spot meter capabilities found in many DSLR and Instrument cameras. In all of these situations, the reading must be adjusted to the subject's characteristics.

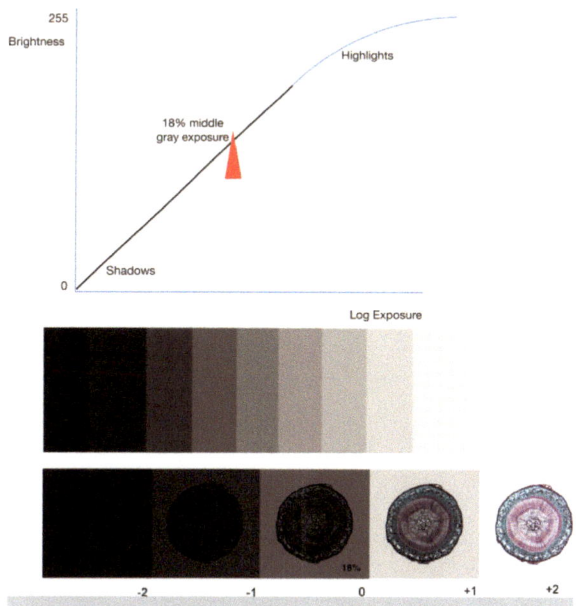

Figure 6.4 A characteristic curve shows how tones will be reproduced when using a ten-step, grayscale image. A ten-tone scale remains an important photographic tool for calibrating a camera's performance. A camera's ability to record a large tonal range is influenced by bit depth and the sensor's dynamic range. The black region of the scene is displayed at the bottom of the curve and the brightest regions on the shoulder of the curve. Located at the bottom of this figure, are five exposures range ranging from −2 EV, the correct exposure at 18 percent gray level and then +2 EV on the right. The correct exposure for this sample would be +2. Images courtesy of Michael R. Peres.

Shutters

Regardless of the quality, a camera must have a shutter and the camera must have a mechanism to control or activate the shutter. A shutter is one of the most fundamental components of a camera. The camera's shutter opens and closes, blocking light entering the camera. This allows or eliminates light traveling to a sensor. When the shutter is opened, light will strike the sensor for a pre-determined length of time. That specific time is referred to as a shutter speed. As computers and microprocessors have become integral parts of cameras, electronic shutters have become the norm and are designed to deliver any time increment. They are not limited by the constraints that were typical of mechanical shutters. In the film cameras, most had mechanical shutters that were operated at specific times such as 1/60 sec, 1/125sec and 1/250 sec times. Each time was a doubling or halving of the adjacent time. This was typical of mechanical timing technology. Contemporary shutters are capable of producing a continuous range of shutter speeds seamlessly from very long (as long as 30 seconds) to very short times, possibly 1/8000 of a second. When the shutter is closed, the pathway to the sensor is closed, which blocks the light from reaching the sensor.

Types of Shutters

There are three types of shutters used in contemporary cameras. A leaf shutter, a focal plane shutter, and an electronic shutter. There are also capping shutters that can be situated in front of a lens but are not common and are used in technical applications such as high-speed photography. In film cameras, and now contained in all DSLR cameras, are focal plane shutters. DSLR camera also have electronic shutters. Large format film cameras and some medium format cameras primarily used leaf shutters. Smart phones use only electronic shutters.

Focal Plane Shutters

Modern DSLR cameras have a focal plane shutters (mechanical component) and an electronic shutter. A focal plane shutter might be considered to be like a slit shutter. It is called a slit shutter because as the shutter opens and it behaves like a slit as it travels across the sensor. When the shutter is activated, the first curtain or gate opens and, at the appropriate time, the second curtain or gate begins to close delivering the correct increment. This allows only a portion or the entire sensor to be exposed to light for a predetermined specified time.

When using a focal plane shutter and electronic flash, there is only one time where the entire sensor is completely opened to light using the shortest possible time increment. This time is called the sync speed and should be selected for use with an external electronic flash. It might be indicated by an X or with a lightning bolt symbol ⚡printed on the camera's shutter selection controller. If the shutter opens using a time duration that does not expose the entire sensor when using electronic flash, a partial exposure will occur. This is revealed by the presence of a dark unlit region in the image. See Figure 6.6. Once the first curtain has traveled across the sensor, the flash is discharged. If the shutter speed is shorter than the sync speed, the second curtain begins to close before the first curtain reaches the other

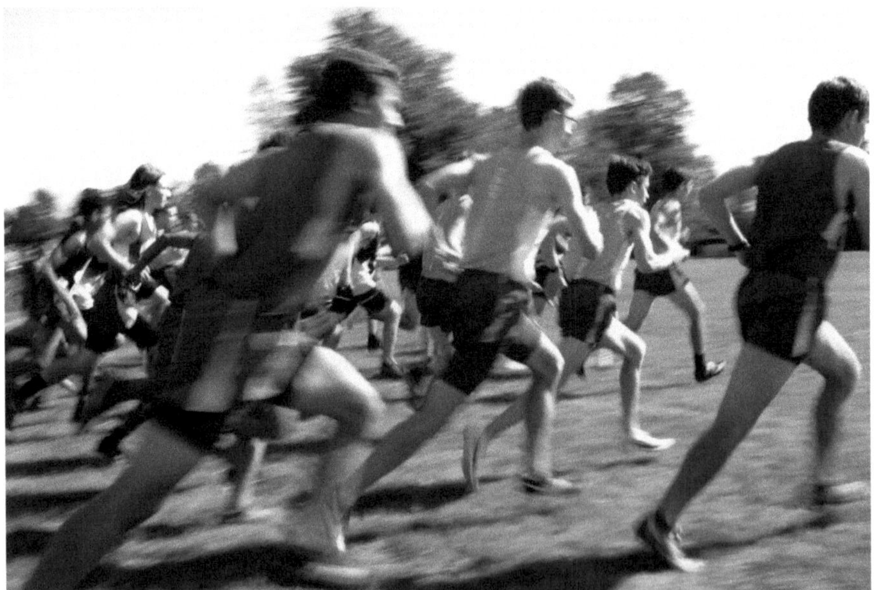

Figure 6.5 Image blur results from using a relatively long shutter speed with fast moving subjects. Subjects can be frozen in space or allowed to move through the frame based on the time increment of the shutter. Because of this potential, a subject can change shape as a consequence of the shutter's travel direction relative to the subject's travel direction. Image courtesy of Michael R. Peres.

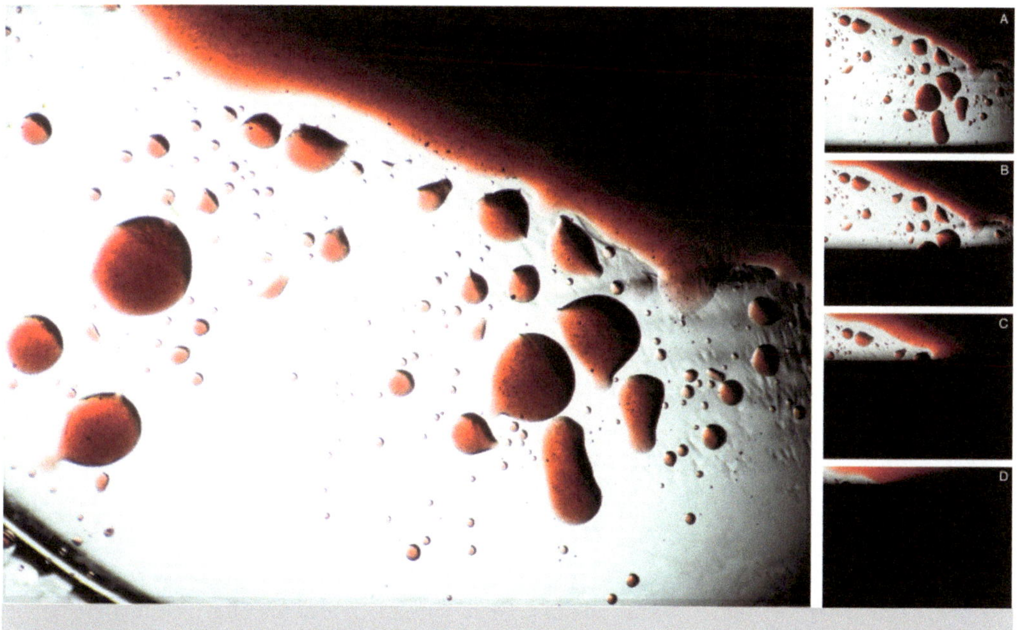

Figure 6.6 When a shutter is operated using a time where the entire sensor is not fully exposed to the discharge of the electronic flash, a shadow from one of the curtains will be cast onto the sensor. In this example, the camera's sync shutter speed is 1/25th sec visible on the left. Image (A) used a 1/500th, (B), 1/750th sec, (C) 1/1000th sec, and (D) was /1/1600th second. Images courtesy of Michael R. Peres.

side. The degree of this error will be revealed in how large or small the shadow is. Using the camera in auto flash mode will control the camera's shutter speed for use with flash. When using external flash equipment and operating the camera in manual mode, it is possible to select and use both a correct and incorrect sync speed. Relative to shutter's mechanism for opening, they can have two travel directions, vertical or horizontal. Vertical traveling shutters will have a shorter sync speed than a horizontal shutter because it travels a shorter distance.

It is possible to use long shutter speeds for different results, too. Longer times will create exposures that contain the flash exposure and an additional exposure from ambient or secondary lighting as well. This exposure creates images that have both frozen and blurry features. Using shutter speeds that are slower than the sync speed is referred to as "dragging a shutter". Dragging a shutter produces ghost like images caused from the ambient light exposure created when subjects move within the frame during the flash exposure. It is frequently used to communicate subject movement for creative applications and is a departure from highly accurate images. Dragging a shutter is not routinely used in technical and scientific photography. An example of stop action photography is included in Figure 16.17.

More on Sync Speeds and Focal Plane Shutters

An electronic flash is an important tool for science photographers. Flash illumination can be useful for stopping motion of moving subjects. Using a flash requires that the shutter and the discharge of pulsed light be synchronized.

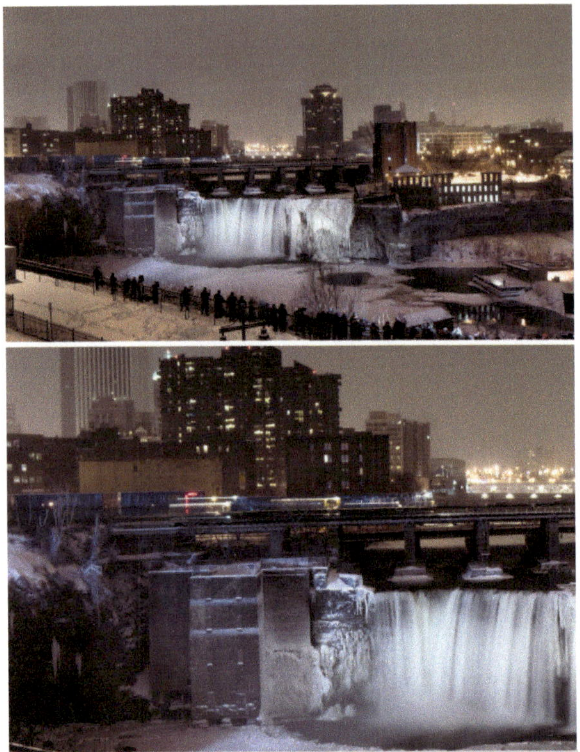

Figure 6.7 This photograph features the use of a flash and ambient exposure of RIT Big Shot No 28. It features the High Falls located on the Genesee River in downtown Rochester, New York. To create the illusion that the train was traveling from left to right, the flash equipment was discharged while the train was backing up. This caused the image rendering of the train's lights to move into the train or towards the left on the sensor. If the train was traveling forward (right), the image of the lights would be located in front of the train. Image courtesy of the RIT Big Shot, http://bigshot.cias.rit.edu.

Syncing the flash to the front curtain is most common. In this fashion, the flash is discharged when the first curtain reaches its maximum opening. If this sync speed is chosen, the entire sensor is exposed to the light discharge prior to the second curtain closing. Electronic flash is useful because it will freeze the motion of the subject or eliminate image shake/vibration in an exposure. It is challenging to use flash on a microscope or other special applications, but the minimizing of vibration affects can be very useful for improved image quality. It is possible when using front curtain sync to use a slower-than-the-suggested shutter speed. In this instance, the rear curtain does not close immediately and may remain open for a longer predetermined time. With front curtain sync speeds, the blur will appear in front of the subject when the flash fires. Many DSLR cameras may also refer to this as the fill flash setting. This problem can be overcome by opening the shutter separately from the discharge of the flash. Start the exposure using the ambient light and then when desired, manually discharge the flash before closing the shutter. In this approach, the blur will be located behind the sharp and well-defined flash exposure. These techniques are useful in public relations photography.

A rear curtain flash sync describes a condition when the second curtain triggers the flash when it begins to close. When this occurs, the ambient shutter exposure will occur first, causing the blur to be behind the subject when the electronic flash is discharged. The blur in this condition will appear to follow the subject and will appear more natural to viewers. Many DSLR cameras come with rear curtain sync settings which can be selected in the camera menu. This type of discharge is might also be called a tail flash.

Leaf Shutters

Different than a focal plane shutter is a leaf shutter. Leaf shutters are located within a lens and not located in the camera itself. Focal plane shutters are located at the focal plane of the camera, which is where the image is focused onto the sensor. For this reason, focal plane shutters are considered more practical than leaf shutters since only one shutter is needed per camera. Focal planes shutters allow one shutter to be used with all lenses. Conversely, one leaf shutter is needed per lens because the leaf shutter is located within the lens itself. Some models of Phase® medium format camera lenses are equipped with leaf shutters. When photography used film, leaf shutters were preferred by studio photographers when using large format cameras. At the time, leaf shutters could sync at any shutter speed and they were cheaper to make than large focal plane shutters for large format film cameras. Since the leaf shutter is located within the lens, it opens from the center of the lens, moves to the periphery and then back to the center as it closes. For this reason, the whole sensor is exposed to light regardless of the shutter speed. The time increment from closed to the maximum open and then back to closed again constitutes the exposure time or shutter speed. By opening and closing in this manner, the entire sensor is exposed to the entire exposure for the whole duration of the event. This allows for flexibility for the synchronizing of high-speed events using shorter duration lighting. Leaf shutters were common in large format camera lenses used on 4 × 5, 5 × 7, 8 × 10 and 11 × 14 film cameras. DSLR cameras have greatly disrupted the use of leaf shutters.

Electronic Shutters

Electronic shutters are used in smartphones, mirrorless, and DSLR cameras. Electronic shutters operate by turning on and off pixel sensitivity to light. Electronic shutters allow one sensor to record still and moving images. Focal plane shutters are also used in DSLR cameras but primarily to provide an absolute block to light to the sensor. When making still photographs, a sensor's electronic shutter operates in the global mode. The global mode allows the whole sensor to create a simultaneous capture. The focal plane shutter controls the duration of time that the sensor is exposed to light. Focal plane shutters are located in front of the sensor and operate in harmony with the electronic shutter.

Electronic shutters are controlled by the camera's processor and require power to operate. In some ways, electronic shutters were invented when digital photography was invented. Their origins are directly linked to the evolution of the sensors including charged couple devices (CCD) and complementary metal oxide semiconductor (CMOS) sensors. The photosite on a sensor is called a pixel. Pixels are unique in the way they work. When a pixel is exposed to light, unlike film, the resultant exposure signal must be moved from the pixel before the pixel can be used again for a new exposure. This feature is sometimes called the rolling shutter feature. Moving the charge from the pixel allows the pixel to be refreshed so it can be used for the next exposure. Electronic shutters are much different than focal plane or leaf shutters because they are part of the sensor. There are no moving parts in electronic shutters. This creates very immediate responses and triggering. When a mechanical focal plane or leaf shutter is used, there is a small delay when triggering the shutter. The delay in an electronic shutter is only influenced by the delay in the triggering of the shutter function. Sometimes when a shutter is depressed there is a delay in the actual opening of the shutter. Early digital cameras had a significant shutter lag before the camera would operate and make an exposure. This delay has been greatly eliminated although it still may be present in inexpensive cameras. When partially depressing a shutter, many of components of the exposure such exposure determination, white balance, and focus functions are operationalized. When pressing the shutter button completely, the exposure is made and saved to the camera's memory card.

How the exposure is moved off the sensor has also changed over time. Not so long ago, a CCD's image was moved from the sensor in what would be characterized as a global movement. When some cameras operate in the still picture mode, the focal plane shutter opens and closes, and the sensor records the data. This data is then moved all at once. This is important and allows the sensor to be ready to receive the next exposure. Once a sensor is charged after

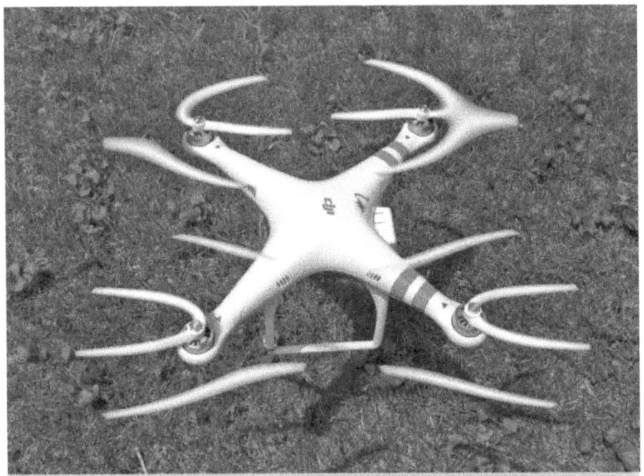

Figure 6.8 The effect of a rolling shutter is evidenced in this picture showing how the rotating blades of a drone's propellers were recorded. Because of the way the exposure is moved from a sensor when photographing a moving object, objects can become distorted as evidenced in this photograph. The propeller's shape changes because of this operation in the sensor. Image courtesy of Ted Kinsman.

an exposure to light, the data must be moved all at once to either a buffer memory or to storage media. Using a camera for video, the sensor becomes what is described as using a rolling shutter. In this type of operation, data is moved line by line immediately after exposure. The data is moved to a buffer area in the camera where the image is recreated into a contiguous file that may or may not have ever existed on the sensor in its totality. When exposures are created this way, there is the potential for objects that are moving in certain ways to have shape or other distortions based on speed and travel direction of the subject relative to the direction of the data's movement. Stationary objects do not experience this type of distortion.

Shutter Induced Image Artifacts

Because DSLR cameras use a mirror, the physical movement of the mirror and shutter during the exposure can cause image shake. Mirror induced camera shake will lead to images that are not crisp or clearly delineated. This artifact will be most evident in close-up photography and other applications where an image's size is larger than an object's size. When the image size is smaller than object size, the effects of image shake can sometimes be "absorbed" into an image and not visible. The vibration is always present but not noticeable when the object is reduced, and is typical in most types of photography. When the image size is larger than the object, the effect of camera or environmental vibration will create images that lack definition and this is easy to see. Using short shutter speeds can actually increase the apparent loss of image detail caused by the camera's mirror. Using longer or slow shutter speeds will actually minimize the effects of mirror shake. When using longer times, the vibration influences will be smaller relative to the time of the overall exposure and minimized. Using a longer shutter speed can be achieved by lowering the overall brightness of the lighting. Using neutral density filters can also be an effective way to create a longer exposure time and not affect color temperature.

Image shake can be introduced by the camera's mirror or from other sources: elevators, subways and trains, computers, or by fans located in lighting equipment. Eliminating or dampening vibration should be a high priority for scientific photographers. Vibration reduction tables are expensive solutions but are excellent tools for isolating vibration from the environment. Many, if not all high-end DSLR cameras have the ability to lock up the mirror after focusing has been completed. Mirror lock-up can be helpful in minimizing image shake produced from mirror movement. Many new DSLR cameras are now being equipped with vibration control (VC) technology, which has been a part of lenses for some time. Mirrorless cameras do not cause vibration since their electronic viewfinder does not use a mirror.

It can be practical to trigger a camera's shutter using a remote trigger or by tethering the camera to a computer. It is also possible to use a self-timer to take a picture. The simple act of pushing the shutter release button can be source of camera shake. Instrument cameras also do NOT use mirrors for viewing and so these types of cameras are not vulnerable to image

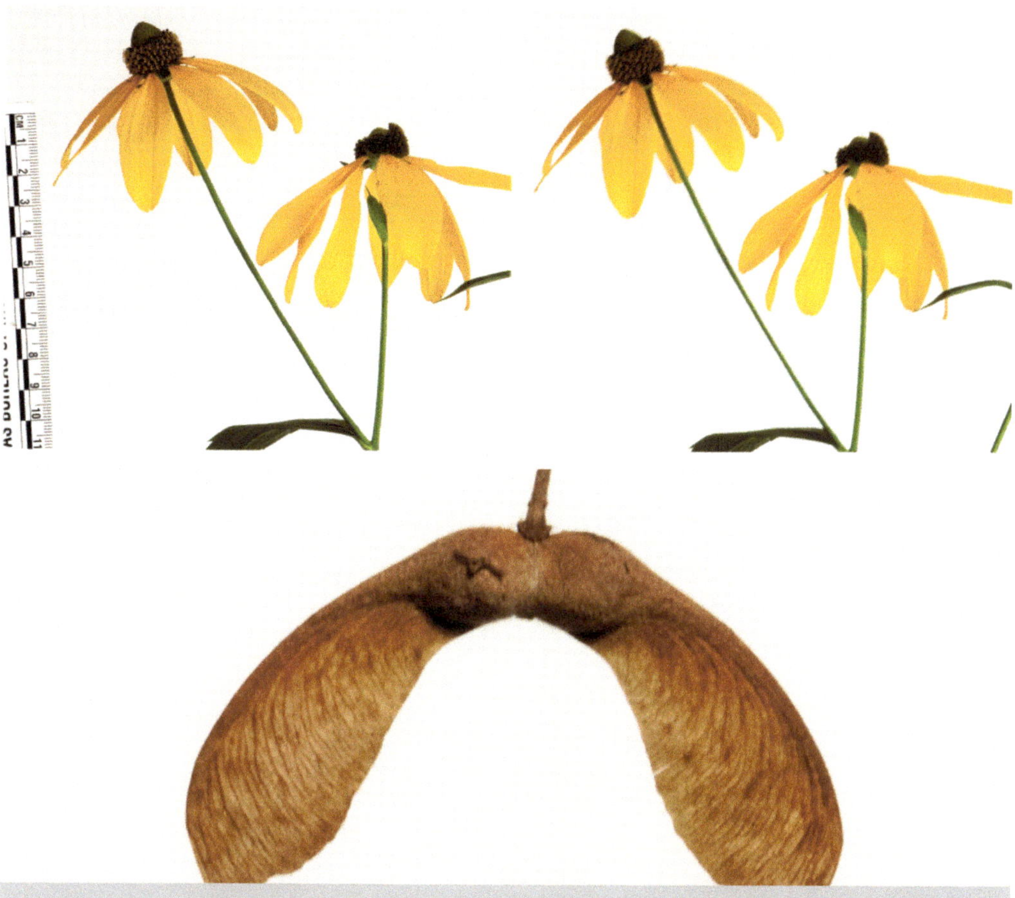

Figure 6.9 This image reveals a lack of sharpness caused by image shake. In this photograph the shake is a result of tripping the shutter using a finger and the effects of mirror vibration. Overall sharpness of the image on the top right is lessened and the effect of vibration can be observed in the enlargement of the paperbark maple seed key image. Images courtesy of Michael R. Peres.

shake from the camera itself. Any imaging systems is vulnerable to environmental sources of image vibration.

Mirrorless Cameras

The simple fact that a camera's mirror and its mechanisms can be a source of vibration in SLR cameras inspired the creation of cameras without mirrors. Leica film cameras have never used mirrors. This type of a camera is called a rangefinder camera. Because they have no mirrors, they are quieter than SLR cameras, however, the image and the scene are not identical in composition. This is because the lens and viewfinder are offset. This error is called parallax. Using a mirror and reflex viewing system requires space in the camera body but allows for precise composition. Without the need for a mirror, cameras can be made smaller and can use various sized sensors as well. Mirrorless cameras do not have optical viewfinders. The image is viewable on the camera's LCD panel and allows both video and still recording to occurring simultaneously. Working in bright lighting can be challenging since the LCD image might not be bright enough to critically see. Composition is accomplished using the LCD display on the back of the camera. Mirrorless cameras use interchangeable lenses. Mirrorless cameras can be used effectively in the laboratories using live view mode as well. Almost all mirrorless cameras have sensors that are similar to those used in DSLR cameras.

Suggested Reading

Langford's Advanced Photography (The Langford Series) 8th ed., by Efthimia Bilissi and Michael Langford. Focal Press, 2011.

The Digital Photography Book: Part 1 2nd ed., by Scott Kelby. Peachpit Press, 2013.

The Manual of Photography 10th ed., by Elizabeth Allen and Sophie Triantaphillidou. Focal Press, 2010.

Chapter 7 Light and Lenses

Making photographs requires equipment. Cameras, lenses, lighting gear, other optical elements such as filters, computers, monitors, and software are required. Each device plays a role in the image attributes that will be visible in the result. This chapter will explore how lenses, filters, and light itself all contribute to image formation and their influences on image attributes.

Light

At its most basic level, photography is described as writing with light. Choosing how to use and manage light is one of the most fundamental elements of photography and the imaging chain. Light or radiated energy informs what can be observed and subsequently what can be recorded in a photographic exposure. Achieving success when making scientific images requires a complete practical and technical working knowledge of light and applied optics.

Making things visible requires the creation of contrast, magnification, and resolution. The ability to see an object – whether large or small – requires the presence of light. The light itself can create or diminish contrast. Additionally, the use of lenses and other optical elements when operated correctly will lead to the creation of well-defined images.

Light Sources and Electromagnetic Radiation

Light comes from various sources or emitters. A radiation source has a physical size and the light it produces has the following characteristics: brightness, directionality, and a spectral (color) quality. These characteristics play a role in how an object will behave and appear when illuminated. These behaviors should be exploited when forming images.

Radiant energy includes visible light and is produced by many sources. Radiant visible light might contain other spectral emissions such as infrared or ultraviolet. One foundational characteristic of visible light is brightness. If the light comes from lighting equipment and not from a natural source, the light's brightness is adjustable. All light will have a physical

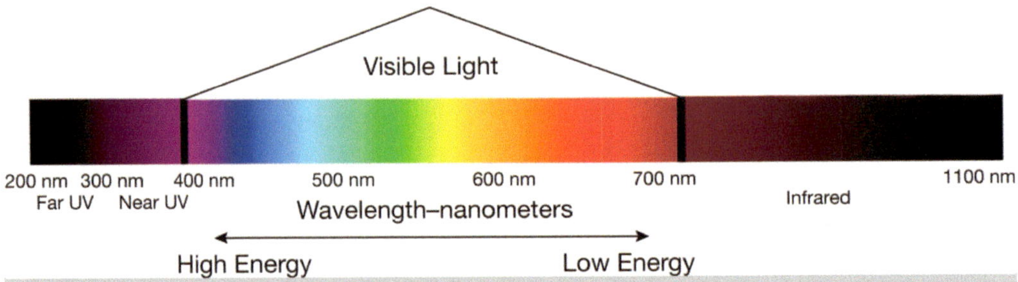

Figure 7.1 Visible light, a portion of electromagnetic radiation, can be visualized on this chart. UV is of a shorter wave energy than visible light and has more energy.

size such as the sun or a 2mm fiber optic light guide. The light will also travel from a specific direction relative to the object. Directionality will affect the creation of texture and help reveal surface information. Radiated light will interact with a material and will sometimes become modified by this interaction. This modification can be evidenced (described) as a reflection, transmission, or absorption. After light interacts with a material, the light energy carries information about the material and its characteristics. Before being "collected" by the eye or camera, the light may interact with other light also leaving the sample, from other sources, or from the same light source itself. These interactions result in collisions called interference. Interference can play either a positive or negative role in the formation of image structure and contrast. The light ultimately collected by the eye or imaging system contains structural information about the subject including its size, surface or internal structure, color, and more.

Describing electromagnetic radiation (EMR) and how radiated energy behaves has been a challenge for physicists for centuries. Even today, no singular model for describing radiant energy travel has been universally accepted because none explains all aspects of energy behaviors by themselves. Many historically important theories describing energy travel have been proposed including the particle and wave theory. Albert Einstein suggested the quantum theory, which utilizes components from both.

In 2015, scientists at École Polytechnique Fédérale de Lausanne created the first photograph that revealed both the wave and particle features of light. It is universally accepted that light travels away from the source of origin but what is not accepted by all is how it travels. Sir Isaac Newton first described light travel in the 1600s using his corpuscular theory. More than 300 years later, Max Plank and Einstein each added their ideas to this conversation, trying to better explain how light travels. In the quantum theory, light is considered to travel in separate packets or photons. The use of singular photons can effectively be used to describe some aspects of energy travel. Using particles explains straight-line travel. Photons are also useful in explaining photographic exposure. A pixel well collects photons. A minimum number of photons is required to create an exposure that registers as an electrical signal in the pixel. This causes the sensor to electronically go from an absolute black level (0 brightness value) to a shade of the darkest grays (1–5 brightness value). The use of photons, however, does not adequately describe speed changes when light goes from one medium into another, or when trying to describe a light's spectral composition or color.

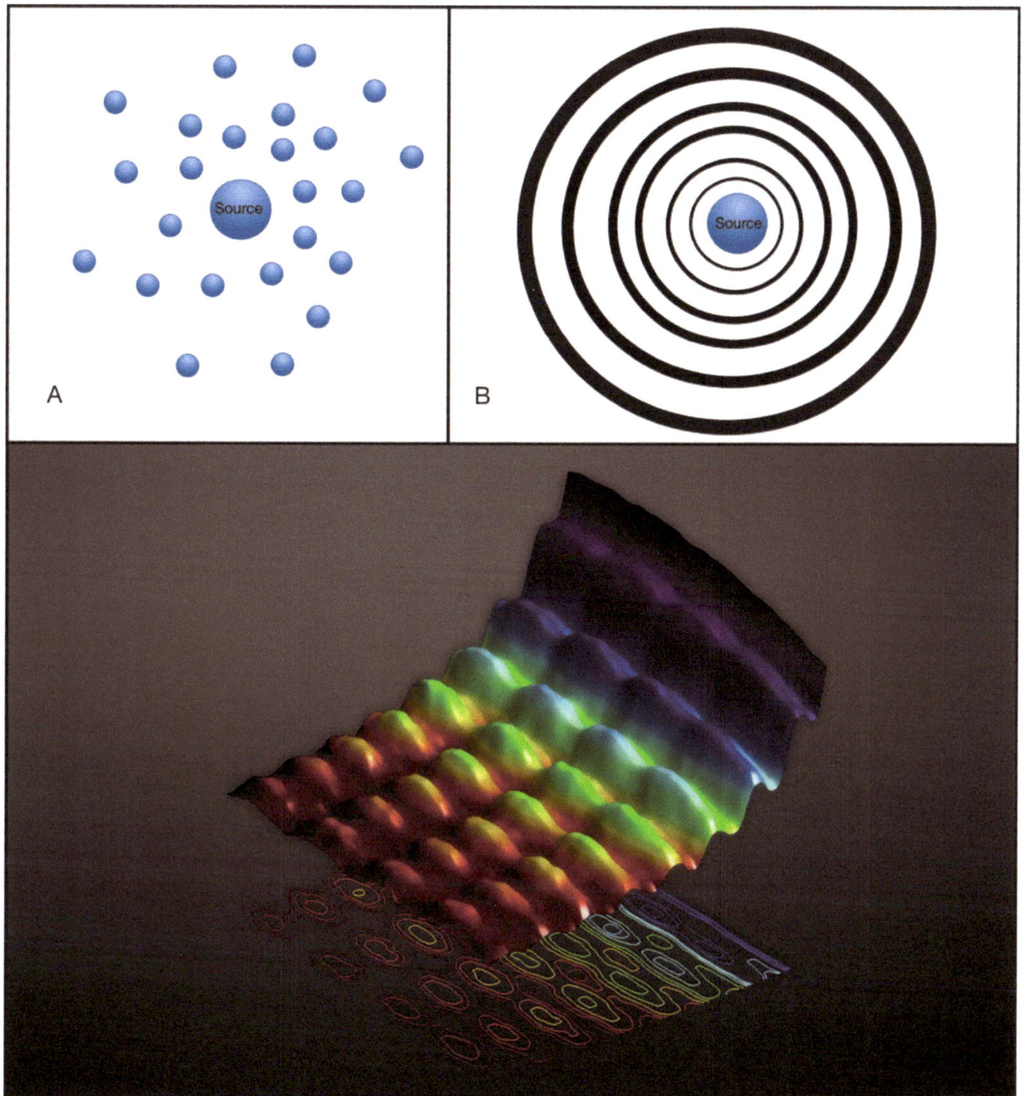

Figure 7.2 Image (A) represents radiated energy particles similar to raindrops leaving a source. The energy will move in straight lines until the particles interact with objects or other materials. Image (B) represents waves leaving a source and traveling in sync away from the source by oscillating up and down. They can collide and can lead to interference that creates or diminishes contrast. Since the days of Einstein, scientists have been trying to directly observe both aspects of this light behavior at the same time. Taking a radically different experimental approach, École Polytechnique Fédérale de Lausanne scientists in 2015 were able to take the first ever photograph of light behaving both as a wave and as a particle. Image (B) courtesy of Professor Dr. Fabrizio Carbone, Laboratory for Ultrafast Microscopy and Electron Scattering, Institute for Condensed Matter Physics, École Polytechnique Fédérale de Lausanne, Lausanne, Switzerland.

Light can originate from a small, point source or from a large, diffuse source. The light source determines the characteristics and quality of the light. A point light is very small and considered a direct light. Direct light produces a dark and well-defined shadow and can be used to reveal texture more effectively than diffused light. Light that is diffused can be shined through a semi-transparent material or reflected off a non-directional surface such as a wall. This modified light is described as diffused light. Clouds create diffused light. This type of light is soft and not good for revealing surface characteristics. Direct sunlight at high noon on a cloudless day represents direct light.

Describing light as particles that travels from a point source, is helpful for explaining some light behaviors. Consider how a mirror, or other flat shiny surface affects light travel and behavior. A photon that is in motion and traveling away from a light source that interacts with a smooth shiny surface will be reflected at an angle equal to the angle of incidence. The reflection will be nearly identical in brightness to the source if the reflection occurs from a mirror. This type of reflection will produce a specular reflection. The size of the reflection will be proportional to the size of the light. If the surface of the subject in the photo is textured, the reflection will be more scattered, less predictable, and create a diffuse reflection. Specular light creates high contrast and clearly delineated surface details. Diffused light creates less contrast and reveals color information more accurately.

When light is described as a wave, the brightness of light is characterized by its amplitude. Waves can be observed naturally in the ocean or in other water bodies. Waves have height (amplitude) and carry both positive and negative energy. The elevation of the wave describes the brightness (energy) and demonstrates the direction of travel. When the wave moves, it pushes forward, and following the raised portion of a wave is the region that is below the mid-line or surface that is called a trough. The region in the middle of a wave is the normal resting or zero energy level. The part below the elevated region of the wave has negative energy. Illustrated in Figure 7.3 are the specific regions of the wave. Crests and troughs are used to describe a wave's amplitude or brightness. Waves have travel directions and exhibit a vibrational magnetic component that is shared in a 90 degree angle to the amplitude.

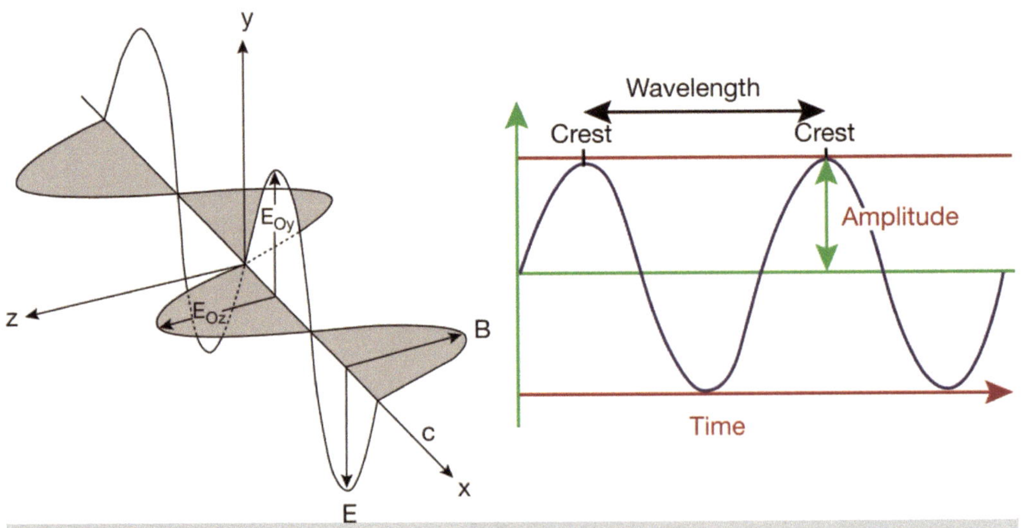

Figure 7.3 Waves are an excellent vehicle, useful for explaining brightness and color. One wavelength is equal to the distance from one crest to the next crest or from one trough to the next trough. Waves also have a travel direction and a magnetic component that is 90 degrees to the brightness component.

Waves also have a frequency of cycles per second in their emissions. Wavelength describes the physical distance from similar points on the wave. This can be evidenced as a crest to crest or a trough to trough. This distance is measured in nanometers (nm) or angstroms (Å) and is also used to describe the color of the light. For example, pure red light would contain wavelengths ranging from 600–700nm. Human vision can respond to radiated energy that is 400–700nm. This region is known as the visual spectrum of light. Short wave light, such as 400nm, has higher energy than long wave energy characterized by lower energy. White light describes a full spectral light source. White light will typically contain a blue component, a green component, and a red component in varying percentages. It is possible to a have source that emits 33 percent of each primary spectrum, red, green, and blue but this is often not the case. A light's spectral composition is described as its color temperature.

Emitters of Continuous Spectrums

Electromagnetic radiation comes from a wide variety of sources including the sun; incandescent sources such as tungsten and tungsten halogen bulbs; fluorescent lighting; metal halide sources; LED, OLED, neon, laser, and HMI; and electronic flash systems. All of these sources might be chosen for use for any number of reasons. A few other light sources that are not often considered include meteors, stars and other astronomical subjects, the auorae, bioluminenscent organisms, lava and volcanoes, nuclear reactions, chemiluminescence, fluorescence, CRT and LED displays, electric arcs, and gas discharges. All of these sources contain different spectral emissions.

Typically, the energy emitted from all these various sources is described using energy (watts) and wavelengths (color temperature). Wavelength and energy are related in the following way. Short wavelength energy will have high energy and long wavelength energy will have lower energy. Gamma rays have more energy than radio waves, but radio waves will have longer wavelengths than X-rays. Shorter wavelength energy produces greater optical resolution and sharper pictures.

The main light source is the sun. It emits all known electromagnetic radiation. The sun is used frequently for general-purpose photography, but is ineffective for laboratory and technical applications such as photomicrography or other highly specialized work because it is always changing. Important properties of light include intensity or brightness, direction, spectral composition, speed, and degree of polarization.

Continuous and Discontinuous Spectrums

When a source emits various spectral elements, the emission can be continuous or discontinuous. A continuous spectrum includes the emission of wavelengths smoothly over a broad range of wavelengths. Incandescent sources such tungsten halogen bulbs or the sun would be demonstrative of a source that emits energy in a continuous fashion. There are other continuous sources including an electronic flash which is a short duration source. A discontinuous spectrum source conversely would be characterized by emissions that are not continuous but include discrete and individual wavelengths that, when added together, form a specific color temperature. An example of this would be a fluorescent light bulb,

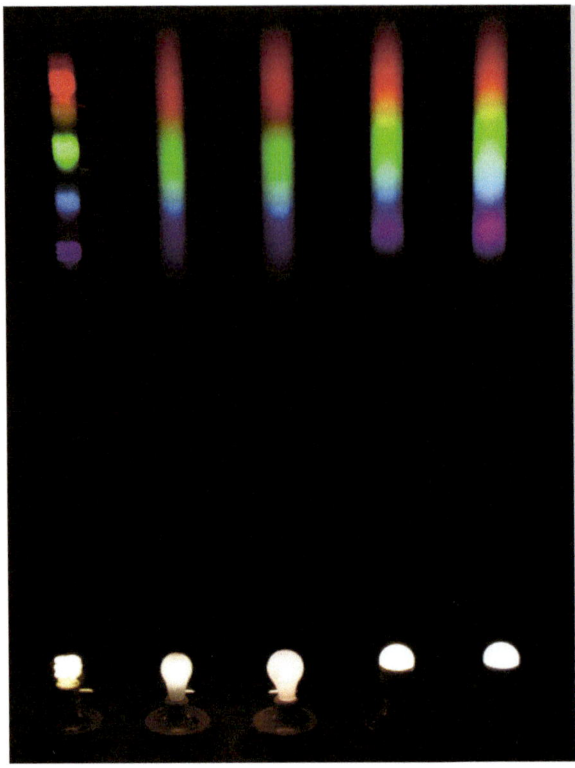

Figure 7.4 Tungsten, fluorescent, and LED lamps all will have different spectral emissions. A tungsten lamp is considered a continuous source, while fluorescent lamps are considered discontinuous sources. This is evidenced by the bulb on the left which has an emission of lines or chunks of color and is not continuous. The compact fluorescent bulb had a published color temperature of 2850K. The second and third bulbs from the left are halogen bulbs and were operated at 2850K. The fourth bulb from the left is a 3000K LED bulb and the bulb on the right is a 5000K LED bulb. The bulbs were photographed through a diffraction grating that is useful in revealing the spectral composition of the light from each bulb. Image courtesy of Nanette Salvaggio and Josh Shagam.

which is evidenced by its line spectral composition. Fluorescent bulbs can be warm or cool white or feature many color temperatures. An example of a single wavelength source of non-incandescent visible light would be the sodium-vapor lamp commonly used in most street lighting. These lamps emit a very intense yellow light, with over 95 percent of the emission being 589-nanometer light with really no other wavelengths. This makes for very bright monochromatic light which is very effective for night-time safety lighting.

Color Temperature

Color temperature is used to describe color characteristics of light with applications specifically to photography, astronomy, video, publishing, astronomy, and home interior products. Light sources are characterized by color temperatures described in degrees Kelvin. A Kelvin number is assigned to a visible color that corresponds directly to a black body radiator that has been raised to specific temperature using a Kelvin scale. A tungsten source is often said to have a color temperature of 3200K. Daylight is composed of sun and skylight and it has a color temperature of 5000K. These color temperatures have been assigned after comparison to the color emission that is correlated to a black body radiator. Color temperatures above 5000K would be characterized as containing more blue light and referred to as cool. Color temperatures that are below 3200K contain more reds and yellows, and would be characterized as warmer. Cameras require and use this information when establishing a white balance. Cameras can be operated in either auto-white balance mode as well as using pre-set values that include daylight, tungsten, fluorescent, or custom sources amongst others.

Temperature	Source
1700K	Matches
1850K	Candles
2700–3200K	Incandescent sources
3000K	Soft warm white fluorescent bulbs
4100K	Moonlight
5000K	Daylight, cool white fluorescent bulbs
5500K	Noon daylight and electronic flash sources
6200K	Xenon sources
6500K	Cloud covered daylight
6500–10,000K	LCD monitors

Continuous vs Pulsed Emissions

When light is emitted from a source, it can be continuous such as sunlight or an incandescent bulb, or it can be discharged in a short duration discharge such as lightning or as an electronic flash. Continuous sources allow for use of a wide range of shutter speeds for making exposures. If a subject is moving during a long exposure, it creates a blurred image. Continuous sources are common. Fiber optic light sources, metal halide bulbs, and other similar sources would be considered to have continuous emissions. A short duration or pulsed source is quite different. Pulsed light sources are used to produce a single continuous discharge or periodically repeated bursts of light lasting from a fraction of a second to milliseconds. Pulsed light sources can be divided into two categories. The first category includes sources that use light radiation from a low-temperature plasma source such as a condensed spark discharge, in gases, or exploding wires. The second source is characterized by the short excitation of phosphor as a result of the passage of an alternating current electric through it or by irradiation using an electron beam. Optical frequency oscillators or pulsed lasers are also considered pulsed light sources.

Electronic flash sources or strobe lights have important applications in short duration imaging applications. The power of a pulsed source is often shared in watt/seconds, guide numbers or sometimes as joules. A typical hand held flash might have a guide number of 160 GN. At full power, an approximate duration of 1/300 sec is required to fully discharge the light. Reducing the power will shorten the duration. Less power allows for more frequent discharges and shortening the battery recycling times. Below are hypothetical durations for an electronic flash.

Full power	1/300th sec
½ power	1/600th sec
¼ power	1/1200th sec
1/8 power	1/2400th sec
1/16 power	1/4800th sec
1/32 power	1/9600th sec
1/64 power	1/19,800th sec

Light Behaviors

Three things happen to radiated energy (light) when it interacts with a new material. It can be reflected, refracted, or absorbed. These three outcomes are influenced by the angle at which the light strikes the surface of the new material and by the material itself. Because photography requires optical elements (lenses) to form images, being aware of how these behaviors negatively affect image formation is helpful. A reflection is produced when light strikes a surface and the light moves away from that material at the same angle at which it struck the surface. The light can also be transmitted when entering a new material. Depending on the entrance angle, the light is often bent when entering the new material. This behavior is called refraction. If light strikes a new material at a unique angle to its surface, the light can be absorbed. These behaviors occur every time you make a photograph. They occur at air/glass boundaries or in situations where two materials of differing refractive indices are adjacent to one another. This is evidenced when photographing in water or glycerine for example. All of these behaviors affect image formation. There could be degradation or improved edge definition. Monitoring image contrast is a tell-tale sign of how reflections, refraction, or absorption are affecting an imaging system. These behaviors can play useful and important roles when choosing and using lights. For example, direct and on-axis light will create a specular reflection or bright spot of itself on the subject. This can be useful for revealing important characteristics about the subject itself.

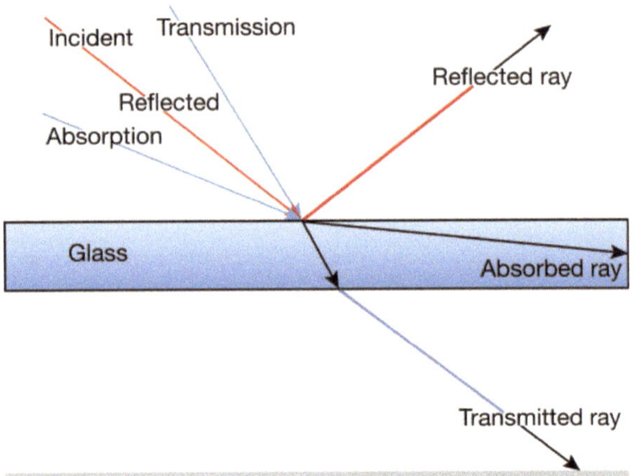

Figure 7.5 When light strikes a piece of glass three things can happen, depending on the angle of incidence. If the angle is great, transmission occurs. If the angle is low, there is a reflection, but if the angle is just right, the light will become trapped in the glass causing flare. This angle is called the critical angle.

Making technical pictures requires awareness of the behaviors light and many other strategies for image optimization. A good

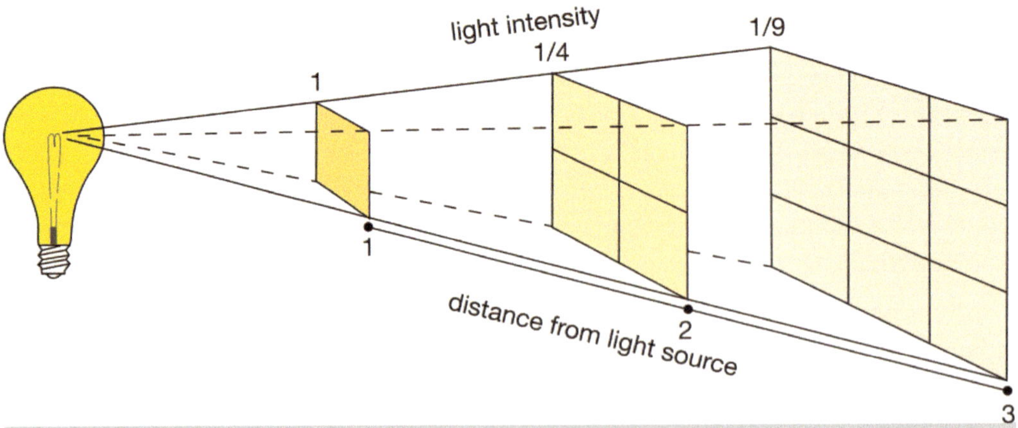

Figure 7.6 A light source will have a brightness associated with it. The brightness can be increased or decreased as needed. A light's brightness will fall off in predictable ways. The light fall off is described using the inverse square law. For every doubling of the distance, the light will be ¼ of the amount.

science photographer anticipates how to control the light when it changes mediums and what occurs when these situations arise.

Reflection

A reflection might be the simplest type of energy travel to understand. A reflection is evidenced in a situation where a beam of light strikes a surface of a material and is reflected back at an angle equal to the angle of its incidence to that surface. You can see a reflection in Figure 5.5.

The surface of a material can be either shiny or diffuse. If the surface is shiny, the reflection will be specular and if the surface is textured, the reflection is said to be diffuse. The angle is measured from a hypothetical perpendicular axis to the reflecting surface called the normal. A reflection can occur anywhere or off any surface. Reflections can occur on the surface of a subject or on other optical surfaces such as within instruments that have many multiple lenses or elements. Reflections can be produced from windows or from overhead lighting. Reflections can find their way into photographs of shiny or wet samples produced in laboratories or other environments such as fish. To minimize and control the presence of random reflections, it is practical to extinguish all room lights, pull room drapes, or manage other extraneous lights when photographing.

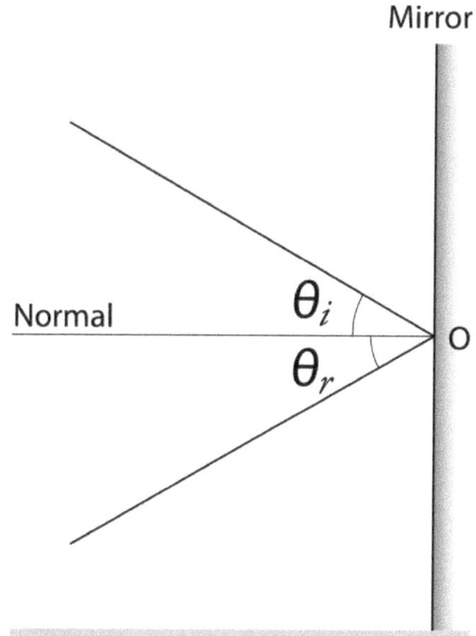

Figure 7.7 When light strikes a reflective or shiny surface, it is reflected at an angle equal to the angle of incidence. $\theta^i = \theta^r$ of reflection.

Absorption

When light becomes trapped in a material, it is often because of the angle of incidence to the object's surface or characteristics of the surface itself. This angle is referred to as the critical angle. Absorption occurs when the light cannot escape the material on the side opposite to its entry. See Figure 7.5. Simply stated, absorption results in the creation of non-imaging forming light called flare. Flare light will lower image contrast and create diminished image visibility. Flare frequently is created by poorly managed reflections. An example of flare might be evidenced in what happens when driving into the sun with a dirty windshield. The windshield will become hazy because the dirt provides a surface for the sunlight to be focus onto. The elements from the scene will appear washed out and without detail or contrast against this flare light in the windshield. Light absorption occurs in lenses with dirt or fingerprints, in prisms, or other optical elements in the system, or how light is shone onto a subject. Random light that interacts with any of optical elements in the imaging system can effectively change the contrast of the system. This change can be subtle or significant. Keeping lenses clean and free of dirt and oils will add greatly in contrast control but does not mitigate the need for ambient light management.

Transmission

The transmission of light through a material with a different refractive index plays an important role in image formation. Some would argue that refraction alone is at the foundation of an optical system's performance. Refraction is the bending of light as a consequence

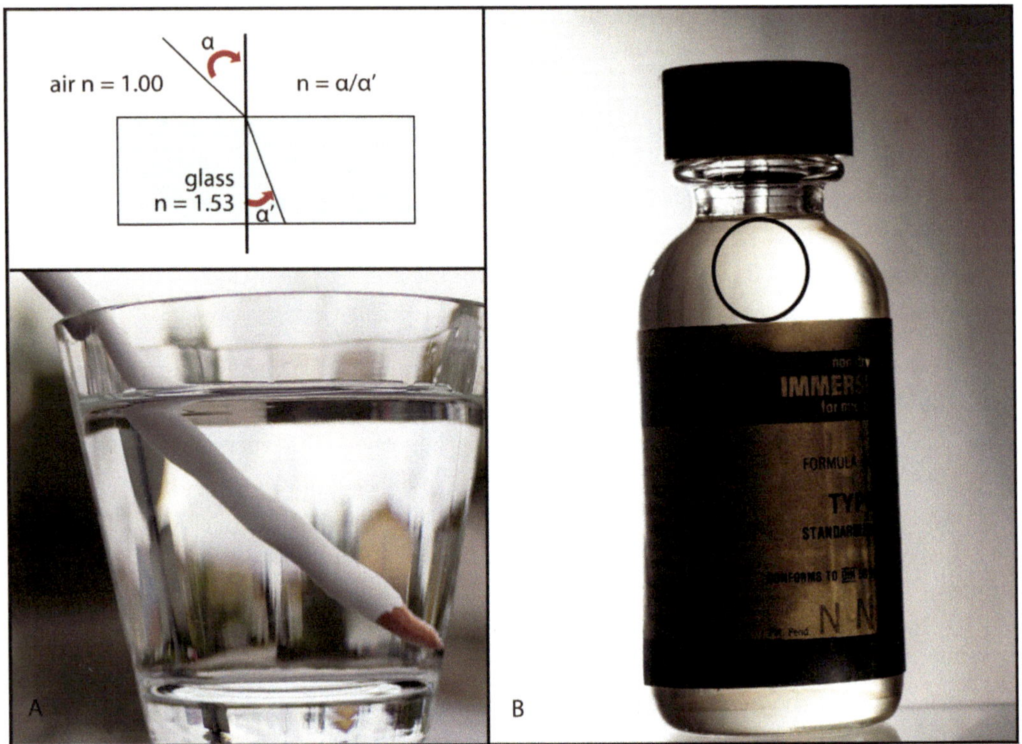

Figure 7.8 When light encounters a material of differing refractive index, it experiences a speed change or refraction. In this graphical drawing, what happens to an incident ray after interacting with a new material is represented. The light is slowed down and bent towards the normal of the material's surface. When the light leaves the glass, it speeds up and is bent away from the normal. It will travel in the same direction as the incident ray, but will be displaced. In image (A), the light's displacement is evidenced by the appearance of the pencil, which looks to have been bent in the photograph. This occurs because the refractive index of the water is different than air and glass. In the image (B) the applicator wand disappears in the oil (evidenced by the circle) and is visible in the air in the region above the oil. This invisibility is because both of the materials have the same refractive indices. Images provided by Michael R. Peres.

of a speed change that occurs when the light enters a new medium of a different optical density. This occurs when light goes from air into glass, or when light enters water from air, or other optical materials such as immersion oils. Refraction also explains dispersion where each of a light source's spectral components are slowed by differing amounts. These speed changes result in the separation of light into the primary spectral components of red, orange, yellow, green, blue, indigo, violet (ROYGBIV). This outcome is commonly described as a rainbow or dispersion.

All materials have different refractive properties. These properties are shared as a material's refractive index. A refractive index is a number that predicts the light bending ability – or optical density (compared to air) – of the material. R.I. affects the speed of light in a material. Air has a refractive index (R.I.) of 1.00 and is considered to be a vacuum to light and causes no speed changes. Particulate matter in air such as particles may affect light travel. Water is denser than air and has a R.I. of 1.33 and glass has a R.I. of 1.53. Refraction plays a significant role in lens design and its corresponding focal length or magnifying power. Shorter focal lenses will have greater refractive power and longer focal length lenses, less refractive properties.

Dispersion

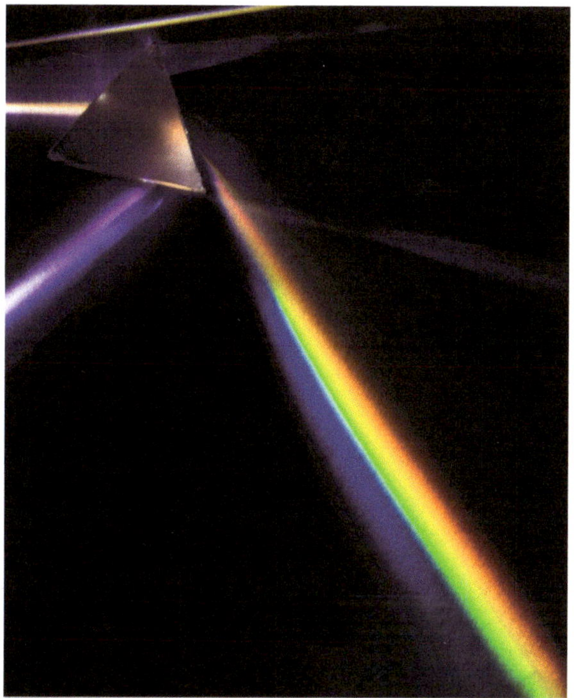

Figure 7.9 Light is refracted differently within a prism that is made of flint glass. The red, green and blue components are spread out as a rainbow. The light source in this example was a mercury-vapor lamp. Image courtesy of D-Kuru.

Dispersion occurs when light is broken up into its components by the refractive properties of an optical material. A prism provides an excellent example of dispersion and so do diffraction gradients. When light enters a glass prism from air, the speed of each portion of the visible spectrum will change differently. Consequently, some wavelengths will be more affected and others less so. This variance of speed results in the bending or refraction differences of light containing RGB. Dispersion is considered a negative quality in an optical system. Lenses that are not highly corrected will disperse light in higher amounts than more highly corrected lenses. Lenses that are highly corrected are characterized as being apochromatic and will have lower dispersion potentials. High quality macro lenses are frequently APO lenses. Dispersion leads to flare and will result in a general lowering of the image contrast. In microscopy, lenses will often have an Abbe number associated with them. An Abbe number is an indication of the degree of correction for an objective lens and is based on the glass's dispersion potential. A higher Abbe number infers the lens has less chromatic aberration. Having high dispersion is not a positive characteristic of a lens except when designed into soft focus lenses used for portraiture or other mood producing applications.

Interference

Since energy travels in waves, there can be situations where collisions of waves occur. When waves travel in phase, or in sync, they are described as coherent. Waves can also travel out of phase and are described as incoherent. When waves collide, interference occurs. Interference occurs when two or more radiated energy waves are present simultaneously at the same location in space. This occurrence produces a new wave that will be the sum of the parts of the contributing waves. Interference can either be constructive – sometimes called additive (positive) – or destructive (negative). In constructive interference, the wave energy will be greater as a sum of the two energies and will contain more energy or be brighter. Effectively this increase in wave energy creates more brightness and sometimes produces a color change. When destructive interference occurs, the total energy of the waves is reduced. This will diminish image brightness and also create a color change. In microscopy, interference is an important element used for image formation. It is used to create or preserve image contrast required for visibility. Phase contrast microscopy and differential interference contrast techniques take advantage of this behavior, making the invisible visible without the use of staining.

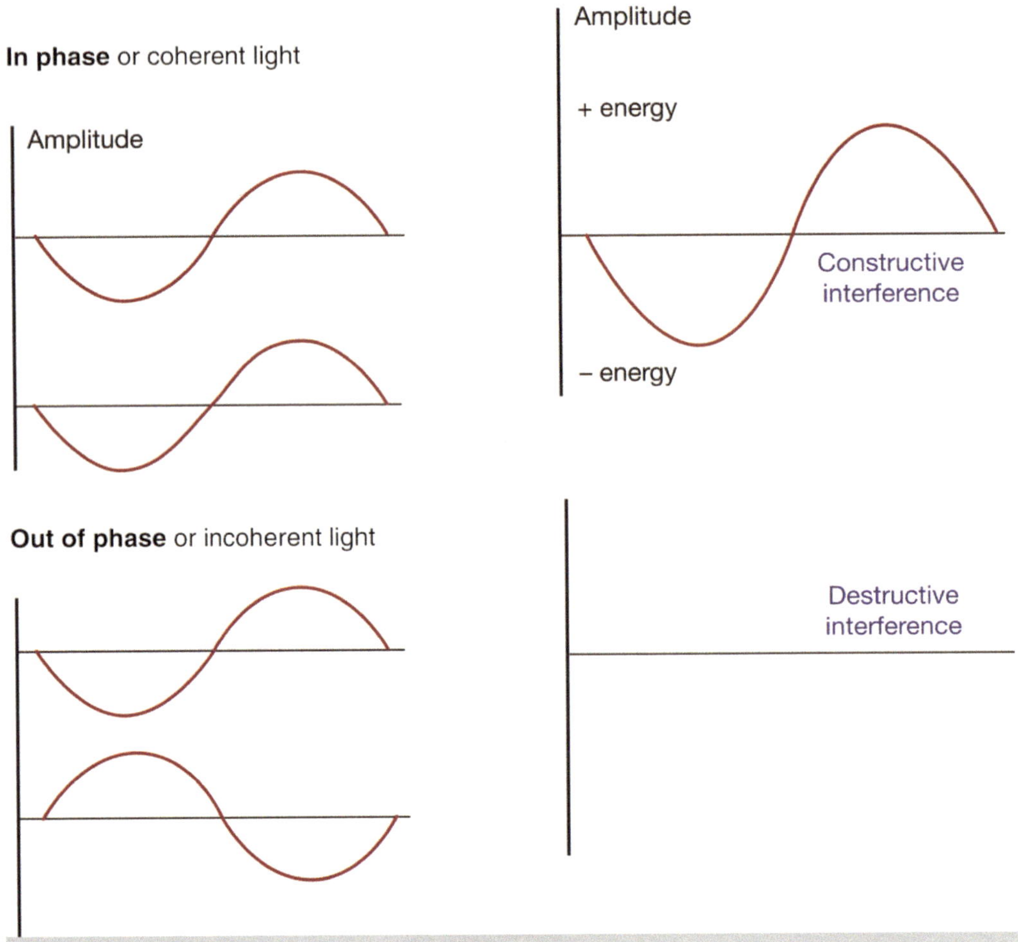

Figure 7.10 In this illustration, waves are traveling in and out of phase. When the waves collide, interference occurs. When waves are coherent and interference occurs, it creates more brightness (amplitude). When they are out of phase, interference causes energy loss and diminishes brightness. Interference plays an integral role in image formation and contrast.

The Lens

A lens may be the most important element of an imaging system. Sometimes called a photographic lens or objective, lenses will define the quality of image attributes. Lenses are coupled to cameras using specialized mounts. Theses mounts are of two types – bayonet or c-mount. Almost all lenses contain apertures. Most lenses used for video, still photography, or other imaging applications are similar in their basic functions. What is different across all the various types of lenses is the degree of correction, the focal length, whether the focal length is adjustable (zoom lens), the light transmission capabilities e.g. f number, and angle of view. Lenses can also be corrected for vibration reduction.

A photographic lens has a focal length with a corresponding angle of view. The focal length is influenced by image magnification (or reduction) associated with the lens. The focal length is typically shared in millimeters (mm) and describes the lens's angle of view and associated circle of good definition. A circle of good definition defines a lens's ability to cover a sensor's diameter. For example, if a sensor were 24mm × 35mm, the diagonal of the image sensor would require the circle of illumination to be approximately 43mm. This is calculated using Pythagorean theorem. The normal lens for this example would then be approximately 50mm or 2 inches. Any focal length lens in this example that is shorter than a 50mm lens would create a greater angle of view and would be characterized as a wide-angle lens. Any lens with a longer focal length used on that sensor would be characterized as a telephoto lens.

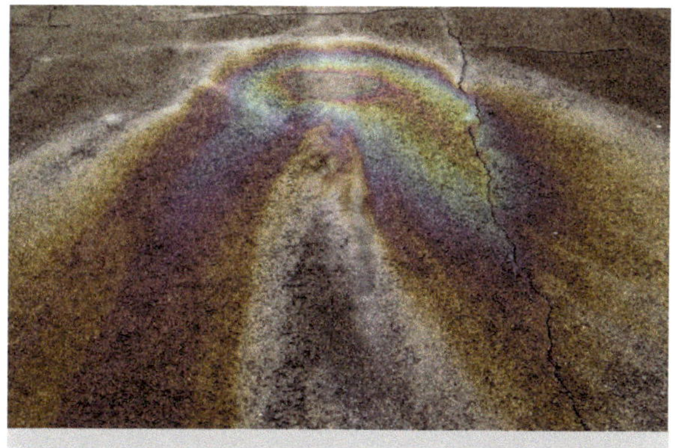

Figure 7.11 Interference is evidenced by the various colors revealed in the oil on the wet roadway. The oil interacts with water and creates numerous thin layers. As light enters the material, it bounces back and forth between the many parallel surfaces resulting in interference and produces this rainbow type appearance. Image courtesy of Michael R. Peres.

Apertures are placed into lenses to influence light transmission and control image attributes. Apertures affect image brightness and consequently a photographic exposure, but apertures also influence image characteristics such as DOF. In modern lenses, apertures remain fully open until the shutter is depressed. This allows for brighter images to be formed when viewing, composing or framing, and is useful for focusing. If the aperture was manual and closed during composition and focusing, the image would be dimmer and difficult to see. When the shutter is activated, the aperture blades close to the aperture setting. This feature is characterized by automatic lenses. A lens aperture that closes when the *f*-stop is adjusted is called a pre-set or a manual lens. Photographic enlarging lenses are manual lenses.

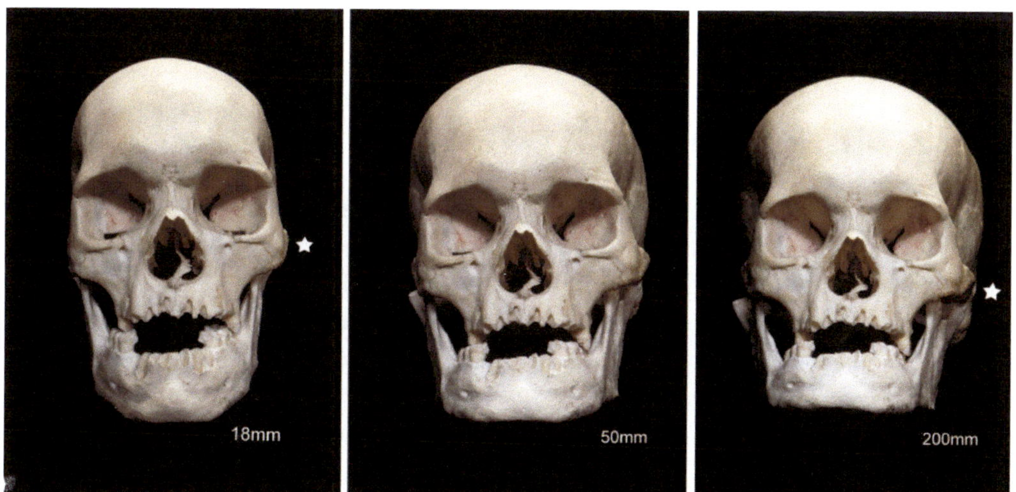

Figure 7.12 In this composite photograph, a skull was photographed using an 18mm, a 50mm, and a 200mm lens. To the extent possible, the skull remained the same size and the camera used the same point of view to make all three photographs. As can be seen, there is a shape change in the skull in the three files. The working distance of each picture was different. A star was placed to help illustrate the changes that are evident in the wide-angle photograph and the telephoto views. The malar and temporal bones have increased separation and visibility in the 50mm and 200mm views when compared to the wide-angle photograph. Images courtesy of Michael R. Peres.

The lens's maximum aperture is determined by dividing the focal length of the lens by the largest clear diameter when the aperture is fully open. Larger apertures are faster (collect more light) and function better in low light applications. Faster lenses are more expensive. All apertures will encroach into a lens's optical pathway. For example, a very fast lens with a high light gathering ability could be characterized by a f-number of *f*/1.2. On microscope objective lenses, numerical aperture (N.A.) is used to describe light gathering ability. Just like with photographic lenses, there are a range of numerical apertures. In some cases the N.A. can be more than 1.00. A lens with less light gathering ability might be characterized as having a f-number of *f*/4.0 and a microscope objective could have an N.A. of .1, for example. More on numerical aperture is covered in Chapter 16.

In addition to affecting light transmission, apertures play an important role in controlling the range of focus of an image, referred to as depth of field or DOF. Depth of field is an important characteristic of an image. Less DOF can be used to optically isolate an image's area of interest and a greater DOF can be used to reveal more of a sample by increasing the image's range of acceptable focus. Choosing the correct aperture is influenced by many considerations and related magnification.

Lenses for Scientific Applications

Some photographic historians consider pinholes to represent the original lens because they were used in a camera obscura. A camera obscura forms an image because of the way light behaves when it travels through the pinhole and is focused onto a wall or paper. Once lenses were discovered in the 1600s, they became valued and useful because image brightness and definition was significantly better when compared to images made from pinholes. Revealing details, structures, and contrast of challenging objects was greatly enhanced using a properly corrected lens and other optical elements.

Lens Fundamentals

Choosing the right lens begins with the evaluation of where and how the lens will be used and the desired image attributes needed in a specific situation. Considerations include long working distance, focal length, and other attributes such as flat field. Lenses are constructed using various types of glass including crown and flint glasses. There are other elements in a lens including coatings that manage image contrast, color fidelity, and affect image resolution. Lens elements will have thicknesses and curved surfaces. The lens's internal elements will be carefully arranged and the position of an aperture also considered to influence image depth of field (DOF), depth of focus, and image brightness. Depth of field is located at the subject and depth of focus is located at the sensor. Modern lens designers use many methods to improve lens performance including adding aspheric surfaces, adding minerals such as fluorite to the glass, and using gradient elements that create improved performance. Modern designs allow for the use of fewer pieces of glass. Nano-crystals are also being integrated into the newest lenses. Lenses designed solely for DSLR cameras are nothing short of remarkable. Image sharpness is experiencing a period of unparalleled excellence, which includes the use of special elements that direct light effectively onto a sensor's surface and more directly and uniformly into the individual pixels. Many lenses are also vibration reduced.

A simple lens, or a single element, can be categorized as either being positive or negative. A positive lens is thicker in the central region and may also be called convex. This lens produces real and inverted images of a subject that is focusable. Negative lenses are thinner in the center and are called diverging lenses. This type of lens produces a virtual image that appears to be larger and visible in front of the lens. Almost all photographic lenses contain both positive and negative lenses. A lens made of many individual lens elements is called a compound lens and must always remain positive for use with cameras.

Lenses all have common shapes. A simple positive lens can have two convex surfaces or one surface that is convex and the other surface flat. A biconcave lens is a single element lens with both of its surfaces curved inward towards the center. This lens is thicker at the edges than in the central region. A meniscus lens is a single element lens that has both spherical surfaces curved facing in the same direction.

Lens conjugates describe locations in front of and behind a lens. Conjugates predict where a lens will focus an image or where an object should be located. Lens conjugates play a role in design and refractive considerations, but more importantly how the lens is used. A lens will have a principle focal length and this is the location where an image is focused when the lens is focused at infinity. There will also be a front focal point for a lens when focused at its near point. When a lens is used for reduction applications – the object will be larger than the image – and the distance from the scene to the lens will be large and the distance from the lens to the sensor will be small. When the image size equals the object size or life-size, the front and rear image distances will be equal and be two times the focal length. When the image size is greater than the object size, the object distance will be small and image distance will be larger.

Image size will be proportional to focal length when used for traditional photographic applications. In this relationship, an increase in focal length will produce a larger image. Depending on the focal length and maximum aperture required, various lens configurations can be used to increase image size. A 100mm lens will make an image that is larger than a 35mm lens if the photographer does not change working distance. A telephoto lens uses a compact configuration to achieve a longer focal length using different lenses in a specific order. In a telephoto lens, a negative lens is placed behind the front positive component. This increases image size and makes for a small and lighter weight optical design.

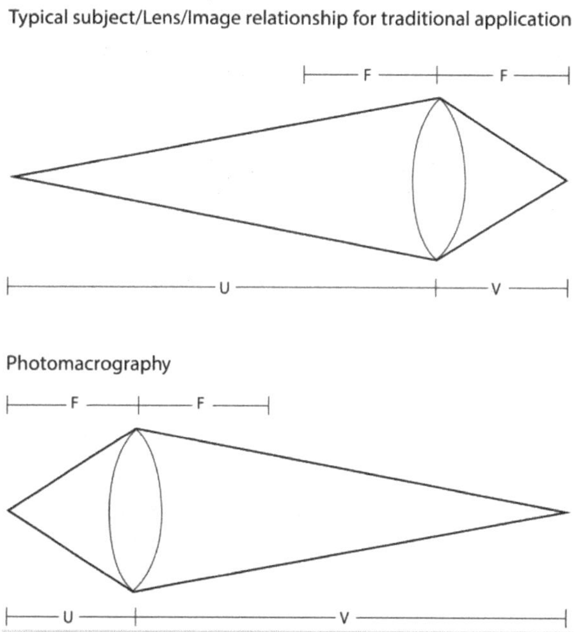

Figure 7.13 In this illustration, the typical lens conjugates found in traditional photographic applications that create images that are smaller than the object is shown in the top view. The lower view shows the lens conjugates required in macro applications where the image size is larger than the object. U is the object distance, V the image distance, and F the lens focal length.

Teleconverters

A teleconverter is a precision optical accessory that can be mounted in-between compatible lenses and a camera body. It will increase the focal length of the primary lens. This configuration can create a longer working (object) distance for a photographer and maintain the same magnification. If a teleconverter is used in combination with a 200mm close-focusing lens, for example, it will yield a life-size (1:1) image and does so while maintaining a 71cm or 28 inch minimum object distance. This amount of working distance is advantageous for many biomedical, industrial, and natural history applications.

Working Distance

Working distance is defined as the distance from the front of the lens to the subject. It plays a role in image size or magnification. Using a larger working distance allows for freedom in arranging lighting and photographing subjects that are challenging. In photographs where the object will be larger than the image, changing lenses and working distances will influence image perspective. Perspective describes how the elements in a photograph appear in the two-dimensional rendering of a scene. Foreground and background magnification differences are included in this process as well as how straight lines are rendered. Images can have a strong or weak perspective based on the location of the camera in a scene. Subjects located within the scene will have different magnifications and locations within the 2D frame if the location where the camera has been placed is modified. As an imaging system's magnification increases, the working distance will decrease unless a specially designed lens is used.

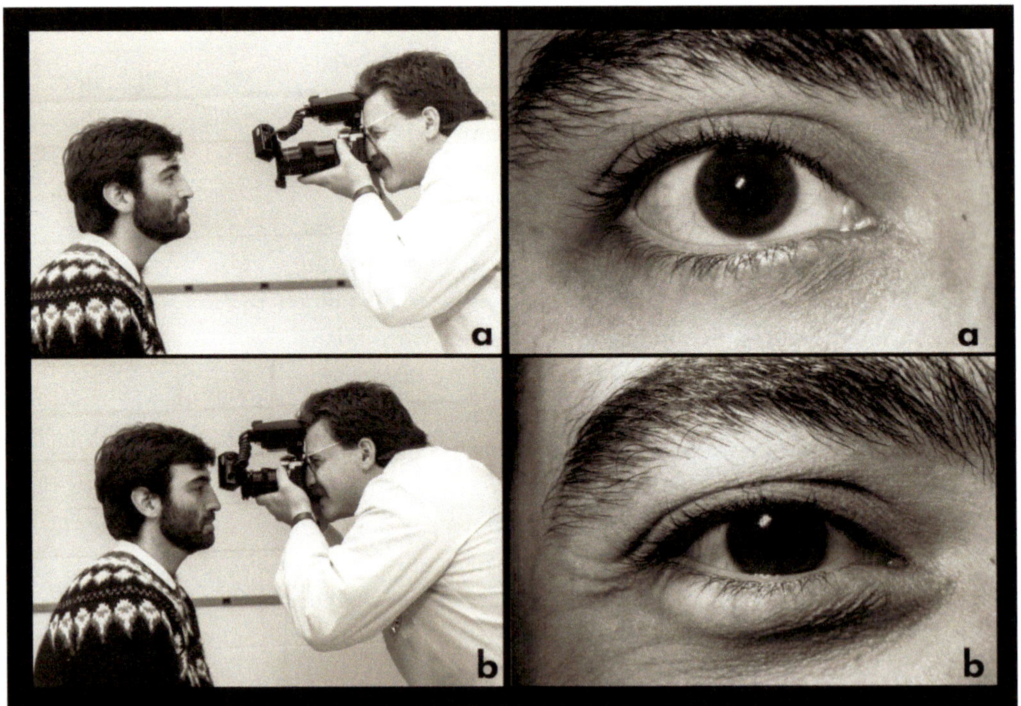

Figure 7.14 In this illustration, a 50mm (top left) and 100mm (bottom left) lenses were used to create the corresponding images on the right side of the figure. The working distance when photographing at 1:1 is 2× the focal length of the lens. While the images are relatively the same magnification, there are differences as a result of the camera's distance to the subject.

Close-Up Lenses

Many relatively inexpensive compact digital cameras are fitted with zoom lenses that are adequately corrected for use in moderate close-up distances. However, DSLR cameras will have a wide variety of quality close-up or macro lenses, both in fixed focal lengths and zoom lens types as well. A zoom lens with close-up abilities is often described as macro zoom. True macro lenses are designed to produce magnifications of 1:1 or greater.

Expensive macro lenses have elaborate internal mechanisms that allow groups of lens elements to move independently as the lens focusing is extended to close distances and create excellent life-size (1:1) or slightly larger images. These lenses will always have reproduction ratios inscribed on the barrel, possess modest maximum apertures (*f*/2.8 to *f*/4), and come in various focal lengths used to achieve various working distances. An image magnification of 1:1 will be 1:1 regardless of focal length. Important optical characteristics are designed into macro lenses including excellent image definition. They have been highly corrected for color and spherical aberrations as well. Their most important feature is their ability to produce edge-to-edge sharpness when used with shorter working distances, especially when used photographing flat subjects.

Supplementary Lenses

The designs of most fixed focal-length camera lenses have been optimized for long camera-to-subject distances and often will have a minimum focus distance of approximately 50cm or 127 inches. This distance can be reduced, and the image size increased by using a supplementary or close-up lens. A close-up lens is a high quality simple magnifying lens. These lenses will either be a single lens or an achromatic doublet that attaches to a lens like a filter. These lenses are sometimes called plus lenses or plus diopters. Using a diopter is the easiest way to achieve close-up images. It is always best to attempt image creation without the addition of lenses when possible, since any additional lenses to a primary lens can degrade optical performance. The optical power of supplementary lenses increases as the diopter number increases for example +1, +2, up to +20. The diopter number of a lens is the reciprocal of its focal length in meters which is expressed:

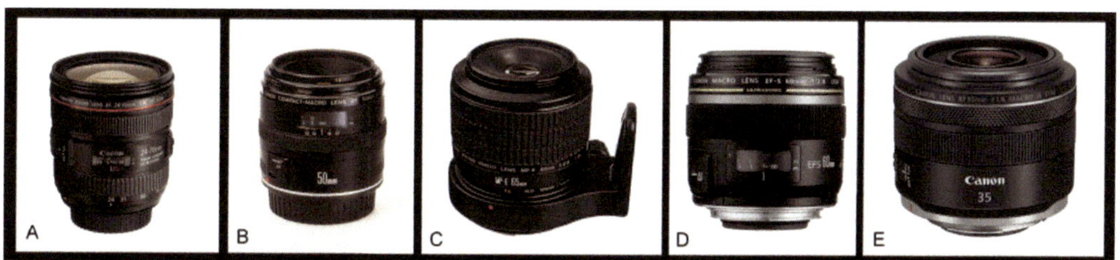

Figure 7.15 Five different Canon lenses that can be used for close-up or macro work. The left image (A) features a Canon 24–70mm macro zoom, (B) a Canon 50mm (discontinued) prime macro lens, (C) a unique Canon 65mm macro lens that produces 1–5×, (D) a EF 60mm macro with image stabilization, and (E) a RF35mm prime macro lens. They have been designed for different applications and should not be considered equal. Images courtesy of Canon USA.

$$D = 1000\text{mm/f or f} = 1000 / D \quad D = \frac{1}{f} \text{ or } f = \frac{1}{D}$$

So, if using a +2 close-up lens on a focal length of .5 meter (500mm) the lens will have a focal length of .2 meter (200mm). Alternatively, it is possible to convert all values to diopters and then convert their sum to focal length. For example, a 50mm lens has a power of 20 diopters. By adding a +2 supplementary lens, the sum becomes 22. The combined focal length is 1000mm. When divided by 22, the new focal length of the lens becomes approximately 45.4mm.

It can be practical to know the combined focal length of the camera lens and supplementary lens if it is required to calculate the lens-to-subject distance. This is possible by using the following lens equation:

$$u = \frac{v - f}{f}$$

where u is object distance and v is image distance and f is the focal length.

Supplementary lenses can be combined to obtain even greater magnification. However, when combining close-up lenses, the strongest should be closest to the camera lens. Image degradation will increase dramatically when more than two supplementary lenses are combined. Supplementary lenses can also be used in combination with extension rings and bellows in order to minimize the total amount of lens extension. Shorter lens extension produces shorter exposures. Shorter exposures may, for many subjects, offset the theoretical advantage of using a camera lens with extension tubes alone. Using a small lens aperture may minimize the aberrations introduced when using simple, positive close-up lenses.

Mirror Lenses

A mirror lens is a special lens and is comprised of a catadioptric optical system where refraction (lens) and reflection is combined into an optical system or lens. Mirror lenses are sometimes called CAT lenses, and have long focal lengths. Because of the way the image is created in this lens, there is a big reduction in the physical length of the optical assembly because of the use of the folded optical path. Mirror lenses typically have focal lengths of 250mm and beyond 1000mm. CAT lenses are much shorter and compact than telephoto lenses and lighter as well. Because of the number of lens elements in this type of lens, chromatic aberrations are well managed by the catadioptric lens design.

Because CAT lenses have only a central element and use a reflection mirror as part of the folded optical design, they do not have an adjustable aperture or diaphragm. This means the lens has a fixed f-number. One characteristic image artifact of this lens is the annular shape of defocused areas of the image, appearing as a doughnut-shaped iris blur, now called a bokeh. These shapes are the result of the lens entrance pupil. Bokeh is modern word that replaced circles of confusion, however both describe the same out of focus effect.

Telecentric Lenses

A special lens often used in technical photographic applications is a telecentric lens. Telecentric lenses produce orthographic images. The true advantage of these lenses is that the image magnification remains the same independent of an object's distance or position in the field of view. These lenses are used for making semiconductor chips, optical lithography applications, and for metrology. Because their lenses maintain magnification, they are also very useful in machine vision systems. These lenses can be very expensive.

Photographic Filters

Filters are important optical elements used in imaging. Filters modify light through absorption, transmission, and interference mechanisms. Filters can also modify light using refraction or diffraction methods. A filter that is part of an image-forming system must be of high optical quality. Filters that are used for illumination systems can be of a lower quality since image formation will be unaffected by their presence except for heat absorbing filters. Filters come in many materials, sizes, and can be made of gelatine, cellulose acetate, polyester, plastics, glass, or multilayer dielectric coated [MLD] materials.

White light is composed of various percentages of blue, green, and red light. Frequently there is also some UV and IR. Absorption filters are common. In this type of filter, absorption occurs, and the remaining light is transmitted and modified by the filter. The amount of light absorption that occurs in a filter is typically wavelength dependent. Often absorption filters are composed of primary colors, which include red, green, and blue. A red filter will transmit red light and absorb green and blue light. All filters remove light and lower image brightness. Exposure compensation will be required to varying degrees when a filter is present in an imaging system. Primary colored filters were very common in film photography and now have found use in Bayer filters located directly on digital image sensors used in DSLR cameras. See Chapter 8 and 19 for more on this topic.

There are two types of absorption filters, broad and narrow band filters. A broadband filter transmits a wide range of wavelengths. For example, a #25 red filter would transmit the energy that is ~600nm or longer. 400nm–599nm wavelengths will be absorbed by this filter. Broadband absorption filters are used to separate red, green, and blue colors required to form full spectrum color digital images. The filters are located on pixels used in CCD and CMOS sensors, and are actually C, Y, and M combinations. Red, green or blue filters were frequently required for use in film photographic applications to create differences in black and white image tonality since reds, greens and blues all reproduce as medium gray. Red filters will darken blue and green subjects and lighten red colored subjects. Green filters will darken red and blue subjects and blue filters will darken green and red elements of a scene.

A narrow band filter might transmit 440–460nm wavelengths, for example. Narrow band filters frequently are manufactured as interference filters and are not absorption filters but serve the same purpose. Narrow band filters are key elements of exciter filters and cubes that are used in fluorescence and confocal microscopy. Using modern technology, it is now possible to produce a filter whose transmission contains just two wavelengths.

Polarizing Filters

Light is believed to travel in waves that are oscillating in all directions. To help visualize this, consider the light to be oscillating north and south and containing a second component that is oscillating east and west. Using special polarizing filters, the light can be modified to oscillate in one direction only. When this condition is achieved, the light is described as polarized. Polarization occurs naturally as reflections from flat surfaces. These might include the surfaces of water, leaves, roads, surfaces of automobiles, windows, and in the atmosphere at 180 degrees opposite from the sun.

A polarizing filter is a type of filter that has rows of aligned structures that transmit light waves traveling in only one plane and blocks the light that is oscillating in a direction 90 degrees to the filter's orientation. Linear polarizing filters can be used to darken blue skies by removing the light polarized (aligned) by the atmosphere, and to increase image contrast of a scene by removing of surface reflections from glass, metal, wood, and plastics. The light that makes up a reflection has been reflected by a material and is traveling in only one orientation.

Linear polarizing filters cannot be used for autofocusing cameras or with in-camera metering systems because they remove the polarized light before it reaches the light meter. A circular polarizer is required for these applications. Circular polarizers contain a linear polarizer as the first element of the filter that orients the light in one direction. The linear polarizer element is followed by a quarter-wave plate, which modifies the now linear polarized light. This modification allows the light to travel in a circular motion and remain polarized before entering the camera. This type of light allows the meter and auto focus to operate normally.

Neutral Density and Other Special Filters

ND filters are used to reduce image brightness in the red, green, and blue spectrums equally. ND filters have no color component and are used to remove brightness only. ND filters typically come in various densities. A filter of a .3 increment will remove 50 percent or the light that goes through the filter. Filters are sold with numerous densities including .1, .6, and .9, for example. Microscopes routinely come with ND filters built in. In landscape photography, ND filters are often used to create long exposure times. This is desirable for making moving water appear smooth in a photograph. In microscopy, ND filters are used to make the illumination comfortable for viewing or imaging.

Interference filters differ from absorption or polarizing filters and are created using multiple layer dispersion (MLD) coatings designed to create constructive and destructive interference, leading to the production of specific wavelengths. Their transmission will depend on the number of layers, their thicknesses, and chemical composition. A dichroic filter contains an MLD coating to transmit primary colors and reflect complementary bandwidths or wavelength ranges. Dichroic filters are used as mirrors or beam splitters in fluorescence and confocal microscopy imaging systems. These filters reflect unwanted color and transmit complementary colors.

An infrared filter is a visually opaque filter that transmits only a very small percentage of the far red light, but transmits the near-infrared region up to 1000nm. Near infrared would be characterized as 700–1000nm. Certain other filters transmit ultraviolet radiation and block the visible spectrum. Another spectrally related filter is a haze filter that is a pale yellow filter used to absorb scattered UV and extreme blue light produced by atmospheric scattering. A camera barrier filter is used for fluorescence photography and is used to absorb the excitation energy allowing only the fluorescent wavelengths to enter the lens and be recorded. More on fluorescence can be read in Chapter 15. This filter would be located somewhere before the fluorescence light enters the camera or before it leaves the microscope.

Aberrations

All lenses and optical materials are vulnerable to flaws that can affect optical performance. These shortcomings are called aberrations. Aberrations can be broken into categories that affect image definition, shape, brightness, and color. The most common aberrations are curvature of the field, barrel and pincushion, coma, spherical and chromatic, and astigmatism. There are other aberrations.

Curvature of the Field

Lenses that have this aberration form images that are curved and not flat. General photographic lenses are designed to photograph a world that is equidistant from the front surface of the lens. Curvature of the field is the natural result of using lenses that have curved surfaces. Often a microscope's objective does not focus in the center and the periphery concurrently because of variances to working distances across the sample. Consequently, the image produced by the lens used in this example will exhibit focus in the center or the outside of the image, but not in both regions at the same time unless corrected. When a microscope is being used for inspection purposes, this is not a problem because the focus can continually be changed. When photography is required, this aberration is a problem. The lens must be corrected for use with flat field applications. See Figures 11.2 and 16.5.

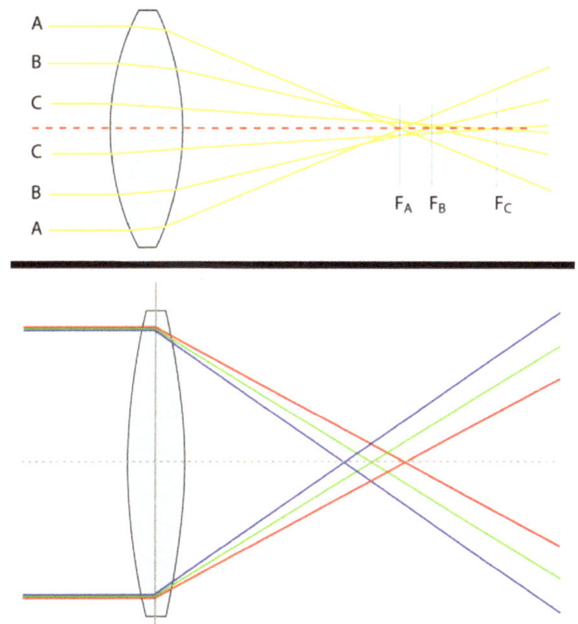

Figure 7.16 Top: Spherical aberration refracts the energy traveling centrally and peripherally differently, leading to images that are lower in contrast and noticeably soft. Portrait lenses by design contain spherical aberration. Bottom: When chromatic aberration is present, red, green, and blue light is refracted to different points of focus, creating colored halos and leads to softness around hard edged objects.

Spherical and Chromatic Aberrations

Spherical aberration is a condition where a lens refracts light differently in the middle of a lens than in the periphery. This is a refractive problem and forms an image that seems low in contrast and poor in definition. A lens with spherical aberration will focus the central and peripheral image points in different image planes and will make the image appear softer and less defined. The resultant image will lack inherent contrast and will display haziness. Spherical aberration can be managed using a lens's aperture

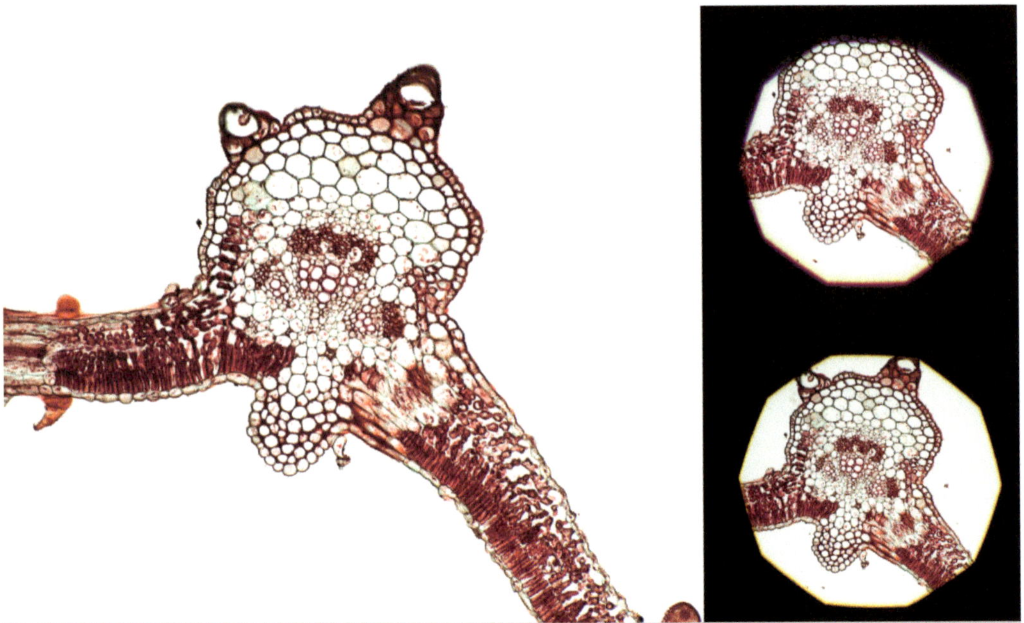

Figure 7.17 The appearance of chromatic aberration can be seen in the edges of the blades of the field stop in images produced from two different substage condensers. These two images reveal the quality of two different condensers. On the top right, the Abbe condenser created an interesting condition creating red on one half of the diaphragm and blue on the other half. An achro-aplanic created a much "cleaner" image with the appearance of less chromatic aberration.

or a microscope's aperture diaphragm. Portrait lenses are designed with spherical aberration, which is useful to create a mood and softness.

Chromatic aberration is a condition where a lens focuses red, green, and blue light at distinctly different image planes. In this condition, a lens will refract shorter wave energy (blue) to a shorter focal distance than the longer waves, e.g. red. The blue will come to focus first and be the largest image, while the red will focus furthest from the lens and will exhibit the lowest magnification. Chromatic aberration can often be observed in a microscope. It would be evidenced at the edge of the focused blades of the field diaphragm that has been centered and remains partially visible in the eyepieces. Unfortunately, stopping a lens diaphragm down does little to resolve chromatic aberration. Regrettably, when photographing, there is only one solution when using a lens that has significant chromatic aberration and that is to photograph using monochromatic light such as green.

Image Depth of Field

The range of sharp focus in a picture is referred to as its depth of field. It is a function of image magnification and influenced by the lens aperture setting. Assessment of the acceptable range of focus an image exhibits is subjective and viewer dependent. As the image magnification changes so does the possible depth of field. Higher magnifications will produce a smaller range of focus. Using less magnification leads to more DOF. When the magnification

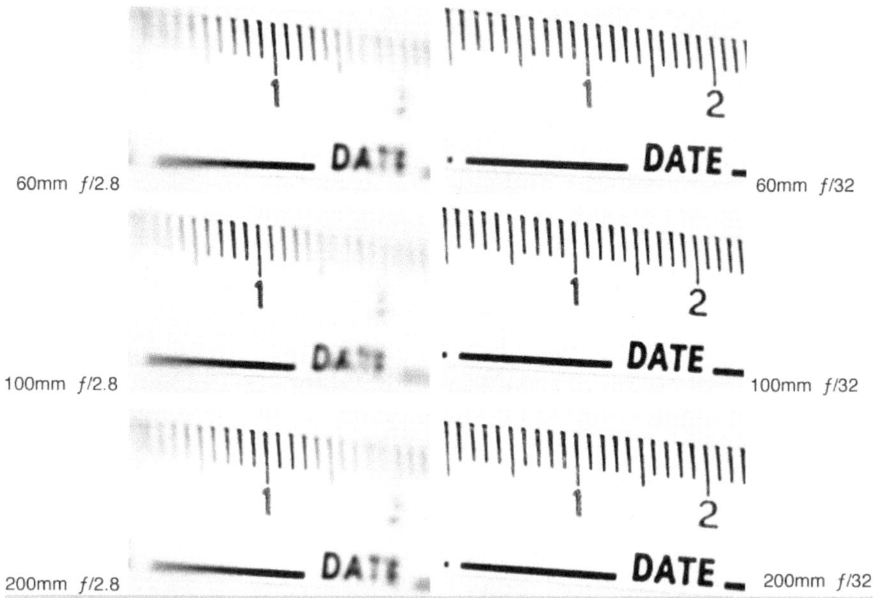

Figure 7.18 Three lenses of different focal lengths were used to create an image of 1:1 magnification. On the left is a series of photographs made at the maximum opening and on the right the corresponding photographs were made at the smallest apertures for each lens. In each pair, the resultant images are identical in the DOF. Sometimes there is a misperception that longer working distances or wider angles will make more DOF but this illustration shows this is not true. Images courtesy of Michael R. Peres.

of a scene is reduced, there tends to be more depth of field behind the subject than in front. In close-up photography though, the DOF distribution is more equal in the front of – and behind – the subject. The depth of field is the distance in front of and behind the object plane that is acceptable. DOF can be controlled by use of the lens' aperture. A smaller aperture will create an image with a relatively large DOF. DOF is located in the image. Depth of focus is located at the rear focal plane. In an imaging system that has a small DOF such as a microscope, the system will have a large depth of focus. If there is a large DOF, there will be a shallow depth of focus.

Diffraction and Image Sharpness

An image's depth of field will increase as a lens is stopped down, but unfortunately there is a fundamental limit to how far a lens can be stopped down without degrading the image sharpness. The behavior that decreases resolution or sharpness is called diffraction. Diffraction occurs when using a small aperture. Diffraction is evidenced when there is the bending of light as it squeezes through a small opening. In everyday photography, an image's DOF can be increased by using a small aperture setting, but as the magnification of 1:1 is achieved or surpassed, the use of very small apertures will produce diffraction. This results in the lessening of image sharpness. Diffraction effects are dependent on a sample's characteristics. Some objects have a potential to create more diffraction effects and others not so much.

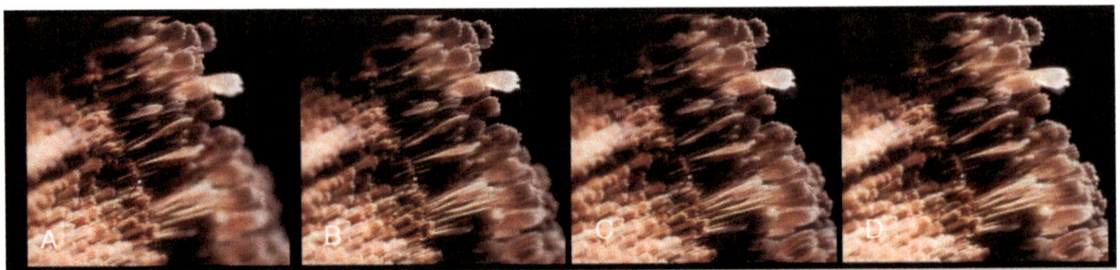

Figure 7.19 A butterfly wing exhibits varying degrees of sharpness based on aperture setting. The image on the left was photographed at ×5 with a maximum aperture of *f*/2.0. It is the sharpest. The middle images exhibit more DOF and are still sharp but have lost critical edge definition because of diffraction. The image on the right has the most range of focus but is the least "crisp" of the three. Images courtesy of Michael R. Peres.

Resolution of the image is influenced by aperture selection and assumed to be diffraction limited. This means that points on either side of the specimen focal plane will be blurred by the combined effects of optical and diffraction blur.

Suggested Reading

Applied Photographic Optics 3rd ed., by Sidney Ray. Focal Press, 2002.

Basic Photographic Materials and Processes, 3rd ed., by Nanette L. Salvaggio. Focal Press, 2008.

Photographic Materials and Processes, by John Compton, Ira Current, Richard Zakia, and Leslie Stroebel. Focal Press, 1986.

View Camera Technique, 7th ed., by Leslie Stoebel. Focal Press, 1999.

Chapter 8 A Deeper Dive into Digital Devices

This chapter was written to demystify digital technology and take a close look at the functions of devices used in digital imaging systems. As an applied person, I have found that acquiring a working knowledge about digital tools has greatly enabled my work. I have also found that sometimes misinformation creeps into an organization's digital workflow affecting the "how to" or "why to" do something. Over time, doing things in ways that are not ideal can become the standard operating procedure because, "that is the way things have always been done here" and that is not good.

Sensors

Sensors in digital cameras are the foundation of the imaging device. There are two types of sensors used in contemporary cameras. Complementary metal oxide semiconductors (CMOS) are found in the majority of today's cameras because they are less expensive to produce and consume less electricity. In the earliest days of digital camera manufacturing, charged couple device (CCD) sensors were more common. In active pixel CMOS sensors, the electrical design has been engineered to use power more effectively. CCDs require ten times more power than a CMOS sensor. The CMOS sensor's design lowers the power consumption required for image processing functions on the sensor. By lowering the power consumption, less heat is produced, and subsequently there is less noise produced by the sensor. A sensor is the capture device in a camera and creates the image by collecting the light that falls on its surface. Located in the sensor are photo sites or pixels.

Pixels

The basic element of a sensor is a pixel. It is the photosensitive element of the sensor. Pixels are the basic picture element of an image and are the smallest component of the picture file that can individually be processed. A digital camera's sensor – sometimes called an array – might contain millions of photo sites that are used to record light values used to display an image. When the camera's shutter button is depressed, the recording of a photographic

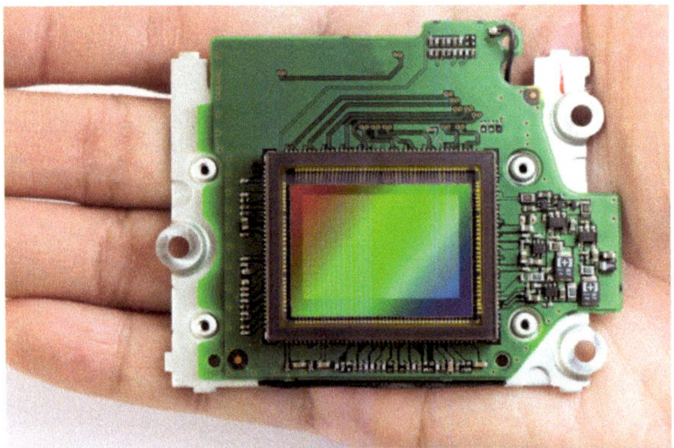

Figure 8.1 A CMOS sensor. It is interesting to consider how miniature these devices are and how powerful they are. Image courtesy of Shutter Stock.

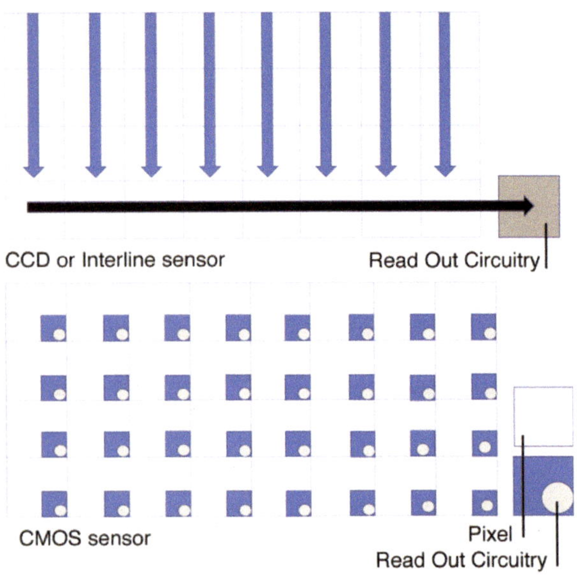

Figure 8.2 This illustration represents the pixel arrangement in a CCD and CMOS sensor. Notice how the pixels in the CMOS sensor are smaller to account for the placement of the circuitry. Image courtesy Michael R. Peres.

exposure begins. During the exposure, each pixel or photo site is uncovered to radiant energy and is exposed by the photons that fall onto its surface. The pixel collects – and stores – the photons that fall onto its surface. Once the exposure has been completed, the camera's shutter closes. Within each pixel, the software calculates how many photons were captured during exposure. The quantity of photons captured in each pixel are then characterized as image brightness levels or bits, which are displayed as other details such as contrast and color on the display.

Pixels are organized on a sensor as either full frame or interline. CCD sensors use full frame pixels. In this type of array, the entire surface of a pixel on the sensor is light sensitive. Interline sensors are designed to have a pixel and image storage mechanism adjacent to the light sensitive region on the pixel. In a full frame sensor, once an exposure is completed, the signal is transferred after exposure line-by-line of the pixel array. The transfer of exposure data occurs at a different time than during image capture. The data transfer in a full frame sensor is slower than an interline array. CCD sensors require a separate shutter for exposure. The design of a full frame sensor allows for higher ISOs and higher resolution that generally produces a better signal-to-noise ratio.

Conversely, interline pixels have smaller light sensitive areas because some of the pixel's surface is used for electronics and transporting charges. An external shutter is not required for this type of pixel. CMOS sensors operate at lower ISOs than full frame pixels, have lower resolutions, and have faster frame rates.

Bit Depth

Producing images with a large number of tones is a function of having a larger bit depth. More bit depth means more information about the subject will be recorded in an image. A pixel's tonal scale begins at 0, a zero brightness and ends at 255. The most basic tonal scale is characterized as being an 8-bit image. A bitmap describes a pixel-by-pixel description of an image that is observable on a monitor or camera display. When the resolution is high enough or appropriate for display, the image is seen as a continuum and its image structure (pixels) is not visible in the image. However, when there are insufficient pixels contained in the image file, the image pixels will be visible. When the image reveals pixels, the condition is called bitmapped. This occurs when the image display does not have the adequate number of pixels required for a smooth display at the proper size for viewing. Bitmapping occurs because either the size of, or number of active pixels, is inadequate. Bitmapping is visible in output in both screen and print applications (see Figure 4.9).

Single Shot Area Array Sensors

There are two types of configurations used in contemporary sensors which are sometimes called arrays. One array produces simultaneous capture and one creates a scanned capture. Single shot or instant capture sensors are by far the most common and are used in DSLR cameras and many other types of cameras. In an area array sensor, pixels are organized in a grid of horizontal and vertical rows like a piece of graph paper. There are a specific number of pixels found in each direction. In a linear or scanning array camera, a single row of pixels is located in only one orientation. There is a predefined number of pixels in that single row of pixels. The number of pixels is designed into the sensor and manufactured to that specification. The linear array – sometimes called a pixel wand – is moved across the device's field of view using a precision low vibration motor. A flatbed or film scanner is an example of a linear array device. Most pixels are rectangular shaped pixels but other shapes have been used in some devices.

Today's DSLR and smartphone cameras are equipped with area array sensors. The sensors record brightness values initially as a grayscale image. This data is used to create color images. The pixels are covered with red, green, or blue colored filters on top of each pixel. This type of sensor configuration uses a Bayer pattern, invented by Eastman Kodak scientist Dr. Bryce Bayer in 1975. On an RGB sensor, each pixel captures only one channel of the scene as either the red, green, or blue component of the image. In this approach, the missing RGB data for each pixel is created or de-mosaiced mathematically by the camera's software by referencing neighboring pixels. This process is called color interpolation. In a typical Bayer pattern array, the sensor will contain twice as many green pixels as red and blue pixels. This is because human vision is most sensitive to green light. The use of an area-array sensor has several advantages. Most importantly, this type of array can be used for both moving and stationary subjects and used for both continuous and short duration lighting where exposures can be anything from a few milliseconds to minutes. There are, in most cases, no mechanical parts associated with this type of sensor. This sensor can also be made in various different sizes and can be easily integrated into most types of cameras, including security cameras for example. Pixels have different sizes depending on a number of engineering factors. An average pixel is 7μm in width and with a few exceptions pixels are primarily square.

A disadvantage of an area array sensor is the simple fact that each pixel records just one channel. Because of this, there is mathematical reconstruction of data not captured in the other two channels. The creation of information by referencing neighboring pixels is called de-mosaicing and creates false data. This process can produce image artifacts. These artifacts are evidenced as color aliasing or color moiré effects. To minimize the occurrence of these artifacts, most cameras use an optical anti-aliasing filter in front of the sensor. The use of an anti-aliasing filter creates some loss of image sharpness because an anti-aliasing filter is a type of soft focus filter. Another disadvantage of this method is that additional time is needed to create and process the data required for image construction. Because the image result is influenced by the color information from surrounding pixels, images from different cameras may produce different color results.

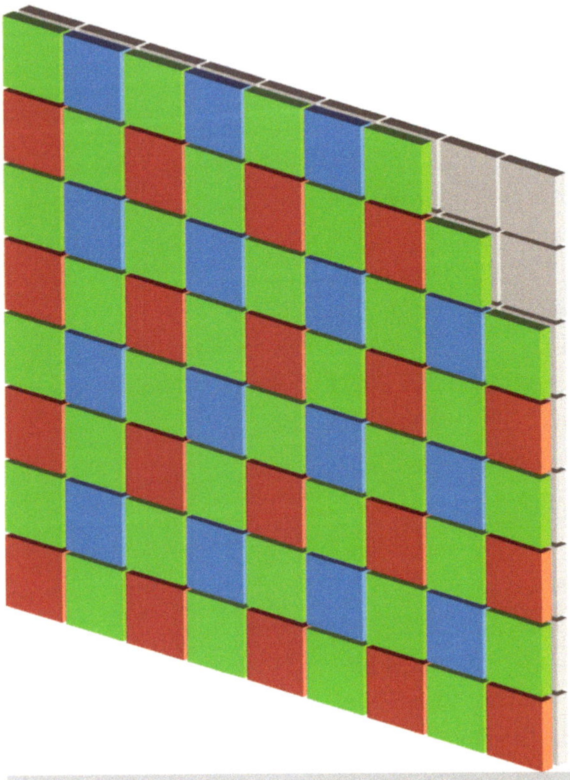

Figure 8.3 This illustration features the appearance of a one-shot sensor that uses a Bayer filter.

Area array cameras can be operated in single shot mode if there are filters on the pixels or using a multi-shot format if there is a filter wheel located in front of the sensor. Multi-shot sensors were frequently used in instrument and studio cameras in the early stages of digital photography. There are currently no multi-shot cameras made that use a filter wheel but the Hasselblad company produces a multi-shot technology that moves the sensor approximately 1 or 1.5 pixels with a precision piezo actuator. This allows the camera to capture more color information and data than single-shot capture forms by having multiple samplings of object information.

Initially, multi-shot cameras were used to create higher resolution color images when sensors were very expensive to manufacturer and a higher resolution was required. The final image from a multi-shot sensor was the result of multiple image capture frames of a static object that were recombined into one larger file. Separate red, green, and blue pictures were made and then a new file was created from those individual files. These sensors created no moiré patterns. In this type of camera and file, each channel contained all of the color data for a given sample. These files had advantages for image processing and frequently produced cleaner and better outcomes of difficult subjects, and numerous multi-shot cameras were used in light microscopy that took advantage of this. The disadvantage was that this camera could not be used for moving subjects.

Hasselblad is not the first or only company to move the sensor in a camera. Special multi-shot cameras were initially engineered to move the sensor small distances after making the first exposure. This created slightly different fields of view for each image capture. These slightly different images were then stitched together and a new file with more pixels was created. This method is still a feature in high-end digital cameras used on photomicroscopes. In this type of camera sensor, for example, a sensor has 1300 × 1000 pixels in its single shot mode but when using multi-shot capture mode, it creates files that contain more than 4000 × 3000 pixels. A multi-shot camera is slower to operate than a single shot camera and cannot be used with moving subjects.

There are area array cameras that do not have a Bayer filter on the sensor and they are characterized as monochromatic cameras. These cameras are used for low light applications and are most often found in fluorescence microscopy and astronomy. Since filters remove light brightness, having a Bayer filter on a sensor reduces image brightness. A naked or chip without a filter will operate more effectively in low light. Color data can be added to the file using what is described as pseudo or false coloring. Colorization is accomplished using image editing software. Astronomical and scanning electron microscopy images are routinely false colored.

More on Linear Array or Scanning Cameras

Scanning cameras were the first imaging systems capable of creating high resolution files with enough pixels that were required for large format printing. These cameras were based on scanner technology. A scanning camera uses a single row of pixels located on the linear array or wand that is moved across an imaging field. During the exposure, the array travels across the image area recording and creating an image line by line. The linear array sensor often contained three rows of pixels. Each row of pixels was covered with either a red, green, or blue filter or stripe that captured each color in a single pass. The significant advantage of scanning cameras was the ability to make a high resolution imaging system at a relatively low cost. Scanning cameras were used most often in studios where subjects did not move and very large files were needed. There are very few cameras of this type still being made but many can be purchased on eBay®. The Better Light® scanning back camera is an example of this type of camera and sensor. They were used primarily with large format cameras and were very popular for a short time.

It should be obvious that a scanning camera can only be used with stationary subjects. Any movement by the subject or movement of the camera during an exposure will produce images containing distorted features. This happens because the subject is not recorded simultaneously at one time and its pieces and parts will be in different locations at different times during the scan. Another disadvantage is the long time required for capturing the entire image as it scans across the field of view of the lens. Bright lights are needed for this type of camera. Flatbed scanners have light sources built into the scanning wand. When using this type of camera with studio lighting, the lights must keep at a constant and high level of brightness during the exposure. If light brightness changes during a scan, the image will exhibit different brightness bands across the image. Tungsten and tungsten halogen lights can change brightness several times per minute because of fluctuations of current to the light. This is called flicker and if tungsten lights are used with this type of a sensor, brighter and darker exposure bands will appear in the image. Fluorescent lights can be more problematic and will add color variations in addition to the darkness gradient. HMI lights are suggested for use with scanning back cameras. These lights do not create flicker.

Sensor Sensitivity (ISO), Binning, and Gain

Adjusting the sensitivity of a sensor can be useful in low light applications. A camera's baseline sensitivity is established in manufacturing and is not adjustable in a smartphone camera. A sensor's sensitivity is adjustable in mirrorless and DSLR cameras. Instrument cameras such as those used on photomicroscopes can also have their sensitivity increased but this is accomplished in a different manner.

In DSLR and mirrorless cameras, the sensitivity of the sensor is referred to as ISO. ISO stands for International Standards Organization, an organization that oversees standards in various industries including imaging products. ISO ratings include the sensitivity of camera sensors. When you change a camera's ISO setting, this adjusts the camera's sensitivity to light. It is possible to raise or lower the sensitivity. Today, it is common to see ISO settings ranging from 50 to 16,000 or higher. These settings are direct indicators of the sensor's sensitivity to light. A lower setting makes the sensor less sensitive and a higher number makes the sensor significantly more sensitive to light. Today's cameras can have very large ISO ranges. There is a downside to increasing a sensor's sensitivity and that is the possibility of creating noise in the file.

ISO is one of three contributing factors in the creation of a photographic exposure. Shutter speed and aperture selection are the other two components of exposure. Shutter speed is selected according to scene brightness and subject motion-stopping requirements, while lens aperture is selected for the depth of focus needed. ISO is changed when there is too little light to make a useable exposure and is controlled in the camera's menu.

Instrument cameras also have a baseline sensitivity not described as the ISO. In fact very few instrument cameras share their base sensitivity as a useable number. Increasing the sensitivity in this type of camera is accomplished using gain. Gain is shared as a factor or number. A setting of zero indicates no gain has been introduced for capture. The gain setting can be adjusted upwards when it is nearly impossible to make an exposure using time alone. It is possible to increase the gain significantly. Electronic amplification of the sensor results in the sensor working harder to record smaller amounts of light. When the sensor is operating, it generates heat. Heat is not a useful by-product of sensor operation and often results in noise. Noise is created when pixels are operating in challenging situations and are not efficient in their operation. Heat produces random exposures on some pixels in a sensor when in use. Noise can cause image data to be diminished. If the signal (good exposure)-to-noise ratio is not high enough, the image will not be easily discernible from the noise.

Sometimes gain alone is not adequate to make an exposure. When the noise levels compete with the sample's signal strength, gain is no longer a useful method and binning is required. Binning combines pixels to make a pixel group which is more sensitive to light because the pixel group has a larger surface area. Pixels can typically be grouped as 2, 3, 4 or more. Like all things, the disadvantage when using binning comes as a result of down sampling the total number of pixels available in an image. If a sensor has a resolution of 1200 × 1000 pixels and a binning of 4 × 4 is selected, this results in 16 pixels behaving as one. When the sensor is set up this way, it will create a file that has 300 × 250 pixels, not 1200 × 1000. Binning is used when light levels are so low there is no other way to make a usable exposure. It is an important strategy to consider when working in challenging situations. An image is better than no image and a smaller file is better than no file.

Noise: Evaluating Dark, Shot, and Noise Sensor

Electrical devices generate heat when operating and when sensors are making exposures, they produce heat. Heat can cause pixels to behave as if they were exposed to light. This artifact is sometimes visible in an image and is called noise. Noise can come from a variety of conditions. When the sensor is operating, this noise is called shot noise. Shot noise occurs when exposure times are long. Long is a loosely defined term in contemporary cameras. In the not-so-distant past, a long exposure time of less than 10 seconds would produce noise in an image. All cameras are not similar in how they make and manage noise. Modern cameras with improved sensors manage the creation of heat and noise much more effectively. It is common to make noiseless images at exposure times approaching one minutes with some of the best DSLR cameras. All digital cameras remain vulnerable to noise creation. DSLR cameras have a noise suppression filter that is accessed in the settings menu. Mirrorless cameras also have a noise reduction setting. Noise reduction filters soften images so they should be chosen with care. By softening the image, there is a blending of noise or pixel information in the image making it less apparent. Some technical instrument cameras will cool a sensor during operation which minimizes the creation of heat. They may also have a noise reduction filter. Cooled chip cameras can be very effective for low light imaging applications and are found in fluorescence and astronomy but can be expensive.

All sensors – because of electrical currents – make noise, even when not being used to make exposures. This type of noise is called dark current noise. Engineers have worked hard to improve a sensor's performance when the camera is on but because sensors are electrical, some heat is always present when the sensor is electrified.

It is possible to evaluate a camera and sensor for its performance capabilities which would include noise production. This can be a very useful test to run to determine how your camera will perform in challenging situations. Testing a camera for noise and other image attributes can be undertaken in the following way.

Sensor Evaluation

Evaluating a camera for image quality and performance including noise production in particular, can be a tedious but useful task that provides practical information about a camera's overall potential and performance. When evaluating a camera for noise and other "sensor created" outcomes, it is important to review the data produced from the sensor with no image enhancement selected and through the use of a standard target. The reason for the evaluation is to determine how image capture attributes are affected in various situations. This will lead to an effective way to manage file capture attribute settings including noise for use with a high bit depth and RAW file formats because of its better signal-to-noise ratio.

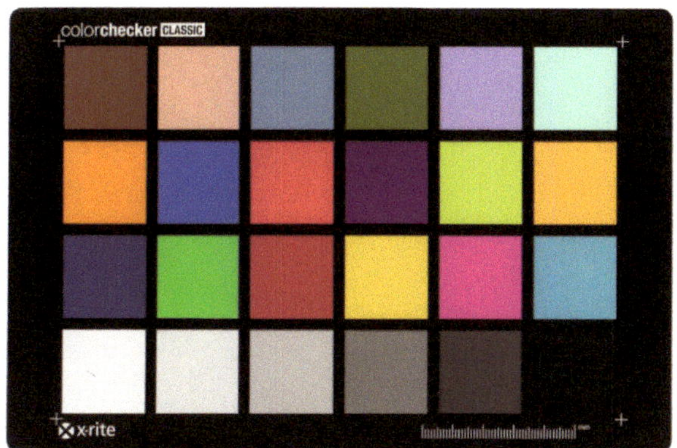

Figure 8.4 A Macbeth Color Checker is a standard accessory needed for camera and imaging system quality control. It is a required tool for color management of image processing. Image courtesy of X-Rite.

To begin the evaluation of a camera's sensor, it is important to ensure all capture settings such as image sharpening, noise reduction, and color enhancement are turned off. Noise reduction and other at-capture choices will affect in-camera image processing algorithms and it is important to remove their influences for performance and sensor assessment.

To set up an evaluation, as mentioned, a standard target or sample is required. For close-up photography, a small color checker can be useful as the target but when a system magnifies images of very small objects, small targets are not common and available for commercial purchase. A good sample target simply should exhibit various color patches and include fine structural details as its resolution component. A useful target should include black, white and gray regions and also contain at least some primary colors. When possible, a target that has fine details oriented both vertically, horizontally, and diagonally is preferable.

To start, all testing parameters must be kept identical to get valid image data outcomes across the various settings that will be evaluated during the testing. This means everything that is done during the test should be maintained for the camera settings, except for ISO or other pre-test choices. Other things to keep constant include use of the same target; use of identical lighting; using identical framing; identical optical settings on the lens as well as diaphragm settings; and performing the same image processing.

Once the target and location are established, set the camera's sensitivity to its lowest ISO, e.g. 100. A slight modification to these recommendations can be made and should be based on the type of camera and what lighting equipment is used. It might be ideal to use manual exposure mode for evaluation but using automatic exposure mode will also work. In addition to turning off noise reduction and image sharpening, it will be important to select Adobe RGB 1998 or another wide-gamut color space.

It is practical to use the RAW file format for the evaluation, but using TIFF files will also be informative. Do not test sensor performance using JPEG files since data will be lost when the file is compressed. Images should be captured using Auto White Balance and RAW images can be converted to 16-bit using the image processor's pre-processing feature.

The test will require the use of uniform and even lighting. The lighting should impart no biases to the target.

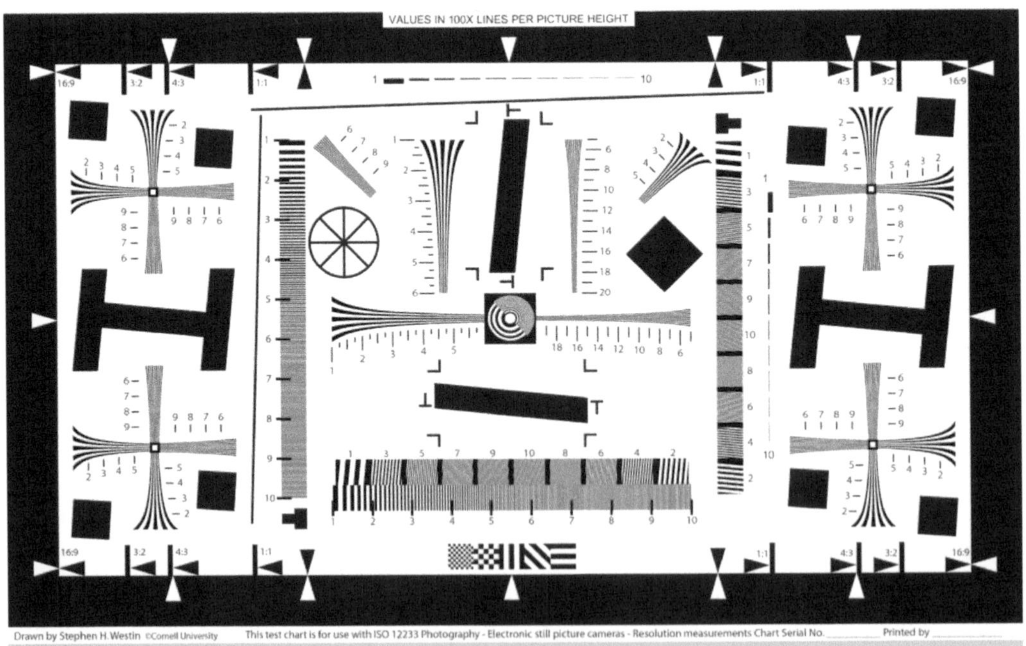

Figure 8.5 A standard test target such as a color checker can be useful when evaluating a camera's performance. Shared above is a re-drawn ISO 12233 Test Chart resolution target. This graphical image was created by Stephen H. Westin and can be downloaded at: https://www.graphics.cornell.edu/westin/misc/res-chart.html

1) Place a sample or target in front of the camera. Compose and focus the image. Fill the frame or choose a composition that will remain the same for each of the next steps. Focus the image.
2) Set the camera to the lowest ISO and make an exposure that achieves the proper time and aperture setting for the Normal (N) outcome or a 0 exposure placement using the Auto mode. This exposure is the proper exposure to achieve an 18 percent gray tone in camera.
3) Change the ISO to a setting that is 2× the initial capture. For example, if the setting was 100, change the ISO to 50, etc. Make an exposure that achieves the proper time and aperture setting for the Normal outcome or a +1 exposure using the Auto mode.
4) Proceed to make exposures using ALL the various ISO settings on the camera.
5) Using Adobe Photoshop or other image processing software, open all the files and evaluate the noise or granularity of the images as the ISO is increased. Degradation of sharpness, color and other image attributes may be evident on initial inspection in the shadows and dark tones.

More on Bit Depth

Within a pixel, a bit is the smallest unit of data that is identifiable. As the basic unit of binary information, a bit can be off and create a zero (0) value or be turned on and create a one

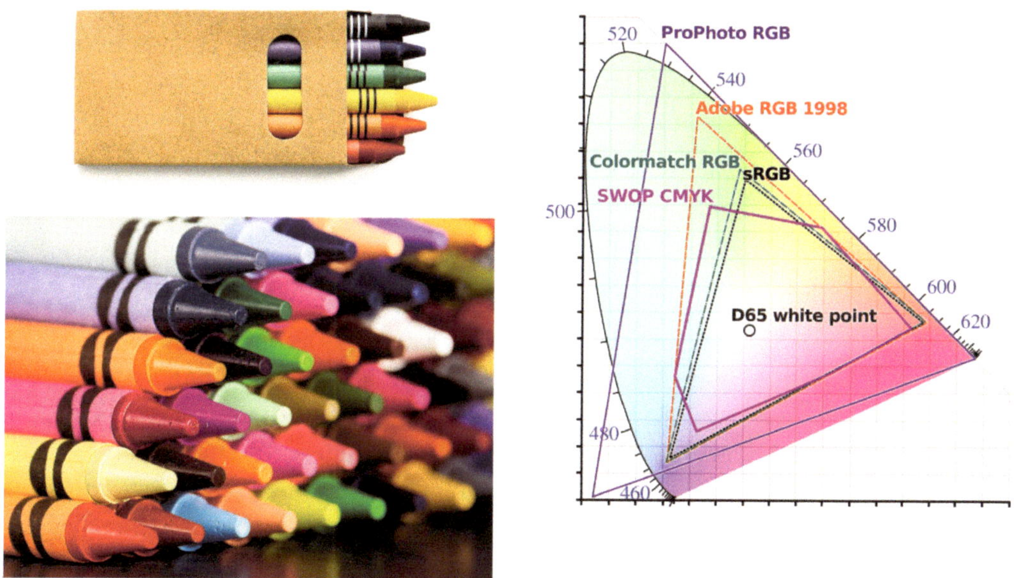

Figure 8.6 A sensor's ability to discriminate colors is influenced by what color space is selected for capture. A useful analogy for understanding color space might be found when considering the number of crayons found in different sized packages. A package with six crayons will provide a user with only a few colors when compared to a package with more choices. The CIE diagram shown in this illustration (right) shares the approximate color gamut of sRGB, Adobe RGB and ProPhoto RGB. It is easy to see how the color spaces fit into the range of colors. Crayon images courtesy of Shutterstock; CIE diagram reproduced as Creative Commons file. Original by C.M.G. Lee and based on http://commons.wikimedia.org/wiki/File:CIE1931xy_blank.svg, using data from Blatner, D. and Fraser B. (2004) *Real World Photoshop CS*, http://canopuscomputing.com.au/gallery2/d/8631-1/Gamuts_comparison-B_F.jpg.

(1) value. Pixel resolution is important but a file also contains color and tonal value information. This is described as the bit or color depth of an image. Imaging devices will create superior image data with greater bit depth. Brightness resolution refers to the number of shades of gray that an image can discriminate that can be derived from each pixel. The more bits, the greater the tonal range will be included in the file and this in turn will lead to greater information useful when processing the file.

Since sensors have become so good at collecting photons, the ability to discriminate more tones can be accomplished by increasing the bit depth when possible. The term bit depth implies that there are a finite number of tones a pixel can manage. Bits are expressed an exponent of 2 and an 8-bit image would be described as 2^8. This bit depth will produce a file with 256 discrete tones for each R, G, B color for a total depth of 16,777,216 colors.

Some contemporary sensors can make 10-, 12-, 14-, or 16-bit images. Increasing the bit depth increases the data available and assists greatly with image processing. Increasing data may increase a sensor's dynamic range but more importantly provides more shades of grays, colors, and data from a subject. When a file has more data, more can be made visible or revealed when image processing. For the record, processing data changes data and sometimes the data becomes more visible and at other times it can be diminished because of noise or other artifacts. It is a trade-off. More bit depth also leads to bigger files. Less bit depth leads to smaller files and less data. Bit depth can be selected in some cameras at capture but during image processing it can be changed from JPEG to RAW or TIFF during operations. When files are down sampled, data is removed, but when files are made bigger, interpolation or false data is created. Capturing the first image using the highest bit depth is most important. Increasing bit depth during image processing is considered by some to create false data. Changing the bit depth after capture is more about how the image will be displayed and seen rather than adding false data.

Color Space

A color space is a mathematical model or description of colors as recorded, described, and/or displayed. Color space describes a set of colors and the corresponding numbers that have been created to describe them. A PMS color descriptor is part of the Pantone system of color space description and is an example of an analog model. Highly structured mathematical descriptions of color were used for ink mixing and choices in the graphic arts field. This information is now found in the Adobe RGB 1998 color space. sRGB files are examples of a color space used for web images and monitor viewing only. There are three values used to describe RGB images and there are four values found in CMYK system, used primarily in the graphic arts applications.

Carefully select a color space. You can easily establish a detailed workflow for capturing color, processing color, and optimizing color for output. Color space is also described as color gamut. A color gamut will be part of what defines the color quality and accuracy in reproduction of an image. For example, Adobe RGB 1998 and sRGB are two different color spaces and they are both part of the RGB color model. The Adobe RGB 1998 space will

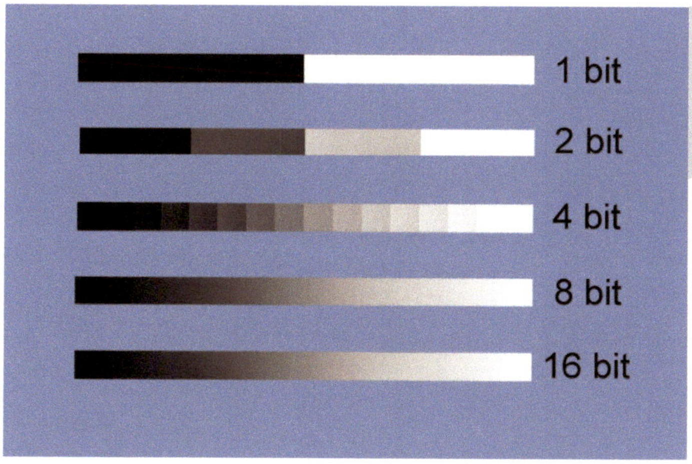

Figure 8.7 Bit depth allows are more tonal resolution in an image. This illustration shares visually the advantage of a larger tonal scale. The obvious improvement between 2- bit and 16- bit is significant.

display more colors and have a wider gamut than sRGB. sRGB is a smaller color space and will compress or suppress more color information. A modern and the largest color space is ProPhoto RGB.

The most common reference standards for color space that include recording and display are CIELAB or CIEXYZ color spaces. These systems were designed to mimic the colors that the average human can see. When deciding what color space to select for a camera, it is always recommended to select the widest gamut available because this will create files with the most information.

Gamma and Contrast

Some instrument and technical cameras allow a sensor's capture contrast to be modified. No DSLR or mirrorless camera can adjust its gamma at capture. Adjusting capture contrast is accomplished using the gamma control and is found in the camera's operating software. Gamma is an important contrast characteristic of all imaging systems. Gamma defines the relationship between a pixel's numerical value and the tonal display from the image's actual luminance when displayed. Without using gamma, all of the shades of gray recorded by a digital camera would not appear as expected. This can cause the display to not exhibit a proper tonal map of a subject. Understanding what gamma is and how it works can improve your ability to create more accurate exposures that can be useful in subsequent image editing.

Using gamma settings effectively is important for imaging because human vision does not perceive light in the same way digital cameras do. When twice the number of photons hit the sensor, the sensor will make twice the signal, which is a linear relationship. Human vision is non-linear and we perceive a doubling of light to be only marginally brighter. This lack of separation is increasingly less at higher light intensities. When compared to cameras, human vision is more sensitive to changes in dark tones than changes in bright tones. Gamma and tone mapping is what differentiates the light sensitivity of the human eye from that of electronic systems. Once the file has been saved as a digital image, the gamma information has been encoded and the tonal values will be displayed more closely to what would be expected by a viewer when the gamma is properly established. Human vision is flexible and digital capture systems are not.

No DSLR cameras allow for the selection of different gamma settings but technical cameras routinely offer this feature. A gamma setting of 1.0 reflects an average contrast display or recording setting. Gamma is displayed graphically and predicts what changes in brightness will be caused by a subsequent change in exposure creating a subject's tonal values. Gamma or $\Delta D = \Delta Exposure / \Delta Brightness$. When an average contrast subject or situation is encountered, a camera gamma setting of 1.0 would be useful. If the sample or situation contains high contrast such as in fluorescence microscopy, a lower gamma setting is preferable. A high contrast setting would be useful for low contrast samples.

Monitors also have gamma settings. These settings are a bit different than the numbers useful in the camera software. A monitor's gamma range would typically begin at 1.4 and go up through 2.2. An average setting was 1.8 for Windows operating systems and approximately 2.2 for Mac computers. Today, both operating systems recommend use of 2.2.

Scenes and subjects will contain varying levels of brightness within them. The difference between a high brightness compared to a region of low brightness describes a scene's brightness range or dynamic range. A sensor can record data more effectively when the exposure comes from a narrow brightness range. A high brightness range creates challenges for a sensor to record the entire range in a single exposure. High dynamic range imaging or HDR methods have evolved to manage this problem. You can read more about HDR in Chapter 12.

White Balance

When photographing and during image processing, a photograph's color is displayed as the result of global adjustments to the intensities of the recorded red, green, and blue color information. The ultimate imaging goal is to render color correctly. The color of an object is affected by the lighting where the object is recorded and viewed. Human vision compensates for different types of light allowing a white object to appear white regardless of the light sources with various color temperatures. Digital cameras can display color correctly but require setting its display to neutral colors of objects photographed in different types of lighting. This allows images to render white objects as white. The method used to create neutrality is referred to as setting a gray balance, neutral balance, or white balance. When adjusting color balance, this changes the overall mixture of colors in an image that are used for color correction and generalized acceptable versions of the image's color balance.

Image color data produced by a sensor must be converted from the acquired values to new values that are appropriate for color of the light used for capture and subsequent reproduction and display. An image's color balance is changed in image editing applications usually by directly modifying the red, green, and blue channel pixel values without respect to any color reproduction.

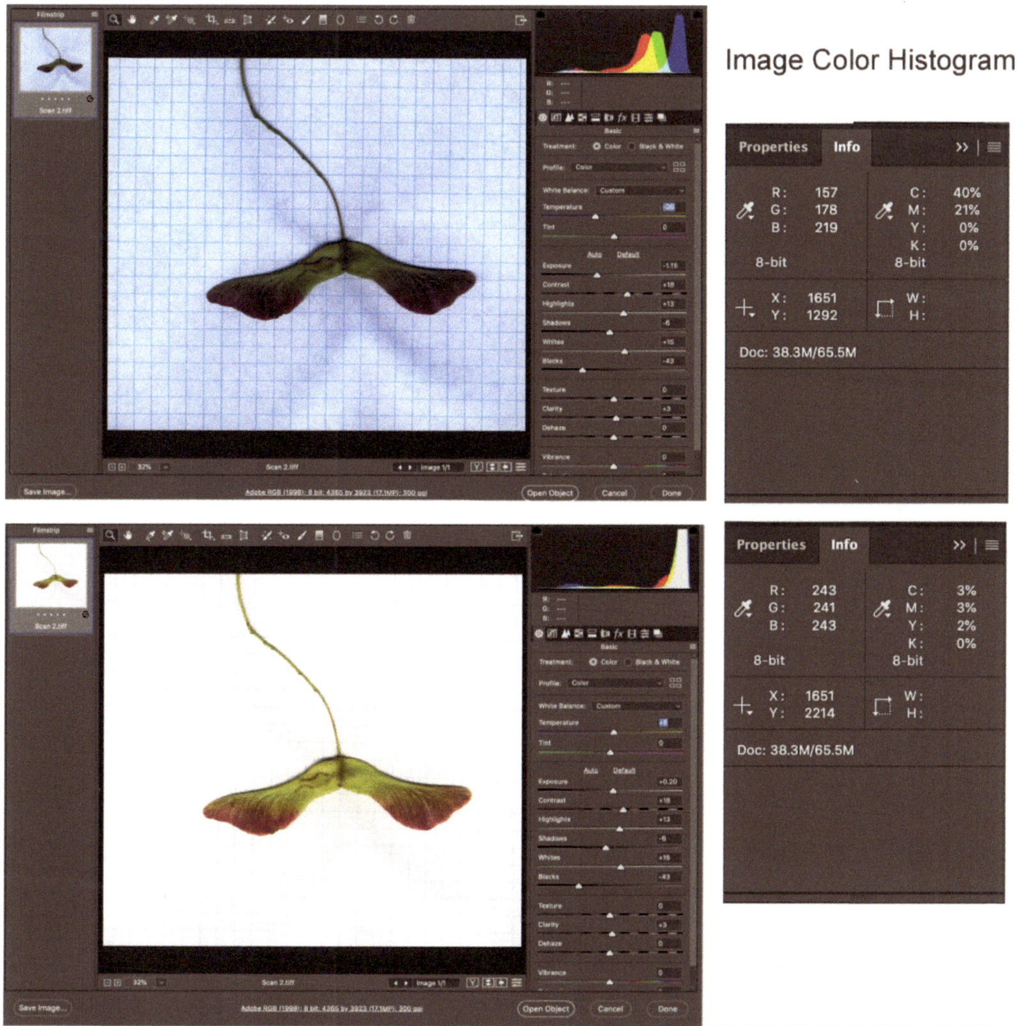

Figure 8.8 In this figure, you can see how I adjusted the color data to have a white point of approximately 240 in each channel. This is evidenced in the info window as well as the histogram.

Cameras can set white balance in several ways. You can use auto, manual, or select from several preset choices called custom. Operators can set white balance using the auto mode and the camera will read the scene's various color temperatures and select a setting from its database of pre-programmed choices. This method works well when a scene has lighting that is comprised of one type of light and there are prominent neutral subjects in the scene. The results when shooting the auto white mode are also affected by changes in the lighting. In most cases the most accurate color rendition requires using the manual white balance setting. Manual settings include incandescent, fluorescent, electronic flash, cloudy, open shade, sunny, and custom Kelvin color temperature choices.

Using the incandescent setting is recommended for tungsten bulbs or when using fiber optic lights or photomicroscopes for example. There are many contemporary laboratory instruments that are equipped with LED lights. Choosing the fluorescent setting will prevent common green casts typical in photographs made using fluorescent lights. The cloud setting will add warmth to the color temperature and the electronic flash setting is used with 5500K lighting produced from electronic flash equipment. The Kelvin color temperature setting is best for total control where the operator sets the degrees Kelvin required to control how the camera will render the hue and intensity of colors in the scene. When using the Kelvin setting, you can fine-tune the process creating total control of color reproduction. Using the camera in the Live View mode will allow you to see in real time the results of a specific Kelvin color temperature change. It can also be useful to place a white or gray card in the scene to use for white balancing a scene.

Capture File Formats

One of the most important decisions a scientific photographer makes prior to making camera exposures is what file format should be selected for saving. There are a number of file formats to choose from. All of the choices have advantages and disadvantages. Decisions about selecting an appropriate file format should be primarily decided upon based on which file format will preserve the most data and which one creates the ideal workflow going forward.

The first decision is to select a file format that is either lossy (compresses captured data) or lossless (does not compress data). A file format that maintains the full data without compression is lossless and would include TIFF, BMP or RAW files (TIF or TIFF files are described using either abbreviation). There are also formats that offer various degrees of compression and are lossy such as JPEG, JPEG 2000, or PNG. There is no reason to choose a file format that results in compression at capture in science. Compression will compromise the integrity of recorded data right from the start of the imaging session. Data will be changed or lost using compression file formats. There is really only one decision to make and that is to select a lossless file format. That being said, many wedding photographers and event photographers use JPEG because of advantages for speed and portability of this file format in image processing. In some situations, thousands of files are created. Rarely should a science photographer ever use a JPEG file format. Technical cameras will also have proprietary file formats which have advantages but only using the work station where the proprietary software is installed.

Digital File Structures

As a side bar, there are two broad categories of digital file structures, raster and vector. A raster image describes an image where the individual pixels can be addressed. It is sometimes called a bitmap file. The other digital structure is called vector and is characterized by a file not having discrete pixel locations in the image. Vector files can be enlarged or made smaller by dragging or compressing the picture as a whole unit because of the way it is mathematically described. Vector images are very different than raster images. Very few if any capture devices create vector images. Vector files are used for desktop publishing, graphics applications used for output, but not for capture. Adobe Illustrator® creates vector images. Raster image files are converted to vector files when used in software products such

Microsoft Powerpoint®. An EPS (encapsulated postscript) file is an example of a vector file format. EPS files can be opened and converted to bitmap files using Photoshop.

In all types of image files, the size of the file is directly related to the number of pixels, the color space, and bit depth. Image files can be compressed or they can remain in a file format that does not compress or change captured data. Compression of a file is accomplished using mathematical methods to approximate data or create a facsimile of the original image's data that uses fewer bytes. This process results in the creation of files that are smaller in size than the original captured file. Compressed files can be opened up to their full resolution and in many cases look as they did prior to compression by using decompression algorithms. Always consider that different cameras and software will create different compression results. It is common for two images containing the same number of pixels and color depth to have different file sizes when differing amounts of compression are applied to the file. Because of the manner in which pixels, color, and bit depth, are evaluated by algorithms, the results are influenced by the variances in the original image subject leading to differences in file sizes after compression. Within some compression mechanisms, an image's complexity may also result in smaller or larger compressed file size. This sometimes results in a smaller file size for some lossless formats when compared to lossy formats. For example, a simple image that has a large region of a similar color or dark region such as a background in a photomicrograph, may be lossless when compressed into GIF or PNG formats. These two file formats may result in smaller file sizes than a lossy JPEG format might make.

There are several file formats that save data files without compression and are ideal for science. These would include TIFF, BMP, RAW. Any compression of recorded data can cause data loss no matter how insignificant. There was a time when storage and storage media was a consideration because storage was not prevalent and expensive. Today that is not a concern and selecting a file format without compression is a more important priority. Proprietary file formats do not compress data and have the advantage that all of the capture information from the equipment itself is exported within the file as metadata. A proprietary file format requires the camera's software or special image processing software to be opened.

The TIFF format is considered by many to be the most universal file format. TIFF stands for tagged image file format and produces files containing data that is exactly like the data that was recorded by the sensor. TIFF files can be read across all platforms including LINUX, MAC and Windows operating systems. They are also backwards compatible. That is to say, a TIFF file can be opened and read using any device regardless of the age of the operating system or when the image file was created. TIFF files can also be opened using many image processing software products without special plug-ins. Plug-ins are special pieces of software that are sometimes required to read various file types. TIFF files will rarely have issues when shared across various platforms, devices and environments. Most instrument cameras can create TIFF files, however not all DSLR cameras create TIFF files directly. The higher end models will.

The ideal file format for use in DSLR cameras is the RAW file format. The suffix of RAW file formats will vary across camera brands and might be NEF, CRW, or ARW, depending on the camera brand. These file formats save data in an uncompressed manner and are typically opened using a RAW file convertor after capture or using Adobe Lightroom®. RAW files are very useful and provide great advantages over other file types for scientific photography. Their architecture allows data to be easily pre-processed and saved which is very useful with challenging subjects and situations. This in turn creates files that are superior for data collection and preservation. RAW files accomplish this by using a mathematically enhanced structure that allows creation of a greater bit depth and color information maintained using the finite exposure range of the sensor. RAW files must be opened and read using what is called a pre-processor or plug-in called a RAW file convertor. This convertor allows the file to be pre-processed, revealing more than if the image was processed using Adobe Photoshop directly. RAW files will be slightly smaller in file size when compared to TIFF files. RAW files are ideal in science because they themselves are never altered and the data display changes accompany the file in an XMP or sidecar file. If a scientist photographer were ever asked to go backwards in time about subsequent image processing steps, a RAW file could be re-opened without image processing information by eliminating the side car file.

When using a sensor with a Bayer pattern, neighboring pixels determine how to create the color for all pixels, but when using a RAW file, the de-mosaicing process is more "cleanly" handled by the RAW file pre-processor. This leads to more precision and control of the photographic process but takes more time. The bit depth of RAW files is also typically greater. A JPEG or TIFF file is frequently 8 bits per channel at capture, creating 256 tones per pixel

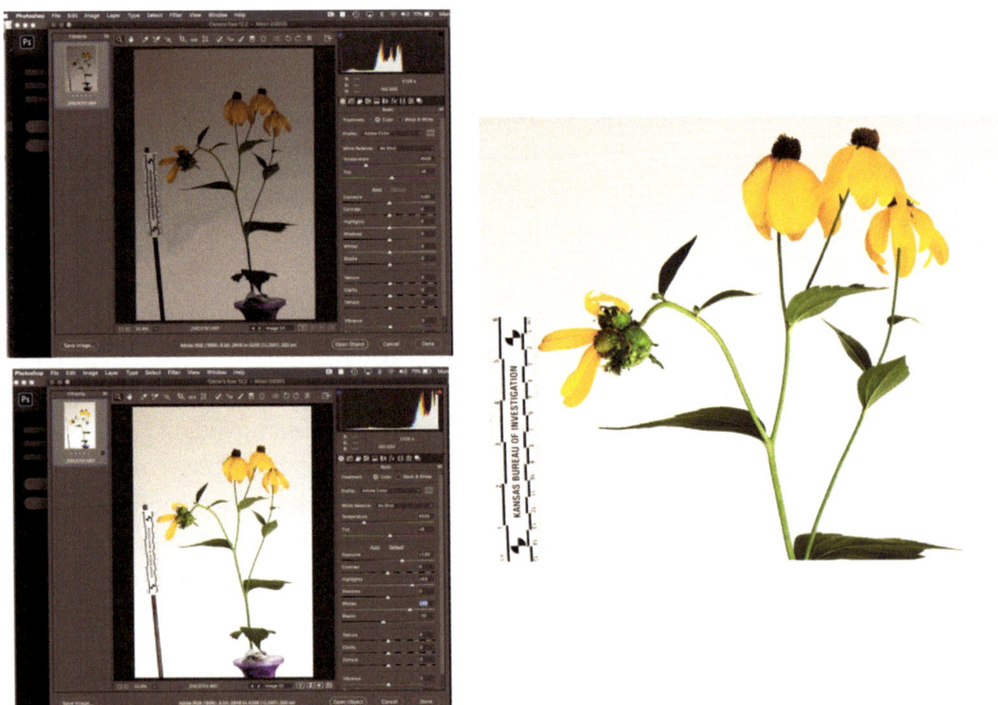

Figure 8.9 The dialog window of a RAW file convertor shown for the captured file and processed file. The Adobe RAW file convertor dialog box allows many image attributes to be optimized before opening in an image editing program. This pre-processing is beneficial to optimize the pictorial information display in a captured file prior to performing additional image processing. Adobe product screenshot(s) reprinted with permission from Adobe.

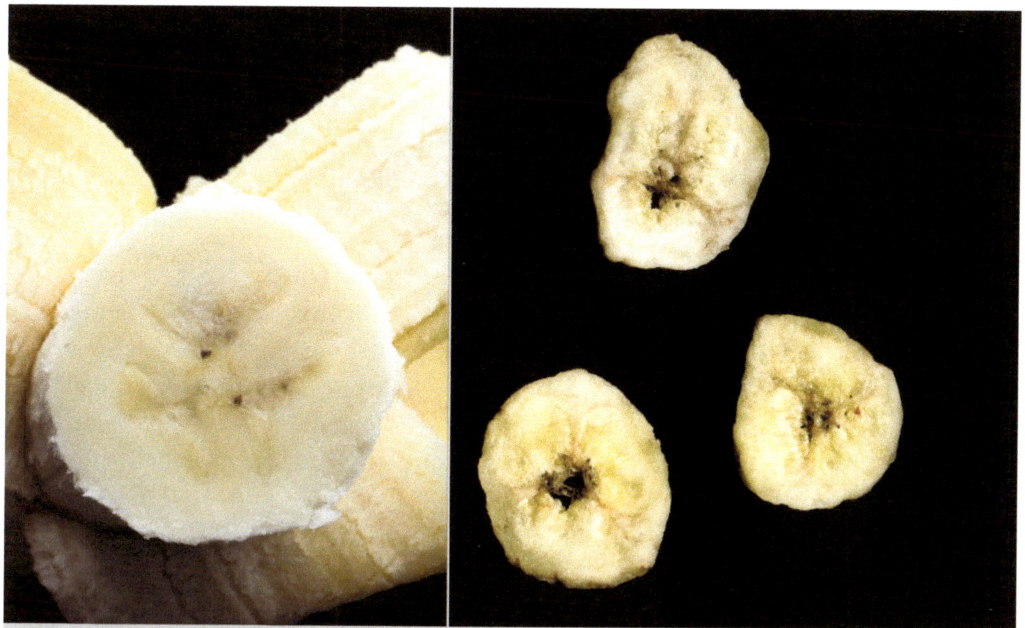

Figure 8.10 A banana might serve as an effective reference when considering file compression. A banana, when fresh, has a certain appearance and nutritional content that is rich in vitamins and minerals. When the banana is dehydrated and becomes a chip, it becomes lighter because of the loss of water, but still contains nearly all of the nutritional value. Attempting to rehydrate the chip will NOT allow the chip to become what it was as an object. The chip's physical structure will forever remain different. Similarly, compression can cause data loss or changes in a digital file by changing the file's data. Images courtesy of Michael R. Peres.

but it is common for a RAW file to contain 12 bits per channel, leading to more than 4000 RGB tones per pixel. RAW file formats are not typically found in instrument cameras.

Other File Formats

Most scientific images can be photographed by using TIFF, JPEG, and RAW files but there are other file types. More is shared about image processing file types such as PSD in Chapter 19. There are, however, a few other file formats that some cameras produce. BMP describes a bitmapped file format. It is frequently found within the Windows operating system environment. These files are uncompressed and create larger files than compressed file formats. PNG describes a compressed file format called portable network graphics. Its origins came from computer scientists as a free and open source choice that in many situations has replaced the popular GIF or graphic interchange format. PNG files support 8-bit images and can display up to 16 million colors. Because they use lossless compression, image processing can be accomplished without the creation of artifacts. PNG is particularly useful for web applications where progressive display modalities are used. PNG does an excellent job when text and photos are combined. GIF is an 8-bit format, however can display only 256 colors. Images that are not complex and do not require millions of colors can most benefit from a GIF file format. It is ideal for web use and images with large areas of white or common color.

GIF files are universal and can be opened in almost all image processing and display software. Frequently GIF files are used for animations and moving subjects. They have found their way into numerous social media applications.

In-Camera Digital Filters

Scientific photographers must consider the appropriateness of having a sample's image attributes maintained, enhanced, or reduced at capture. These choices occur when selecting for use in-camera filters or image modifications. The most common of these "at-capture" attributes would include color reproduction, noise, and sharpening. There are various reasons why an operator would choose, or not choose, to use any of these filters. If JPEG files are created at capture, performing sharpening can be helpful or enhance capture artifacts but if using RAW, the best practice is to de-select sharpening, noise reduction, and color enhancement at capture.

Sharpening

Almost all camera sensors have an IR cutoff and anti-aliasing filters on top of the Bayer filter on a sensor. These filters are required to manage how the certain subjects are recorded without making artifacts. There are a number of cameras that no longer use anti-aliasing filters. While these filters are only a few microns in thickness, they adversely affect critical image sharpness. In-camera sharpening is an image-processing algorithm sometimes chosen because like all sharpening tools, it increases the localized pixel contrast that makes things appear to be more sharp. This localized increased image contrast creates an image that appears to be more delineated. Having files that are electronically enhanced sometimes produces some digital noise. Over time and when future image processing is undertaken on files with digital artifacts, edge defects can actually degrade the appearance of image detail. The initially captured image will typically exhibit a smoother gradient of tonal data per pixel without sharpening but when sharpness is boosted at capture, images can appear exaggerated. More importantly, if the images are to be used to create a computational or derivative picture, sharpening should NEVER be selected at capture. Creating files that have edge enhancements at capture when blended can appear over processed or embellished. Perceptually, viewers will respond more favorably to images that are sharp and crisp. Over-sharpened images will appear image processed and may not be well received for a number of subjective and individual reasons.

Different cameras provide operators with various levels of sharpening choices at capture. A sharpening filter can be applied using auto-mode or by selecting the amount of sharpening that is desired. It can also be turned off and provide no sharpening at all.

Color

Similar to sharpening, an image's color saturation can be managed at capture in some cameras. Since viewers have expectations about what colors are believable, choosing how to record the correct color saturation without embellishment is an important decision a photographer makes. Cameras may have a variety of commands to control the amount of color that is captured. An example of these settings could be standard, neutral, vivid, monochrome, portrait, landscape, or flat.

The choice of how to record color is sample dependent and allows an image to have more or less color data. Decades ago, the Kodak Research Labs performed exhaustive testing to evaluate what consumers desired in color reproduction when using Kodak color negative films. After years of conducting surveys, they learned people wanted to see color that did not offend them. So that being shared, individual preferences play an important role in choosing to improve color saturation at capture or not. In science, not enhancing color at capture should be the basic starting point and the choice to add more color would be selected when sample visibility challenges exist.

Noise Reduction

There are reasons why noise reduction should be selected for use, but it should be used carefully. Noise filters actually soften images. This image softening solution creates images that will appear less crisp which might be the opposite outcome of what is expected. In general noise reduction, an image is softened while sharpening will increase contrast making images appear sharper. Scientific photographers may have tight tolerances for the amount of noise reduction or sharpening that are selected. Noise reduction is accomplished by actually blurring the image data to varying amounts. The amount of blurring blends image defects that are caused by pixels exposed by heat but not by photons. There are many blurring filters that can be selected in post-production.

Various cameras will have different noise reduction settings. These settings could be off, low, normal, and often a camera's default, high. These choices allow the operator flexibility in managing the compromise between subject detail and noise levels in a JPEG file for example. Many cameras do not share information at what ISO or at what long exposure times noise will be produced. There are a number of ways to reduce noise using Photoshop, and a popular noise reduction, Topaz®.

Digital Artifacts – Aliasing

Low and medium resolution computer imaging systems normally create and display images with smooth edges. When the frequency structure of an object matches the pixel resolution correctly, no aliasing occurs. When a sample's fine detail is greater than the number of pixels, aliasing can occur. Aliasing is evidenced in an image as jagged edges or banding. Aliasing is often a result of under sampling and is created when the detail exceeds the rendering ability of the sensor. I am sure you have seen images where clothing exhibits strange moiré patterns because the frequency of the fine patterns and details of the object are more detailed than the sensor can resolve. Aliasing can be the result of spatial or frequency sampling. In either outcome, it is not desirable, and produces images that do not appear correct. It is difficult to predict with certainty when image aliasing will occur. Anti-aliasing filters are located on sensors to help manage this image artifact. Unfortunately these filters make images softer at capture.

Connecting Devices

Cameras, scanners, card readers, and any number of peripherals including flash drives, need to communicate with a CPU. Almost universal at this time is the Universal Serial Bus (USB). USB has become an industry standard initially developed in 1994. At that time, there were all types of cables, connectors and communication protocols used for connection to power supplies, between computers, and for use with other devices. Users demanded a standardized connection for all things connected to imaging devices. This included digital cameras, card readers, printers and other devices such as monitor calibration tools. There were many connection technologies, including FireWire and serial ports. In 2000, USB 2 was released. USB 1.1 was capable of transfer speeds up to 12 Mbit/sec while USB 2 was able to transfer 480Mbits/sec. In 2008, USB 3 was released and could transfer data at up to 5Gbit/sec. In 2016, USB-Type C found its way into computers and other devices and is the most common connection technology. Today many devices communicate wirelessly by using Bluetooth. A physical connection will always be significantly faster using USB.

Memory Cards

In some cameras digital image files are stored in the camera's buffer camera memory, which is a temporary location before being moved to a memory card or to the CPU. Some cameras cannot record an image without a memory card in the camera and others may operate with or without a card. A memory card is capable of reading and writing both when located in a camera or when located in a card reader attached to the CPU. Different names have been given to memory cards including Smartmedia®, memory sticks, or compact flash. Each of these devices is a different storage technology. Differences are evident is size and speed. Many cameras can take both compact flash or SD or XD Media. Media continues to become smaller in physical size and faster. It is not uncommon to see 32 or 64GB cards that have speeds of 32X. Some important camera companies including Sony and Nikon are now using XQD media. The two main benefits of the new XQD format are its extremely fast read/write speeds (current and proposed for the future) and high capacity storage limit. The downside is its cost, which is causing slower adoption in the market.

Chapter 9 Lighting for Natural Science Photography

Light and Lighting

Technical considerations about light were covered in Chapter 7. Brightness, color temperature, and other characteristics of energy travel were shared using a theoretical point view. Creating effective lighting for natural science photography requires more than simply having a working technical knowledge about light and its related behaviors.

Creating effective lighting for scientific photographic documentation does not have to be overly complicated. Selecting and using basic equipment in conjunction with some practical strategies can lead to surprisingly good outcomes. Unfortunately and frequently, artificially lit photography looks amateurish. Often an on-camera flash is used and the results are simply awful. Short working distances often associated with close-up photography, for example, in combination with electronic flash lighting leads to very direct and harsh (contrast) lighting. Pictures illuminated with direct flash at short working distances will display very "white" whites with NO detail, and black harsh shadows also without detail. Ambient lighting can also play a role in creating "not so effective" lighting. Bad lighting simply creates bad pictures. Good lighting makes the characteristics of subjects visible and helps make the nearly invisible visible. That outcome is the ultimate goal of scientific photography.

For science images, the use of artificial lighting needs to be carefully considered. Many practical suggestions about how to modify light or how to make the light work for a specific subject will be discussed and illustrated in this chapter. In many of the book's other chapters, approaches used for specific applications are covered in greater detail.

Science photographs need first and foremost to share scientific data in highly precise ways. Science images need to exhibit a neutral point of view and the use of good lighting should not be a random accident or an afterthought. Effective lighting should not embellish, amplify, or distort characteristics of a subject, but rather should reveal things about the subject that otherwise might not be visible. The lighting should never create exaggerated perceptions about subjects or characteristics. Science pictures must create science facts and not science fiction.

Sometimes creating factual pictures can be challenging because – unfortunately for science – photography is an interpretive process. Just like the focal length of a lens can affect the shape of an object and the aperture will influence the range of focus in a picture, lighting will define and reveal features of an object or not. Lighting can change the emphasis and make common subjects look dramatic. It can distort the relationship of the elements contained within the frame or the lighting can make the subject appear normal as we would expect to see it.

Where the light is shone onto a subject, or how the quality of the light is modified to create an emphasis is part of the process. It is useful to consider how adjectives are used in writing and how similar they are to lighting, which can add emphasis. Adjectives and light are descriptors. The characteristics of a sample can be diminished or amplified by

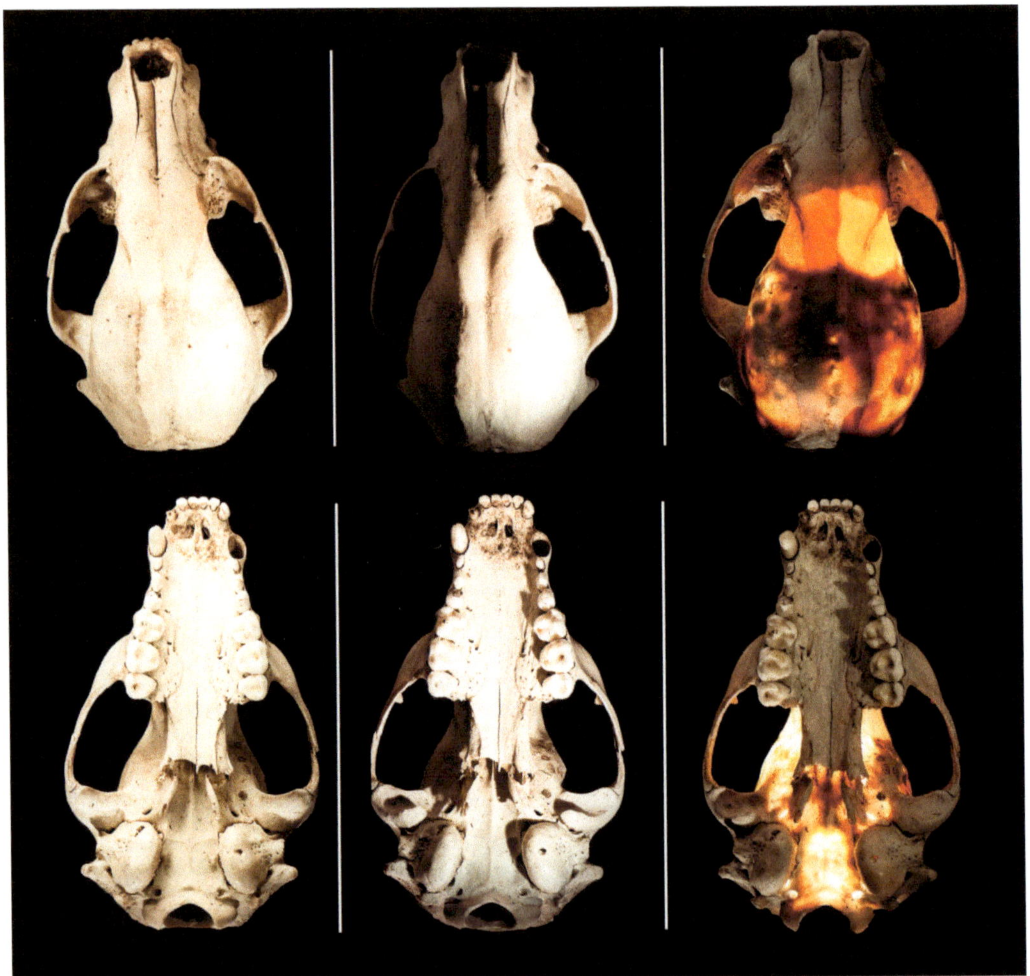

Figure 9.1 This skull demonstrates how appearances can change when using different lighting. The images on the left are examples of flat light that comes from the front. The middle images demonstrate the influences of directional (side) or medium contrast lighting with some fill lighting. The images on the right use primarily backlighting. In each view, the skull reveals different characteristics. Which way is best? It depends. Images courtesy of Michael R. Peres.

a direct consequence of where and how a light is shone onto, or at a subject. Producing effective lighting is a process that starts with an analysis of the characteristics of the subject, deciding what needs to be revealed, and then creating lighting that accomplishes that goal.

The subject itself plays a foundational role in the development of effective lighting. A subject can be opaque such as a rock, or shiny like a coin. Subjects can be translucent like a fish embryo or really just about anything. Subjects can also be relatively flat or possess textures or topographical features. So, the sample's surface characteristics or natural visibility will dictate where and how to light. The subject's dimensionality or its reflectivity will influence how diffuse (broad) or how direct the light should be. There are numerous variables to consider.

Global methods for lighting and modifying a light used for artificial lighting will be explored in this chapter. Various types of lights will also be covered. When discussing artificial lighting, it might be practical to break the subject into small tabletop set-ups for small things, and approaches needed for larger subjects.

When creating good lighting, typically two lights should be used, or a least a main (key) light with a fill light or card when possible. The main light influences what is featured, and the fill light determines how much contrast is created by the lighting. Using a small and direct light will create very black and sharply defined shadows and reveal texture. Using a diffuse and possibly large light, will emphasize color and produce softer shadows.

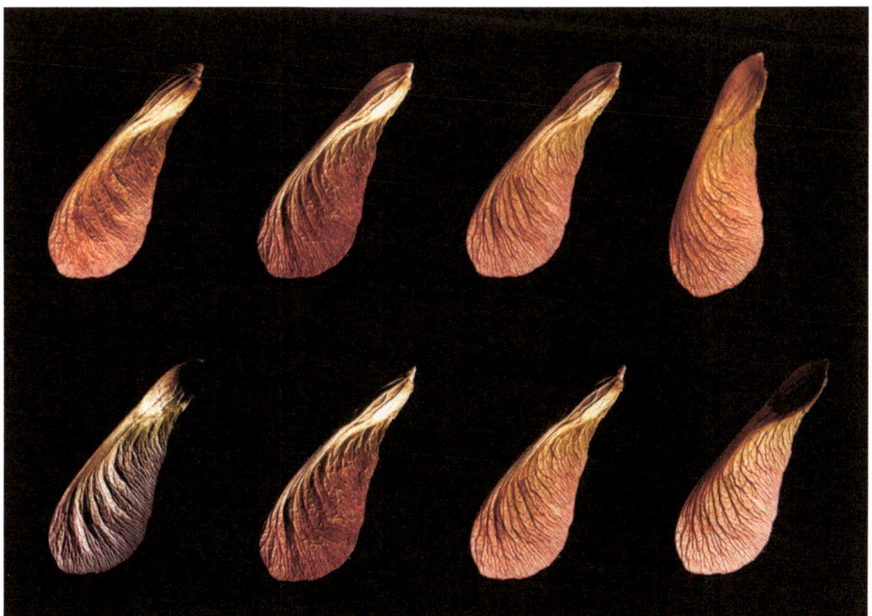

Figure 9.2 Light might be considered similar to adjectives we use in language. Lighting allows various features of a subject to be revealed and create an emphasis. In this composite photograph, the seed was photographed in eight different ways. Images courtesy of Michael R. Peres.

Tabletop Lighting

When making tabletop pictures, smaller lights are useful. There are four common types of light sources that might be considered for use. Continuous lighting equipment includes tungsten, LED, fluorescent, and HMI types. HMI lights (Hydrargyrum Medium-Arc Iodide) are arc lamps and are very bright and expensive. They are popular for video productions and have advantages that include being up to five times as efficient as incandescent bulbs. Because they use less power, they also produce less heat, and have a color temperature of approximately 5600 which is near to daylight.

For short duration lighting, electronic flash equipment is desirable. Small strobe lights can be useful for location photography and large powerful electronic flash lighting equipment is typically used in studios with larger subjects.

Types of Lights for Small Subjects

LED light cubes and fiber optic lights are special lights and are excellent for small subjects. An important part of a fiber optic light is its light guide(s). Fiber optic lights deliver light to small and precise locations using a light guide. This is accomplished using glass fibers. A light guide is composed of hundreds – if not thousands – of microscopic glass fibers that begin at one end of the light guide where the light enters the guide and terminate where the light departs the guide at the other end. They are pipes for light. The fibers, on average, are slightly larger than a human hair. They are fairly robust but remain vulnerable to breakage when used roughly. These fibers cannot be bent into a 90 degree angle or dropped on hard concrete floors. Dropping a light guide will certainly break fibers. As fibers become damaged, their ability to transmit light will become diminished, and the light's brightness will go down.

Achieving an even and homogenous light, useful as a white background can be accomplished using a light box or by other approaches. Even and neutral light is frequently a key element to creating images with scientific veracity. When using multiple lights, it is important to achieve equal brightness across the entire region that will be located in the background. Using diffusion materials can be useful to spread the light evenly across the background. Care should be taken to minimize the spreading of the light used for the background from falling onto a sample. Making a white background can also be accomplished using a light box or by using a laptop's screen to create a uniform white background. Small objects can be located in front of the computer screen.

A fiber optic light source itself contains a light housing, sometimes called a lamp box, and contains a very bright light. The light is directed to the fibers at the light guide port. The light's brightness is adjustable. Fiber optic lights can be outfitted with a variety of light guides and are sold in a range of models ranging from inexpensive to very expensive. The cost for a unit will be influenced by how robust it is and whether the brightness can be reduced by voltage or through the use of a graduated neutral density filter (tungsten halogen models). LED light models use a pre-stepped brightness range controller. More expensive units will also come equipped with fans that are dampened for vibration and may have the ability to

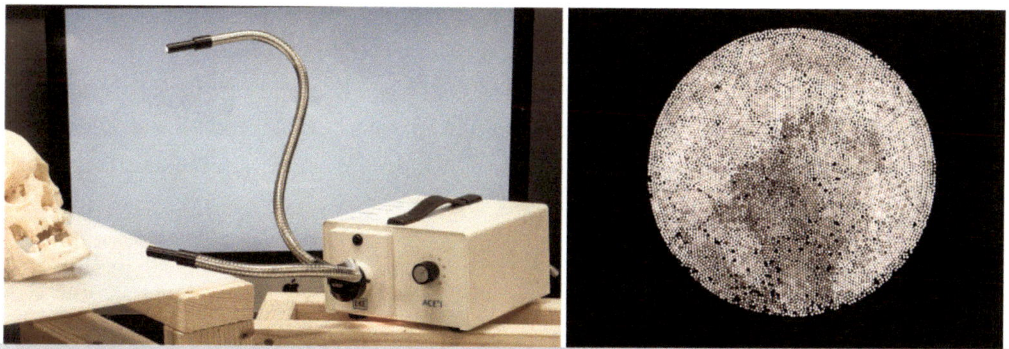

Figure 9.3 A fiber optic light is a very important tool for scientist photographers. It can produce highly precise lighting that can be delivered to specific locations on a sample and easily modified. The nearly invisible microscopic fibers carry light and behave much like a pipe. When the fibers break though, they do not transmit light. In this photograph, an illuminator on the left is equipped with a bifurcated gooseneck guide, and on the right is a close-up photograph of the light guide revealing broken fibers evidenced as black dots. The white dots represent fibers that carry 100 percent of the available light. The diameter of this light guide was 5mm. The gray dots represent fibers that are having problems. Images courtesy of Michael R. Peres.

use brighter bulbs. The heat and air currents from a lamp and its housing can affect delicate biological subjects by causing them to move in air currents from the fan or worse yet, dehydrate them when illuminated. Illuminators with robust fans will not be ideal for photography because of the heightened vibration they create that can affect image sharpness.

It is easy to check a light guide for damage by first removing it from the lamp housing. Point one end of the guide towards a light and place a finger also on that end. By moving the finger, it is possible to peer into the guide to determine the percent of fibers that are not transmitting light. Broken fibers will appear as black dots within the guide and fibers that are transmitting will be visible as white dots.

Fiber optic light guides come in a variety of types and sizes. They are unique to specific lamp housings and there are various types. There are a variety of common light guide styles and a single or bifurcated rigid gooseneck is by far the most widely used. The light guides can be encased in rigid coaxial cable or flexible casing, which allows the light to be positioned. Rigid guides that allow guide positioning can be very practical. The following is a short list of types of light guides. Each light guide has typically been optimized for a specific application.

- Straight and single fiber optic light guides
- Dual branch bifurcated fiber optic light guide
- Multiple gooseneck light guides
- Quartz light guides
- Ring lights
- Line lights
- Back lights

A ring light and line light can be useful for lighting when the working distances are small and access to a sample can be a challenge. Ring lights make omni-directional lighting that

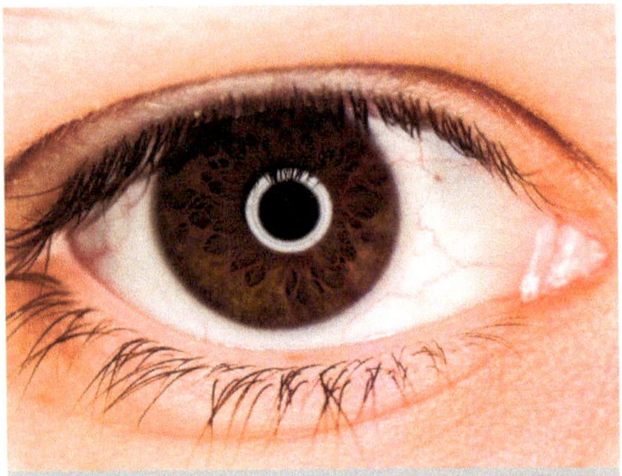

Figure 9.4 Shiny objects present unique challenges for photography. The use of a ring light will create a circular reflection of the light on a shiny subject. The reflection sometimes is a significant distraction. In this image of an eye, the ring light's circular reflection is noticeable and evident. Image courtesy of RIT Biomedical Photographic Communications.

has low contrast. Ring lights are useful for non-shiny objects where color and textural surface information is the subject. When used with reflective objects, a circular reflection of the light will be evident in the photograph. Line lights are great for showing texture when placed at a low angle to the subject's surface. This light will not produce as much contrast as a direct or point light.

Small LED light panels and light cubes are also an excellent and contemporary choice for tabletop lighting. They come in a variety of sizes including small squares. These versatile lights offer some excellent features, including wireless controls using Bluetooth apps, adjustable light levels from off to 100 percent, waterproof and durable housings, daylight color temperature, a 60 degree beam angle, 750 LUX, and 1/4" 20 screw built in for tripod use.

Electronic Flash Equipment

Small or large electronic flash equipment produces discharges that are frequently between 1/1000th and 1/200th of a second. Electronic flash light has a color temperature of approximately 5500Kelvin. Because its light is short duration, using flash lighting is useful when photographing moving objects. See Figure 16.17.

Flashes are routinely integrated into many cameras including smartphones and DSLRs. Some DSLR cameras have an accessory bracket called a hot shoe for mounting small flash units onto the camera. Large studio electronic flash equipment is significantly larger and has much more power. It is sometimes unwieldy to take on location. All studio flash equipment will have lighting heads with flash tubes and are operated and connected to a powerpack. Or they can be a mono-block type where the power and flash tube is a self-contained unit ideal for location work. Discharge of the electronic flash lighting must be synchronized with the

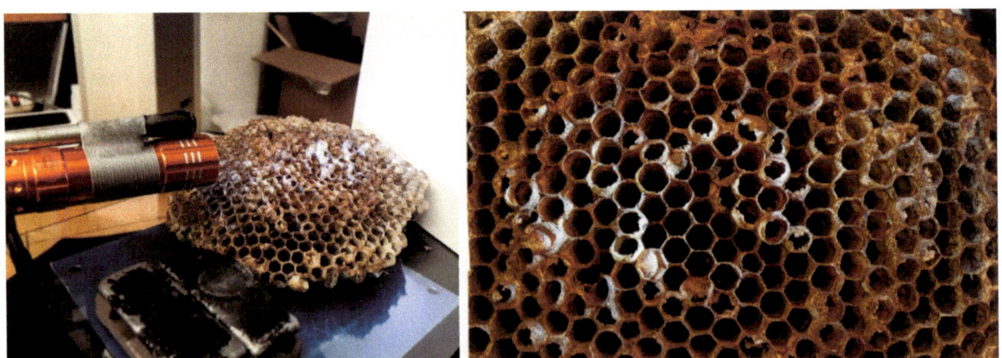

Figure 9.5 Using small lights can be helpful when photographing small objects. There are a number of specialized lights including Lume Cube®, however, using an LED flashlight can also be effective. A flashlight's size and portability make it unique and it can easily be adapted for close-up work. Image courtesy of Michael R. Peres.

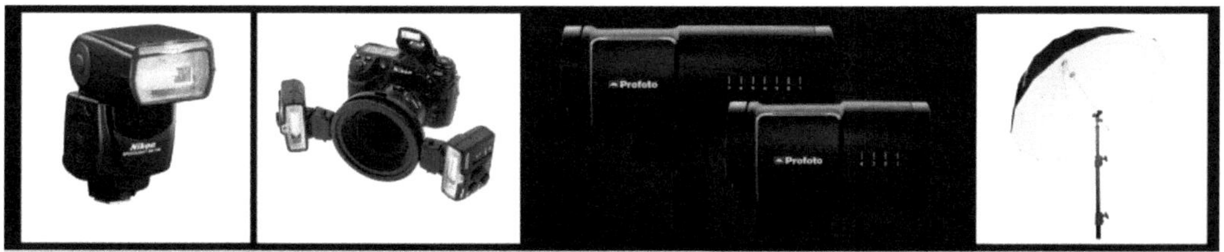

Figure 9.6 In this composite photo, various electronic flash equipment is featured. On the far left is a Nikon Speedlite, an on-camera flash, second from the left, a Nikon Macro flash ring. The third photo from the left features two Profoto B-10 studio flashes. The B-10 PLUS is the larger unit, and on the far right is a Westcott white umbrella used for light modification. Images courtesy of Nikon USA, Profoto, and Westcott, Inc.

camera using a cable, signal, or can be set off using a light trigger called a slave. Flash units are characterized by their power. For small portable units, power is shared as a GN or guide number. Lower numbers such as 60, for example, represent lower power sources and 160 would be associated with more powerful units. Studio equipment is characterized in watt seconds or joule ratings. 500 watt/second is not very powerful when compared to units that might be 3200 watt/second. The more powerful the system, the more the cost, but typically there will be more features such as stroboscopic and burst.

Modifying Light

Good lighting is the result of knowledge, skills, good samples, and using the right type of light sources and modifiers. Depending on the subject, effective lighting can be pretty easy or challenging to achieve. The KIS method – "keep it simple" – is best whenever possible. It is easy to make things complicated and the lighting should operate in the photograph without becoming the subject of the photograph. It is practical and often the best way to use only one light if possible. The physical size of the light relative to the sample's size is important when making artificial light photographs. Small light sources make more contrast and directionality. The larger the light source, the less contrast it will create. It is recommended to use a light that is proportionally sized to an object's size. As a rule of thumb, small subjects benefit from using small lights and larger subjects benefit from larger lights.

The term "quality of light" describes how the light achieves, or does not achieve, effective lighting on a particular subject. Quality refers to the characteristics of the light. Direct or diffused illumination describes a quality of light. Direct light is very harsh and creates hard shadows but also emphasizes texture. Harsh light is created from a direct and undiffused source. It can come from the small on-camera flash or from a fiber optic light shined onto a subject without diffusion. The size of the light has a direct relationship to the harshness of the light. The quality of light is one feature of light and another is the directionality.

When making lighting decisions, it is important to consider what is important to reveal and then how the light itself will interact with the sample, or surface of the sample. Color, texture, shape, topography, cracks, or other characteristics of the sample that are important can be made visible using lighting. Lighting can be shone from above a sample, and is called

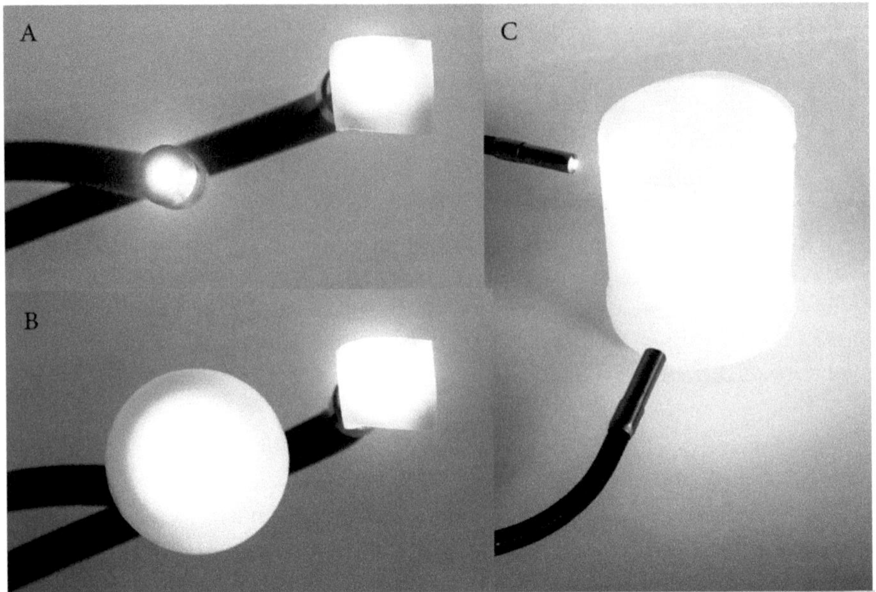

Figure 9.7 This composite photograph shows how to increase the size of a fiber optic light guide. In photo (A) a Scotch® tape loop has been added. In (B) a ping pong ball and a tape loop were added to the light guide, and in (C) a milky plastic liquid bottle was employed as a large and excellent diffusor. Images courtesy of Michael R. Peres.

reflected light, and light that is shone from below translucent or transparent samples is called transmitted light. Some samples will require both transmitted and reflected light to show surface and internal structure characteristics.

Making White and Neutral Backgrounds

Subjects can sometimes be transparent and benefit from using transmitted light. Producing uniform transmitted light is not overly complicated but there can be challenges when smaller lights are used. A light box or light panel – if available – will be the most effective solution for making even lighting. Masking the size of the illumination field can be helpful. It is also possible to evenly shine lights from two sides onto a white board. Using ping pong balls on the ends of the fiber optic light guides can be helpful to spread light out more evenly. Creating uniformity across the entire field of view is very important. If the light is not uniform, false data or image artifacts might be created in the image. While the background is not the subject in science pictures, it will play a role in the image's chance of success. A well-managed background is a fundamental expectation in science pictures. Irregular lighting and the formation of shadows will lead to random regions of darkness or lightness in the background and will compete with the sample for attention.

Making Contrast

Contrast is an important element of a photograph. Increased contrast will make pictures look sharper even though they may not be. Creating more contrast can be accomplished in several ways. One of the easiest ways to increase contrast is to move the light further from the subject. While the brightness will be diminished, the size of the light will become proportionately smaller relative to the object. Smaller lights create more contrast on the subject. The location of the light relative to the object can also modify the contrast. Having a light proportionally sized that is brought closer to the subject will make less contrast even when located at a 90 degree angle to the surface of the subject. This technique might be called edge or texture lighting. If light is raked across the surface of an object it makes more surface contrast than if the light comes from the same axis as the point of view of the camera. A small light used from an acute angle, will make strong raking light and produce shadows that will be very black. The reason to use side lighting would be to show surface structures or texture.

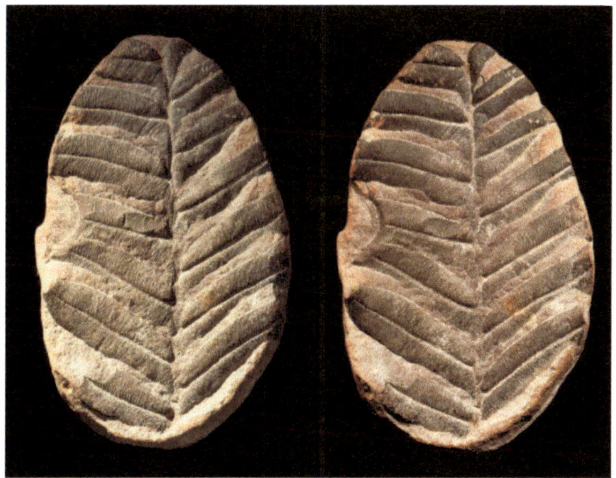

Figure 9.8 Flat and non-directional light reveals colors more effectively than side lighting. Light that is diffuse is of low contrast and is shown in the image on the right. Light that comes from the side creates texture lighting. This is shown in the image on the left. Images courtesy of Michael R. Peres.

Light Ratios

The brightness ratio light when considering the brightest regions of a photograph when compared to the darkest shadows of the scene is called the luminance ratio. When light is added to the shadow side of the scene, this light is called fill light. Adding fill light will lower the brightness ratio and image contrast. The brightest light used in a photograph typically comes from the main light. It is the dominant light for the photograph and using other lights adds fill light to control contrast and make shadow detail more visible. The main and fill lights have a ratio to one another. A fill light is most frequently located at or near to the camera's point of view, but not always. A lighting ratio of 1:2 means the brightest part of the picture will be 1 *f*-stop brighter than the darkest region. A ratio of 1:3 will make the brightest part of the scene 2 stops brighter and so on. The higher the lighting ratio, the higher the contrast will be visible in the image. A lighting ratio of 1:1 describes lighting with an equal brightness across the entire field of view. Digital cameras can typically reproduce a 6+ *f*-stop range from the darkest tone to the brightest tone without being overly taxed. Once the brightness range exceeds this range, good detail cannot be recorded in both the darkest and brightest regions simultaneously without making small adjustments. HDR or high dynamic range methods are required for creating images with details in both the brightest and darkest parts of the picture when the ratio is too great for a single exposure. See Figure 12.6.

Fill light can be added using light from a second light of similar color temperature or by using a small white card, or other reflective surface to reflect light back onto the subject. The brightness of the fill is based on how near or far from the subject the card is placed. By varying the angle of the reflector to the camera and the distance to the subject, the amount of brightness reflected into the shadow can be adjusted. Fill cards for small objects can be held in place using many ways including simply holding it, using modeling clay, a third arm (a soldering tool accessory), or a lab stand with black tape. I sometimes use aluminum foil as well.

Reducing Contrast

To reduce the contrast of a light itself, the overall size of the light must be made larger. Lights contain both bulbs and sometimes accessories such as reflectors affecting the size of the light. This in turn affects the softness or harshness of the light. Some lights may also use a convex-like mirror behind a lamp making it focusable. Moving the light closer to a subject can make the light larger proportionally to the sample. Changing the light's location will have limited effectiveness but should be considered as a possible first step. Moving the light closer to a subject will increase the light's brightness and a shorter exposure will be required. A better approach is to broaden the light by shining the light through a piece of diffusion material, or by shining the light onto a semi-reflective surface such as piece of white cardboard of appropriate size located near to the sample. The cardboard will reflect the light back onto the subject. Making the light source larger will reduce contrast. Shining a low contrast light using a low lighting angle to the subject's surface is effective in revealing a sample's surface characteristics and maintaining details in both the highlight and shadow regions of the scene.

A favorite trick for enlarging the surface of fiber optic light guides is by using 3M Magic® tape; or using white plastic bottles originally used for milk or medicines. See Figure 9.7. Ping pong balls are also excellent light diffusers. It is useful to poke a small hole into the ball where the logo has been printed. Keep the hole smaller than the diameter of the light guide. Once the hole has been established, the ball can be slid over the end of the light guide. Relative to an object's size, this approach can produce excellent and well-managed contrast and uniformity of the light. Highlights or bright reflections can be easily managed. This method is practical for semi-reflective metal objects such as coins, circuits, and many other things. The quality of the light's contrast when using this method will be affected by the subject's size, surface characteristics, the sample's texture, and the placement of the light. A lights to subject distance will always play a role in a lighting outcome.

Shining light through a white milky plastic bottle is effective for making a broadly diffused light on small objects. There are commercial products that can be purchased but I find homemade light modifiers work just as well. It is best to cut the bottom and top of the bottle off leaving only the cylinder of the vessel for use as the diffusion material. The height of the tube will be determined by the focal length of the lens. Once positioned over the subject, the lights can be positioned at short distances away from the diffusion cylinder. This type of lighting behaves much like a light tent and subjects will be surrounded and bathed in diffuse and 360-degree light.

Axial Lighting

Axial lighting is a unique type of light that makes pictures appear as though the light is coming directly from the camera's lens itself and is on axis with the camera. To achieve axial illumination, a beam splitter is located at a 45 degree angle to the lens between the lens and the subject. When using axial lighting, 50 percent of the light that is shone onto the beam splitter will be reflected down to the subject and 50 percent will be transmitted

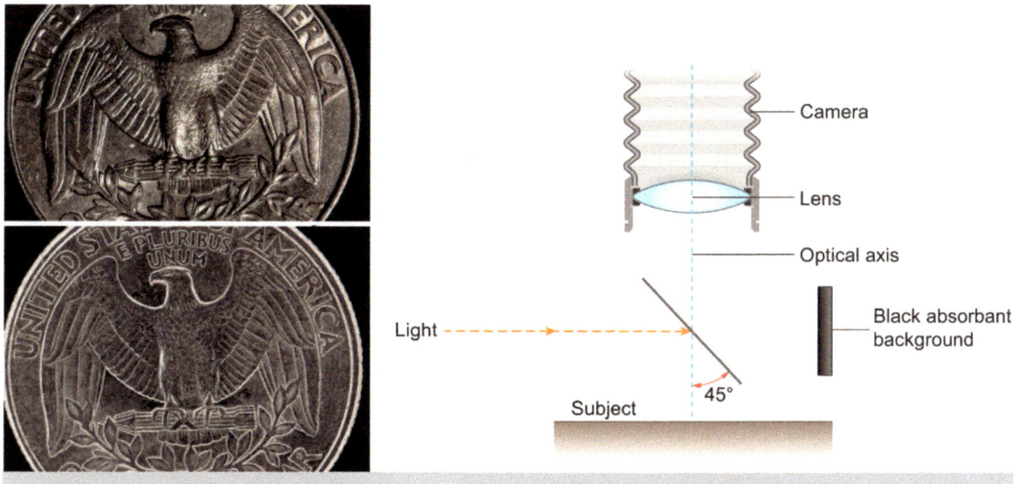

Figure 9.9 Axial and diffused lighting can create very different results. In this composite, the top left image of a U.S. quarter was illuminated using axial lighting. The bottom left image was created using diffused lighting. More surface characteristics are revealed in the axial and higher contrast lighting. The image's structure exhibits more crispness and the image appears to have some relief when compared to the flatter and more diffused non-directional lighting used for the bottom image. Image courtesy of Torey Miller.

through the glass and away from the subject. Axial illumination produces lighting that is almost without shadows and exhibits a high degree of contrast on the surface of the subject. It is very effective for shiny objects such as coins and circuits. It reveals surface texture and structure without the creation of directional shadows that obscure detail.

Axial lighting can be tricky to set up. I frequently use pre-cleaned microscope slides or cover slips, depending on the size of the subject, as a beam splitter. I also use a small fiber optic light guide for this work. The beam splitter is located above the sample and positioned at a 45 degree angle to the surface of the subject. The camera lens is located directly above the beam splitter. There are no photographic challenges when using axial lighting but there are a few things to keep in mind about the lens and aperture settings. Because of where the light comes from (the same pathway in which the image is formed), it is crucial to photograph using a nearly wide-open aperture. When an aperture is closed down, it increases image DOF. If the DOF of the system is too large, the "light" reflecting off the beam splitter can become defined and create flare in the system. The diameter of the lighting coverage is also affected negatively when a small aperture is selected. The circle of illumination will get smaller as the aperture is closed. Never use autofocus for technical photographic applications, especially in axial lighting set-ups. It is also very important to ensure the beam splitter's alignment achieves the same height across the sample from left to right. It can be helpful to locate a small piece of black material on the side opposite of the light itself. The surface of the beam splitter that is away from the light may reflect the ceiling or other things back into the lens. It is also better to work with the room lights dimmed or off.

Laboratory Glassware

Glass presents interesting problems for photography because it has 360 degrees of reflective surfaces. Glass can be colored or neutral. It can contain liquids or be empty. Because of the importance of controlling specular reflections, it can be very useful to broaden a light or to light from behind the subject. As previously discussed, broadening a light can be accomplished using a white card. That remains an excellent strategy to consider, however, it will create a large and soft white reflection on the surface of the glassware. A better way to light glass is using white line and dark line lighting. When making dark line lighting, first place a white or light gray background behind the subject. It does not need to be huge but simply should fill the camera's viewfinder. Place the glassware on a sheet of glass that is elevated and above the table. Allowing light to enter the glassware from the bottom will produce better lighting. Once composed and focused, shine a small spot of light directly behind the subject. This will make the subject appear backlit and display dark edges. It is important to attempt to keep the raw lighting from striking the glassware. Place the light in the front of the camera and use light baffles or cardboard to block any direct light from striking the glassware.

There is also "darkfield" lighting used for glassware, also called white line glassware lighting. It is useful to reveal glass inclusions and edge definition. In white line lighting, a piece of black material is located behind the subject and needs to be large enough to cover the field of view of the camera's lens. Two lights located outside out the field of view are shone through the subject from behind using an oblique view to the camera lens. It is best to locate pieces of black cardboard or other black material to act as a curtain or baffle to the lights. This creates slits of light that come from behind the glass subject. Light coming from this direction will become trapped within the vessel and cause the edges of the glass to glow. When permissible, it is important to clean the glassware carefully using lint-free materials. It might also be advisable to use a dusting brush as a final treatment prior to depressing the shutter. Fingerprints and other particulate material can quickly become the focal point of a photograph if prominent in the composition and made visible by this type of lighting.

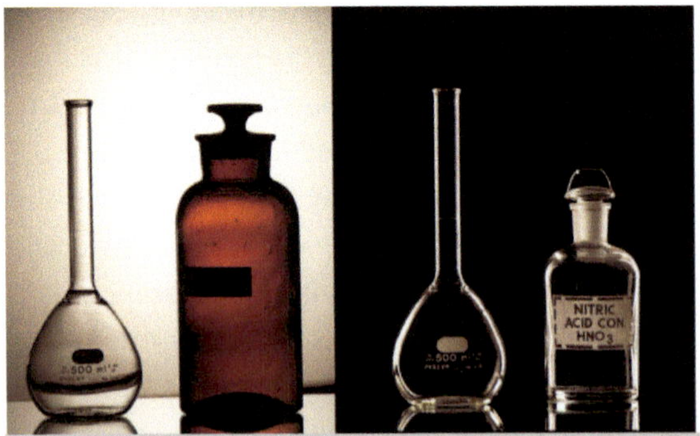

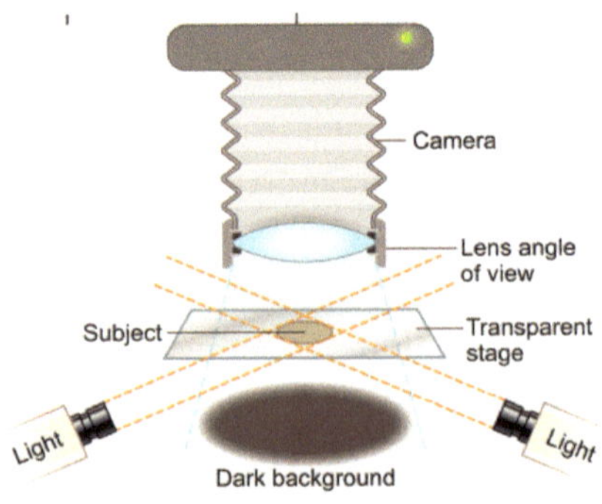

Figure 9.10 Laboratory glassware can be a challenge to light. Managing stray reflections is one of the primary challenges. The glassware image on the left used dark line glassware lighting and the image in the center used white line lighting. White line lighting is similar to darkfield lighting used in microscopy. The illustration (right) shows how to establish darkfield lighting. Images courtesy of Michael R. Peres.

Metal and Tent Lighting

Shiny objects like glassware can be difficult to light. Similar to glass, highly polished metal or mirrors have entirely reflective surfaces reflecting anything that is in front of them. A shiny surface can be a very tough subject to light and photograph for sure. Lights, cameras, photographers, and anything else will be reflected back to the camera. Depending on the size or prominence of the reflections, they can be terribly distracting and certainly are not a positive element in the photograph.

Better lighting can be achieved by making a very broad light, which also will reflect itself in the subject, but will be more pleasing and less random across the highly reflective surface. Sometimes, a better approach is to light the mirror-like surface using what is called a tent. A tent used for photography is almost always going to be white. A large and semi-rigid system is ideal. White cardboard, for example, or something such as a white shower curtain or liner might work equally well. The subject's size will influence how large a tent is needed. A light or lights can then be shone through the inside of the tent, which will be reflected in the highly polished and reflective surfaces of the subject. On occasion, it can be effective to locate rectangles of black paper or other appropriate materials to create desired reflections in strategic places. These other tones may be useful to break up the possibly monotone tent reflections. Black reflection may also create the perception of shape in the subject. See Figure 14.5.

Immersion

Wet samples can often benefit by being photographed in liquid. The objective of keeping a sample wet is useful for managing surface specular reflections. When a light is shone onto a wet object, the various topographical structures will produce many and irregular reflections. These reflections will be random and may obscure important details. It is possible to place a sample underwater or other appropriate liquid and manage these reflections using one light.

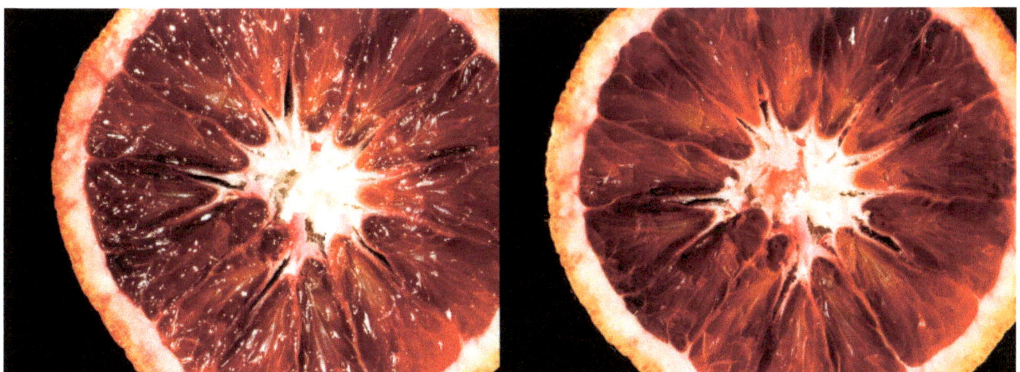

Figure 9.11 A blood orange has a wet surface and is highly reflective. The image on the left reveals countless specular reflections created by the fruit's surface. "Speculars" are the by-product of a wet surface. In the image on the right, the orange was submerged in a water bath and the specular reflections have disappeared. Images courtesy of Michael R. Peres.

Distilled water is a good choice since this water has been de-oxygenated. It may be necessary to hold the sample below the surface of the liquid using some type of a weight. Many objects will float. Using one small light, the surface of the subject can be illuminated without the presence of any specular highlights to compete with any surface details. Immersion management and lighting will create low contrast lighting. It is useful to add contrast back into the picture during the image processing.

Suggested Reading

Basics Photography: Lighting 02, by David Prakel. Bloomsbury Academia, 2007.

Stoppees' Guide to Photography and Light: What Digital Photographers, Illustrators, and Creative Professionals Must Know, by Brian and Janet Stoppee. Taylor & Francis, 2012.

Studio & Commercial Photography (Kodak Pro Workshop Series), by Jack Reznicki. Pixel Press, 1999.

Part III Applications

Chapter 10

Aerial Photography

Keith A. Mancini

Producing effective data rich photographs is a constant challenge for all photographers. In natural science, the ever present need to provide informative points-of-view for photo documentation is very helpful in sharing the most complete narrative of a subject. Aerial photography is sometimes the ideal way to reveal details of a subject from above. Examples of where aerial photography can be helpful includes documenting damage caused by natural disasters, monitoring crops, sharing real estate and properties, and mapping new developments required for inspection.

There are several ways you can create aerial photographs. The technique(s) that is ultimately chosen will depend upon a project's needs, budget, and the equipment that is available. In this chapter, several options for aerial photography will be explored, covering a wide range of considerations and challenges. Some approaches could range in cost from free if using existing satellite imagery and software such as Google Earth, or very expensive should the use of helicopters and high-end photographic equipment be required.

Photographing from an Aircraft

One common technique for aerial photography is by using a small aircraft. The use of aircraft will depend on the specific needs, budget, and access to an aircraft. Obtaining quality images from an airplane is possible using common photographic equipment.

Equipment Recommendations

- Camera – a DSLR or mirrorless camera having a variety of operational controls including manual control of shutter speed choices.
- Memory card – the larger the better since changing cards while the aircraft is in motion is not recommended, especially if the doors or windows are open.
- Lenses – a VR (Vibration Reduction) zoom lens allows for the selection and use of multiple focal lengths without having to change lenses. Note that the longer the lens, the

more potential for amplification of motion blur. Lens filters are useful for protection of the front element of the lens. A polarizing filter also may help reduce glare, as well as darken blue skies. It is not advisable to use a lens hood because it will add wind resistance if the lens is used outside the aircraft. It can also be at risk of falling and causing damage to objects or injury to people below.

- Straps, tethers, and harnesses – secure yourself and all your equipment to the aircraft before taking off. Keep all necessary equipment nearby or have an assistant available to pass your equipment during the flight. You should be tethered to the aircraft either by seatbelt or harness if you will hang outside of the airplane during the flight.
- Gyroscope or gimbal – this device is used to steady a camera. It may be cost-prohibitive to buy. However, one may be available for rental.
- Microfiber cleaning cloth – this is used to clean the windows of the aircraft. The cloth will not scratch the fiberglass windows.
- Blackout material – to reduce internal reflections and glare when shooting through windows.

Choosing the Best Aircraft for the Imaging Requirements

There are two main considerations when identifying which type of aircraft to select because each will have different benefits to making pictures from them. Aircraft availability may ultimately dictate the final choice. Regardless of the aircraft, be sure to have a realistic plan for photographing the subject from above and discuss the flight plan with the pilot prior to taking off.

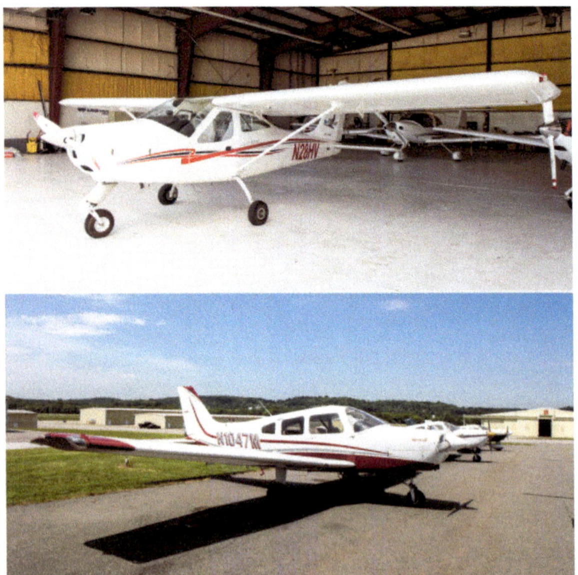

Figure 10.1 Airplanes with high and low wings. Images courtesy of Teri Zgoda.

Airplanes

Airplanes must fly above 500ft. In populated areas that altitude rises to 1000ft. Airplanes are useful for photographing large areas. To remain airborne, planes must maintain a minimum airspeed leaving little time for precise composition of images. Circling the subject is a way to provide more opportunities for photographing. Whenever possible, choose an airplane with high wings to avoid including wings in the frame of the photograph. Photographing will most likely take place through an airplane's window. Be cautious when pointing a lens out of an open window because the plane's speed and wind resistance may make it difficult or even dangerous to operate.

Helicopters

Helicopters fly at approximately 500ft or below and depending upon the surroundings may fly even lower. Helicopters can fly at slower speeds than airplanes and are able to hover in the one

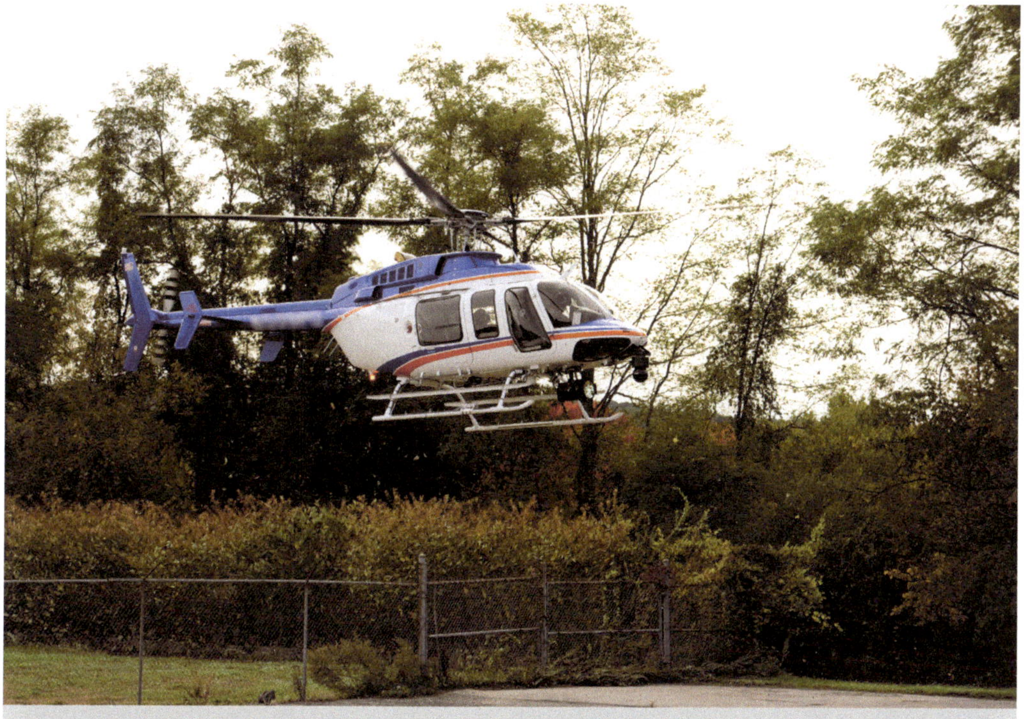

Figure 10.2 A Bell 407S helicopter. Image courtesy of Keith Mancini.

location, which makes them preferable. Depending on the flying conditions and safety, it is always recommended to keep the helicopter in motion and never held in place for too long. Helicopters do not have wings to get in the way of the view, and the doors may be removed prior to flying or to remain open during flight.

Aviation Safety

Safety while flying in an aircraft is of the utmost importance. If you are unfamiliar with the safety procedures in and around the aircraft that you will be using, be sure to discuss them with the pilot prior to takeoff. If the pilot is landing prior to picking you up, be sure to wait for their signal before approaching the aircraft. Never walk behind a helicopter or near the props on an airplane.

When preparing for flight keep in mind that less weight will mean there is more power for the aircraft so be careful to pack and bring only essential items needed for the photography. It is a good idea to prepare a required list of photographs needed prior to the flight and even better to map each out anticipating the flight plan. This will greatly help a pilot unfamiliar with the area being photographed and lead to the making of better results.

Before boarding an aircraft, empty pockets and secure any loose articles of clothing. Once onboard, secure all the equipment to the interior of the aircraft. Loose and unsecured equipment can become a safety hazard in the aircraft if turbulence is encountered. Options for securing equipment includes using anchors, the aircraft's seatbelts, as well as bungee cords or tie-downs for your camera bag. At the very least, use a camera strap on the camera and secure it around your neck. Alternatively, specially designed camera rigs can be mounted to the aircraft and may be great choices.

Time of Day Considerations

It is not always possible to choose the date or time for a flight. However, given the choice, plan the flight during good weather. A cloudy or overcast day may actually be preferable since there will be no glare or high contrast from the sun. When choosing a time, late morning or early evening will provide the best detail because of the angle of the sun to the surface of the ground. This situation creates texture or topography in an image, which also may be helpful to a photograph.

Avoiding Glare

If it is necessary to shoot through the windows, be sure to clean them prior to taking off. It is always best to ask the pilot for permission first. Use a non-abrasive microfiber cloth. Avoid wearing brightly colored clothes when flying to reduce reflections. Dark solid colors are ideal. Other methods of reducing reflections in a window include reducing the light in the interior of the aircraft, blocking the light behind you with blackout cloth, using a camera mask, or the use of a lens hood that attaches to the window like the LENSKIRT™ Flexible Lens Hood.

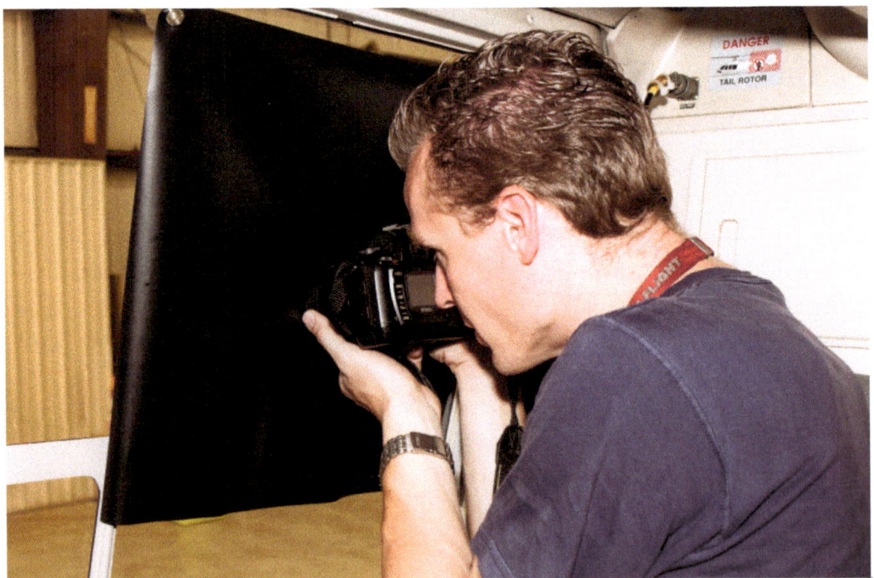

Figure 10.3 Officer Sean C. Lonergan demonstrates a technique for reducing reflection while photographing through a window. A hole is cut in the center of an opaque material, called Savage© Velvetine, which is hung on the window and the camera lens is placed through the hole. Image courtesy of Keith Mancini.

Reducing Vibration

Regardless of aircraft, vibration will be an issue when flying and may introduce significant motion blur into images. There are a few ways to reduce the effects of vibration in photographs. The easiest way is by using short shutter speeds such as 1/1000th second or faster if using long telephoto lenses such as 200mm or longer. If using Vibration Reduction (VR) lenses, be aware that they consume more power when in use and may drain camera batteries more quickly. The use of a gyroscope or gimbal attached to the camera will aid in steadying the camera and significantly reduce the effects of vibration, but they can be expensive.

Exposure

The easiest way to ensure the use of a fast shutter speed is to operate the camera in shutter priority mode. Select the desired shutter speed and allow the camera to choose the aperture. Do not be overly concerned with aperture. The required depth of field is easy to achieve since the ground is relatively flat when photographed from the air.

Focus

Use a single-servo autofocus mode since the relative distance to the ground will not change significantly, even though the aircraft is in constant motion. When shooting from high altitudes or while using a wide-angle lens, it may be beneficial to use manual focus with the lens set to infinity.

Composing

Be sure the pilot is aware of the subject of interest before the flight begins. Ask to be informed when you are getting close. Stay in communication with the pilot during the flight, and during the shooting. It may be necessary to circle the subject several times to get adequate photographs and all of the angles that are needed. In an airplane, it may be necessary to bank turns to keep the wings out of the frame, so be sure to discuss this need with the pilot before that situation occurs. Helicopters can hover, however, depending upon the circumstances, the pilot may prefer to circle the subject instead. If it is necessary to document large areas, consider photographing in a grid or pattern across the scene. When doing this be sure there is plenty of overlap between in the adjacent sections of individual photographs.

Unmanned Aircraft Systems

Unmanned Aircraft Systems (UAS), more commonly known as drones, are a class of radio-controlled model aircraft used for recreation and commercial photography. Advances in electronics and lowering of costs have made it easier than ever to get a camera up in the air.

Drones are small commercially available radio-controlled aircraft, typically operated by a person using a transmitter usually within the line of sight of the aircraft. They have a wide

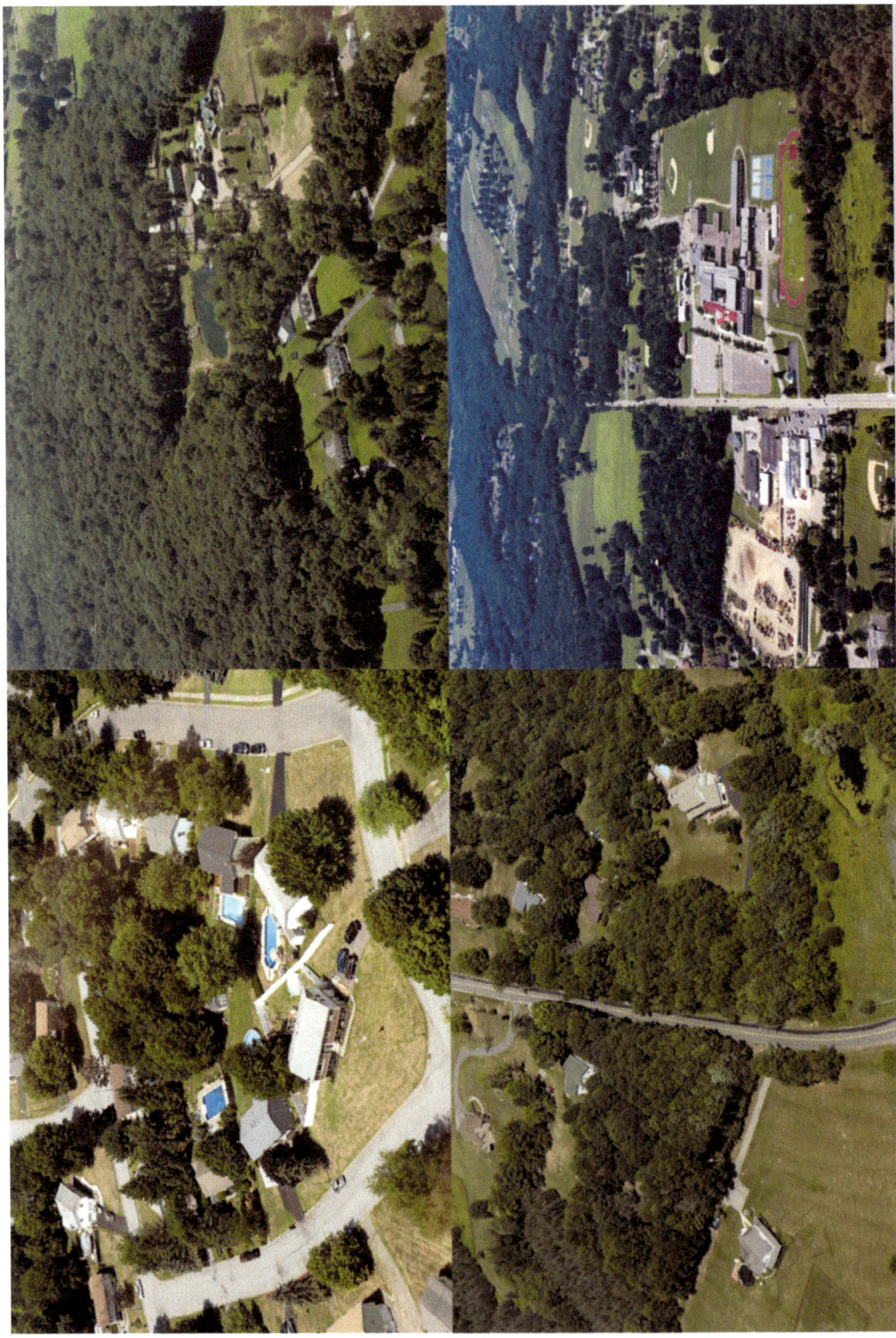

Figure 10.4 Examples of photographs taken from an airplane. Images courtesy of Teri Zgoda.

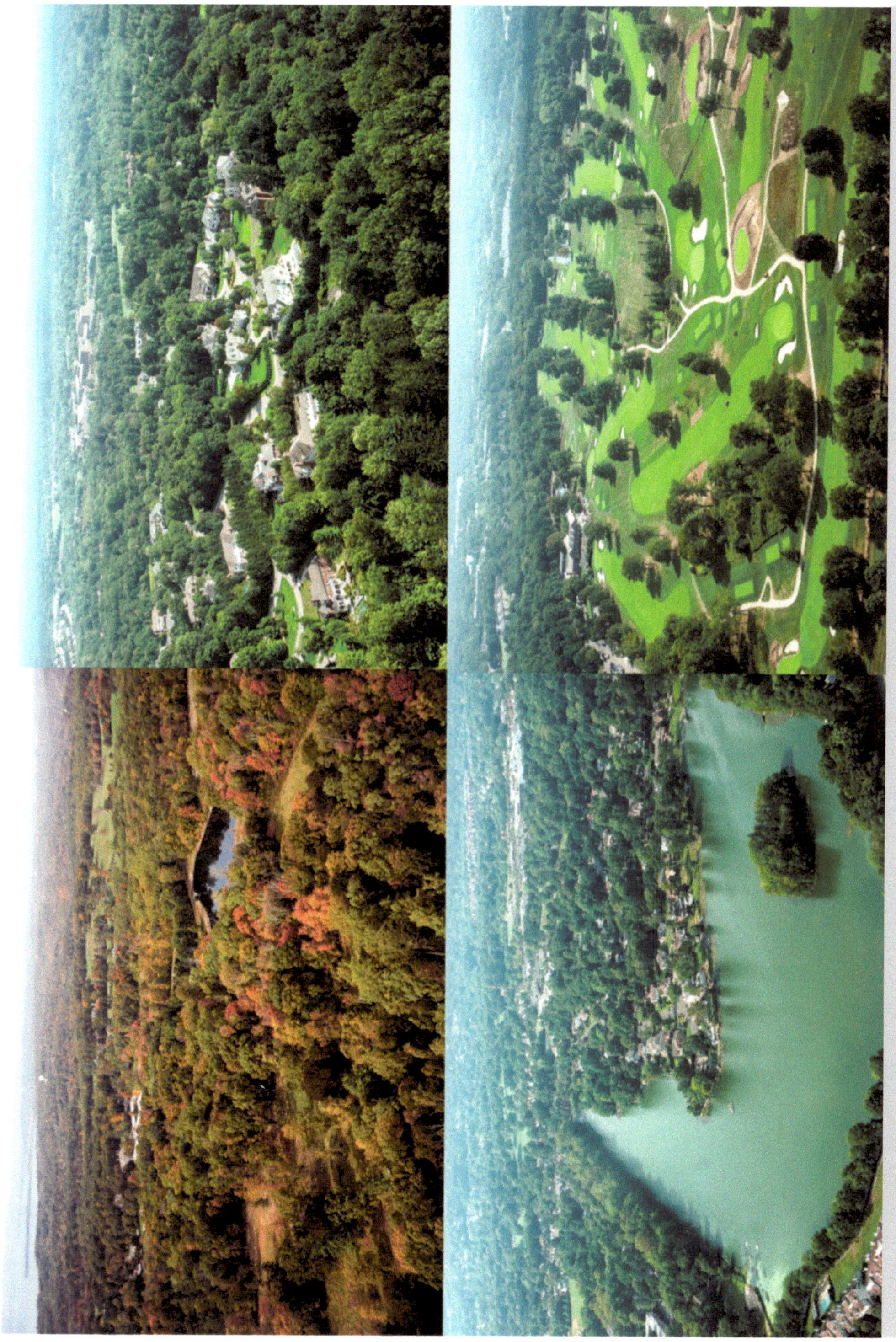

Figure 10.5 Examples of photographs taken from a helicopter. Images courtesy of Keith Mancini.

range in costs from very inexpensive "toys" to very costly advanced systems. Drones may fly with the help of global positioning systems (GPS), or first-person view (FPV) cameras and they can be controlled in real time or be programmed to fly specific routes.

Multi-Rotor

The drones most commonly associated with aerial photography are multi-rotor helicopters or simply called multi-rotors. Multi-rotors have several propellers mounted horizontally to provide the aircraft with lift. Moving the multi-rotors position forward, backward and turning is achieved by varying the rotation of one or more of the rotors. Multi-rotors create a stable platform for aerial imaging since they can hover like a helicopter and multiple rotors contribute to the aircraft's stability. The most common configuration of a multi-rotor is a quadcopter, named for its four rotors, although it is also common to encounter hexacopters (six), octocopters (eight), or other configurations as well. Rotors are primarily found in even numbers since their rotation is alternated between clockwise and counter-clockwise to prevent the aircraft from spinning.

Considering the costs for making aerial photographs using piloted operated aircraft, multi-rotor airplanes are relatively inexpensive. Many of the most popular quadcopters for aerial photography vary in configurations and range in price from $400 to $2000. They are completely ready for use including providing a high-quality camera in many cases.

Flying a Multi-Rotor

Flying a multi-rotor drone is relatively easy. There are essentially four controls used for flight. Throttle, yaw, pitch, and roll. The throttle controls the rate of ascent of the aircraft. The more throttle given to the aircraft, the faster it will climb. When less throttle is used, it will cause the aircraft to descend. Yaw controls the rotation of the aircraft clockwise or counterclockwise changing the direction which the aircraft faces. Pitch is the direction of travel, forward or backward of the aircraft relative to its front. Roll is the direction of travel, left or right of the aircraft relative to its front.

Common controllers or transmitters have two sticks that control the aircraft's movement. Throttle and yaw are controlled by one stick and pitch and roll by the other. Some multi-rotors have a feature called smart mode where the pitch and roll controls are relative to the pilot's orientation. For example, forward will always be in the direction away from the pilot regardless of what direction the aircraft is facing. In addition, some multi-rotors can be controlled via a Wi-Fi connection from a smartphone or tablet application. These controllers typically have a shorter range than traditional radio transmitters, however, that varies depending on the model.

There are many ways to learn how to fly a multi-rotor drone and training is recommended prior to flight. Even though many models fly with the assistance of GPS, it will be beneficial to know how to fly unassisted in case of a malfunction. There are several inexpensive multi-rotor models available that fly using the same controls as larger more expensive models. Practice can be accomplished using an inexpensive model prior to purchasing or flying a

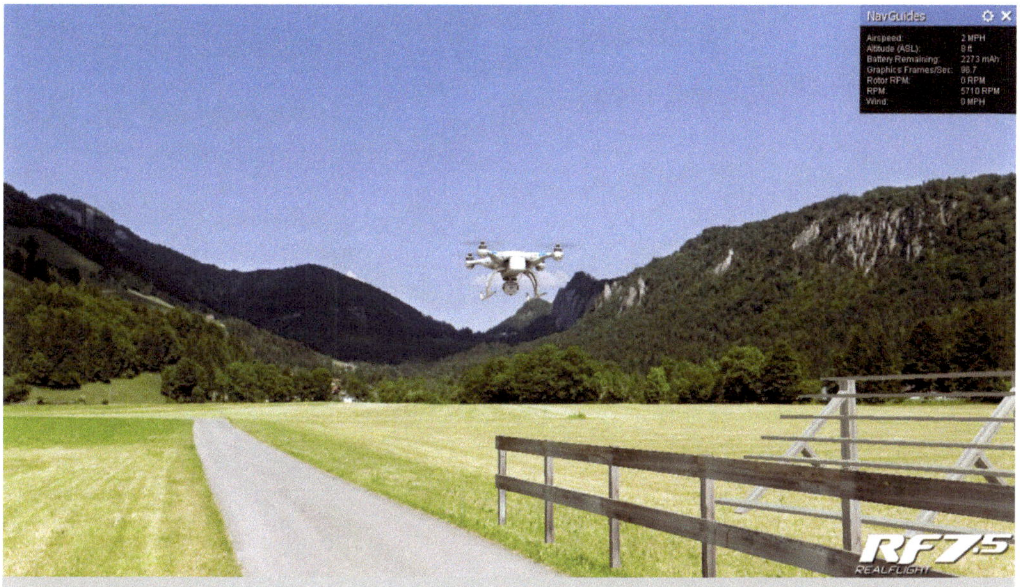

Figure 10.6 A screenshot of RealFlight® flight simulator software. Image courtesy of Keith Mancini.

more expensive model. It is far better to crash an inexpensive "toy" than a much larger and expensive drone. Also, many of those "toys" can endure a crash better than their larger, more expensive counterparts.

Another alternative for training is the use of flight simulator software. Commercially available software like Real Flight are available for approximately $100. The software includes a controller and several model aircraft within the software to practice flying.

Drone Basics

There are essentially three categories of drones: beginner, intermediate, and advanced. Beginner drones are usually compact in size and come equipped with basic cameras, but it may not be suitable for aerial photography. Beginner models typically lack features such as GPS or advanced flight modes, and may only be suitable for indoor flight. Although they may not be suitable for aerial photography, beginner drones can be a fun way to learn to fly.

Intermediate or "prosumer" (somewhere between consumer and professional) models are larger than beginner drones and typically can fly higher, farther, and for extended periods of time. They have the ability to carry larger payloads and use better quality cameras. Intermediate drones are primarily geared toward aerial imaging and include a built-in camera and stabilizer called a gimbal, or they are designed to carry small action cameras such as the GoPro Hero. Intermediate drones have features including GPS, and automatic return to home which automatically will bring the aircraft back to its takeoff spot. Examples of intermediate drones include the DJI Phantom and Mavic series, Yuneec Typhoon, Autel Robotics Evo, and Parrot Anafi.

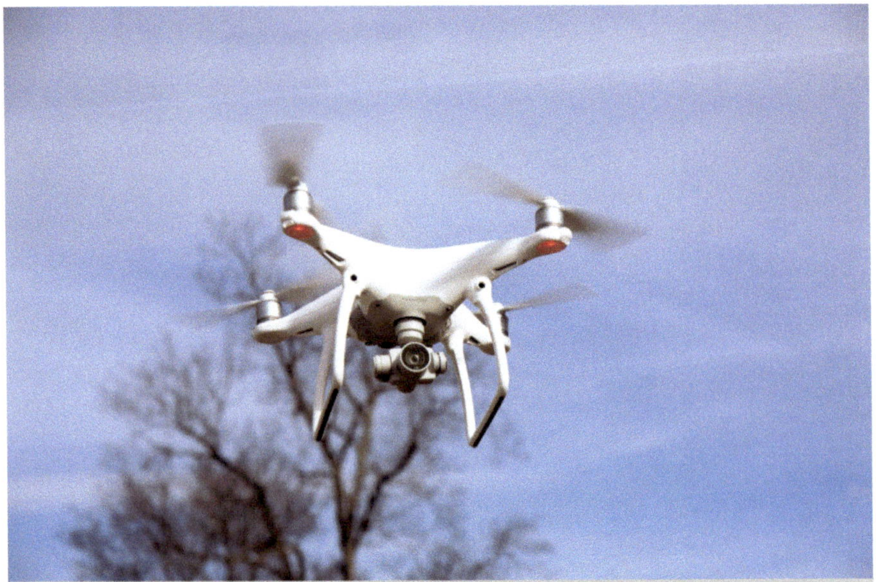

Figure 10.7 The DJI Phantom is an example of an intermediate or prosumer quadcopter. Image courtesy of the Westchester County Police.

The advanced category of drones have similar features to the intermediate drones but tend to be larger and have more powerful motors. As a result, they can carry even heavier payloads such as dual cameras, redundant GPS systems, and additional back-up batteries. These drones can support a flight point of view (FPV) camera for the pilot as well as another independently controlled imaging camera. Professional drones are typically used for digital cinema cameras, however they may also be configured to carry a mirrorless camera or DSLR. Examples of advanced drones include the DJI Inspire and Matrice, Yuneec H520, and Freefly Alta.

Drone Features

The following features are common across most drones. Many drones are being designed for easier flight and as a result, may improve the quality and success of aerial photographs.

Ready to Fly (RTF)

RTF drones come mostly assembled and require very little setup prior to flying. Usually only charging the batteries and calibrating the compass is required for set-up. RTF drones include a transmitter, a battery, and charger. An RTF drone is a great choice to get out and fly relatively quickly.

Global Positioning System (GPS)

GPS greatly improves the accuracy of flying the drone by communicating with several satellites and aiding in the positioning of the aircraft. A built in GPS in the drone allows for other flight modes such as the Return to Home function.

Return to Home (RTH)

RTH enables the aircraft to return to its original takeoff position. This may be activated by pushing the button automatically when the battery is low, or when the aircraft loses communication with its transmitter.

Stick Relativity

The aircraft will always fly in the direction, relative to the pilot, that the control stick is pushed regardless of which way it's front facing.

Altitude Hold

The aircraft will maintain its altitude (height) in the air. This allows for precise control roll, pitch, and yaw.

Waypoint Flying

Using software or a smartphone app, you can set up waypoints on a map prior to a flight which allows for the aircraft to fly a predetermined route. This feature can be combined with camera control which ensures the camera is always pointing in the desired direction.

Flight Boundaries or Geofencing

This feature allows you to set up a boundary that the aircraft will not leave. This is a nice feature for learning to fly, or if you want to keep the aircraft in a specific area.

Follow Me

The aircraft follows the pilot or a subject, typically using the transmitter, a smartphone application, or using image tracking on the aircraft.

Obstacle Sensing

Some aircraft feature sensors that assist with the avoidance of obstacles. Depending upon the aircraft, obstacle sensing may only be frontward facing or sensing in multiple directions including backward and downward. These sensors allow the aircraft to precisely maneuver through close quarters as well as indoors where no GPS is present.

Image Tracking

The aircraft uses its camera to follow a subject. This feature may be used in conjunction with obstacle avoidance to autonomously film a subject.

Drone Camera Options

There are several features worth considering when selecting a camera for aerial photography. Many imaging systems will come with capabilities for video and still image capture. If the goal is to get into the air quickly, this type of system is the best option. There are several cameras that may be mounted onto a drone. Many professional drones like the DJI Inspire or Yuneec Typhoon allow a choice of cameras for specific purposes. Drone cameras vary in quality and features, therefore, it is best to evaluate real imaging needs prior to making a purchase. Specifications such as resolution, and light sensitivity, as well as the ability to capture in RAW format should be considered because these features will vary across cameras. The greatest limiting factor to mounting a camera to a drone will be the payload capability of the aircraft. Be sure to confirm the capacity of a drone before loading a camera and flying for the first time.

Using a GoPro or other Action Camera

An action camera is a small durable video and photography camera typically used by athletes or enthusiasts to document their sporting adventures. An action camera is well suited for mounting on a drone because they are typically lightweight and compact. Despite their size; the photo and video quality of sports cameras can be very impressive. Many models feature 4K 60fps video and 12MP photos. Action cameras may be controlled via remote or through a smartphone application, allowing the pilot to start and stop the video camera or take photographs remotely.

Professional drones have the ability to carry larger cameras due to their larger frames and more powerful motors. These aircraft can carry mirrorless cameras, DSLRs, and even digital cinema cameras. Many have the added options of carrying multiple cameras as well as smaller first person view or FPV cameras. Control of the cameras can be achieved by the pilot or another camera operator. This frees the pilot up to focus on flying the aircraft.

Multispectral and IR Cameras are sensitive to different wavelengths of light and energy and can be used to monitor the health of crops, plan fertilization, pesticide application, and irrigation.

Thermal cameras convert heat into images and may be used for search and rescue, inspection, and public safety.

Regardless of the camera selected, there must be a way to mount the camera to the drone. The camera will need a mount and the mount should provide stability, vibration reduction, and even permit the positioning of the camera while in flight. The simplest way to mount a camera to a drone is with the use of an anti-vibration mount. An anti-vibration mount will separate the camera from the body of the aircraft by using rubber anti-dampening material. The vibration of the drone may cause something called the "jello" effect on the video of some cameras. The "jello" effect is a result of the manner in which some camera's shutters record video.

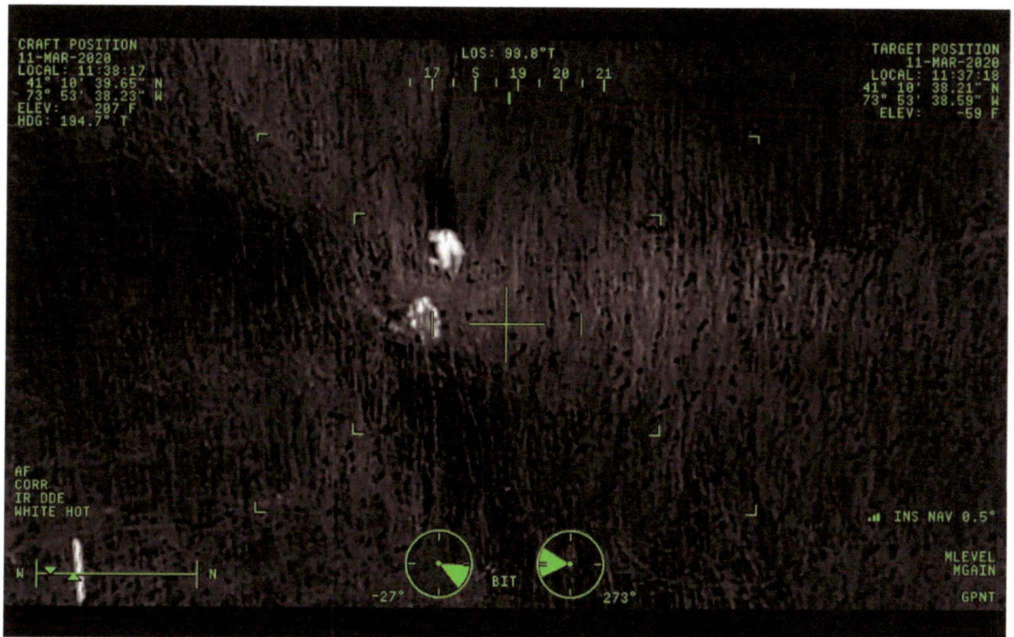

Figure 10.8 A thermal image of a search and rescue mission. Image courtesy of Officer Sean C. Lonergan, Westchester County Police.

The addition of a gimbal will add stability to the camera. A gimbal is a device that steadies the camera and reduces the effects of the tilt or swing of the aircraft. The camera gimbal may be combined with an anti-vibration mount. Camera gimbals are available in two types. Passive which attempts to keep the camera level or active, which uses motors to achieve a higher level of steadiness. Gimbals can stabilize in two or three directions, called two or three axes. Another feature found in some gimbals is the ability to control the camera direction remotely via a transmitter or application.

Some more advanced cameras and many drones come with built-in cameras with a feature called first person view or FPV. The camera transmits video directly back to the receiver, allowing the pilot to see exactly what the camera is seeing. The receiver can be built into the transmitter, an accessory screen, or a smartphone application. The video can be transmitted over Wi-Fi or as an analog signal. Each technology has advantages and drawbacks. Be sure to check the distance of the transmission to avoid flying beyond the range.

Photogrammetry and Mapping

Using a drone and photogrammetry software, it is possible to create accurate 3D maps from 2D images. The drone is used to take multiple photographs of a subject using a combination of vertical and oblique images. Software is available to help plan the flight path and capture the images. The images are imported into the photogrammetry software, which matches common points in the images and creates a 3D model. The model may be used in other 3D software or used to collect data such as measurements, as well as planning and development.

Flight Safety and Federal Guidelines

Drones have become extremely popular in a relatively short period of time, due to their affordability and ease of use. However, stories of aircraft going astray, as well as pilots not following common sense guidelines have raised a lot of concern among the public. It is because of this, that the Federal Aviation Administration (FAA) has adopted several guidelines pertaining to the use of drones. The following are some of their recommendations for the recreational use of drones. The FAA defines recreational use as flying a drone for fun, and not as part of a business.

- Register the drone.
- Fly the drone at or below 400 feet.
- Keep the drone within your line of sight.
- Be aware of FAA Airspace Restrictions.
- Respect privacy.
- Never fly near other aircraft, especially near airports.
- Never fly over groups of people, public events, or stadiums full of people.
- Never fly near emergencies such as fires or hurricane recovery efforts.
- Never fly under the influence of drugs or alcohol.
- Recreational flyers should know that if they intentionally violate any of these safety requirements, and/or operate in a careless and reckless manner, they could be liable for criminal and/or civil penalties.

For commercial, government, and additional information about recreational use of a drone visit the FAA website at https://www.faa.gov/uas/. Regulations for both the commercial and recreational use of drones are continuously evolving, therefore it is recommended to check this site regularly.

Photographing from a Drone

Most drones with built in cameras, allow remote control of the camera via an application or the transmitter. The application displays exactly what the camera is seeing, as well camera settings such as shutter speed, aperture, ISO, and even RAW may be chosen. Capturing images can be as simple as composing the shot, steadying the aircraft and pressing the capture button. However, if the drone does not have a built-in camera, you will have to find another way to capture images. Many modern digital cameras have a feature to remotely connect to the camera and control the camera, this includes several models of action cameras, point and shoot, mirrorless, and even DSLR cameras. If the camera does not have this feature, there are aftermarket accessories available that can add this function. Be aware that the signal used for this communication may share the same frequency as a drone's transmitter and receiver. This may result in interference between the transmitter and receiver and may cause erratic flying or a crash. Be sure to check the frequency of your drone's transmitter and the frequency of a Wi-Fi camera before a flight.

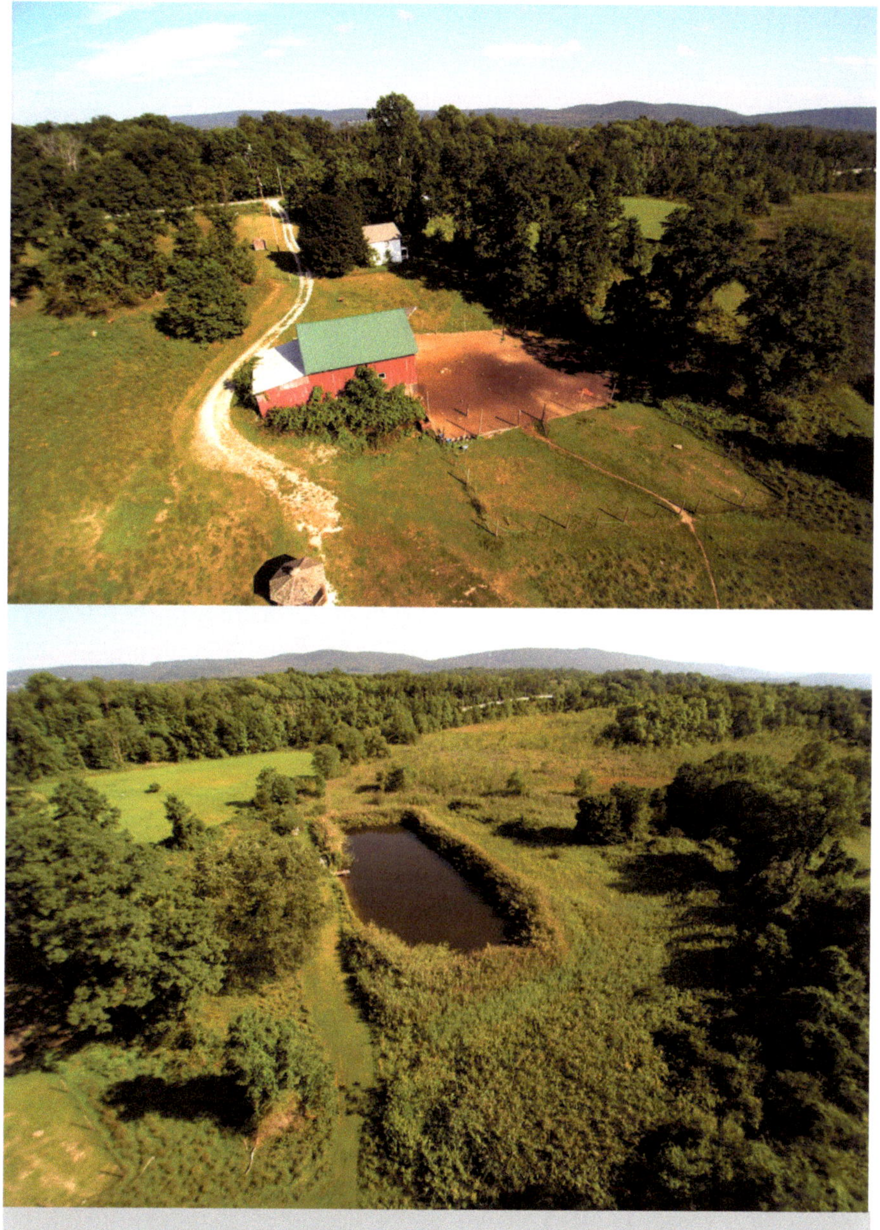

Figure 10.9 Examples of photographs taken using a drone. Images courtesy Joseph P. Mancini.

Exporting Frames of Video

Another option for capturing still images using an aerial point of view is to record video produced during a flight and then export still frames from that video footage. There are several methods to do this and they do NOT require the use of video editing software. Adobe Photoshop, supports working with several video file formats. It is possible to open a video file in Photoshop then save the desired individual frames as still image files. Alternatives to Photoshop include; GOM player and VLC. Both of these products have the capability to export individual frames of video to still image formats.

Saving a Single Frame of Video using Adobe Photoshop

Open the video in Photoshop. Choose the "Motion" workspace, and ensure that the timeline is visible. Using the red slider scrub to the desired frame.

Go to File > Save As, and choose the desired image format. TIFF is recommended as there will be no further compression to the images.

Uncheck the "Layers" option. Name and save the file. These steps can be repeated multiple times for additional frames.

Saving Multiple Consecutive Frames using Adobe Photoshop

Open the video file in Photoshop. Trim the video to the desired frames using the timeline. Click and drag either the beginning or end of the video in the timeline to trim.

Convert the frames to layers.

Click on the Timeline panel menu (located in the upper right corner of the timeline window), choose Convert Frames > Flatten Frames into Clips.

Save the layers to files. Choose File > Export > Layers to Files

In the Export Layers to Files dialog box choose the location, name and format of the files. TIFF is recommended as there will be no further compression to the images.

Saving Single Frames using VLC Media Player

Configure VLC snapshots by going to Preferences > Video

Under video snapshots, choose the folder, prefix and format of the images.

While the video is playing, or paused on the desired frame, right click on the video and choose "snapshot". Alternatively use the keyboard shortcut Command + Option-S on the Mac, or Shift + S on the PC. An image of the current frame will be saved to the selected folder.

Saving Single or Multiple Consecutive Frames using GOM Player

Setup screen capture by going to Video > Advance Screen Capture, or using the keyboard shortcut Ctrl + G.

Select the image format, resolution and the directory where the images will be saved.

Select the frame desired using the seek bar, then press the capture button, or use the keyboard shortcut Crtl + E to save the current frame, or Crtl + C to copy it to the clipboard. Several consecutive frames may also be captured using the burst capture option.

Low Cost Alternatives to Aerial Photography

Satellite Images

In this chapter, I have discussed several options for making aerial photographs. The price ranges for these options can vary greatly from a few hundred dollars and up to a few thousand. However, if the timing of when an image is made is not of concern, it is possible to capture aerial images using Google Earth Pro. Google Earth Pro is a free mapping and satellite imagery software available for download at www.google.com/earth/. To save aerial images using Google Earth Pro, you should first navigate to the desired location on the map. Compose the area using the zoom tool and arrow keys on the right side of the screen. Then follow these directions to save an image.

Select File > Save > Save Image... or click the Save Image button on the toolbar. The image will be saved with all placemarks, borders, or other layer information that is visible in the 3D viewer.

Overhead Photography

The following technique, while not exactly aerial photography, will enable the capture of photographs taken from an overhead point of view. Because the following techniques will put the camera out of an operator's reach, the greatest challenge will be how to trigger the camera remotely. This may be accomplished in one of several ways. The use of a long shutter release cable, a wireless shutter release (remote control), or through a smartphone application connected to the camera via Wi-Fi.

It is possible to use a tripod as a boom by extending its legs to their longest lengths without spreading them, then hold the tripod by the legs and suspend the camera over the subject.

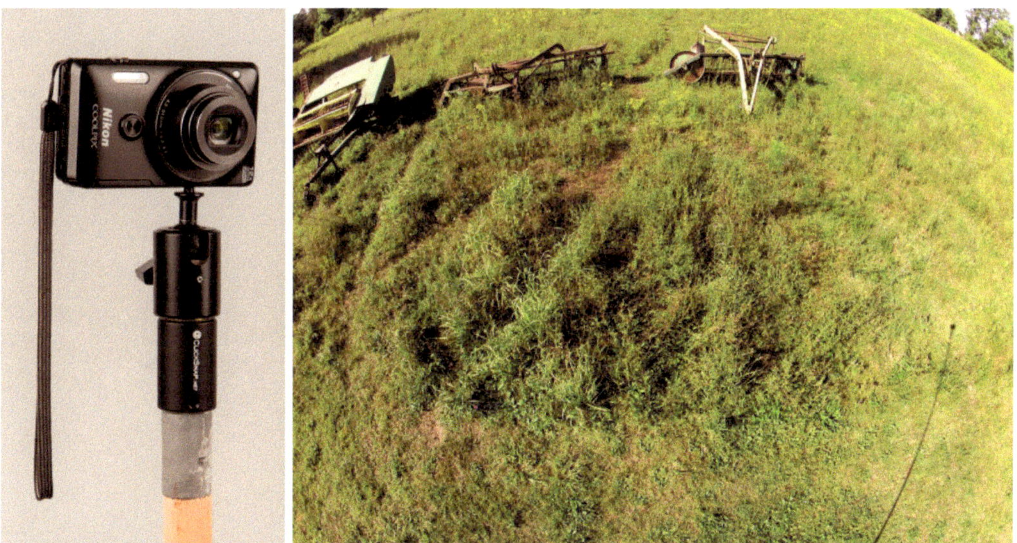

Figure 10.10 A small camera and small ball head can be attached to a painter's pole adapter. An example is shown above. On the left is the system, and on the right, an image taken using this system. Images courtesy of Keith Mancini.

Additional height may be gained through the use of a ladder. Be very cautious using this setup because the system will be top-heavy and can be unsteady.

A second technique is the use of a painter's pole adaptor to mount a camera atop an extension pole. These poles are commonly available at hardware stores in the tool or painting section and come in various lengths up to 23 feet or longer. Again use caution as larger cameras will be very top-heavy and unstable at long lengths.

Several options exist for capturing images from overhead. When there is no budget or when equipment is limited, finding creative solutions to photograph subjects from the air can challenge even the most creative individuals. I have known photographers to use a balloon or kite in extreme situations.

Conclusion

Aerial imaging is a vast and very dynamic field. The continual release of inexpensive unmanned aerial imaging systems is exploding and changing almost monthly. This makes discussing elements of hardware and methods challenging for a text book.

Making high quality images from elevated points of view opens up many unique opportunities for the creation of photographic documentation that can serve many needs. The products, software, and accessories available to users is experiencing explosive growth and the local and FAA policies governing their use are also changing. The subject cannot be completely captured for inclusion in a book because of the dynamic changes that are affecting the industry. I hope you will find the information included to be useful and practical.

Suggested Reading

From Above: The Story of Aerial Photography (150 Years of Breathtaking Imagery), by Gemma Padley and Eamonn McCabe. Laurence King, 2019.

The Photographer's Guide to Drones, by Colin Smith. Rocky Nook, 2016.

Aerial Photography and Videography Using Drones, by Eric Cheng. Peachpit Press, 2015.

Chapter 11 Close-Up and Macro Photography

Overview

Close-up photography and photomacrography – sometimes incorrectly called macrophotography – are techniques that use small camera-to-subject distances. Close-up photography typically produces images that are smaller than subjects. Images will have image sizes of one tenth life size ($1/10^{th}$ or 1:10) down to life-size (1:1) often expressed as an image reproduction ratio (image size:object size). Photomacrography typically describes making images that are larger than subjects and is characterized by ratios from 1:1 to no more than 50:1, or 50 times magnified sometimes written as ×50. Photography above ×50 magnification is normally considered to be photomicrography and is covered in Chapter 16. No definition is absolute.

Defining Close-Up Photography

Close-up pictures play important roles in the photo-documentation of all kinds of subjects. Many applications of close-up photography are found in the medical, botanical, biological, or geological sciences and includes all sorts of things: butterflies, fossils, fingerprints, coins, or postage stamps for example. This chapter will cover foundational considerations for close-up techniques. Close-up techniques are also discussed in Chapters 14 and 17. There are a number of ways to make close-up photographs using simple equipment and/or accessories.

Lenses for Close-Up Photography

Almost any camera can be used for close-up photography, but the use of a quality close-up or macro lens is required to create the best imaging outcome. Many relatively inexpensive compact digital cameras – or now smartphones – are adequately corrected for moderate close-up distances, despite being designed primarily for use at longer focusing distances. Smartphones have also become a common tool used for making marginally successful close-up pictures. There are many DSLR and mirrorless cameras with a wide variety of close-up or macro lenses. These can be purchased in either fixed focal lengths or zoom lens types. A zoom

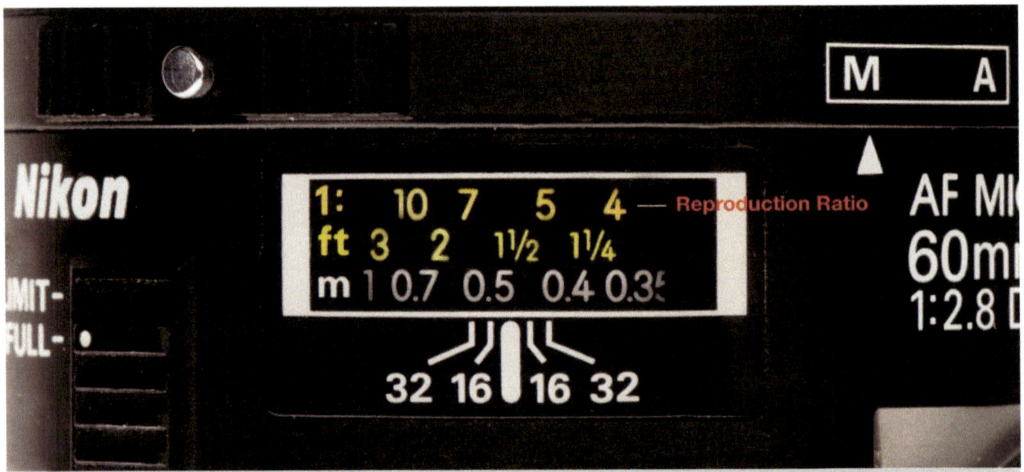

Figure 11.1 In this photograph, the reproduction scale on a Nikon 60mm lens is highlighted. This lens can create a magnification range of 1:1 through 1:10 life-size. The desired reproduction range is established by positioning the magnification adjacent to the 1: visible on the left in the focusing window. Image courtesy of Michael R. Peres.

lens with close-up abilities is often described as macro zoom. It can create useable results, but using a single fixed focal length close-up lens is a significantly better choice for close-up applications. Important optical characteristics are designed into macro lenses including excellent edge-to-edge image sharpness. They have been highly corrected for color and spherical aberrations as well. Their most important feature is their ability to produce edge-to-edge sharpness when photographing at small working distances. This optical design characteristic is called flat field or PLAN.

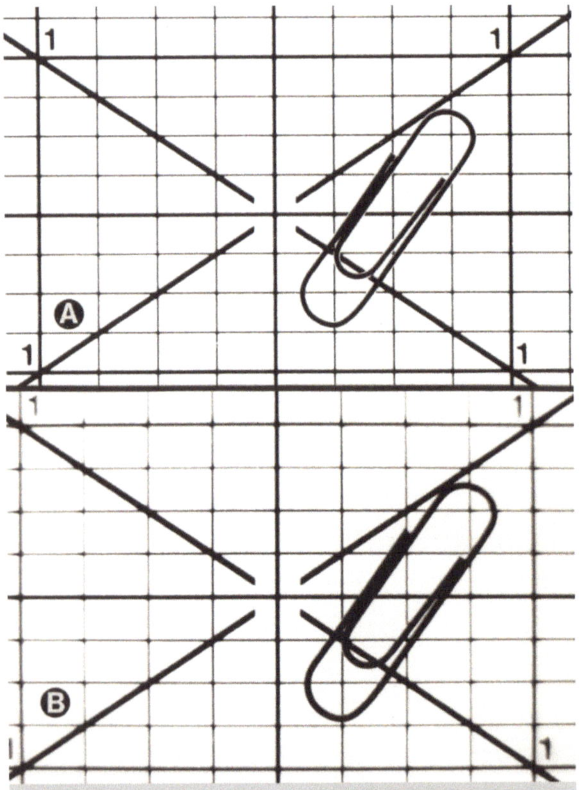

Figure 11.2 This composite photograph reveals the performance difference of a 50mm macro lens on the top (A) and a 50mm normal lens on the bottom (B). There are considerable quality differences in the edge-to-edge definition of image characteristics such as sharpness and subject definition. Image courtesy of Michael R. Peres.

Expensive macro lenses have elaborate internal mechanisms that allow groups of lens elements to move independently as the lens focuses to shorter working distances, creating excellent down to life-size (1:1) or slightly larger images.

These lenses often have reproduction ratios inscribed on the barrel, have modest maximum apertures, and come in various longer – and other than a normal – focal lengths. These lenses all produce larger subject to lens distances when operated at 1:1. Common focal lengths for macro lenses might be 50mm, 60mm, 100mm or 200mm with a maximum aperture of approximately *f*/2.8.

When images are being made at a 1:1 magnification, both the working distance and the image distance will be equal to twice the focal length of the lens. Longer focal lenses provide increased working distances, which is helpful when lighting or access to the sample is important and it is not prudent to be near to heat or infectious diseases, for example.

Getting Closer: Supplementary Lenses

Sometimes there is a need to focus closer than a macro or close-up lens will allow. This is because most fixed focal-length camera lenses are optimized for long camera-to-subject distances and they are usually limited in their minimum or short focus distance. This distance can be reduced and the image size increased by using a supplementary or close-up lens. These lenses will either be a single lens or lens group that attaches to a lens like a filter. These lenses are sometimes called plus lenses or diopters. Note, Photojojo® sells small threaded diopter lenses that are useful on smartphones.

The magnifying potential of supplementary lenses increases as the diopter number increases e.g. +1, +2, up to +20. A +2 close-up lens will have a focal length of .5 meter (500mm) and a +5 close-up lens will have a focal length of .2 meter (200mm).

Alternatively, it is possible to convert all values to diopters and then convert their sum to focal length. For example, a 50mm lens has a power of 20 diopters. Add a +2 supplementary lens and the sum is 22 diopters. The combined focal length is 1000mm divided by 22, or about 45.4mm.

Teleconverters are precision optical accessories that can be mounted in-between compatible lenses and a camera body. Teleconverters increase the focal length of the primary lens and maintain a shorter working (object) distance. If a teleconverter is used in combination with a 200mm close-focusing lens for example, it will yield a life-size (1:1) image while maintaining a 71cm (28 inch) minimum object distance. This working distance can be advantageous for many biomedical, industrial, and natural science applications. Having an increased working distance allows for more freedom in arranging lighting or keeping safe distances from dangerous events or diseases. Teleconverters have optical elements and can degrade the resolution of the primary lens if not of high quality.

Getting Closer: Extension Tubes for Close-Up Photography

Cameras with interchangeable lenses allow for extension tubes or a bellows to be inserted between the camera and lens, which extends the lens to image distance. Increasing the image distance rather than changing the focal length will often produce a better optical result than using supplementary lenses. Extension tubes are cylindrical tubes of various lengths, that can be used alone or in combination to change the image size. Tube lengths may vary from as small as 5mm up to 100mm. It is possible to create large extensions of up to 250mm. The tubes are fitted together with threads or bayonet-type mechanisms and some allow the camera's automatic diaphragm features to be maintained, but typically camera lenses must be operated in manual mode.

Extension-tubes are relatively inexpensive but can be a bit inflexible because they have specific lengths. Making precise compositions is challenging because of this lack of flexibility. A bellows conversely is more costly, but is essential for professional work because it allows for precise control over lens-image distances, leading to more precise image compositions

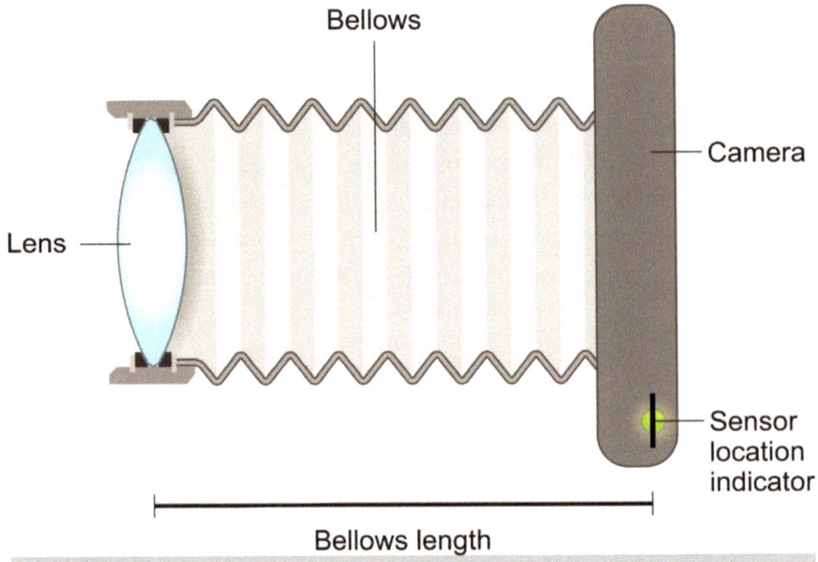

Figure 11.3 The basic parts of a close-up or photomacrography system include a close-up or macro lens, a bellows or extension tubes to increase the lens-to-sensor distance, and a camera. It is useful to know where a sensor is located and this is indicated on many DSLR cameras as a circle with a line in it.

and magnifications. The number of companies manufacturing and selling bellows and extension tubes has significantly declined in the last decade.

Macro Lenses

When using extension tubes or bellows for close-up and macro work, ordinary lenses are not ideal as they are not well corrected for short working distances. A common or normal camera lens has been corrected for relatively long object-to-lens or working distances and uses typically a short lens-to-detector distance. When this type of lens is used with short working distances, the resultant photos will exhibit flaws, since there are a number of performance issues that will become evident when used for this type of application. Optical performance can be markedly improved by using the normal camera lens reverse mounted, so that the rear element of the lens is closest to the subject. Reverse mounting rings are available for this purpose but many of a lens's automated features will be disabled.

Focusing, Depth of Field, and Diffraction

To achieve a specific image size in close-up photography using a DSLR camera, traditional techniques are abandoned. To achieve critical focus in close-up work, first establish the required reproduction ratio on the lens. The image can then be focused by changing the working distance, which involves moving the subject or camera closer or further from one another. This may be done precisely by using a focusing rail. A quality rail has calibration marks, smooth adjustment controls, and will lock in place creating precise control over its operation. The rail should be attached to a tripod or camera stand creating image stability and precision of magnification.

Focusing can be tricky in close-up and photomacrography. A lab jack or focusing rail can be a useful tool for achieving precise control over focusing. A quality lab jack is preferable to an inexpensive one. There will be a lot of wiggle or play in the focusing apparatus when height adjustments are made using inexpensive equipment. These negative movements will cause changes to a photograph's composition. Moving the camera can also be accomplished by adjusting the height of the lab stand or tripod up and down.

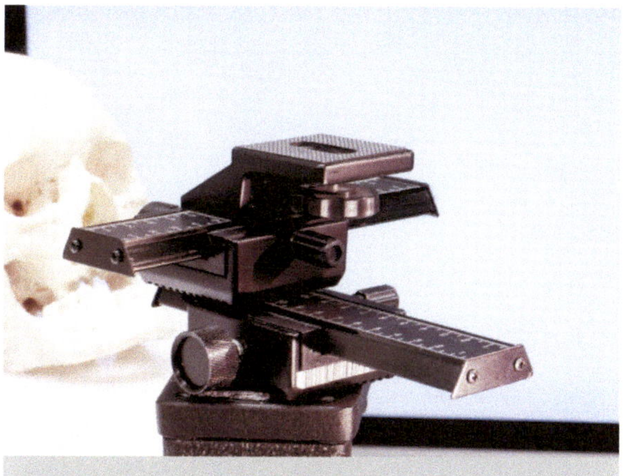

Figure 11.4 This is a focusing rail. Rails can operate in one direction or are capable of moving a camera left and right as well, precisely changing working distance to the subject. Image courtesy of Michael R. Peres.

If a lab jack is not owned, an old lens can serve as a lab jack and can be used for focusing small and medium sized objects. Since the helical focusing mechanism in lenses behaves much like a focusing rail, adjusting the image focus will be accomplished by moving the sample up and down. It is useful to secure the lens's base to the table, if possible, to minimize side to side movements.

The range of defined focus in a photograph is referred to as its depth of field (DOF). Deciding what is in focus can be subjective and viewer dependent. In close-up photography depth of field will be noticeably small and becomes smaller with increased magnification. In traditional photography, an image's depth of field can be improved by using a smaller aperture. As a magnification of 1:1 is approached, using a small aperture (*f*/16–*f*/22) will produce diffraction and degrade image definition. Diffraction effects are dependent on a sample's characteristics and magnification. In general, there tends to be more depth of field behind the subject than in front in more common applications of photography. In close up photography though, the distribution is more equal in the front of and behind the subject. An image's DOF can also be increased using Z-stacking or increased DOF imaging. More is shared about how to achieve increased DOF using computational techniques in Chapter 12.

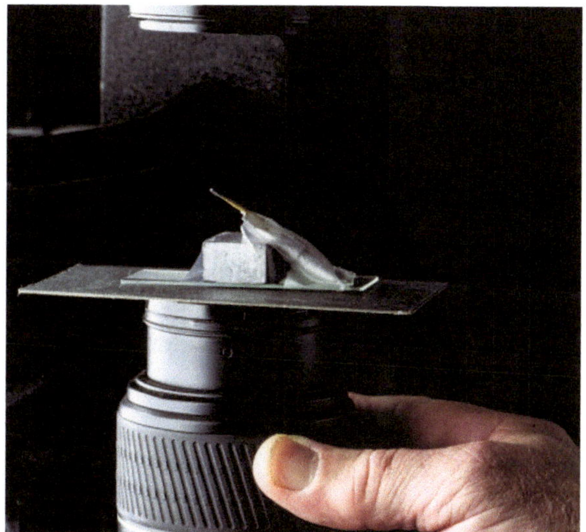

Figure 11.5 Changing the working distance can be accomplished by moving the sample or moving the camera. It is important NOT to change the reproduction ratio on the lens. A lens can be used as a lab jack. Because of the smooth helical focusing mechanism in a lens, small changes in working distance can be made very precisely. It can be useful to cover the top of the lens with a piece of black cardboard as shown in this illustration. Image courtesy of Michael R. Peres.

Creating the Maximum Zones of Focus

Since the DOF in close-up photography is shallow and there is a short working distance, it is useful to align the camera to the sample to maximize the DOF. Once the reproduction ratio has been set on the lens, the camera should be aligned to the surface of the subject. Either the camera or the subject can be adjusted. Figure 11.6 shows considerations for this alignment. When focusing, it is important to roll the focus slightly into the sample. DOF can easily be wasted in the space above the sample. Using a spirit or electronic level can be useful to create surfaces that are parallel. This will create the maximum DOF independent of the lens's aperture choice, and is also affected by image magnification.

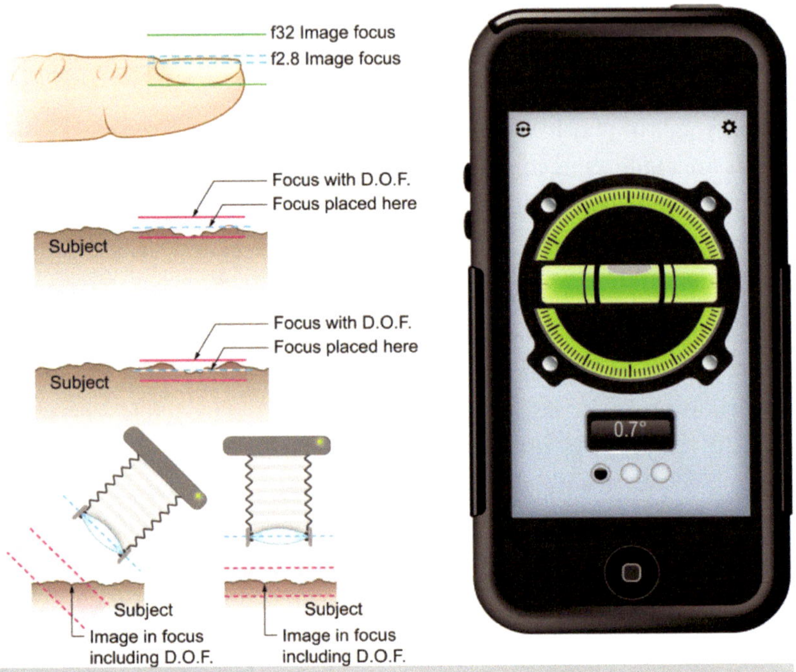

Figure 11.6 Creating and preserving an effective range of focus in close-up and magnified images can be a challenge. Using an appropriate aperture can be most helpful. In the example, the zone of DOF can be increased proportionately by using *f*/32 rather than *f*/2.8. Additionally, rolling the focus into the sample rather than placing it on the top of surface can increase the DOF by a small amount. One of most important tools is camera alignment to subject alignment. Keeping the sensor parallel to the sample surface can be one of the most effective of all three strategies shared in this graphic. On the right is an image of an app from a smartphone for a level. A level can be a useful tool aligning the camera to the subject's surface.

Choosing the Best Aperture

Camera to subject alignment will play a role in the range of focus, but so will the DOF created by the aperture. Using a small aperture will increase the zone of focus, but can also introduce diffraction as the image size increases. Using an aperture that is too open will create a photograph that does not provide enough focus to make an effective image. Figure 7.19 shows diffraction effects on details from the scales on a butterfly wing when various settings of the aperture were used.

Exposure Compensation and Determination

When increasing the sensor-to-lens distance, light will be "lost" in the camera and underexposures will be created if no compensation is added. When using today's sophisticated cameras, calculating a corrected exposure produced when light loss occurs can be accomplished by the camera itself without effort. Contemporary metering systems are highly accurate. As image magnification increases, the amount of light reaching the detector from a given area of the subject is correspondingly reduced. However, exposure evaluation using through-the-lens metering (TTL) equipment is easy and straightforward. Using TTL should be selected as the first choice and your concern should be if the subject is "average" or 18 percent gray in its tonality.

When using hand-held or external light meters, you must compensate the hand-held reading for light loss caused by the increased lens-to-sensor distance. As an example, when using a 1:1 magnification, the rear element of a 50mm lens may be 100mm from the sensor. At this reproduction ratio, there will be two stops of light loss. The amount of light loss for any reproduction ratio can be calculated. There are several equations that can be used to determine light loss and subsequent exposure factor.

A useful equation is Exposure Factor = $(R + 1)^2$ where R is image magnification. In this equation, 1 represents one focal length of any lens used and R is the reproduction ratio expressed as a decimal. The reproduction ratio always corresponds to the increase in extension. For example, if the reproduction ratio is 1:5 or .2, the exposure factor would be $(.2 + 1)^2 = (1.2)^2 = {\sim}1.4$. Once an exposure factor has been calculated it is applied to the measured exposure creating the corrected and adjusted time. To calculate a system's reduction or magnification, divide the image size by the object size. Mag = image size/object size or i/o.

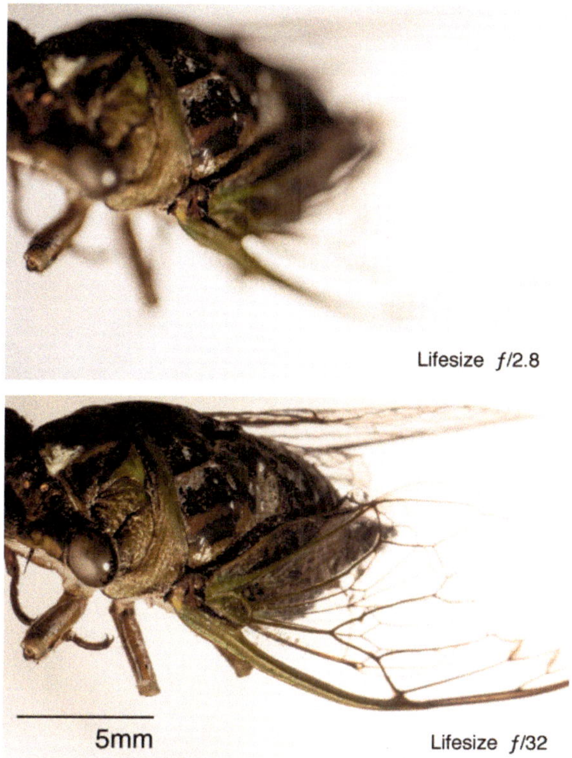

Figure 11.7 This photograph features a cicada where the minimum and maximum DOF was created using a lens magnification of 1:1. The top photo used an aperture of *f*/2.8 and the bottom photo *f*/22. Images courtesy of Michael R. Peres.

Photomacrography

Photomacrography is usually defined as forming images that are in the range of 1:1 (life-size) and up to approximately 50:1 (50 times life-size), but it is more correctly defined by the equipment used to create the magnification. Photomacrography equipment is generally more elaborate and complicated than the equipment used for close-up work. In photomacrography, the lens is further from the sensor than the lens is from the subject.

There are several ways to produce magnified images or photomacrographs using a simple microscope. Photomacrography uses a single stage of magnification. The best lenses for photomacrography are no longer produced. They had special features described below. Many close-up lenses can achieve 1:1 but only the Canon 65mm macro lens creates magnification by itself. Today most of the photographic work done in this magnification range is accomplished using stereomicroscopes. You can read about using a stereo microscope in Chapter 16. Some scientist photographers use a compound microscope equipped with a 1× or 2× objective with varying degrees of success. The resolution of low power objectives for a microscope will be lower than higher quality true macro lenses.

When making magnified images, the object is placed at a distance equal to the focal length of the taking lens. Additionally, the image must be projected to a distance to the detector that is greater than two times the focal length of the lens to create magnification. In this

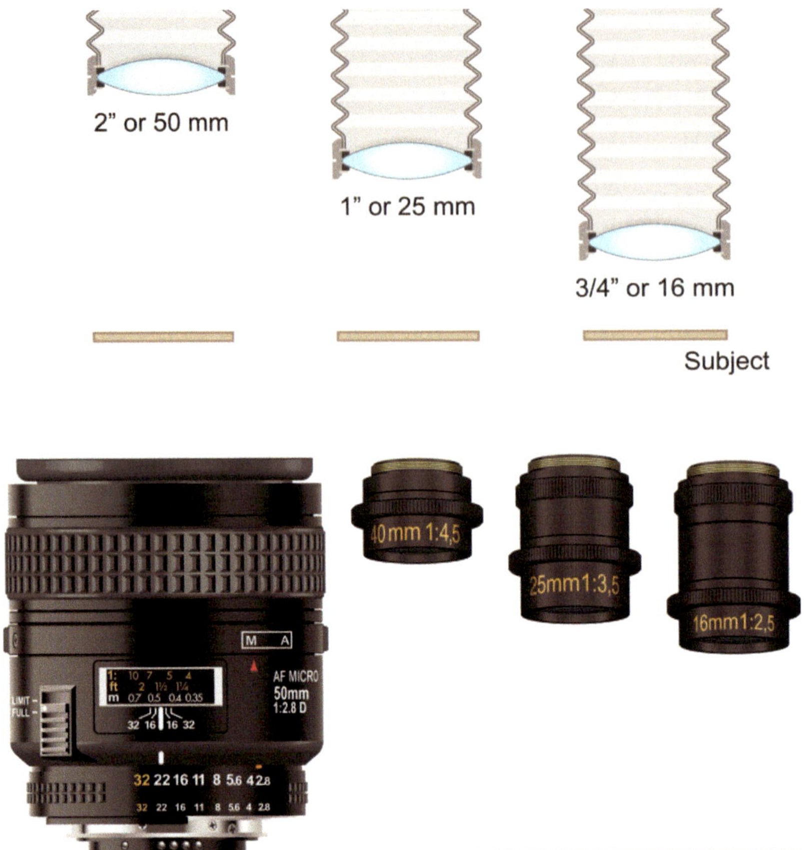

Figure 11.8 A close-up lens looks very different from true macro lenses. On the left is a Nikon 60mm lens and on the right, three Zeiss Luminar macro lenses with focal lengths of 40mm, 25mm, and 16mm respectively. Notice that the macro lenses have no focusing ring and only an aperture adjustment ring. The lens's relative working distances are estimated in the illustration.

magnification range, it is essential to have rigid support of the camera that is capable of moving each of the pieces (object, lens and detector).

Bellows and Laboratory Setups

A bellows is, in effect, a variable length extension tube with a lens board at one end and a location for the camera to be attached (Figure 11.3) on the other. The camera and lens are mounted onto a focusing rail. The bellows extension can easily be adjusted over a wider range than extension tubes, achieving a correct and precise magnification for a sample. Magnification will be limited by the rail's length and the focal length lenses that are used. A bellows is a seemingly simple piece of equipment to operate, but it can be difficult for new users. This happens because knowing where to locate the key elements – sensor, lens and subject – on the rail is confusing. In a "normal" camera operation, two of these three variables are usually fixed. Although some bellows can accept a wide variety of cameras and lenses, many are designed for a specific manufacturer's lens mounts.

True Macro Lenses and Optical Considerations

A laboratory photomacrography camera system described above will usually accept a wide range of apochromatic macro lenses optimized for specific magnification ranges, but versatile enough to be used with a true compound microscope and eyepieces. This link between micro and macro endured for many years because true macro lenses often used RMS (Royal Microscopical Society) threads used for mounting rather than a bayonet mount.

True macro camera lenses are special-purpose lenses designed for magnifications greater than infinity. These lenses may have other names reflecting their size, such as short mount, or sometimes called thimble lenses because they look like thimbles. Macro lenses are unlike close-up lenses. The first and most obvious difference will be that this lens has no focusing collar because the lenses are designed for use on a bellows. Consequently, most macro lenses will have only a lens diaphragm ring, and focus is controlled by changing the working distance, keeping the bellows' length fixed. The aperture range is usually smaller than for traditional lenses used for other technical purposes. True macro lenses will typically have maximum apertures of $f/2.8$ and minimum apertures of $f/16$ or $f/22$. At high magnification, small apertures produce undesirable diffraction effects and poor resolution and are rarely selected by someone who knows the subject.

The diaphragm rings of true macro lenses are seldom marked with conventional f-numbers. The f-numbers on ordinary camera lenses are defined in terms of an object photographed at infinity, which is never the case in magnified imaging applications. Macro lenses are used in short lens-to-subject distances. The effective f-number therefore increases as bellows extension increases. For this reason, some macro lens diaphragm settings are designated by a simple numerical sequence, for example, 1, 2, 3, 4, 5, 6, where each number is a factor of two which creates a one stop exposure value difference. Other lenses use Stolze numbers, which are similar to f-stops where each number 1, 2, 4, 8, 15, 30 is proportional to its adjacent stop by a factor of two.

Other Lenses that Can be Used for Magnifications 2:1 and Higher

Different manufacturers such as Carl Zeiss, Nikon, Leitz, Canon and Olympus produced true specialized macro lenses of excellent quality for years. As previously mentioned, Canon makes a special lens, a 65mm macro $f/2.8$ that creates an image range of up to 5:1. It is very sharp and very useful for creating image sizes that are larger than object sizes. Regrettably it is not possible to use this lens for samples that need to be reduced in image size or magnification that are less than 1:1.

Since camera companies are longer making true macro lenses, they can be purchased from re-sellers on eBay for example. Perfectly serviceable lenses made by Bausch & Lomb® or Wollensak® might also be used but are not as high quality. Also, as a consequence of the evolution from motion picture film to video in the 1980s, wide-angle and normal focal lenses used for 16mm cine cameras can easily be found and adapted for use in this magnification range as well. The 16mm, 25mm and 50mm lenses work well for photomacrography when reverse-mounted. Cine lenses are not as well corrected as true macro lenses, but they can make a satisfactory "poor man's" substitute, especially when reverse mounted.

Setting up a Macro System

1. It is important to estimate the magnification needed prior to starting by measuring the object size and comparing it to the sensor's size. In this way, predicting magnification is simple. When using a full sensor DSLR camera, the sensor will be 24mm or 1 inch in the shorter dimension. The magnification requirement can be calculated using the equation m = i/o, where the sensor represents the image and the object will be measured. For example, if a sample were 5mm in the long dimension, and using the equation m = i/o, a 35mm frame size of a 5mm object would require an image magnification of 7× to fill the frame.
2. The best way to calculate the magnification of a single lens system is by using the equation

   $$v = (m + 1)f$$

   where v is the image distance, m is the magnification of the system, and f is the focal length used. This equation can be used effectively for magnification calculations in close-up photography or photomacrography, independent of the format of the camera system utilized.

   Using the example shown above, approximately 7× is needed. Start by selecting a 25mm or 1 inch lens. Substituting 7 for the magnification and 1 inch for f, a bellows length of 8 inches or 200mm would be required.
3. Using a 1 inch lens, creates 8 inches of separation between the lens and camera sensor. The sensor's location can be determined by located the circle ∅ on the camera body.

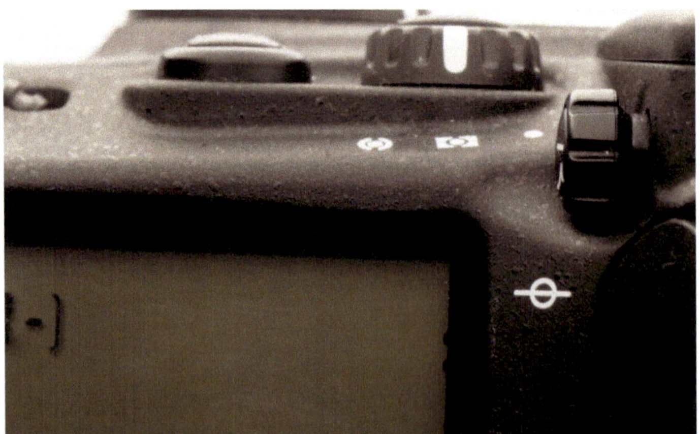

Figure 11.9 Location of sensor within a DSLR camera body useful for technical photographic applications such as photomacrography. Image courtesy of Michael R. Peres.

4. Locate the subject at a distance of one focal length or 1 inch in this example from the lens and change the working distance to focus the image. Do not change the bellows length.
5. The camera can be operated in a tethered or Live-View mode and the exposure can be assessed in the preview window. The amount of light loss in the system can actually be calculated using the method shown below.
6. Focus by changing the working distance.

Exposure Compensation for Photomacrography

As the lens is moved further from the sensor to increase magnification, light loss will occur in the system. This is the result of light behavior and can be described by considering the inverse square law of illumination. A camera with an internal metering system will automatically adjust for the light loss, however, with some equipment, it can be helpful to predict the exposure change when using an external method. Light loss in a macro camera system is referred to as the bellows factor. If using an external meter, the factor needs to be applied to an external meter reading to obtain a correct exposure.

Exposure Factors

There are several equations that can be used to determine an exposure factor, but the two most widely used are:

$$\text{Exposure factor} = \left(\frac{\text{image distance}}{\text{focal length}}\right)^2$$

The following equation is equivalent to the previous one because the image distance is the lens focal length plus any extension, making the equation:

$$\text{Exposure factor} = \left(\frac{\text{extension}}{\text{focal length}} + \frac{\text{focal length}}{\text{focal length}}\right)^2$$

but extension/focal length is equal to R, and the second term is 1, which brings us back to:

$$\text{Exposure factor} = (R + 1)^2$$

This equation was previously discussed in close-up photography. The $(R + 1)^2$ version is easier to use if scales or rules are substituted in the specimen plane needed to determine reproduction ratio. The (image distance/focal length)^2 version is easier to use if a graduated bellows is used to determine image distance (remember to add the focal length to the bellows extension reading to get the entire image distance if the bellows scale reads "0" at the infinity focus position). The exposure time indicated for a selected *f*-stop with a hand-held exposure meter is multiplied by the exposure factor.

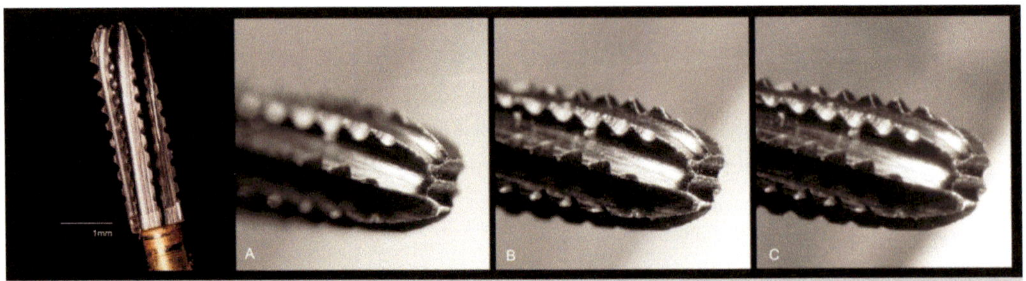

Figure 11.10 This illustration reveals the DOF and sharpness of a dental drill burr (left) photographed with a magnification of approximately 5×. Photograph (A) was made with the lens wide open at *f*/2.0. It shows the lens's least amount of DOF and an effective resolution. Photograph (B) was made at *f*/11 and reveals a modest gain in DOF at no real expense to resolution. Photograph (C) used the lens's smallest aperture, *f*/22 and reveals what happens to an image's sharpness that is negatively impacted by aperture-induced diffraction. On the flip side, there is a considerable increase in DOF. Image courtesy of Michael R. Peres.

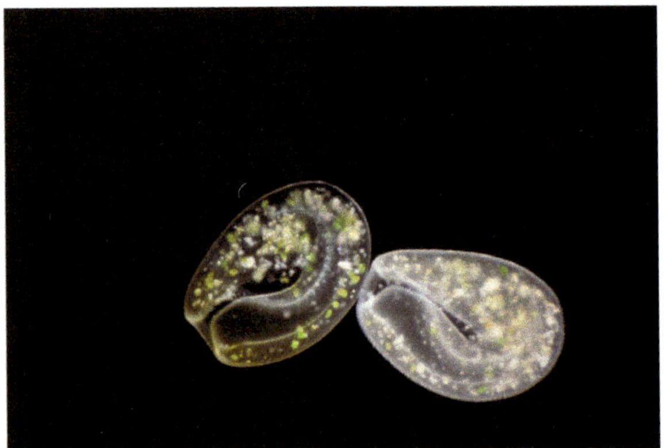

Figure 11.11 A swimming *Bursaria truncatella*. The photograph was made using a simple microscope equipped with a 25mm lens. A camera magnification of 4× was used. Darkfield illumination produced from an electronic flash was used to maximize image contrast and produce a dramatic and very sharp result. Image courtesy of Michael R. Peres.

Depth of Field

In a photomacrographic camera system, relatively no DOF is produced. Depth of field increases as a lens is stopped down, but unfortunately there is a fundamental limit as to how much a lens can be stopped down before degrading an image, even in the plane of sharpest focus. This softening of the image is the effect of diffraction caused by the bending of light as the light passes through small openings. This process leads to a lessening of image sharpness. More on diffraction can be read in Chapters 7 and 16.

Resolution of an image is affected by many factors. Image resolution is largely assumed to be diffraction limited with data points either side of the specimen focal plane blurred by the combined effects of optical performance and diffraction blur. Higher magnifications will be more affected. In practice, any aperture can be used when magnifications are less than ×5, but as the magnification goes beyond that, image degradation becomes obvious both in the viewfinder and on the subsequent image file. In theory, all lenses will form their highest resolution when operated wide open or at their maximum opening, *f*/1.4 for example.

Case Study: Photographing Ice Crystals (Snowflakes)

Photographing snowflakes requires certain conditions such as cold temperatures and the presence of well-formed crystals. I am most interested in photographing what are described as dendritic flakes that are approximately 1–3mm. This sized image falls into the realm of photomacrography. I use a simple microscope for this work.

Working Outside Below 32° Fahrenheit

Photographing snowflakes must be accomplished in temperatures that are below freezing. I must work outdoors or in unheated structures located outside. To photograph snowflakes, all equipment must also be kept below freezing. This will preserve the ice crystal for as long as possible. I place snowflakes onto clean microscope glass slides for ease in handling when using a simple microscope. I have learned from experience that pristine snowflakes have a short lifecycle. They can melt or sublimate which can cause evaporation. My microscope is located in my garage. I have also found in practice that when photographing, temperatures really need to be below 30°F since there is a microclimate in my garage that is a degree or two above the outside air temperature. When temperatures are above 30°F, crystals melt too quickly for photography. Acclimating equipment to the cold takes more time than might be imagined. Isolation from the wind and falling snow is also very important. A garden shed or a tent might be another good space if it is large enough to accommodate a stable table and other required pieces of equipment.

Catching and Transporting the Flakes

The best way to catch and identify good snowflakes for photography is through the use of a piece of black velvet placed in a baking tray. Cardboard will also work. Black velvet is ideal because it allows for easy identification of the clear flakes, but also provides for the easy lifting of flakes using a sowing needle taped to a pencil. Carefully pick up the very small delicate ice crystal and transfer it to a 1 inch × 3 inch glass slide. I put a static charge on the needle by rubbing it in one direction a few times before use. Efficiency is important. By the way, do not breathe heavily onto the slide once the snowflake has been transferred because you may accidently blow it away, or worse yet melt it.

Achieving Magnification

Snowflakes come in many shapes and sizes. I find working in the ×3 to ×8 range is a good fit for photographing most ice crystals. Achieving this magnification can be accomplished using many types of equipment. I use a simple microscope. My system uses a standard medium focal length macro lens attached to a bellows. I prefer a 25mm lens. If you wanted to use a 60mm macro lens, approximately 8 inches of bellows (separation) from the sensor is required to produce this magnification. A shorter focal lens will provide higher magnifications. A stereo microscope will also work and so will a compound microscope equipped with a 1× or 2× objective. Most compound microscopes come with illuminators. Coupled with the secondary magnifier of a microscope, the magnification will be (sometimes) too high for photographing an entire flake.

Attaching a Camera to a Microscope

Coupling a DSLR camera to a compound microscope can be difficult or easy depending on many factors. If you use a camera with an attached lens, such as a smartphone, place the smartphone camera's lens at the eye point produced by the photo ocular. When this is accomplished, the camera's lens will relay the image from the microscope to the sensor. I usually suggest setting the camera to its widest focal length setting when using this technique.

Hand holding any camera will be very difficult if not impossible if good results are needed. The alignment of the camera is critical. Most of the time, I use a tabletop copy stand and I suspend the camera over the instrument. If you do not have a vertical stand, you can use a tripod and locate the microscope on the ground. I have also cannibalized an old enlarger by removing the enlarging head and replacing it with a wooden camera plate. The plate has a ¼–20 bolt/wing nut for use as a camera attachment. This stand has worked quite well when considering its cost. If you are fortunate enough to be using a photomicroscope, the camera system may already be integrated. If you are using a DSLR camera and a bellows, suspend the camera system over an illuminator.

Illumination

Lighting is a critical element required for the production of interesting photographs of all types. Almost exclusively I use a fiber optic illuminator for snowflakes. This light is used to

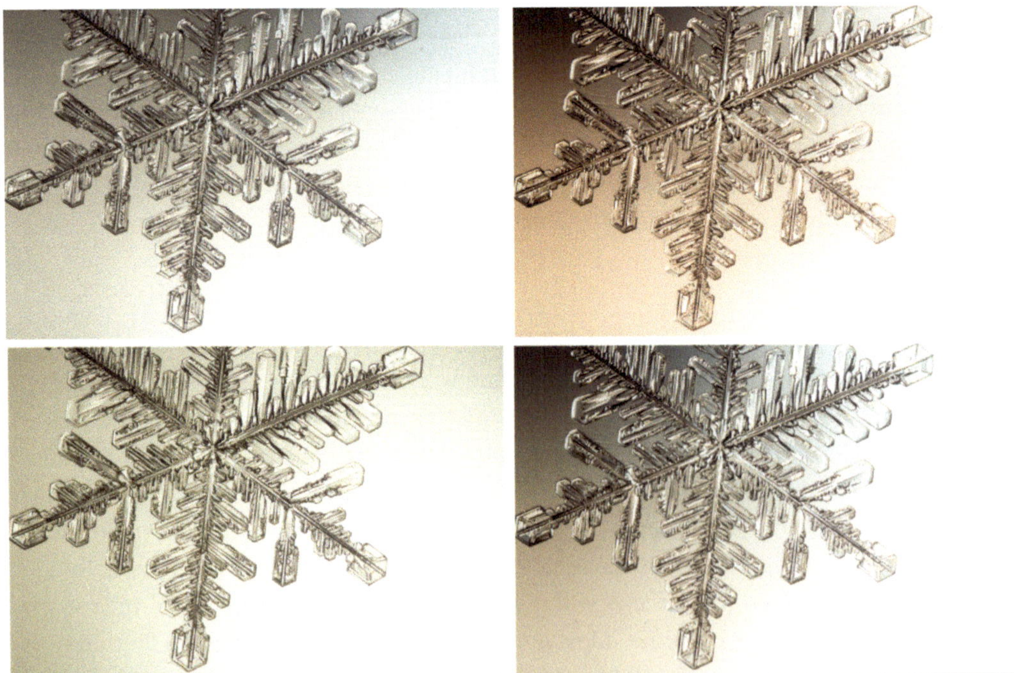

Figure 11.12 This composite photograph shows how small changes to the colored material held below the crystal can change how the various internal surfaces can be emphasized. Images courtesy of Michael R. Peres.

supplement the microscope's built-in illuminator and will lead to images with interesting internal reflections. The snowflake's structure is much like a diamond with its many facets. Producing light that creates internal reflections and refracts in the flake's numerous facets is my ultimate goal. I sometimes use color gels and color backgrounds as well as monochromatic strategies. One of the things I enjoy the most is learning how to light the flakes.

Be sure to trigger the camera using the self-timer.

Suggested Reading

Biomedical Photography, by John Paul Vetter. Focal Press, 1992.
Close-up Photography and Photomacrography, by Henry Louis Gibson. Eastman Kodak, 1970.
Handbook for Scientific Photography 2nd ed., by Alfred A. Blaker. Focal Press, 1989.
Macro Photography, by Rob Sheppard. Pearson Education, 2015.
Manual of Close-Up Photography, by Lester Lefkowitz. Watson-Guptill, 1979.

Chapter 12 Computational Photography

Sometimes a photographic outcome is inadequate for any number of reasons. Sometimes there are technical problems or sometimes the DOF for example at a given magnification is inadequate. An image with a very shallow depth of field or a camera that produces lower digital resolution than needed can be a source of frustration and disappointment when images fall short of expectations. While those realities are possibilities using single shot capture, digital technology provides methods that can effectively fix many image shortcomings. These methods use many "component" pictures that are used to form a new derivative picture from the component pictures.

Photographs that are the result of using other photographs are described as computational photographs. While the term computational photography is not universally used to describe this processes in particular, the term suggests an image is the by-product of processing other photographs to make the new one. Making computational pictures using today's powerful imaging tools can create unique and highly informative images both for artistic or scientific applications. This chapter will share the best practices used for increased depth of field (DOF) imaging; wide-field high-resolution imaging; and high dynamic range (HDR) photography.

Making images with more DOF, that have more digital resolution, or that contain a wide dynamic range of exposures, requires the production of "frames or image slices" as they are sometimes called. It is common in some macro and microscopy applications to use many files. It is not far-fetched to need more than 50 individual frames to form the very best images. Making files for derivative pictures requires careful attention to the initial file's creation, the image processing, and folder management. Since all of the files will be used to create the new file, it is imperative the files are all processed in the same way using batch processing. All image processing software operates in slightly different ways, however, using a repeatable Best Practices Workflow should be the goal.

Batch Processing Files Using Adobe Lightroom

It is impractical and not best practice to process individual files prior to using them in a computational photograph. It is more practical to process the files the same way and in a batch that will be used in the computational photograph rather than processing each file individually. Batch processing is a real time saver. When the images are processed in the same way, their characteristics will also be similar if their initial capture attributes were the same. Reducing the number of pixels in a group of images is an example of an image-processing function that lends itself to batch processing. Opening 100 files, reducing the image size, and then Saving As, all in individual steps, will take more time than expected. Running batch action on a folder of images will accomplish the task very quickly. There are many global image processing tasks that work well in batch processing. These include auto-contrast or auto-brightness, for example, if the circumstances warrant those corrections. A word of caution. Care should be taken when choosing to use auto "anything". The results from auto-processing may actually be disappointing. It is practical to test how batch processing works when applied to the files if variability exists across the image slices or if there were various types of subject images.

Steps for Batch Processing

1. The first step used for batch processing with Adobe Lightroom is to edit one of the photos that will be used in the computational photograph.
2. Once the original file is edited, it is possible to apply those same edits to other image files that will be used to form the derivative photo.
   - Highlight the edited image.
   - Using Control/Command, + Click on any other images that are to be edited.
   - With the required photos selected, choose Settings > Sync Settings from the menu. You can also click the Sync button in the lower right of Lightroom or click CMD/CTRL + Shift + C.
   - Make sure the proper settings are checked. It is not practical to batch process cropping or dirt removal.
   - When all image files have been identified, run Sync on the selected files. The desired changes will be applied to all the photos that were selected.

Batch Processing Files Using Adobe Photoshop

To image process a folder of images using Adobe Photoshop, you must create an action or macro as the first required step of batch processing. The sequence and specific image processing steps must be selected and recorded when making an action. Making an action creates a series of recorded commands that can be saved for future application. Once the steps are created and saved, name the action.

1. To program an action in Photoshop, open the Actions panel found in the > Window pull down menu. It can be also be found as an > Automate function in the pull down menu below File.
2. Once launched, the steps used should be recorded using an option located in the dialog window that starts the recording function. Once initiated, activate the Record process. It is recommended to have a file open prior to the starting of the recording sequence. If the Open File step is recorded into the action, it will need to be deleted prior to running

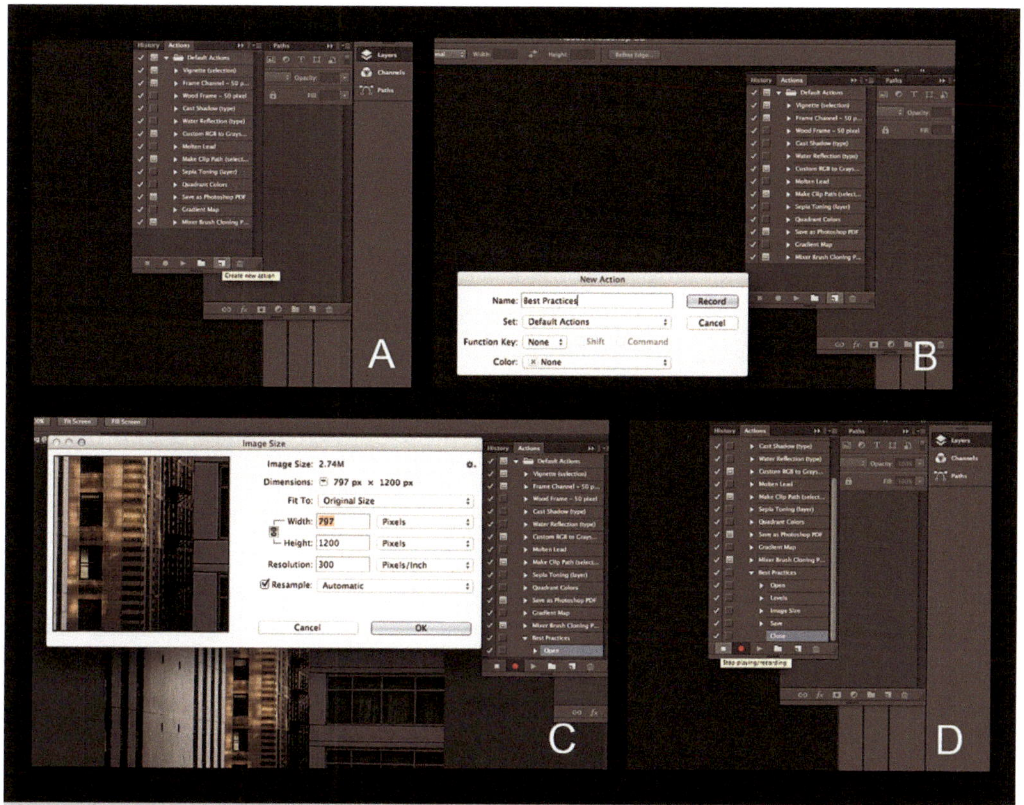

Figure 12.1 Creating an action for batch processing images requires creating many programmed steps. The steps that were chosen are identified in the action's dialog box. The following composite picture shares what would be visible in each window. Step 1: Create the action. Step 2: Name the action. Step 3: Record the steps in the action. In this example, Step 4 would resize the image. Step 5: Stop recording. Adobe product screenshot(s) reprinted with permission from Adobe.

that action. Removing a step in an action is easy to accomplish by selecting it and then deleting that step.

3. Once the recording has started, the image processing steps are programmed sequentially by the software. In sequence, each of the various adjustments are recorded by the action. As you record the steps, it is recommended to always use Save as needed to preserve native files. It is also recommended to save files into a new folder where they can be segregated from the native files. Once the image processing steps have been identified and concluded, push the Stop recording function. The action must be named. When finalizing an action it remains important to program the Close file step, as well.
4. Running an action is relatively easy once the action has been created. Running an action begins by going to the Automate function located in the > File pull down menu. Once in the Automate section, select > Batch.
5. Choose the action that has been recorded for the image processing steps that are needed.
6. Select the Source folder where the images to be batch processed are located.
7. Select the output folder where the saved files will be re-directed.
8. Start running the action by pushing the OK button and Photoshop software will batch process all the files in the source folder using the steps programmed into the action.

Increased Depth of Field Imaging Methods

In close-up photography and other magnified imaging techniques, a limited amount of DOF is produced by a lens. The focus range is influenced by the system's magnification and the aperture selection. As the magnification of the system increases, the image's DOF will be reduced. Some images may have only a few microns of DOF. There are only a few if any optical methods that are useful for increasing an image's DOF, but frequently these lead to a loss of image sharpness because of the influences of diffraction – see Figure 7.19. Closing the aperture of the lens or the aperture diaphragm of a microscope can increase an image's DOF but these images will be degraded and possess less optical resolution.

A practical approach for creating images with more DOF uses increased depth of field techniques sometimes called focus stacking or making a z-stack. This method is also referred to as z-stacking. As its name implies, it produces images with greater DOF than possible from a single image by itself. Increased DOF imaging blends individual image slices – by first organizing them on top of each other using software. The image slices are placed into individual image layers similar to a stack of cards. Once the files have been placed into layers, the software seeks out contrast differences between like pixels in each of the image slices. Where the sharpest focus in adjacent pixels (edges) exists, that data is maintained and all the rest is minimized. These regions are isolated and used to make a new derivative image from the high contrast regions of all of the individual image slices. There is a lot of computing required to identify and process the out-of-focus regions and then re-blend the in-focus regions into a new and completely focused image.

This method – like anything – requires careful attention to detail, and even then, it may NOT work as expected. There are several recommended software products used for increased DOF imaging including Adobe Photoshop, Helicon Focus®, and Zerene Stacker®. It is also possible to use Image J and GIMP. Research grade photomicroscopes also come equipped with software that can produce increased DOF images. Increased DOF images can be programmed into the instrument's camera software. The automation feature in the software changes the pre-programmed working distance and then operates the image processing features to create truly excellent outcomes.

Helicon Focus, Zerene Stacker and Adobe Photoshop software are commercially available. I have found that each of these products works differently and leads to various results when used on different types of samples that have different magnifications. I have found it necessary sometimes to try each product on difficult samples because one will invariably work better than another. The software is not overly complicated to use and the knowledge needed to operate one is easily transferred from one product to the other.

Making Image Slices

It is important to create a photographic and lighting approach where nothing changes between exposures except for where the focus is placed in the individual frames. It is important for all of the exposures to be identical as well as all of the various image attributes. To minimize any deviation in exposure and other image attributes, it is best to use the camera's manual

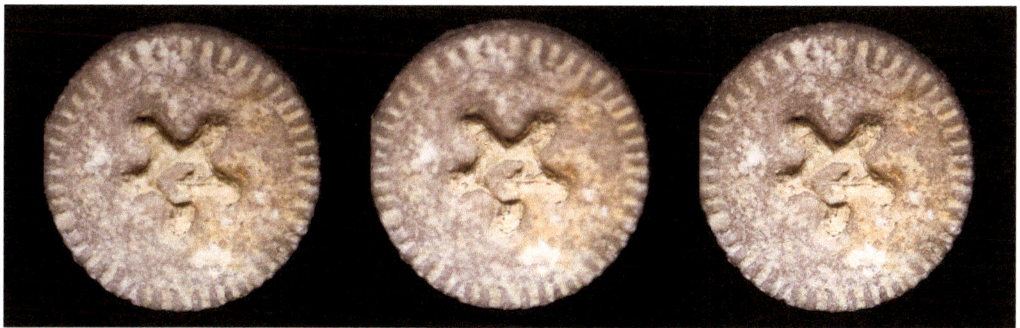

Figure 12.2 This composite photograph contains an increased DOF image on the right. The focus in the left and center images was placed at a different plane by changing the working distance by several µm. The subject was 4mm in diameter and features a fossilized worm segment collected in the Mazon Creek bed, Illinois. Images courtesy of Michael R. Peres.

exposure mode and set the *f*-stop, white balance, shutter speed and corresponding ISO to the same settings. It is also important not to adjust the lighting or brightness of the light(s) during the photography. Because found light is sometimes changeable, photographing with ambient light can be difficult because of exposure and other variances. Subjects that move cannot be easily z-stacked. It is also important to focus the lens using manual mode during each of the individual exposures. I also recommend using the lens's optimal aperture which is often 1.5 to 2 stops from wide open. Never use the aperture completely closed down (e.g. *f*/32) for this work. The optimal aperture will create the maximum optical resolution. A lens's optimal diaphragm setting can be determined from the lens's specification data sheet.

Focusing Rails

Once the correct exposure has been determined and an optimal aperture selected, the making of the image slices for the increased DOF image can begin. It is possible to move the sample up or down. There is no advantage to either travel direction. The more individual slices that are made, the better the results. Producing too few image slices can create problems because the software may struggle to form data where insufficient pixel data is present in the files. It is also important to try and move the sample or the stage using the same distance for each image slice to the extent that is possible when working manually and without automation.

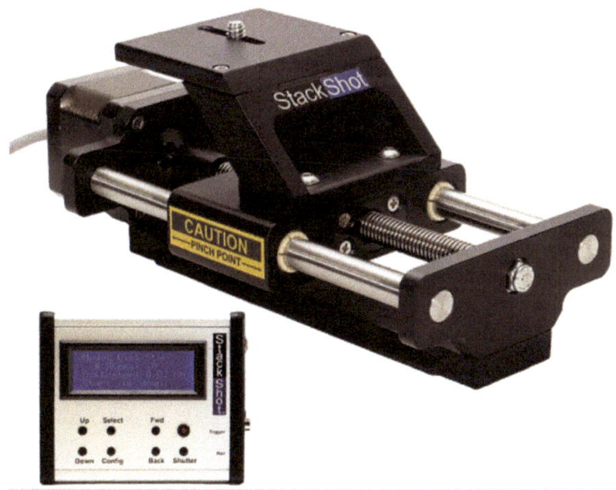

Figure 12.3 Pictured above is the Cognisys® Stackshot system. It is an excellent tool designed to precisely control focusing distances using a close-up rail. It produces tiny and precise distance changes automatically after being programmed into the controller. The focusing rail is capable of moving the lens/camera system by as little as 2µm and can cover up to a 10cm distance. A USB port and a computer is required for operation of this device. Image courtesy of Cognisys.

A focusing rail is an excellent accessory for this task. Many focusing rails are calibrated and include printed distance increments. When making each of the individual frame's exposures, NOTHING can change in the set up except the working distance. If any structures in, or on the sample move, a ghosting image defect will be created in the new derivative file. Sometimes, nearly transparent objects lead to unsatisfactory results. When making individual image slices, it is important to create overlap of the in-focus areas. It is important to capture all of the exposures needed when shooting because it will be IMPOSSIBLE to recreate a replacement image in the same precise manner – if needed – for integration into the stack of the previously captured images. Image acquisition should include regions above and below the subject. It is recommended to capture image files using a RAW file when using DSLR cameras.

Processing Files Used in Computational Images

Once a set of image slices have been successfully recorded, they should be moved into a named folder. Image optimization is suggested prior to creating the z-stack image. It is fundamental to the success of the final image that all the slices can be processed including the same adjustments to each file. Depending on what image processing software is used, it is recommended to run an action on the files. Adjusting them all together leads to matched image characteristics and color balance. It is also useful to pre-sharpen files a small amount. Once all of the image characteristics of each frame are matched, the files should be saved as TIFF files. I do not recommend using unprocessed RAW files for stacking. Your results might exhibit flaws.

Another consideration for building an increased DOF file might be the individual image's capture resolution. Many contemporary DSLR cameras are capable of recording more than 7000 pixels in the long dimension. Depending on the software and number of slices, the amount of cumulative image data may exceed your software's and computer's RAM capabilities needed to process such a HUGE file. Unfortunately the largest number of pixels that can be handled by the computer can be determined only by trial and error. If a file has become too large, it may cause a computer to freeze.

How many pixels are required for a specific output has been a prevailing question for practitioners of digital and scientific photography since its beginning. A current HD monitor displays 1900 pixels, while a VGA monitor will display just more than 1000 pixels. Output should always drive input. Downsizing large files can be very helpful when preparing work that will be displayed on only a display and not in print.

Making a Z-Stack Using Adobe Photoshop

The individual files that will be used for the stack need to be imported into Photoshop as layers.

1) Select all of the required images using Adobe Bridge or Lightroom. In Bridge, under the pulldown > Tools menu; select > Photoshop; > Load Files into Photoshop Layers.
2) Once the files have been loaded into Photoshop layers, the images must be aligned prior to blending. All of the various files in the layers must be made active by manually clicking each layer while holding the shift key or under the > Tools menu, select all layers. Once the layers are selected, and under the > Edit menu, > Auto-Align Layers.

Once aligned, the > Auto-Blend Layers feature should be executed by selecting the > Stack Images option. In that window, check the seamless tones and color dialog box. The software will then work through each file isolating the pixels that have the highest contrast needed to build the new file from the individual frames.

In a relatively short time, the stacked image with increased DOF will be available for preview. The final image is created from the individual layers and each file still can be reviewed within an individual layer by clicking on the eyeball in the software window if desired. If the result is satisfactory, the file's layers can be flattened using Save As. It is possible to perform some touch up on the file using Photoshop.

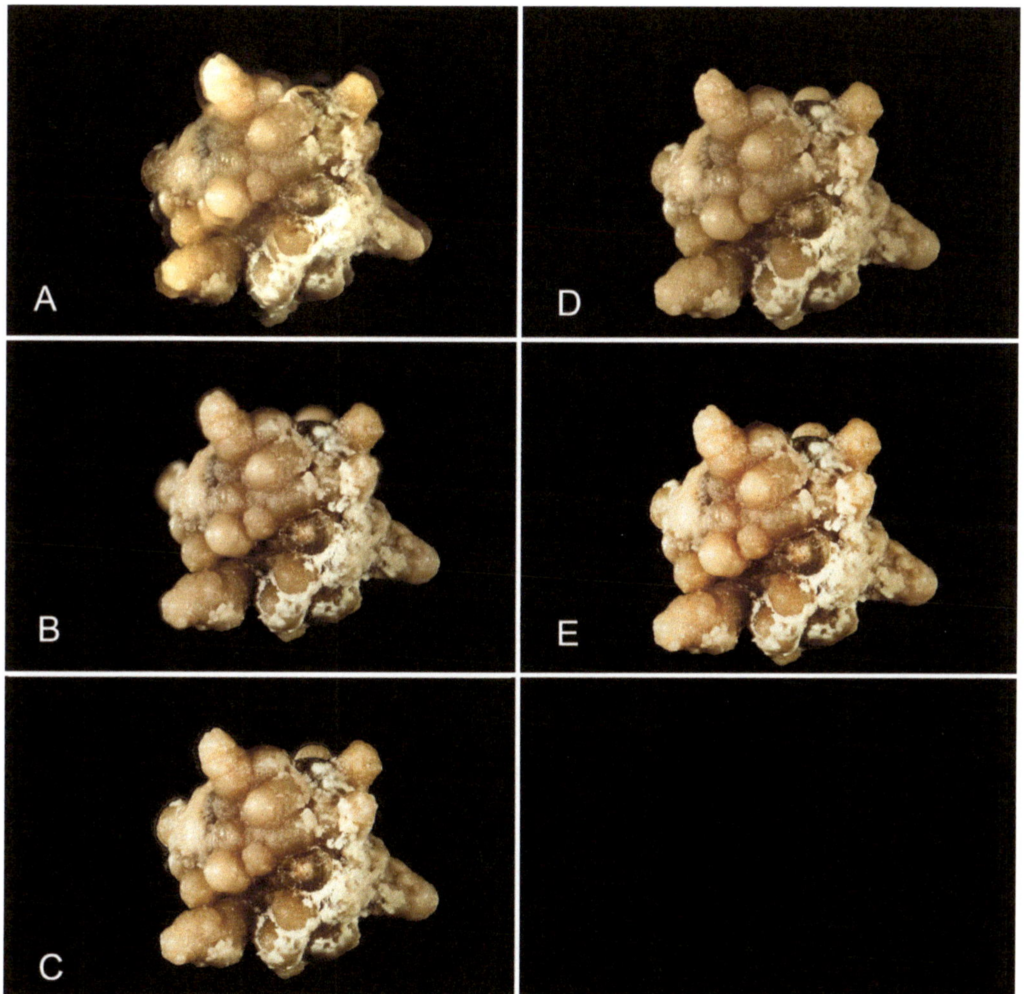

Figure 12.4 This composite photograph of a kidney stone shows the results produced by Adobe Photoshop, Helicon Focus, and Zerene Stacker. Image A reveals artifacts using Photoshop. Image B used Helicon Focus in the Depth Map mode and also contains artifacts. Image C used Helicon Focus in the Average mode and shows improvements. Image D was created using Zerene Stacker in the PMAX mode and in Image E, Zerene Stacker was used in the Depth Map mode. Image D created the most successful result. Camera capture image files courtesy of Staffan Larsson.

Using Helicon Focus

Using Helicon Focus software can be easier and faster than Photoshop. There is one command to add the image files using the "add image" command or the files can be added by dragging them into the program's load window. Once the files have been added, the files can be rendered using Density Map or Average stacking modes. The stacked file will appear in a window while it is being constructed. Some subjects will benefit from using the Density Map method while others will do better using the Average selection. A visual inspection of the stacked file will reveal which mode is best for any given type of subject. Do not sharpen or perform other image processing using this image stacking software. Image adjustments to the stack file will be better managed using image processing, including sharpening. Once satisfied with the stacked file, save as a TIFF file and flatten if warranted.

Using Zerene Stacker

Using Zerene Stacker is also faster than Photoshop. You can "add image" files using the add image command or you can add them to the software by dragging the files into the program's load window. Once the files have been added, the files should be aligned before they are rendered using DMap, PMAX, or BOTH modes. Similar to the recommendation when using Helicon Focus, do not sharpen the files or perform other image process using this image stacking software. Image adjustments to the stacked file will be better managed using image processing including sharpening. This software can Save its output images as 16-bit TIFF files.

Creating Wide Field High Resolution Files or Image Maps

When photographing using low power objectives or when creating an image that will be scaled or zoomed, the optical or digital resolution in a camera by itself may be inadequate. This reality leads to results that are unacceptable for publishing or large scale display.

It is relatively easy and fast to make a single photograph of an entire sample using a low magnification lens where the image may not delineate structures within the sample because of poor optical resolution. The need to create files that are highly resolved and interactive is not new but as tools continue to evolve, it is easier to make widefield high-resolution files. Producing high-resolution digital files is quite simple but takes care and patience. The images produced from widefield techniques are scalable and allow viewers to zoom in and out, or move around the image – much like Google Maps. This allows users to see more information within the image and allows the images to be published onto Internet sites in a fashion similar to Google maps where the images are tiled.

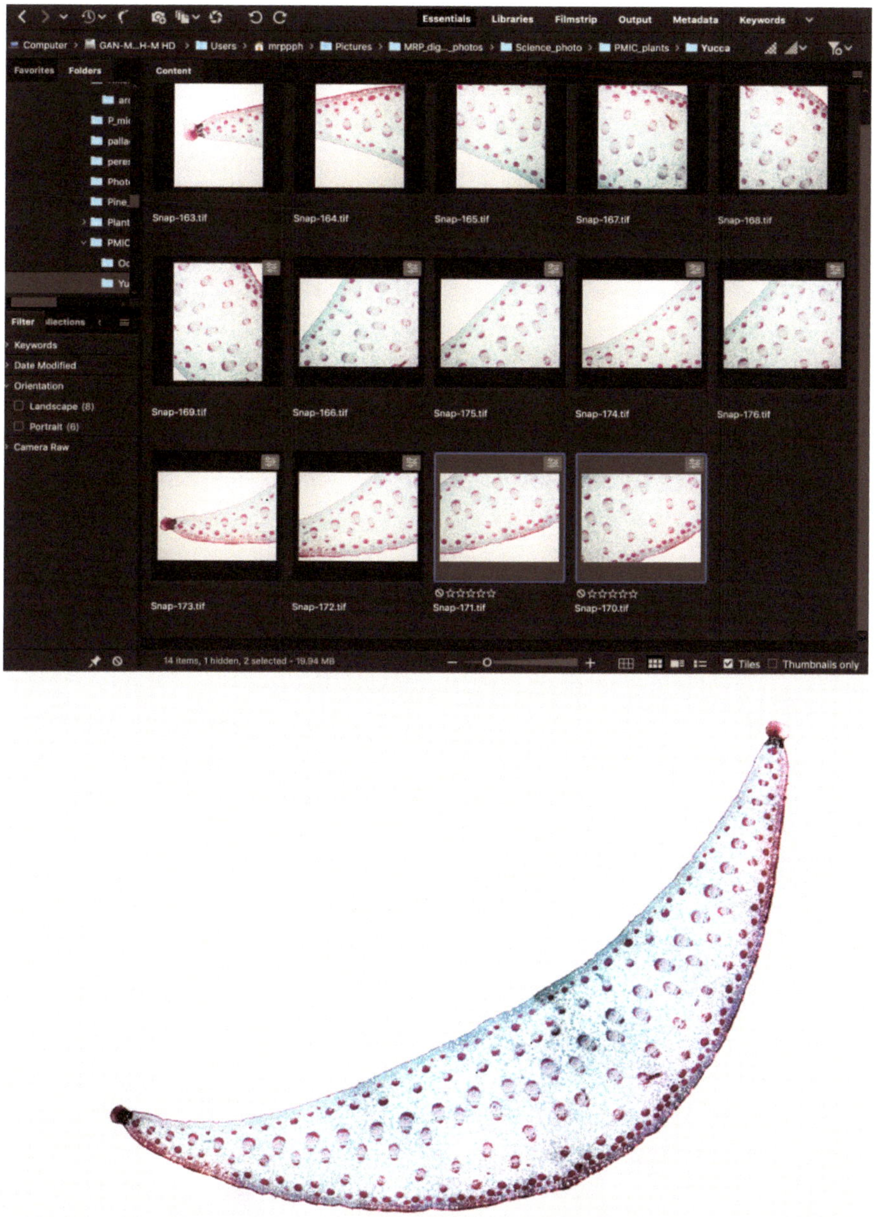

Figure 12.5 This figure shares the screen appearance of files ready to be imported to Photoshop's Photomerge function. When shooting files for use in a map, it is important to overlap regions of each of the individual frames. Staying organized during image recording is tedious, imperative, and requires paying careful attention to location on the sample. Some photomicroscopes have an automated stage that is programmable, precise and easily manages this process. The subject in this photograph is a Yucca leaf. Adobe product screenshot(s) reprinted with permission from Adobe. Composite image courtesy of Michael R. Peres.

Methods

Producing individual files required to make a widefield high-resolution image is not particularly difficult. Creating the individual files needed for the process needs to be carefully managed. Each field should contain some overlap to the adjacent regions included in each frame in all directions. In a process that is derivative, the end results are influenced by the sum, and quality of the individual parts.

The numerical aperture of an objective or the aperture of a macro lens will play a significant role in the optical resolution of the image. The higher the magnification of the objective, the higher an objective's numerical aperture will be. The higher the magnification of the image, the narrower the field of view of the lens. This becomes a problem in this type of work. Images made from objectives with a wide field view creates low optical resolution and images that are more highly magnified will be well resolved but will not include the entire sample.

To create both outcomes, individual images photographed using a more highly resolved lens are combined to form an improved image. It is important for all of the exposures for each frame to be identical in all of their image characteristics. It is vital to minimize any deviation in exposure and other image attributes when making the individual frames. Using the camera's manual exposure mode, white balance settings, and corresponding ISO when determining the best exposure attributes is recommended. It is important not to adjust the lighting, apertures or diaphragm settings, or brightness of the light(s) during the imaging. Only the field of view and re-focusing (if needed) should be changed. When photographing fields, I overlap the images by approximately 15–20 percent. It is practical to consider the sample as a grid and work back and forth across the sample when making image files. When using an instrument camera and related software, all aspects of the process of image slice formation including focus can be programmed and managed by the software and microscope.

Using low power and low N.A. objectives will create fewer image tiles but the wide field image will be less resolved. Using a very high N.A. objective with high magnification will require making many more individual image files and require significant computing power to create the derivative image. Additionally the camera's digital resolution can affect an imaging system's processing power. Files from big sensors will tax even the most robust computers. It is possible to make image files that are larger than 1GB without much effort.

Once the correct exposure has been locked in and other camera or microscope settings established, the making of the image tiles can begin: photographing left to right, and then shifting down or up, and repeat, then process. There is no advantage when considering how to photograph the frames of the "grid". The only requirement for imaging is that the individual frames MUST overlap each other. Creating too few image tiles can tax the software to stitch the files. Having too few individual frames may create false data when insufficient pixel data is available to overlay. It is also very important to move the sample or stage the same distance for each tile to the extent that is possible when doing this by eye/hand and without automation.

When making the individual frame's exposure, NOTHING can change in the set up except the working distance. If any structures in or on the sample move, ghosting or other image defects will be created in the new derivative file. Images should include the regions outside of the subject itself. The ideal image capture quality file should be a RAW file when using DSLR cameras or TIFF when using an instrument camera.

Image Processing for Mapping

Once the images tiles have been successfully formed, recorded, and moved to a named folder, image optimization is required prior to building the high-resolution wide field image. It is fundamental to the success of the final image that all the frames are processed for the same adjustments. More about these considerations can be found in the increased DOF section.

Creating the Image Map

Creating an image map is easily accomplished using Adobe Lightroom or Photoshop software. There are other software products such as Panorama Maker®, PTGui®, Image J, or Fiji – an image processing package that bundles Java, Java3D and plugins to make a widefield image.

If using Photoshop, the new file can be built using the Photomerge function. This function can be launched using the automate menu bar located in the > File menu. Once loaded, the procedure is pretty straightforward. The software will bring up a dialog box that allows the images to be blended using an Auto, Perspective, Cylindrical, Spherical, Collage or Reposition mode. Check the Blend Images Together feature. Once the images are added, simply click OK and the software will produce the computational photograph. I prefer to use the Reposition mode of the Photomerge function.

High Dynamic Range Images

Many times, objects or scenes will have a luminance ratio that is beyond the sensor's ability to record details in both the darkest and brightest regions. A camera's sensor will have a finite exposure range. Often this range might have an approximate brightness ratio of 120:1 or approximately a 6+ *f*-stop range. Once the brightness exceeds this range, the sensor cannot record useful exposures in both the brightest and darkest regions simultaneously. The middle gray tones are not typically affected by the dynamic range. You can read more about a sensor's dynamic range in Chapter 5.

High dynamic range photography is a method used to create a derivative image built using two or more files that each have different exposures. The human eye can see into dark shadows and bright highlights almost simultaneously. Because a sensor creates a static capture, this is not possible. HDR images record sample information from both the darkest and brightest regions in the sample. Fluorescence imaging is a type of application that can challenge a sensor to capture the entire brightness range in one exposure.

Making Photographic Exposures

The various exposures – the ones above and below the recommended ones and needed for an HDR file – can be produced either using a manual or automatic exposure mode. Many cameras will have an HDR exposure setting and will create multiple exposures very quickly and efficiently without effort. The camera can be programmed to make what is called a bracketed series of exposures. This exposure range is typically −2, −1, 0, +1 and +2 *f*-stops. In an ideal situation, nothing about the subject can change during the image-recording phase of the process except the exposure. Since the images are being blended, it is important to change exposures using time only. Never use aperture, as image attributes such as DOF are affected.

Blending the Images

Once the images have been created, creating the HDR file can be accomplished using a number of image processing software including Photoshop, Image J, GIMP, PTGui, and HDR studio. These programs are simple to use. Many of the same concerns resident in stitching files or increased DOF application should be addressed when pursuing global image processing prior to running the image processing program.

Blending of images with various exposures is accomplished by what is called tone mapping. Generally "tone mapping" applies an algorithm to an computational image based on how images were recorded and stored. The derivative image will be influenced by bit depth. In HDR image processing, the image data is derived from a digital file rather than a scene. In this operation, all the variances resident in similarly located pixels are averaged to create a new file lowering contrast across the subject's dynamic range. Tone mapping is the process by which the software accomplishes this blending of files of differing exposures.

Creating the HDR file is easily accomplished using Adobe Photoshop software. Once Photoshop has been launched, the file can be built using the merge to > HDR Pro function. This function can be found in the automate menu located in the > File pull down menu. Once the files are loaded, the procedure is pretty straightforward. The software will bring up a dialog box that allows the images to be loaded simply by clicking OK. The software will produce the computational photograph. Use the align images features as well.

Other software products will have a different workflow, but in the end will accomplish many of the same tasks.

While useful for making images that contain information in the highlight and shadow regions of the subject, there may be shortcomings for HDR images. Exposure and image brightness is often not a singular element of an image file. Image contrast and color data is also associated with exposure. Because of these linked image attributes, HDR files will often display a unique color palette and shadow characteristics. They may also exhibit a glow around the highlights. HDR images that are overworked have a characteristic look and this appearance may be inconsistent with technical imaging expectations. Many times, this appearance is bad and infers the object is not displayed in its correct tonality. It looks like a computer image and because of this appearance, HDR might not be appropriate

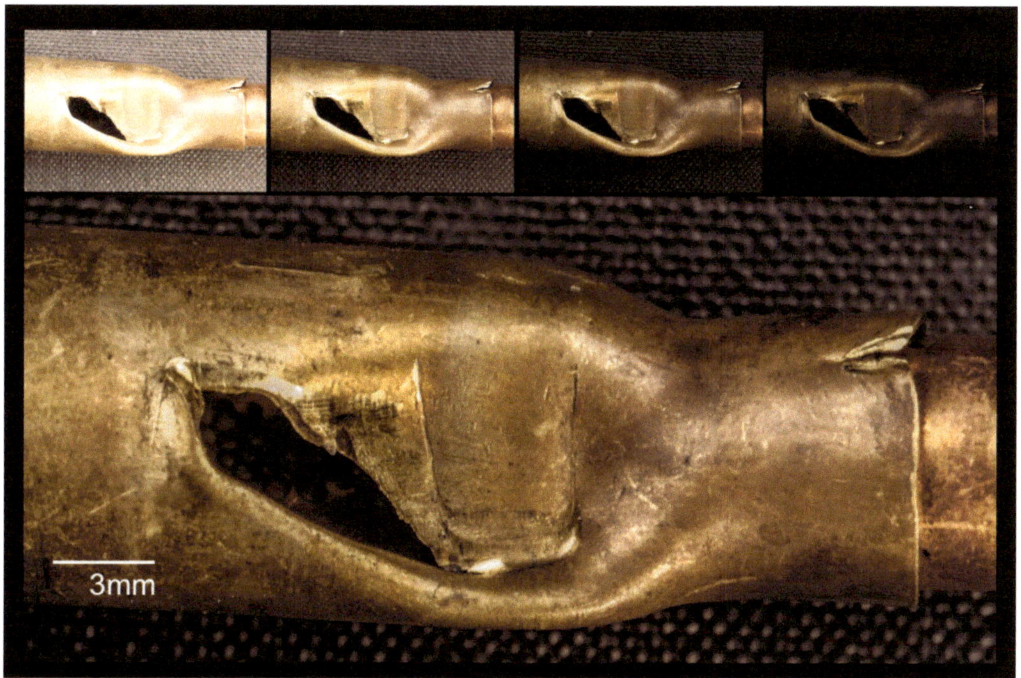

Figure 12.6 This HDR photograph features a 7.62mm cartridge fired from an M143D 6-barreled Gatling gun, often called a minigun. The illustration reveals damage to the shell as a result of a feeding malfunction. The HDR image was beneficial for recording detail in both the highlights from the bent metal, and the interior of the cavity where unburned powder is visible. This would not be possible using a conventional single shot photograph. The four exposures were made using *f*/25 at 1/4sec, 1/8sec, 1/20sec, and 1/50sec. Images courtesy of Adam Hartley.

for scientific applications. Testing should be done to determine the appearances of HDR images. Fluorescence photomicrographs may appear completely false and simply not be acceptable.

Suggested Reading

Creating HDR Photos: The Complete Guide to High Dynamic Range Photography, by Harold Davis. Amphoto Books, 2012.

The HDRI Handbook 2.0: High Dynamic Range Imaging for Photographers and CG Artists, by Christian Bloch. Rocky Nook, 2013.

Chapter 13 Field Photography

Adrian Davies

Introduction

Photography of natural science subjects in their habitats is the best way of conveying the most information about the them. It allows viewers to see where they live and reveals aspects of their behaviors or ecology.

However, photography in the field brings with it its own set of problems such as variable weather, temperature extremes, or inconsistent lighting, for example. Daylight's brightness, quality, and color, varies throughout the day, and the environment in which you may be working offers a multitude of potential problems and hazards. For the recording of subjects in a consistent manner, such as moths caught in a moth trap, it may be necessary to use artificial lighting such as electronic flash or LEDs to achieve a collection of images with consistent lighting, which perhaps were made over a longer period of time.

Various mobile phone apps including Photographers Ephemeris® and PhotoPills® provide useful information about things such as the time and direction of sunrise or sunset, moonrise and moonset, and other useful information for field photographers. There are other apps that provide up-to-date information on weather and tide times as well.

Ethical Considerations

One of the basic precepts of nature photography is that the welfare of the subject must always be considered above creating an image. See Chapter 4 for more about this. Photography should not be undertaken if it places the subject at risk. Make sure, when photographing plants, that you do not trample other specimens, or place tripod legs on them. In the UK, many subjects are specially protected by law and require licenses before photography can be undertaken. The Royal Photographic Society in the UK publishes a Code of Practice for Nature Photographers which deals extensively with birds, mammals and other animals, plants and legal aspects. It can be downloaded from their website. Similar documents, "Conservation Photography" and "Principles of Ethical Field Practices", are available from the North American Nature Photography Association which not only deal with ethics but also various other aspects of photography and conservation.

Another major issue surrounding the ethics of nature photography resides in approaches to image enhancement and manipulation. This has been an issue almost from the birth of photography, and there are many well know examples of images that have been retouched or composited from a number of separate individual images into a derivative and completely different result.

Image processing software today is capable of a huge range of processes, increasingly using artificial intelligence technology to produce perfect outlines around a subject and drop this image into a different environment, for example. There have been many examples of manipulated images being entered into competitions, subsequently winning, and then being disqualified. The effects can be long lasting. One particular example, from 2006, showed a herd of endangered Tibetan antelopes running across an arid plain while in the background an express train speeds over the recently completed Qinghai–Tibet rail line. The image was used extensively to show that the controversial railway was having no effect on wildlife. The photographer, Liu Weiqiang, said later that he had gone to extraordinary lengths to get the image. "It took only several seconds for the Tibetan antelopes to pass in front of me, but I had waited for eight days".

However, the image was subsequently shown to be a fake. It was composed in Photoshop. It caused huge embarrassment in China, Liu was dismissed from the *Daqing Evening News*, and its editor was forced to resign. Perhaps worst of all, however, was the impact of the disclosure on China's claim that the railway had had little impact on the wildlife of the Tibetan plateau. "The truth is probably the opposite of what the picture was trying to claim" Su Jianping, a zoologist at the Northwest Institute of Plateau Biology in Xining, said. "The antelopes are shy, and their migration patterns are being disturbed by the trains. There is no such thing as harmonious coexistence between trains and antelopes".

For the purposes of natural sciences photography, it is vitally important that the integrity of the image remains intact, and nothing is done to alter the content or context of the image. The boundary between enhancement and manipulation still remains a gray area!

Image processing software can also be used to analyze the content of images, to count the number of birds in an image for example, or measure the area taken up by a fungus. Adobe Photoshop has a basic analysis tool which can be used to count objects or produce linear measurements. More sophisticated software is available to perform a huge range of analyses. These include Image J, a free to download open source piece of software available for Mac, Windows and Linux systems. Users can write their own macros for automating a huge number of processes. One example is a toolkit for producing multi-spectral images showing how animals perceive objects. This software can be downloaded from www.empiricalimaging.com.

Equipment Recommendations

Cameras

To photograph a wide range of subjects in the field you will probably need a camera with interchangeable lenses, either a digital single lens reflex (DSLR) or mirrorless camera. You will need lenses with a wide range of automatic or semi-automatic settings for use with Aperture Priority and Shutter Priority exposure modes. The camera must also have the capability of being used in a completely manual mode. Most cameras have auto-focus capabilities (often highly complex), though you may, at times, want to switch this off and use manual focusing.

Mirrorless cameras are generally much lighter and more compact than DSLRs, and are likely replace the DSLR camera in popularity in the next few years. They have a large range of features including high resolution video, focus stacking for close-up and macro photography, and a special feature by Olympus, Pro-Capture. Using Pro-Capture, when the shutter release button is half depressed, the camera starts recording images at up to 60 frames per second, but does not write them to the memory card and keeps the last 14 frames in a memory buffer. When the release button is fully depressed the camera starts recording images to the card, including the buffered 14. This can record images of birds about to take off or whales about to leap out of the water.

Today's image quality is generally excellent and any camera having 20 megapixels (MP) or more will give superb results if enlarged to an 11 × 14 inch A3 print (297mm × 420mm), or used full page in a journal or book such as this one. For photographing animals in dark rainforests or birds, with long telephoto lenses, you should use a higher ISO setting. The quality of images shot with high ISO settings have improved significantly in recent years and ISO settings of 800, 1600 and above will yield excellent results.

When travelling, and where weight is a major issue, you may consider using zoom lenses such as 24–70mm or 70–200mm rather than carrying three or four fixed focal length lenses. Longer zooms can usually be used in conjunction with tele-converters to increase their focal length.

Trail Cameras or Camera Traps

Camera traps have become established as an excellent way of monitoring and photographing animal populations in an area. This often reveals surprise results such as the photographing of a black leopard caught by a camera trap in 2019 by the British photographer Will Burrard-Lucas in Kenya. It was the first authenticated image of the species in Africa in nearly a century. These cameras work in daylight, or at night using infrared (IR) technology and can be left out in the field for many months at a time. They are frequently placed overlooking pathways in forests or near water holes

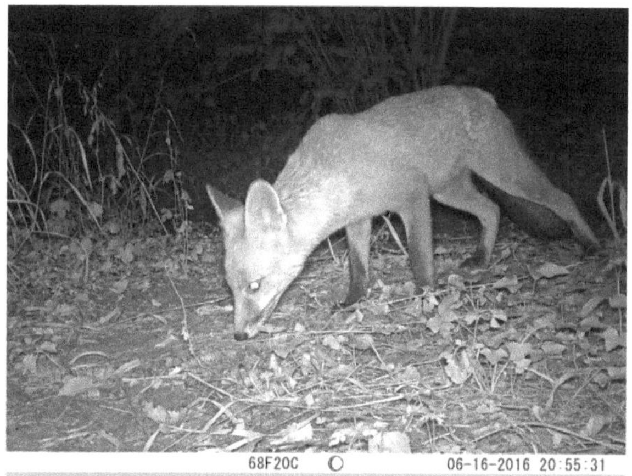

Figure 13.1 Using an IR camera trap, the presence of a red fox (*Vulpes vulpes*) in a woodland was confirmed. Note the date, time, and temperature records onto an image file. An animal's retina reflection from its tapetum lucidum can become an issue, from the invisible IR flash because the lens and flash axis are so close to one another. The photograph was made using a Bushnell Trophy Cam HD Essential E2 camera. Image courtesy of Adrian Davies.

Drone Photography

Another type of photography which has grown in popularity enormously over the last few years is photography from drones, which can give an unequalled aerial view of the habitat in which you are working. For more on aerial photography see Chapter 10. Drones are potentially dangerous and their use must be licensed in certain countries such as the U.S. and the U.K. Their use is mostly limited to areas away from human populations. Their use is banned in many nature reserves and national parks around the world. If you intend using one it is worth considering undertaking a training course in their use.

They have been used extensively for a wide range of monitoring activities, from the health of woodland and agricultural crops, counting the number of seals on isolated beaches, to counting nesting birds on sea cliffs.

Camera Supports

A good camera support is essential for an expensive camera's performance to shine. Although most contemporary cameras have image stabilization (IS) or vibration reduction (VR) capabilities, there is still no real substitute for a lightweight solid tripod. Beyond holding the camera steady for timed exposures, a tripod also forces you to slow down and analyze all areas that will be included in the frame. This process will perhaps help you identify out of focus branches or other intrusions into the frame which may be distracting to your final image. Always check your image's attributes using the depth of field preview button. This feature reveals how an image will appear and how the subjects in the background may come into focus and de-emphasize the subject. Some manufacturers recommend turning the image stabilization facility off when the camera is used on a tripod. Check with your own specific camera/lens manufacturer for any of their specific recommendations.

One particularly useful feature on many cameras used for field photography is the built-in digital "spirit level" (such as Nikon's Virtual Horizon feature) which will ensure that the camera is level for landscapes. You can also find a number of smartphone apps for levels and leveling.

Good tripods consist of two parts, the legs and the head. A tripod that goes down to ground level for photographing plants, as well as extending to at least your eye level when standing is ideal. For travelling and field photography a lightweight version will help and will probably be made from carbon fiber. This type of tripod tends to be expensive.

There are two main types of tripod heads. A pan and tilt style and a ball and socket type. Ball and sockets are more compact than pan and tilt heads and far easier to set up for most natural history subjects. Most models will have a quick release plate (e.g. an Arca® Swiss fitting), which stays on the base of the camera, and enables the camera to be mounted quickly onto the tripod.

One particularly useful type of head, particularly when using long telephoto lenses is a gimbal type such as the Wimberley® WH-200. These can be balanced so that the heavy lens becomes effectively weightless and easily maneuverable, allowing you to follow flying birds, for example.

Other supports such as monopods can be used for butterfly and dragonfly photographers, and bean bags are useful for both photographing from vehicles, and for ground level photography of fungi.

Even with the camera on a sturdy tripod or other support, it is still necessary to employ good techniques to ensure the creation of the sharpest images. It is always a good idea to use a cable or remote release if possible and use the "mirror up" mode, a feature found on most top-end DSLR cameras. In this approach, you press the release once to raise the mirror, then wait for a few seconds for any vibrations to settle down before pressing the release again to activate the shutter. Make sure to cover the camera eyepiece when you fire the shutter, to prevent light entering through the viewfinder at the time of exposure. This can lead to incorrect exposures.

Filters

When film was used in the natural sciences, photographers would carry a range of filters including skylight, polarizing, graduated, and possibly colored ones, too. Today, with the image processing capabilities of programs such as Adobe Photoshop many of these filters have become redundant or even obsolete. It remains a good idea to put a good quality skylight (UV) filter onto the front of your lens protecting it from dust or liquids. They are easier to clean than the expensive front element of your lens.

Polarizing filters are useful for reducing reflections from shiny, non-metallic objects such as leaves or from the surface of water. This has the effect of increasing the color saturation of an image. They can also be used to darken blue skies in a landscape photograph, particularly when used at or near 90° to the sun. Polarizers have rotating mounts which allows their effect to be previewed when looking through the camera's viewfinder. Choose the setting carefully because the effect can become too obvious and lead to image falloff into darkness from one side of the frame to the other, particularly with wide-angle lenses.

Carrying the Equipment

A camera bag selected for carrying equipment should be comfortable to carry in the field, hold the equipment safely yet accessibly, be weatherproof, and flexible about how it can be packed. There is no such thing as the perfect camera bag, but good ones are made from waterproof material (some have rain covers in a pocket which can be brought out quickly), and be able to be used as a backpack, easing the weight on your back. Adjustable dividers inside the bag will enable you to customize the layout for a particular set of kit. It is worth considering a sling-type bag, which does not need to be put down on to possibly wet ground when accessing equipment. It is a good idea to see if the bag is suitable to be a carry-on when flying.

Other Accessories

Other useful items to carry when photographing in the field might include a good pocket knife or multi-tool such as a Swiss Army knife or Leatherman, a jeweler's screwdriver set, hex wrenches (also called Allen keys), camera plates, a head light or torch, a length of string for tying back branches, and a small LED light source. It is worth reviewing the contents of your bag periodically to make sure you are not carrying items you no longer need. Make sure to take plenty of spare batteries, chargers, memory cards, and a facility for storing and backing up your images in the field, such as a small laptop PC with external hard drives.

Clothing and Personal Safety Suggested Guidelines

Always wear appropriate clothing for the location in which you will be working, bearing in mind that conditions can change rapidly, particularly in mountainous regions. Wearing several layers is better than just two layers – it is a warmer option, and it is easy to remove a layer if you get too hot rather than trying to add layers. It is always worth having a change of clothing in your vehicle in case you get wet. Lightweight breathable fabrics, some impregnated with insect repellent are good for the tropics, particularly as you will probably need to cover your arms and legs to minimize the potential for insect bites. Wear a hat, and always use sunblock whenever necessary. Wearing the correct shoes is also very important, from solid climbing boots in mountain areas to insulated boots in arctic regions. Carry plenty of water and food with you, tell someone where you are going, and when you think you will return. Remember that mobile phone signals may be unavailable in remote regions. It is worth carrying a GPS device, but also carry a good map and compass and know how to use them. It is also worth considering a small first aid kit. Nothing should be left to chance if planning to photograph in a remote location. Many unplanned things will occur on such adventures.

Modifying Light in the Field

Although natural light is preferred when working in the field, it is, of course, unpredictable in its color, quality, and quantity. Natural light can be perfect, have too much contrast, be too flat (low contrast), or too dim for making successful photographs. The color of daylight will vary throughout the day and the use of a gray card or other white balancing method to achieve the correct color for your subject is a part of the best practice of working in the field with available light. When working in high contrast light, contrast can be reduced with the use of reflectors, diffusers, or by using a fill-in flash.

Often camera sensors cannot record the entire range of brightness that the human eye can see. While we can see detail in all parts of a scene with a brightness range of around 10,000:1, a digital camera may only be able to record successfully around 400:1, meaning loss of detail in highlights, shadows, or both. The use of a weak fill-in flash can greatly reduce contrast. If, for example, a scene has a brightness range of 800:1, try using a weak flash (even the pop-up flash on the camera). If the on-camera flash adds just one unit of brightness, the contrast range becomes 800:2, or 400:1. The fill-in light will contribute mostly to the shadow areas, which in turn will lighten them.

Fill reflectors also are very useful when working in the field, particularly with small subjects such as plants and fungi. A particularly useful type is a circular reflector that folds down to a small package such the Lastolite®. They are available with several different reflective

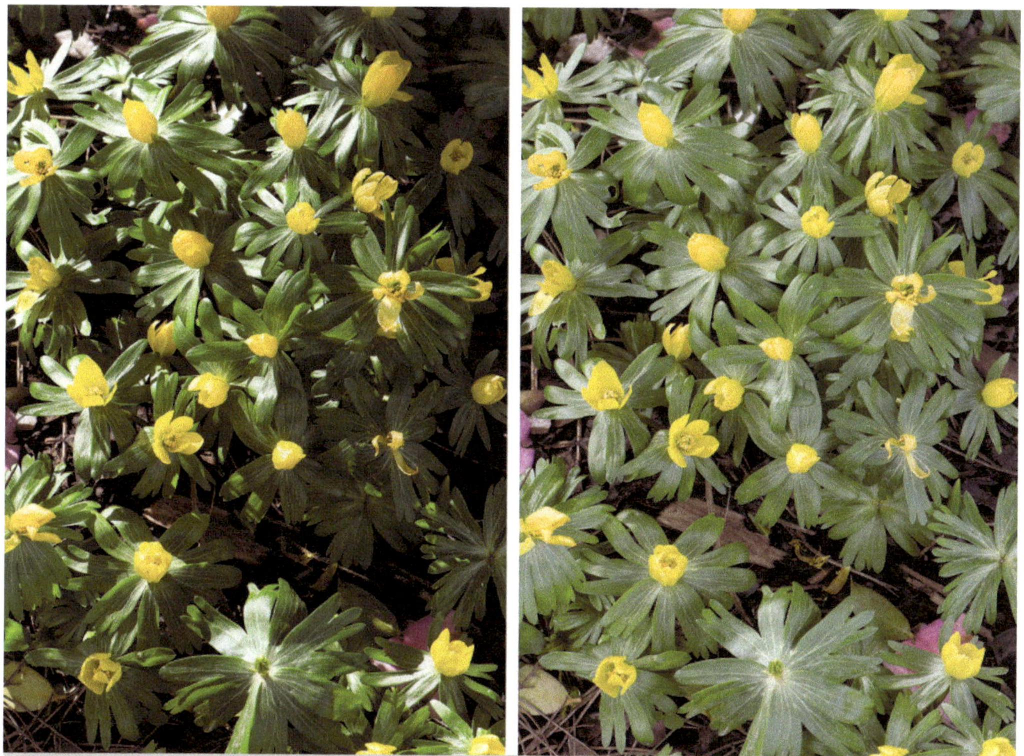

Figure 13.2 Winter aconite (*Eranthis hyemalis*) flowers are shown in full sun (left), and reduced contrast with a 50cm circular diffuser (right), held between the sun and the specimen. An alternative method is to cast a shadow over the subject. Images made with a Nikon D800 with 105mm micro-Nikkor lens and courtesy of Adrian Davies.

surface including white, silver, gold, white/silver and white/gold, and in a range of sizes. For small subjects a 30cm version is ideal and easily folds down to fit into a pocket.
Diffusers are also useful, when the sun is very bright, creating dark shadows and bright highlights. These situations generally will need a larger reflector. A good size for many applications is the 50cm, or 10 feet in diameter variety.

Another way of reducing contrast in an image is to use an exposure blending technique. In this approach two or more images are shot of the same scene using different exposures. One exposure is for the highlights and the other for details in the shadows. The two images are opened in an image processing program such as Adobe Photoshop, and the two images blended together using the layers facility. This technique is also called HDR photography. More on computational photography can be read in Chapter 12.

A Useful Image Processing Technique for High Contrast Scenes

There are several different ways of achieving an HDR result. An example is shared below. Specialist HDR (High Dynamic Range) software, such as PhotoMatix® is also available giving a huge range of options. When using Adobe Photoshop:

1. Shoot two images of a high contrast scene, one exposed to give detail in the highlights, the other to give detail in the shadows.
2. Open both in Adobe Photoshop. Open the layers palette. Click on one of the images to make it "active", and go to Select > All. Copy it, then paste it on top of the other image.
3. In the layers palette you will see two layers, the background, and Layer 1. With Layer 1 active (by clicking on it) add a Layer Mask by clicking the icon at the bottom of the palette.
4. Select a brush from the toolbox (size will be dependent on the size of the image, but you will probably want to give it a soft edge in the hardness control).

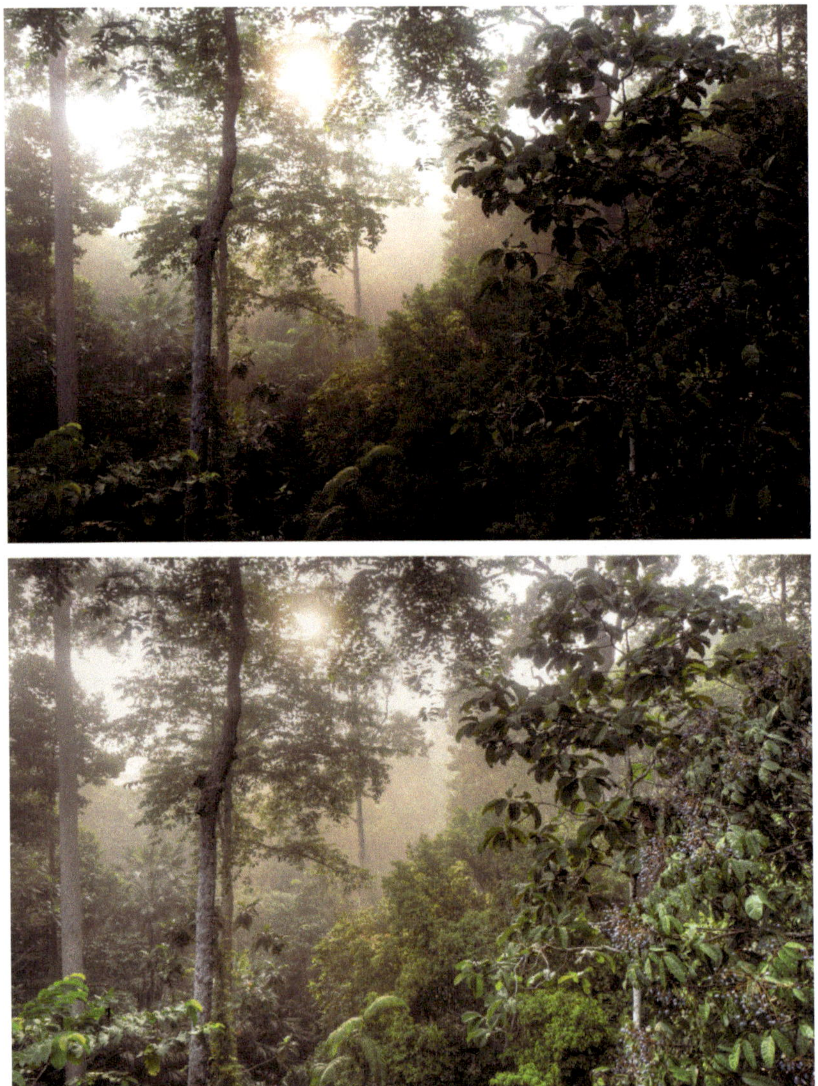

Figure 13.3 A view of the Bornean rainforest from a raised canopy walkway, at sunrise, showing extreme high contrast (top). The final image (bottom) is a composite of two shots, one exposed to retain detail in shadows, the other to retain detail in the highlights, and blended together in Photoshop. A Nikon D800 camera with 16–35mm lens was used. Images courtesy of Adrian Davies.

5. At the bottom of the toolbox the background/foreground icon should show black and white. Select black for the brush colour, and start to paint it over the image. This will effectively make a "hole" in the mask, allowing the image underneath to show through. If the shadows are too dark, try painting them to reveal the lighter shadow tone underneath.
6. You can switch to white to "repaint" the mask. Changing the opacity of the layer (at the top of the layers palette) will yield a lower density "hole" making the tone darker.
7. For more on image processing for natural science applications, please refer to Chapter 19.

Low Light Photography

Much field photography is carried out in low light conditions ranging from dark forests and caves, to sunrise and sunset. Modern cameras give remarkably good results at higher ISO values which can be up to several thousand. Such images may be "noisy" which means they exhibit granularity. In general, the higher the ISO, the more grainy the appearance. Software programs such as Photoshop and Lightroom have excellent noise removal filters, as well as Topaz® Denoise AI software which gives superb results.

Scales

It is very important, particularly when photographing unfamiliar subjects, to convey a sense of scale – is the subject 1mm or 10cm in size? Technical photographers make use of rulers placed alongside the subject, but with subjects such as fungi or flowers photographed in the field for example, a natural comparison can be useful. This might be a leaf, pine cone or other suitable subject, laid alongside the main subject. This can also be a good way of giving an indication of the habitat, if your subject only grows in conjunction with other specific species such as pine trees.

Close-Up and Macro Photography in the Field

Much field photography of plants and insects will involve close-up and macro photography techniques. Normal lenses, without a close focusing facility can be made to focus closer by using either close-up lenses, or extension tubes in between the lens and camera body. More details about this can be found in Chapter 11.

A major problem when using close-up photography in the field – where it may not be possible to control all aspects of the process – is the management of depth of field (DOF) or lack of it. At a magnification of ×1/2, and an aperture of f/8 with an APS-C sensor camera, there is only 2.8mm of DOF and only 1.9mm is in focus with a full frame camera. For this reason, it is very important to place focus onto the most important part of the subject. This is usually the eye in the case of insects. Manual focus is preferred rather than auto focus for close-up photography.

Increased DOF Imaging or Focus Stacking

One technique which has become popular recently with close-up photographers is focus stacking, which was discussed in detail in Chapter 12, where a number of images are taken of a subject at different focus points, then the images are merged together afterwards in a suitable program. The results can be astonishing. While the technique lends itself to the controlled conditions of a studio, it is quite possible to shoot a suitable set of images in the field, particularly with solid subjects such a fungi. The main proviso is that the subject and camera do not move in relation to each other during the shoot, so a tripod mounted camera is almost essential. If there is slight movement between frames the software can usually align the images (in Photoshop this is Edit > Auto-Align Layers). You will need to use manual focusing.

Ecological Monitoring

Photography has long been used to illustrate ecological issues, such as the growth of invasive vegetation over a period of time, the numbers of nesting seabirds on a cliff, the movement of a glacier, or the effects of management in a nature reserve.

For monitoring changes to a habitat, the ideal solution is to photograph the same area, with the same lens and camera, and from exactly the same viewpoint, using a technique known as fixed point photography. If possible, a metal pole or stand should be erected where it can be left in situ for a period of time, possibly several years. A bracket on top of the pole holds the camera and lens in exactly the right orientation. It is best to use a fixed focal length lens rather than a zoom, to ensure that exactly the same area is covered, and a camera with the same size image sensor. If it is not possible to erect a pole, then take a print with you of the previous image to help alignment. Drones are useful for aerial views of habitats.

Photographing Plants and Fungi

Plant photography covers a huge range of subjects, from the smallest mosses and lichens to the largest trees, each of which will require their own special technique. It will be worth carrying a range of lenses to cover all eventualities. Finding the right subject in the right place is a major key to success.

A perennial problem faced by plant photographers is that of wind, where even the slightest breeze can cause delicate plants to sway wildly in front of the camera. It is worth having some sort of support system for holding plant stems still. This might be as simple as a bamboo cane and a couple of clothes pins (pegs), or a more sophisticated purpose-built device such as the PLAMP (Plant Clamp) made by Wimberley for the purpose. These devices cannot completely eliminate movement but will certainly help.

For much plant photography it may be worth turning off the autofocus system of the camera and focusing manually. You can then decide precisely where to focus on the subject (stamens, front or rear of an orchid for example). Depth of field, or lack of it, may become an issue, particularly with trumpet shaped flowers for example, and it may not be possible to use a small aperture to gain the required depth of field. As discussed previously, focus stacking can be used to produce images with almost infinite depth of field, though can be difficult to carry out in the field.

Telling a Story

It is certainly worth shooting more than one image if it is important to illustrate various aspects of a chosen plant. These could include a "portrait", a close-up of an important area of the subject, and an environmental or habitat shot, and then any other detail worth noting. In the example shown here, of the cauliflower fungus (*Sparrassis crispa*), which only grows at the base of pine trees. In this example, the whole specimen was photographed, followed by a close-up showing the texture of the surface, and then a wide-angle view showing the fungus at the base of a pine tree, making sure to include enough bark to enable identification of the tree species, and showing the background behind the tree. Unusually, the same specimen was used for each of the three images. They do not always grow in such a good and photogenic position! Note the leaves and pine needles around the fungus, adding scale and information about the surrounding vegetation.

Many fungi do not last long, so it is suggested to always photograph them when you find them.

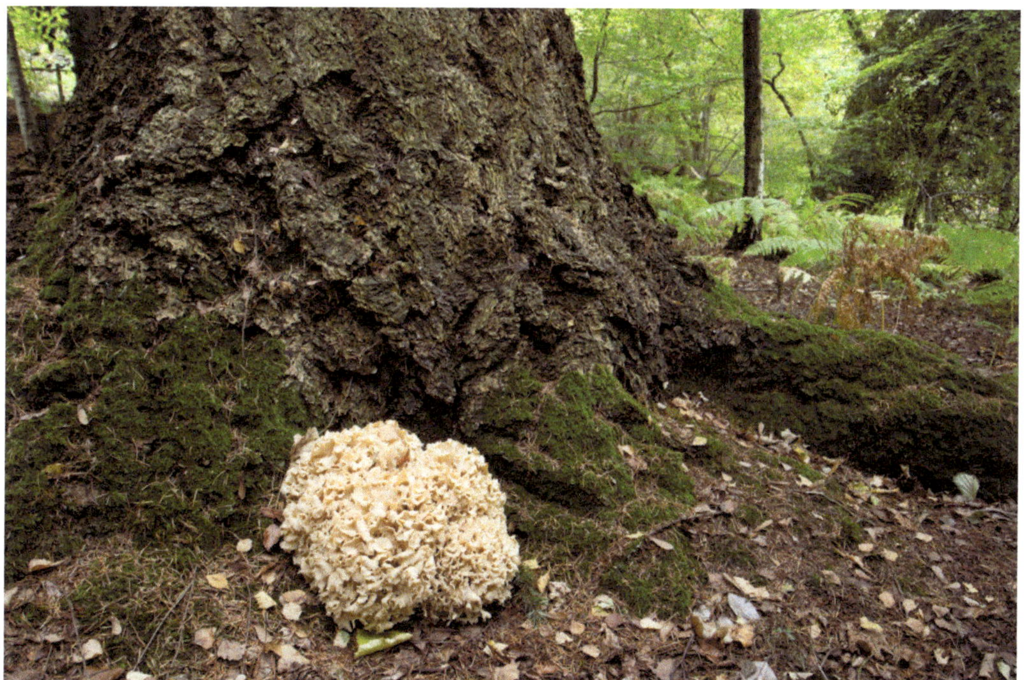

Figure 13.4 A cauliflower fungus (Sparassis crispa) growing at the base of Scots pine tree. Note the pine cone and needles, providing scale for the fungus. Wide-angle: Nikon D800E with 24–85mm lens at 24mm. Image courtesy of Adrian Davies.

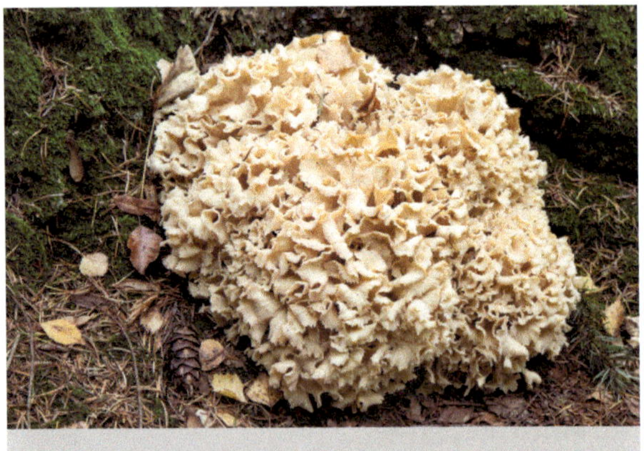

Figure 13.5 A cauliflower fungus (*Sparassis crispa*) growing at the base of Scots pine tree. Note the pine cone and needles, providing scale for the fungus. Whole specimen: Nikon D800E with 24–85mm lens at 85mm. Image courtesy of Adrian Davies.

For "portrait" shots of vertical subjects such as orchids there will often be a large area of plain space occupied only by the plant stem. This redundant space in the image can be filled by shooting through out of focus regions of other vegetation, which can serve to frame the main subject. Such out of focus vegetation is often known by plant photographers as "mush" and should be well out of focus so as not to distract from the subject. Another way of achieving the same effect is to hold a leaf or piece of moss, for example, very close to the front of the lens so that it just appears in the bottom of the frame.

One essential piece of equipment for a plant photographer is a kneeling mat (or perhaps the separate knee pads used by carpet fitters and tilers). There is nothing worse than having a superb specimen in a perfect position and the ground surrounding it is stony. This can make it painful to kneel on and make carefully composed photographs. If you use a matt, a useful tip is to put a large cross on one side and always place that side downwards. Mats can pick up thorns and other items that will stick into you if unlucky!

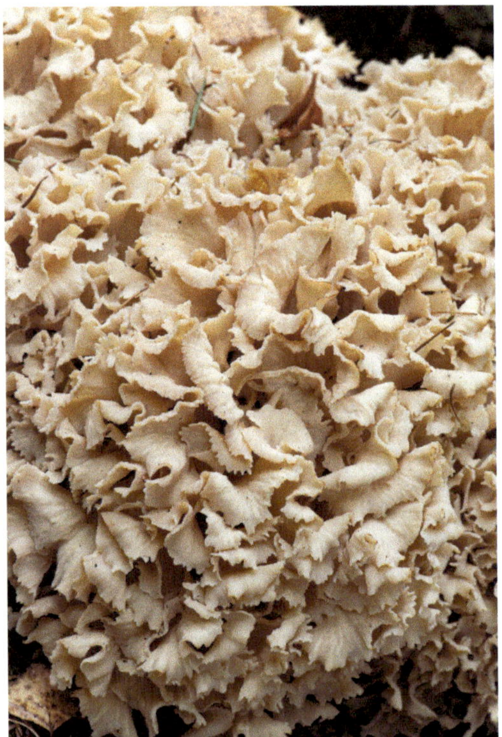

Figure 13.6 A cauliflower fungus (*Sparassis crispa*) growing at the base of Scots pine tree. Note the pine cone and needles, providing scale for the fungus. Close-Up of surface: Nikon D800E with 105mm micro-Nikkor lens. Image courtesy of Adrian Davies.

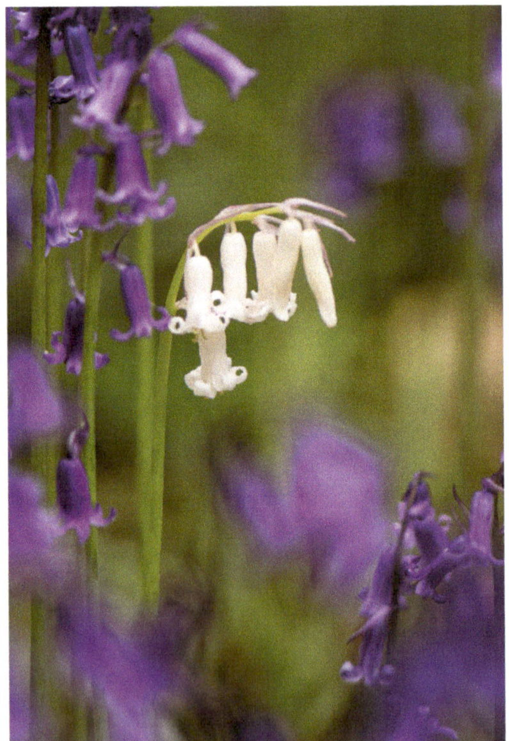

Figure 13.7 A bluebell (*Hyacinthoides non-scripta*), white variety, shot through out of focus blue flowers. A Nikon D7100 with 105mm micro-Nikkor lens and 1.4× tele-converter, 1/250th at *f*/4 was used to make this picture. Image courtesy of Adrian Davies.

Blue Flowers

There are some flower colors in particular that both film and digital photographers have struggled to photograph and produce accurate color. These colors are purples and blues such as the bluebell (*Hyacinthoides non-scripta*), and various gentians and geraniums. In the days of color transparency film, Kodak actually referred to this as the Ageratum Effect, where the color film recorded portions of the ultraviolet and infrared reflections from the flower, giving the photographs a pinkish appearance. To help reproduce these colors, they published a filter "recipe" for correcting it that consisted of several filters used together which unfortunately reduced the amount of light by more than five stops.

Today it is possible to white balance the subject by including a photographic 18 percent gray card or other light balancing target into the image frame at capture. Choosing the right lighting conditions will greatly help color accuracy. Generally, a bright overcast sky will be better than full sun for photographing bluebells.

To use a gray card, shoot two images of the same subject, in the same light and have only one image include a portion of the gray card. In the RAW converter software open both images and select the image with the gray card. Select the white balance dropper and click it on the gray card. This will neutralize the color in the card to be neutral gray. Now synchronize the

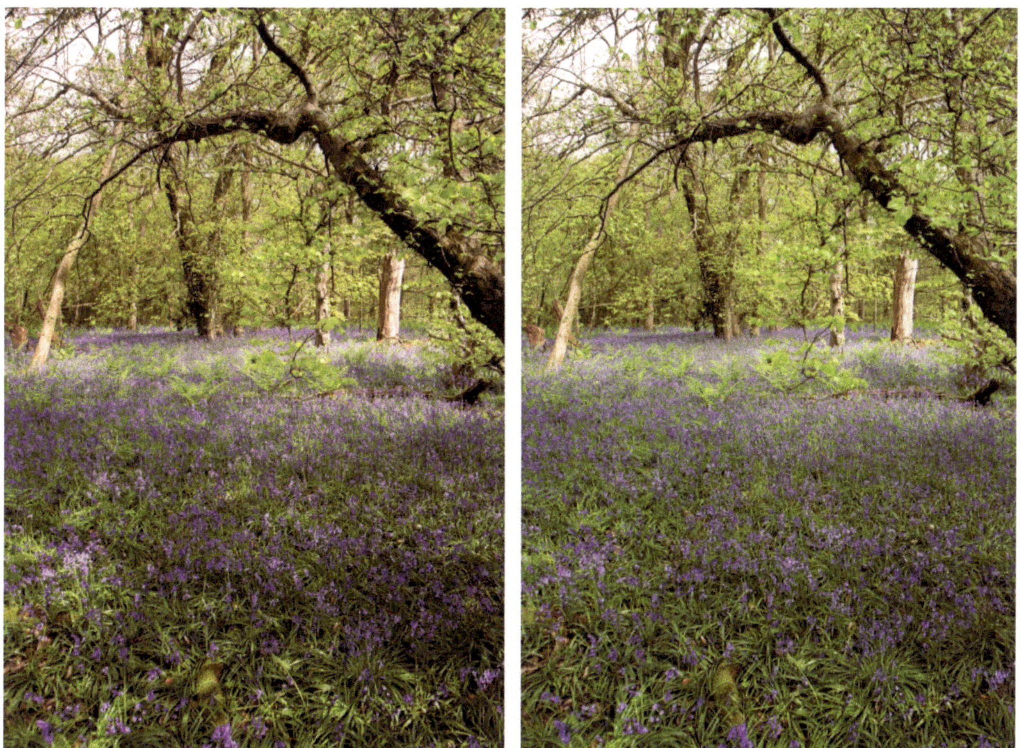

Figure 13.8 A bluebell (*Hyacinthoides non-scripta*) wood in spring, in sunlight (left) and overcast conditions (right). Note the increased color saturation and richer blue color in the overcast version. A Nikon D800 with 24–85mm lens, at 28mm was used to make these pictures. Images courtesy of Adrian Davies.

two images so that the settings from the gray card image are applied to the image without the target.

Insects

Almost by definition, insects are small, and prone to running, jumping, or flying away if disturbed. Lenses with a close-up capabilities are usually required. A 100 or 150mm macro lens is perfect for many species, giving a reasonable working distance and excellent quality. Although a 200mm lens or longer may be necessary for particularly difficult subjects such as dragonflies. A good combination is a 105mm macro lens, together with a 1.4× tele-converter, gives a combined focal length of approximately 150mm. If you are travelling, this approach will give you two different focal length lenses and the converter may also be used with other lenses such as a 70–200mm zoom. Make sure the converter matches the quality of the lens that it will be used with.

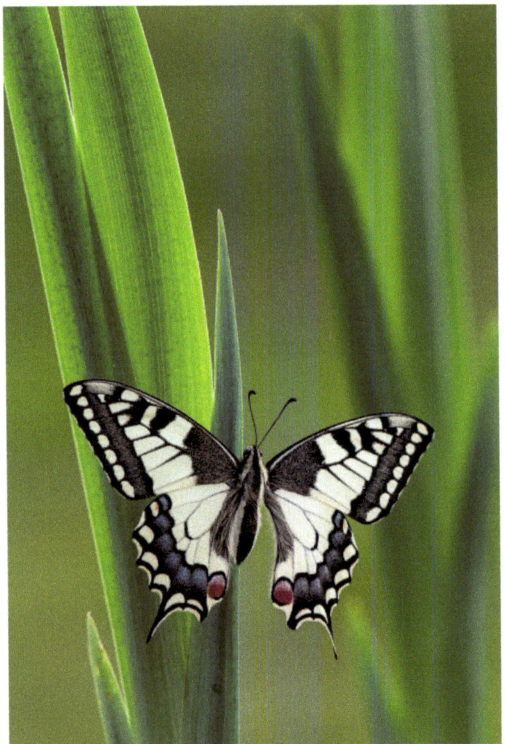

Figure 13.9 A swallowtail butterfly (*Papilio machaon*). By keeping the camera parallel to the main body of the butterfly, depth of field was maximized. The goal of this work should be to keep the background out of focus. A Nikon D300 with 105mm micro-Nikkor and 1.4× converter (giving an overall equivalent focal length of approximately 220mm) 1/100th second at *f*/8 was used. Image courtesy of Adrian Davies.

As with plants, it is worth trying to use a wide-angle lens on the camera and place the camera into the insect's habitat. Doing this will of course necessitate being very close to the insect. Unlike plants, tripods may be impractical for insect photography, and the camera may need to be hand-held. Use the image stabilization feature and take great care when activating the shutter. If you use the auto-focus facility, make sure it focuses on the eye of the insect.

It may well be worth designing a bracket system for holding one or more flash guns to enable easier photography of insects, and provide a consistent light source. A single flash gun, perhaps with a soft box to diffuse the light held above and to one side of the lens will provide a good light for many species. More details of this are given in the section on rainforest photography.

Animal and Bird Photography Using a Long Lens

Most photography of birds and animals will require the use of a long telephoto lens at 500mm or longer. Although some modern designs are much smaller and lighter than previous versions (such as the Nikon 500mm *f*/5.6, weighing just 3.5lbs/1.5kg), long telephoto lenses are still bulky and heavy and require good technique to achieve sharp images.

A solid tripod will help along with a good head such as a gimbal type. Try to minimize vibrations by pressing your eye to the back of the camera and place a hand on the top of the lens to help stabilize it. If hand holding is possible, it is the preferred by many photographers when photographing birds in flight.

Stalking

Stalking animals and birds in the wild will require a good knowledge of their habitats, feeding habits, or other daily routines. This is known as fieldcraft and is best developed over a period of time and through experience.

Wearing rustle free clothing, keeping downwind of mammals, and always being aware of potential hazards can be helpful. Do not try to get too close to animals such as bears, which will have a "flight zone" – a distance which, if entered, will alarm a subject. This will either cause it to flee or possibly attack. Stalking will require a considerable commitment to waiting time and often creates little result.

Working from Blinds (Hides) both Permanent and Temporary

For photographing many birds and animals, a form of concealment will be required. This is known as a hide in the U.K. or blind in the U.S. In the very early days of photography, it was thought that a hide needed to look like a part of the environment to be effective. Photographers Richard and Cherry Kearton, who were bird photographers in the U.K. during the early part of the twentieth century, made dummy haystacks where they hid and covered wooden frames with sheep and cow skins. We now know this is strictly not necessary and effective hides can be tents, vehicles, or permanent buildings.

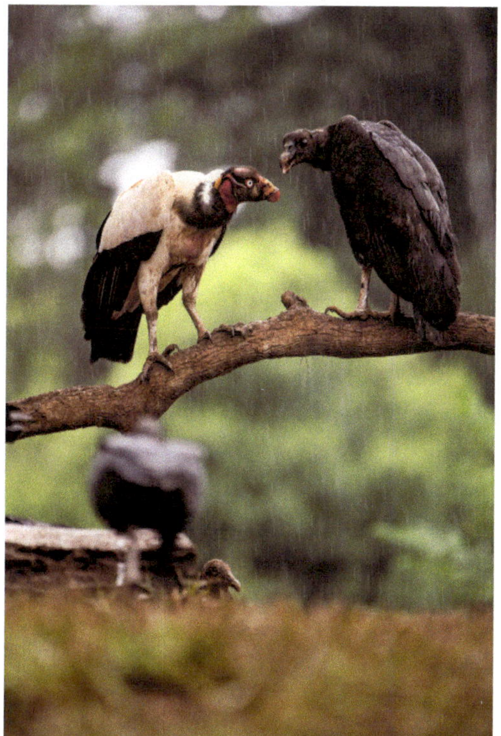

Figure 13.10 A king vulture (*Sarcoramphus papa*) and black vulture (*Coragyps atratus*) were photographed from a specially designed ground-level hide in Costa Rica. Note the rain! A Nikon D810 camera with 200–500mm lens, 1/160th second at *f*/5.6 was used. Image courtesy of Adrian Davies.

Today there are two main types of hides. Temporary mobile hides and permanent hides, usually in nature reserves. Mobile hides are very useful and can usually be carried and erected by one person. They are often dome shaped, and have camera portholes and viewing windows on four sides. If you are using one in the open in windy conditions, it is worth weighing down with rocks. Many will have pockets for this purpose at the bottom of each wall. It is best to introduce a hide slowly into an environment. A small perching stool is also a very useful, if not essential piece of kit.

Permanent hides are usually found in nature reserves and can often accommodate large groups of people. They are often situated near waterholes or overlooking marshes and are often not ideal for photography. This is because they may be too far away or too high, creating a point of view where you are looking down on your subject. Bence Máté, a Hungarian photographer, has built a series of hides around the world, often at ground or water level, allowing unique perspectives of the subjects. The image of the king vultures shown here was shot from a hide dug into the ground at a site in Costa Rica, so that the camera was virtually at ground level. The site was baited each day with cow heads, and attracted dozens of black vultures as well as the rarer king vultures.

Nest Photography

When cameras were bulky and heavy, most photography of birds was carried out at the nest, from a hide. This involved gradually moving the hide towards the nest so that eventually it was within camera range, and possibly cutting away some vegetation to reveal the nest. Unfortunately, this often distressed the parent birds, causing them to desert the nest. Nowadays, very little nest photography is carried out, particularly of small songbirds. Some birds, particularly seabirds such as gannets, terns, and puffins, nest colonially and can be successfully photographed without undue disturbance. See Chapter 4 for more about Eliot Porter, bird photographer.

Baiting

Birds (as well as other animals) can be attracted within a camera's range by using food put out in the same way that birds are fed in gardens. Take great care when using food to entice birds and animals to a site so that they do not become dependent on it as a food source. They will become distressed when the food supply is stopped.

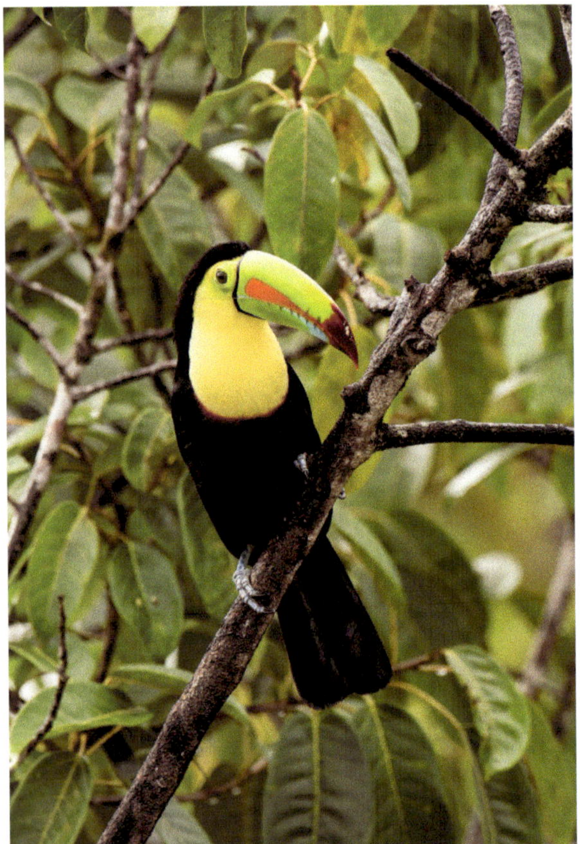

Figure 13.11 A keel-billed toucan (*Ramphastos sulfuratus*) in Costa Rica, baited with ripe bananas, just out of shot. It was photographed using a Nikon D810 with Nikon 200–500mm lens. 1/80th second at *f*/8. Image courtesy of Adrian Davies.

Extreme Conditions

Many habitats can be classed as extreme, either in terms of weather, terrain, or the potentially dangerous animals that can be found there. Extra preparation should be taken when photographing in such conditions.

Rainforests

Rainforests can be difficult habitats in which to work, but potentially the most rewarding, because they are teeming with life. They are dark, hot, humid, and frequently very wet. Lighting conditions are often very difficult. A dense rainforest is very dark, or at best, demonstrates extremely high contrast with shafts of sunlight piercing the canopy (see Figure 13.3).

Such conditions can encourage the growth of mold on equipment, particularly on the surface of a lens. It is a good idea to dry equipment each night by using containers of silica gel to absorb moisture. The best way to do this is to place the lens or camera in an air-tight sealed container such as a Peli® hard plastic camera case along with the silica gel. The silica gel can be re-activated by putting it in an oven for a few minutes to dry it out. On returning from a rainforest trip it is worth getting your camera gear out of its bags and leave out for a day or two to completely dry out. Check all lenses for any sign of fungus, which can be removed if caught early enough.

Potential natural hazards include snakes, constantly irritating biting insects, and possibly leeches, as well as large predatory animals. Even the smallest animals can be very dangerous. In the jungles of South America lives the bullet ant, capable of inflicting the most painful bite known! Wear protective clothing such as leech socks, boots, and long-sleeved shirts and long trousers. It can be a huge challenge to keep yourself dry in the sometimes torrential rain and it is often better to concentrate on keeping equipment dry using a large poncho which can cover both you and your camera. Note, use of a good insect repellent, together with a high factor sunscreen is MANDATED.

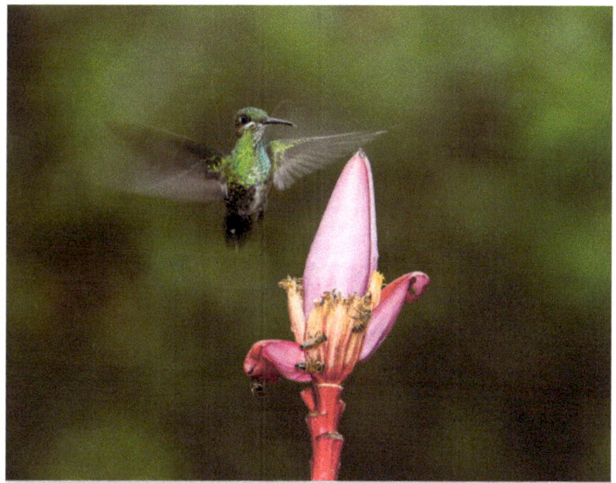

Figure 13.12 A green-crowned brilliant hummingbird (*Heliodoxa jacula*) was photographed in Costa Rica. This image was an attempt to get away from the classic image of a hummingbird, frozen in mid-air almost like a carboard cut-out. A flash exposure was mixed with daylight to create a dynamic image of the bird hovering in front of the banana flower. The flash froze the body of the bird, while the daylight exposure allowed the wings to become blurred. Note also the rain. A Nikon D801 with 70–200mm lens, at 200mm. Nikon SB900 flash gun. 1/80th second at *f*/9 was used. Image courtesy of Adrian Davies.

Rainforests come alive at night with a completely different range of animals. It is always worth going out into the forest at night to see what you can find. Never go alone and always go with an experienced guide if possible, who can help find subjects as well as be able to identify potential hazards.

You will almost certainly need to use flash, although some reserves ban the use of flash. Always check before you set out as you need to be able to operate your equipment in near dark conditions. A head flashlight is a necessity when handling photographic equipment. It is worth experimenting with some sort of bracket which can hold the flashgun and produces consistently lit images of nocturnal subjects. It is worth considering some sort of diffuser to soften the flash and reduce specular reflections from shiny subjects such as wet frogs or shiny beetles.

Designing a Flash Bracket

The primary goal of a flash bracket for night-time photography (and daytime too) is to hold a flash gun and allow for the angle of the flash to be changed in relation to the subject as needed. Some photographers prefer to use two flash guns. One will act as the main light, the other, a weaker flash, used as a fill-in on the opposite side to the main flash. See Chapter 9 for more on lighting. Most nature photographers will have, at some time, designed their own, usually an L shape bracket, which clips on to the base of the camera. The arm of the bracket (this could be rigid, or a length of "goose neck") should be able to move to position the flash to the side or top of the subject as well as acting as a handle for the set up. For many subjects, the flash should be held above and to one side of the subject to give some modelling or shape. A small ball and socket head on the top of the arm will help to angle the flash precisely.

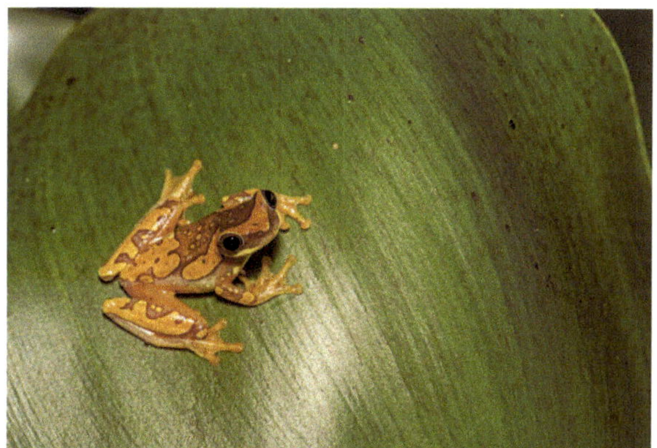

Figure 13.13 This beautiful hourglass tree frog (*Dendropsophus ebraccatus*) was photographed at night in a rainforest in Costa Rica using a single heavily diffused flash gun mounted on a flash bracket. The camera was hand-held, and a head flashlight was used for focusing and composition. A Nikon D801 with 105mm micro-Nikkor lens and hand-held diffused flash on bracket was used. The exposure was 1/160th second at *f*/16. Image courtesy of Adrian Davies.

Diffusing the flash will often be a good idea, to soften the shadow and reduce specular highlights, and various diffusers are available such as various sizes of mini soft boxes. Specialist macro flash kits are available from various manufacturers such as Nikon and Canon, where one of more flashes are held on the mount which is placed on the front of the lens.

Desert Photography

Desert regions are hot, dry, dusty, and frequently windy. Changing lenses should be made as little as possible and keep cameras covered in bags as much as you can. It is worth carrying two camera bodies with different lenses to minimize the need to change the lenses. Lighting conditions will often be challenging with harsh light, though this can be put to good use to emphasize the shape of sand dunes, for example, or backlighting the spines on cacti. Try to keep cameras out of direct sunlight if possible. It may be worth restricting photography to early morning and evening because of lighting considerations and your own personal comfort.

The animals and plants that live in deserts are adapted to survive for long periods of drought and will often have specific adaptations which are worth photographing. The welwitschia plants shown here in Namibia are endemic to the Namib Desert of south west Africa. They are extremely slow growing and may be several hundred years old.

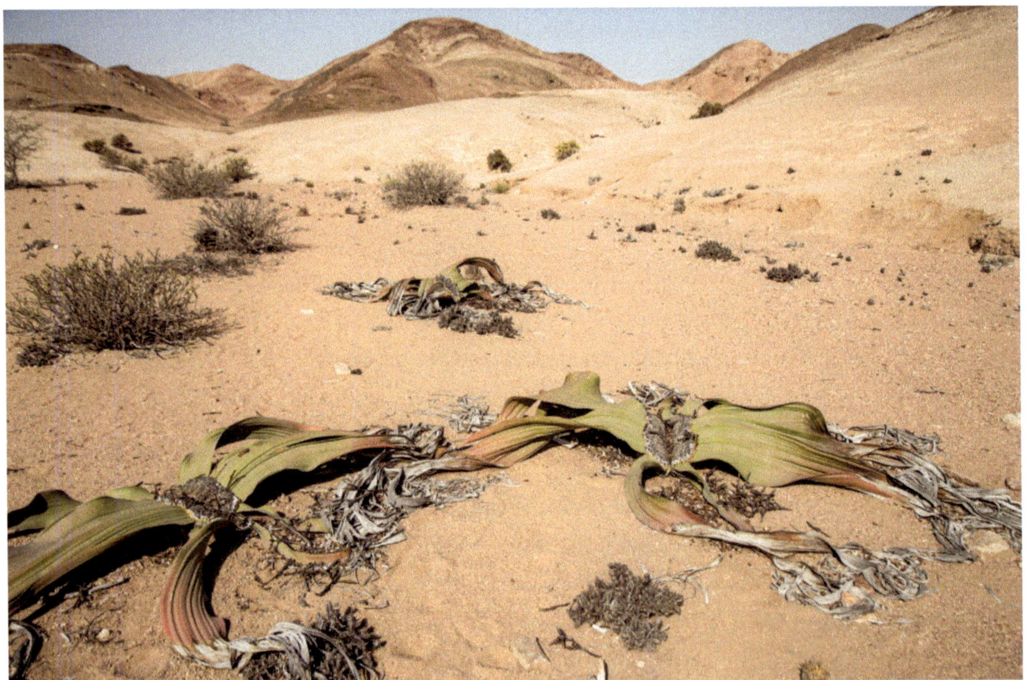

Figure 13.14 Welwitschia (*Welwitschia mirabilis*) is an endemic plant in the Namib Desert, Namibia. This picture was made with a wide-angle lens to show the plant in its desert habitat. Plants like this grow exceedingly slowly and can be several hundred years old. A Nikon D800 with 24–85mm lens, at 24mm was used. Image courtesy of Adrian Davies.

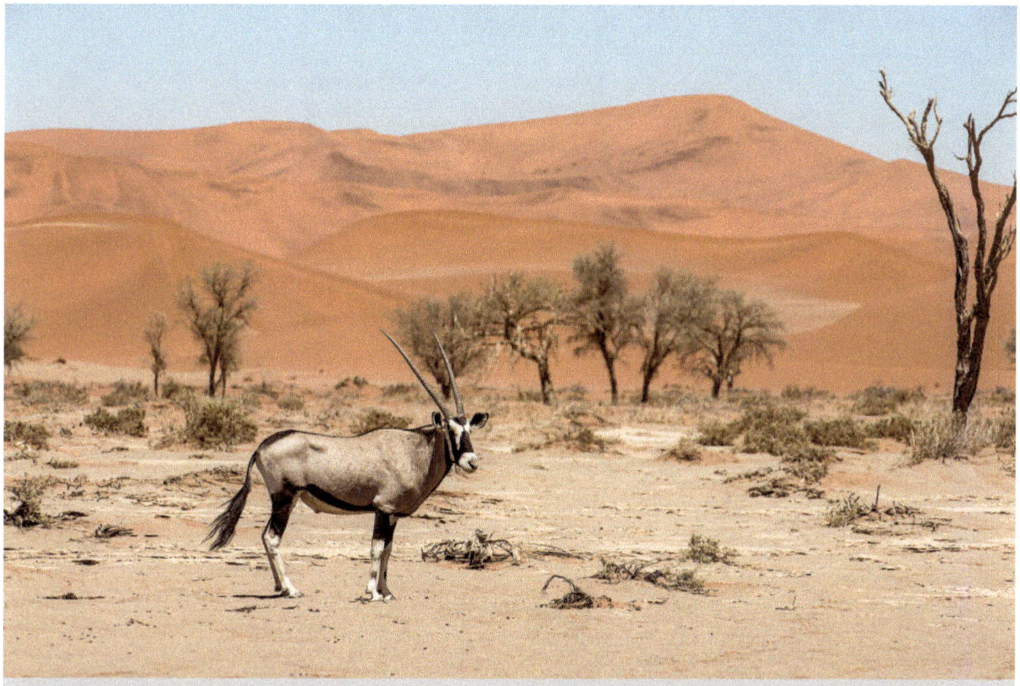

Figure 13.15 An oryx (*Oryx gazella*) was photographed against a sand dune in the Namib Desert, Namibia. A Nikon D800 camera with 70–200mm lens, at 150mm was used. Image courtesy of Adrian Davies.

Polar Region Photography

Working in polar regions brings a unique set of challenges, namely the sometimes incredibly cold temperatures (down to –60°F or more) coupled with strong winds, which causes bitingly cold wind chill effects. The first priority is personal safety and preventing frostbite. Gloves with built-in warming devices are available and well worth investigating as well as boots specifically designed for arctic conditions. It will almost certainly be necessary to operate your equipment with thick gloves and this should be practiced beforehand. A warm hat is also essential. While most modern photographic equipment will function perfectly well at low temperatures, camera batteries in particular can stop working in extreme cold and may need to be kept warm. LCD screens can also be affected by temperature extremes.

If you are working in the Arctic it is essential that you keep an eye open for polar bears at all times. Many guides carry rifles, which are used to frighten the bears away if they approach too closely.

When photographing in snow, it may be worth giving extra time to your exposures using the exposure compensation facility. If using automatic, maybe +1 or more stops will be useful to ensure the snow stays white rather than being recorded as a gray tone.

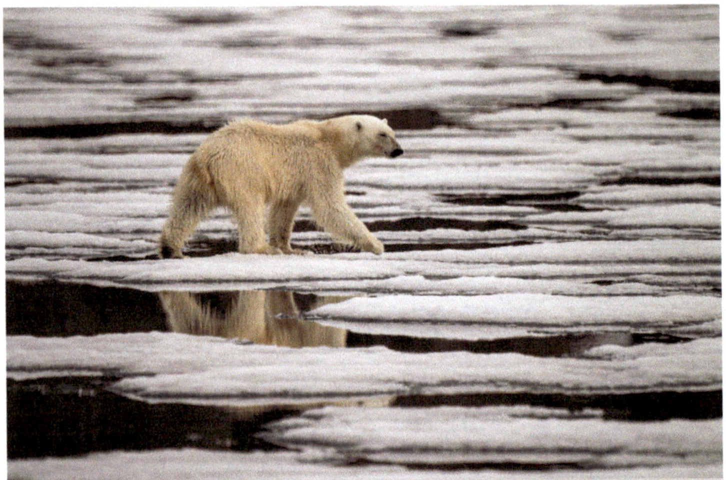

Figure 13.16 This polar bear (*Ursus arctos*) was photographed at Svalbard in the Norwegian Arctic. This particular young bear is very thin, finding it increasingly difficult to find food on the shrinking ice floes. Shot from a boat using a Minolta Dynax camera, 500mm lens. 1/1000th second at *f*/8. Image courtesy of Cecilia James.

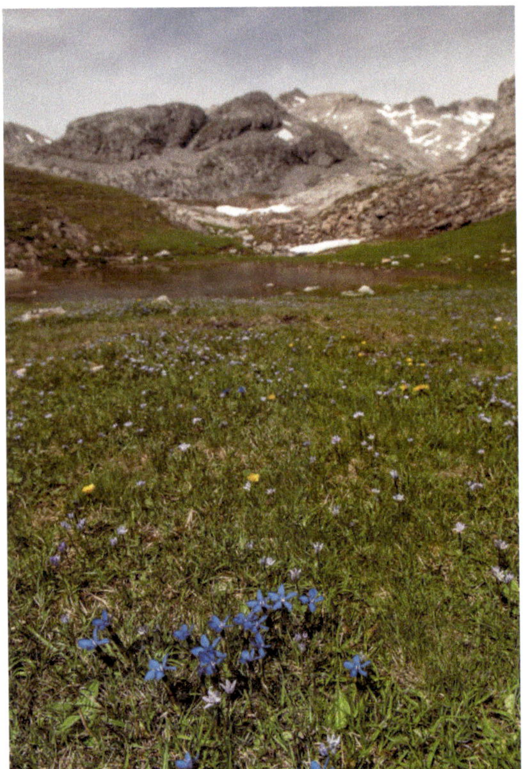

Figure 13.17 Spring gentians (*Gentiana verna*) and other flowers in an alpine meadow in the Picos de Europa, northern Spain. Note the lying snow and meltwater. Altitude approximately 1800m. A Nikon D300S camera with 17–55mm lens, at 25mm was used. Image courtesy of Adrian Davies.

Mountains

Weather conditions in mountainous regions can change rapidly, and it is essential you go prepared for all likely conditions. Mountains have their own specific flora and fauna, adapted to the altitude and extreme weather. Many plants are cushion shaped to retain heat. Wide-angle shots showing plants in relation to their alpine habitat will be useful.

Safari Photography

Although not an extreme condition, safari photography which includes photographing animals and wildlife in open conditions from vehicles is well worth discussing. Unless you are very fortunate you will probably be sharing a vehicle with other people, so tripods are usually not practical. A useful alternative is a bean bag, which can rest on the window ledge and into which the camera or lens is pushed. This creates a remarkably firm support. It is worth having a long piece of string attached to the bag so that should it fall from the vehicle you can retrieve it without having to leave the vehicle. It is also possible to buy window mounts for tripod heads, which clip onto the window sill.

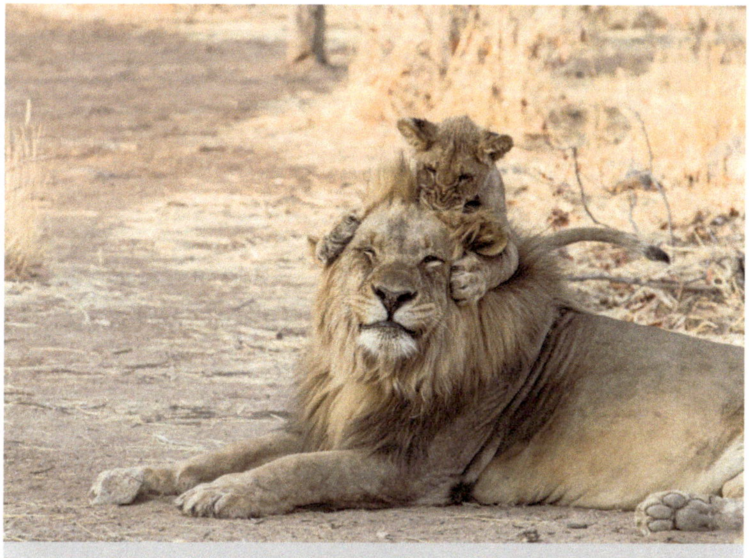

Figure 13.18 A male lion (*Panthera leo*) with a young cub was photographed from a safari vehicle using a camera supported on a bean bag on the window ledge. Etosha National Park, Namibia. A Nikon D800E camera with 70–200mm lens and 1.4× tele-converter was used. Image courtesy of Adrian Davies.

Suggested Reading

Understanding Photography Field Guide: How to Shoot Great Photographs with Any Camera, by Bryan Peterson. Amphoto Books, 2009.

Nature Photography Field Guide, by John Shaw. Amphoto Books, 2000.

Wildlife Photography: Stories from the Field, by George Lepp. Lark Books, 2010.

Chapter 14

Fossils, Minerals, and Geological Photography

Norman Barker

Representing the Specimen

The practice of photographing fossils, minerals and geological samples blends the demanding craft of the scientific illustrator with the expressive aesthetics of a fine artist.

I am continually amazed by the sheer beauty of earthly riches that are available to both the casual collector as well as the advanced scientist. Even though paleontology, mineralogy and geology are considered separate disciplines, there is significant overlap. You can't have fossils and gemstones without geology. Additionally, today in the minerology and fine gem collector worlds, more and more synthetic stones are coming onto the market. Synthetic diamonds are an example of this phenomenon.

One of the challenges found when photographing these subjects is trying to figure out what the goal of the photography is and what characteristics of the sample should be emphasized. This becomes especially true since these three areas intersect and overlap. The serious collector or museum curator will often have examples of fossils and minerals and certainly the geological formations will be a part of the matrix of the specimens that they collect. Practitioners of this activity fall into various categories. They are sometimes referred called rock hounds, gem and mineral collectors, prospectors, or fossil hunters. Regardless of what these individuals are called, they all appreciate the beauty of these earthly treasures.

Many of the photographic techniques that I use can be easily adapted for photographing fossils, minerals, or geological specimens. I have found that the more I learn about geology and earth sciences, the better my chances of success with my photography.

The Plan

Whenever photographic documentation is required, the first step is to develop an imaging plan. In Chapter 2, the term treatment was used and we learned why it was important. It is best to anticipate what could happen and be prepared for serendipity. This process is more complex than simply taking a picture of a specimen. The process should be more

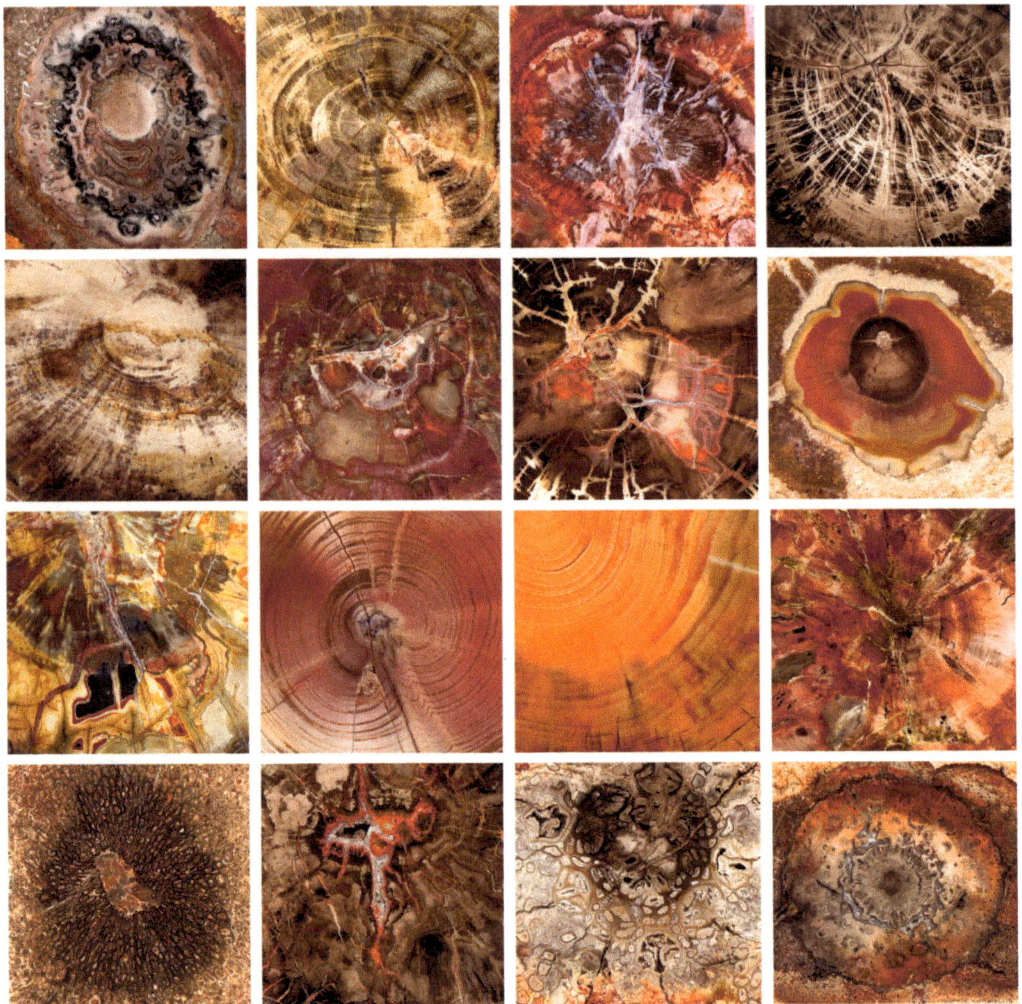

Figure 14.1 This composite photograph features xylem rays from nine different specimens of petrified wood. Each individual photograph was lit and composed to feature the various internal characteristics of the sample. Image courtesy of Norman Barker.

about making an image that reveals attributes about the specimen. Louis Pasteur (1822–1895), the great French scientist said, "In the fields of observation, chance favors only the prepared mind." More than 100 years earlier, the great polymath Johann Wolfgang Goethe (1749–1832) said "We only see what we know." These philosophies being shared, I suggest you should always have a critical look at a specimen, before starting to work. I often do my examinations using a magnifying lens to inspect for regions of interest and look for dust or extraneous information that should not be included. With this visual examination completed, I have a better insight into what my photographic approach should be and what techniques I will need to use before I even look into the camera's viewfinder. Often at times, this information informs my development of lighting, the choice of background, and which camera angle is ideal to help me accomplish my objectives.

Several questions need to be asked and then answered before setting up equipment or operating the shutter. These include:

- What will be your approach?
- What are you trying to reveal and what are you trying to document or describe to your audience/viewer?
- What are the most salient features you want to identify?

Always keep the viewer in mind or consider the possible use for the image when creating an imaging strategy. Foundational to that is what you are trying to communicate with the image. My questions that drive my solutions include:

- What is the purpose of the photograph?
- Will the image be used for quick documentation in the field or become a publishable photographic record?
- What lighting technique will reveal the most?
- What background and angle of view is ideal?

Whether the photograph will be used in an academic journal or selected for the cover of a high-quality magazine, care and concern for creating the best outcome should influence everything you do. Each specimen is unique and comes with its own set of individual challenges and demands. I always make photographs as though they are for publication regardless of their ultimate use. This forces me to exercise great care, discipline, and technique in my practices. If working on a geological specimen, I try to highlight the important mineralogical features that represent the specimen for proper identification in the field. If documenting a beautiful fossil or gemstone, it can be practical to show off the beauty of the specimen for purely aesthetic reasons but it's important to try keep geological science in the representation.

In the end analysis, successful images are really about how effectively light is used. As image makers, sometimes we don't think in terms of how to create the illusion of a third dimension in a 2D image. But that's exactly what you should do each and every time you light a specimen. Sometimes photographers get too caught up in cameras and lenses and forget the most important tool that they have access to – light. This holds true whether the light is natural or artificial. One of the most exciting developments in the last few years is not only in the evolution of the camera but all the new choices in artificial lighting. Ranging from new battery powered light emitting diode (LED) technology to fluorescent portable lighting. The contemporary photographer has many new choices in the lighting gear. Whether working in the field, on location, or in the studio/laboratory, there are methods and techniques that allow for high quality reproduction of new finds or specimens. In the case of those working exclusively with fossils or minerals, the image will often be used purely for aesthetic reasons and must show off the beauty of the specimen with well-considered lighting and careful choice of background. For dealers or traders in geological specimens, the goal might be to form an accurate document of what's in front of the camera, while an academic geologist or paleontologist might be trying to make a high quality image highlighting a particular region of interest for journal or book publication. The web is full of images shared by people trying to sell fossils, gems, and minerals or images from collectors showing off their beautiful collections.

Camera, Essential Lenses, and Tripods

It is amazing how many different high-quality digital cameras are available on the market today. In the near future, we know equipment will continue to improve and change. Unfortunately the one thing all electronic equipment has in common is that it all has a planned obsolescence built into it. I don't recommend falling for advertising where they boast that the latest and greatest is best. Pursuing this strategy will put you on the upgrade treadmill forever and create a feeling there is a need to replace equipment every new release cycle. Your camera should have a 24 megapixel sensor or more and I recommend you spend your hard earned money on high-quality lenses. A single lens reflex camera with a 100 percent accurate viewfinder is also best. This makes a tremendous difference when doing close-up or macro photography. Another important feature to have in a camera is mirror lock-up for use with longer exposures. This enables the mirror to remain in the up position during exposure, and produces less vibration from the mirror slapping up and down during operation. Mirror vibration affects will make soft images.

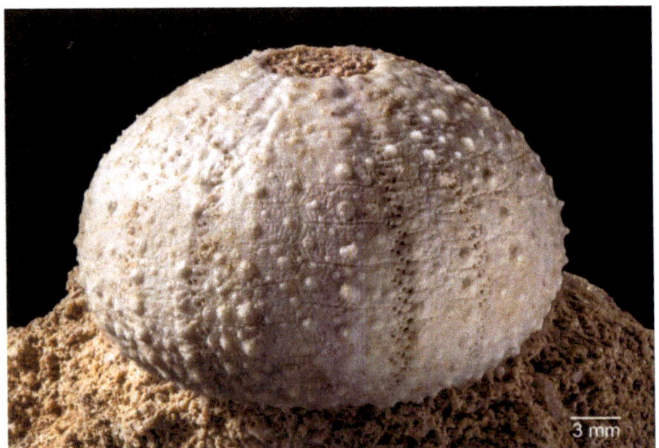

Figure 14.2 In this figure, you can see a very tiny oblong fossilized sea urchin from the Mannum formation in Southern Australia dating back to the Lower Miocene period still attached to its matrix. The tiny specimen is more than 23 million years old and so well-preserved that one can observe the finest of subtle details on its shell, where at one time protective spines protruded out. A total of 25 exposures were made using different points of focus. The images were processed in Helicon Focus® software. Focus stacking requires exacting techniques as well as use of a platform that is completely free of movement or any type of vibration for both specimen and camera. Image courtesy of Norman Barker.

There are also several new mirrorless cameras on the market. Most cameras that fit into this category will have the ability to shoot in RAW format which is highly recommended simply because of the flexibility it provides in post-processing. For ease of use and removing the challenges of image quality assessment using a small LCD screen, the camera should be operated in tethered mode. This enables operating and viewing images directly on the computer's display. Sharpness, content and image quality can be evaluated both in the field or studio. Screen brightness on location in bright light can challenge visibility but using a hood over the screen can mitigate that problem.

A good specially designed macro lens (not zoom close-up lens) is essential because of its flexibility and its capability for use from ×1 (life size) to infinity. This feature is very important, especially if you work on location at a dig site. Prime macro lenses are available in focal lengths of 60, 105, and 200mm, depending on the manufacturer. Chapter 11 provides a great deal about close-up techniques. Some would argue, one do-it-all lens is easier to carry and you don't have to change lenses in a dusty environment. Opening a camera always risks dust and dirt then finding its way onto the camera sensor. This is an important consideration for fieldwork. Another helpful tool is a macro focusing rail that provides very fine movements and more control over focusing. This is an essential tool, especially when making close-up and macro images that will be focus stacked. With advancements in software for focus stacking (extending an image's depth of field) it's amazing the types of images you are able to produce that were impossible to form just a decade ago. These methods record multiple images from different focus points throughout the region of interest. The images are produced using the lens's optimal aperture or sweet spot (sharpest aperture without diffraction) which is usually between *f*/5.6 and 8. The images are combined together in software and a limitless depth of field can be obtained. For more information about this useful technique, see Chapter 10.

Image Shake

Camera vibration can be an exasperating problem when photographing small subjects. The use of a heavy-duty tripod is an essential tool that is needed for the production of sharp images, especially when working with continuous light sources and long exposures. A high-quality tripod can be a pleasure to work with and make the task at hand much more efficient. There are several models of carbon fiber tripods that are heavy duty and reasonable in weight when carried into the field. Do not go low on this purchase. A good tripod will last for many years if taken care of. Another essential tool for the studio is a good heavy-duty copy stand. Be sure it is multi-purpose and can be used for small objects and macro specimens. Having a robust stand will make life easier when you need to look straight down onto a subject such as an old photograph, fossils, or minerals.

Special Lenses

With specialized applications, sometimes more magnification is needed than 1:1 (life size). This is true when documenting microfossils or fractures in gems. With the addition of a bellows or extension tubes you can obtain much higher magnifications depending on what details of the specimen are to be highlighted. Extension tubes are a cost effective way to gain slightly more magnification. They can be purchased individually or bought in sets. They can be used one at a time between the lens and the camera body, or in combination to achieve greater magnification. A bellows for even more magnification in the 2–10× range is very convenient. When additional magnification is needed there are a series of specially made lenses for macro applications. This series of lenses yields excellent results for color correction, resolution and flatness of field. Unfortunately many of these lenses are no longer manufactured. They are available in second-hand markets and eBay. Zeiss made a series called Luminar which were a series of five different lenses designed for different magnifications ranges when used with a bellows (see Chapters 7 and 11). Many of these lenses are still in drawers of university research labs because most people don't know what they are or how to use them. When more magnification is necessary, using a microscope or a specially designed stereo macroscope might be a better choice. This type of apparatus can be very expensive and cost upwards of the $25–35K range depending how equipped. Using high quality equipment is a pleasure to work with, and if you plan on doing this work often it might be worth considering a purchase.

Lighting

One of the most important outcomes for technical photography is accurate color reproduction. This can be a challenge. One thing that has made life much easier in the last several years has been the availability of different types of new and innovative light sources from several manufacturers. LED or light emitting diodes offer a wide range of light types that can change Kelvin temperature from daylight to tungsten and everywhere in between. Many of these units come with battery capabilities for use in field work as well as being easily adjustable and having a very accurate color balance. Highly recommended and a critical tool that is a must for consistent color is a quality color checker. X-Rite® makes an excellent product called ColorChecker Passport®. It is made of plastic, folds up, and fits neatly in any camera bag.

It's critical that once a specimen has been selected, the angle of view determined and the lighting established, the first photograph should be of a color checker and a proper scale (see Figure 14.3D). In this approach, you will always have an accurate reference for color when image processing and a scale for accurate and precise measurements. Using a coin or some kind of field tool is not recommended since some viewers will not be familiar with that item. A small plastic ruler is easy to carry in your gear bag and when processing the image the scale can be removed and a small bar scale can be inserted and is much more accurate. When photographing gems and some minerals, its often the case that owners prefer to leave scales out of images because of aesthetic concerns which they feel will out-weigh the need

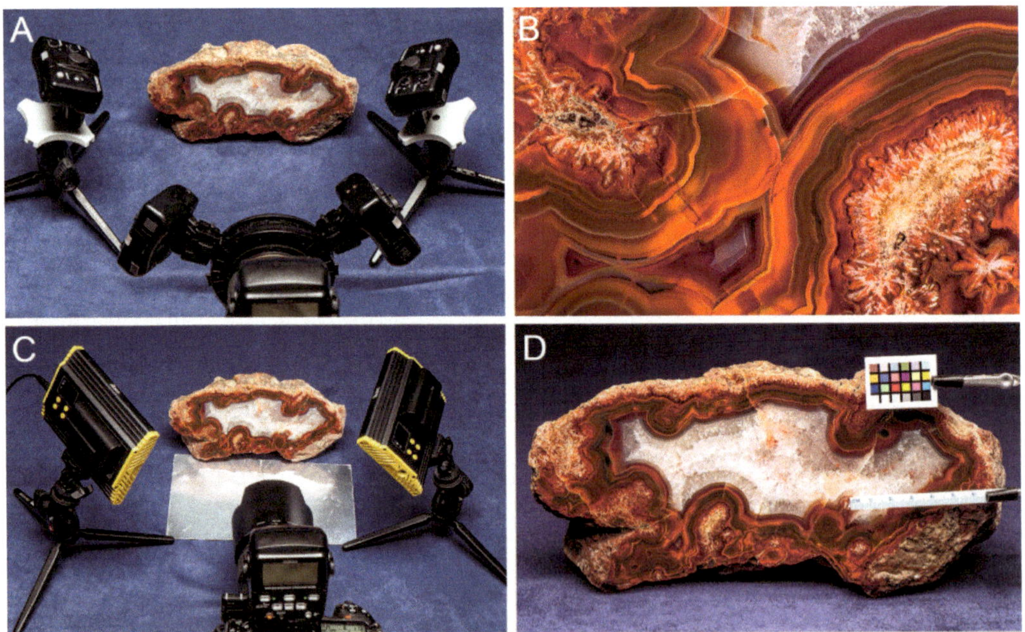

Figure 14.3 (A) and (B) A Nikon R1 Wireless Close-Up Speedlight system in use. (C) LED lighting with silver bounce reflector. (D) An Agua Nueva Agate, from Rancho Agua Nueva in northern Chihuahua, Mexico. I made a first frame reference photograph with a color checker and scale. Image courtesy of Norman Barker.

for measurements. Regardless, I always shoot at least one frame with a scale as a record for me. At some point down the line, these images may be published in a journal and the information of size is important. Even if the scale is not used in the photograph you will have the ability to give the specimens measurements in the legend. For many fossils and geological finds, scale is critical for future study, especially when images are published in an academic journal or book.

Sunlight and Documenting the Site

Sunlight can often be too harsh for photography but it is possible to manage using the right light modifiers (diffusion material, bounce cards and mirrors). With diffusers, the light can be modified but not completely controlled. Also, one of the challenges of natural sunlight is that during the course of the day the color temperature of the light will change. This is not as significant a problem as it used to be when film was used for photography. Where accurate color is needed, always shoot a color checker! There are many times where geological specimens and fossils will need to be photographed in-situ. This would be to show where the specimen was located or how it was found in a particular formation or vug.

Sometimes out in the field, a white bedsheet can help diffuse direct sunlight. It can be folded over itself as well, depending on how much diffusion is required. A cloudy day often works best, but obviously that's something you will not have control over. Always keep in mind that the specimen's inherent features or characteristics will dictate the lighting. The more control you have over the light, the more successful images will be. Often when photographing fossils, gems and minerals, it's effective to set up a photographic studio or table-top set-up giving you more control and access to make changes.

As mentioned, having an understanding of basic geology can be essential to identify sedimentary rocks prized by fossil searchers, such as shales, mudstones and sandstones. Igneous rocks from lava flows offer great collecting opportunities. Agates, topaz and quartz crystals can be among the spectacular finds. The most sought-after igneous rock sites contain pegmatite intrusions that have formed during the final stages of the magma's rock crystallization, and can contain beryl, emeralds, sapphires and other gems. All you need to get started is a rock hammer, protective glasses, magnifying lens and a sturdy pair of boots and work gloves. A highly recommended website is https://geology.com/, an excellent place to start. It provides valuable information and resources for all types of geology and collecting techniques.

Another way to search and document the site is the use of camera drones. This technique gives one a birds-eye view and can help plot your collecting activities. Large areas can be covered and collecting sites identified. Commercially available drones have come down in price and most use a standard Go Pro camera which can capture 4K video. The big problems with drones in geological fieldwork is that in several states and especially on Federal lands there are numerous restrictions. In many places it's a legal gray area and governments are working to restrict their use. In some areas they are outlawed completely, so check local laws before you take off.

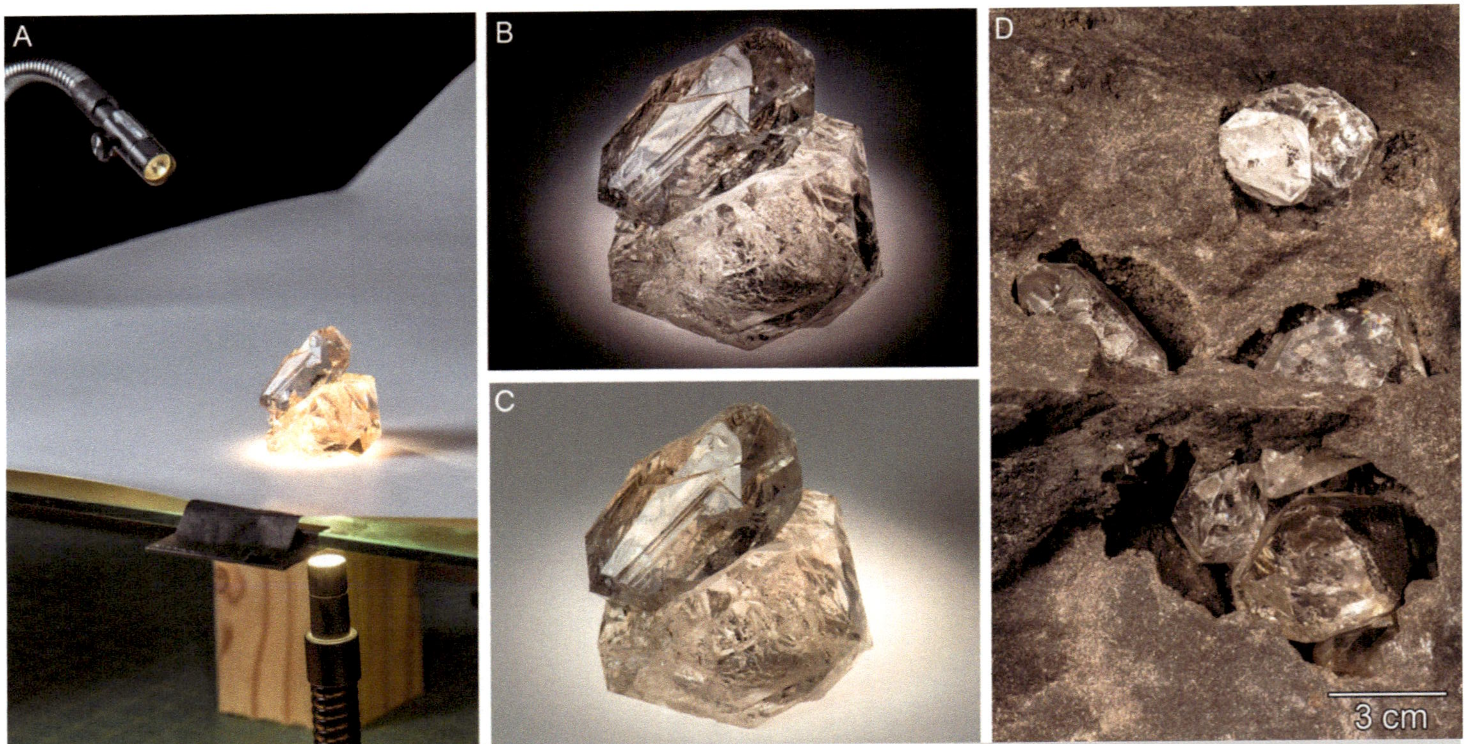

Figure 14.4 (A) Herkimer Diamonds photographed with Savage Translum, a translucent diffusion material that allows for shoot-thru and complete control of backgrounds. Fiber optic illuminated was used on the background and on the specimen. (B) and (C) Two variations of background light. (D) Herkimer Diamonds shown in-situ in a cavity in rock which is lined with mineral crystals, often referred to as a vug. Image courtesy of Norman Barker.

More on Artificial Light Sources: Tungsten, Fiber Optics, Electronic Flash, and LED

In the past it was common practice, (especially at gem and mineral shows) to use a simple inexpensive portable Tensor® lamp purchased at any office supply store as a light source for small object photography. This type of light used small tungsten bulbs and performed satisfactorily with smaller specimens. It was a good light for a small budget. The down side of the light was – as with anything – you get what you pay for. The light offered little control and provided no ability to focus the light, which was a significant disadvantage. Artist vellum could be used as a diffusion material and was available at art supply stores. With this material you could diffuse the light which reduced its harshness. This was much better than the light from the bare light bulb. There are different types of vellum with different thicknesses. If you use this material as a diffusion material, be careful not to put it too close to hot tungsten lights because of the possibility of catching fire to the paper. Photographic supply stores also carry a product called Savage Translum® which comes in rolls or sheets which is translucent and can be cut to size depending on the specimen or lighting set-up. For more on lighting, see Chapter 9.

You can see the effects of Translum as a background for the Herkimer diamond photograph in Figure 14.4. This semi-transparent plastic material can be used to shape or diffuse light. It can be used as a background to project light through and give great control over background light. This technique is useful when shooting gems or minerals because of the infinite possibilities for the use of colored filters. Translum can be used with a wide variety of light sources, flash, LED or tungsten hot lights.

There are numerous tungsten photo bulbs that are still used as well as quartz halogen bulbs. Halogen lights can be very bright. Rated at 750 or 1500 watts, these bright lights produce a great deal of heat. Some fossils and minerals such as sulfur can be very sensitive to heat. The heat might crack or change the color, or the material might dehydrate when exposed to this heat. For larger samples such as slabs, larger rocks, or for minerals, halogen lights are cost effective and provide a lot of bright light. These types of lights cannot be used with a dimmer because the color temperature of the lamp will change with voltage. This will impact an image's color balance. The one advantage of using this type of artificial light is you can observe in real-time what effect the light has on the specimen and observe any changes you make. This is extremely beneficial when trying to get light into cavities or trying to control light that reveals facets in a gem and controls highlights that become overexposed. For smaller specimens, a fiber optic light with a dual gooseneck can provide intense and cool illumination for a wide variety of imaging applications and has flexibility for easy positioning of the fiber optic pipettes.

The ammonite in Figure 14.5 became pyritized during the process of fossilization. Pyrite is also known as fool's gold because of its color and it was often mistaken for the real thing. The shiny glowing gold-like specimen creates a challenge to light with all of its curves and reflective surfaces. Building a small enclosure or "tent" around the subject with diffusion material softens the lights, opens up details, and prevents highlights from blocking up or

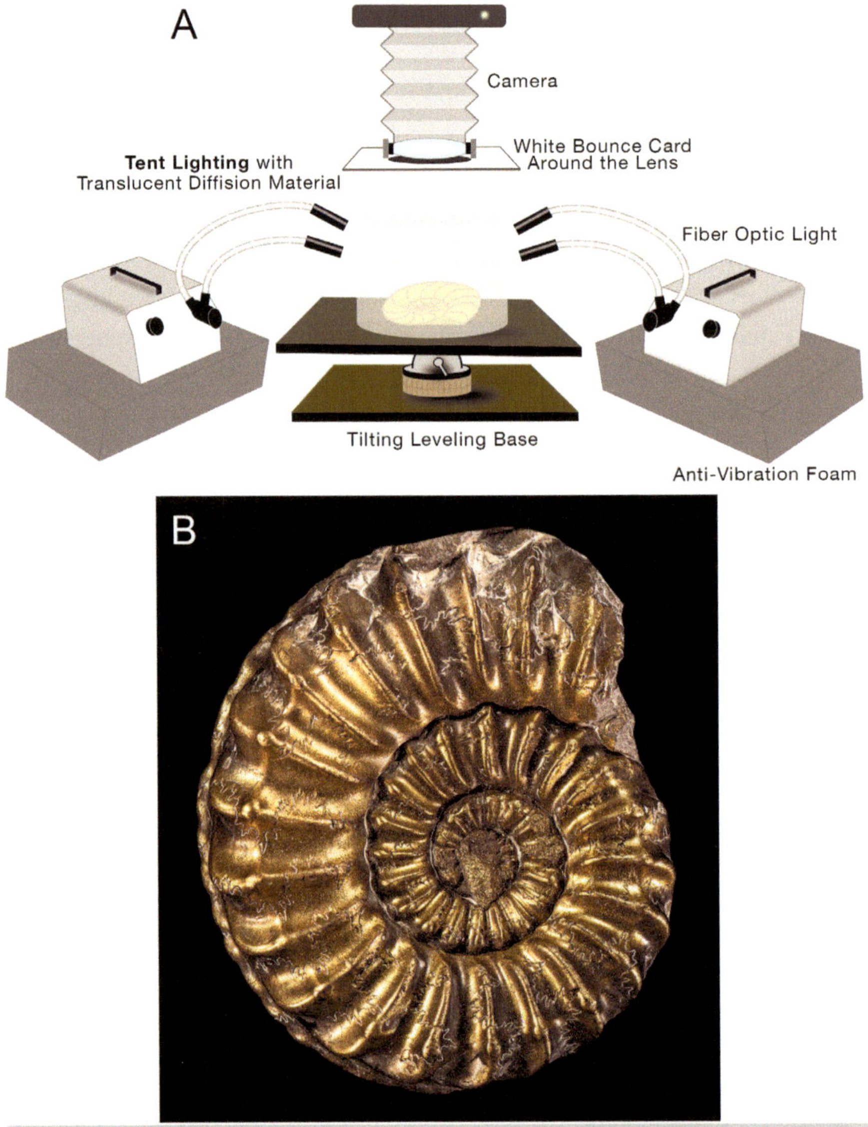

Figure 14.5 (A) Set-up showing fiber optics in use for tent lighting to diffuse shiny or metallic surfaces. (B) A pyritized ammonite photographed with the above set-up. Notice the tilting specimen stage for leveling and white card around the lens for bouncing light back onto the specimen. Image courtesy of Norman Barker.

burning out. Portable light tents are commercially available and fairly inexpensive, and some kits are also available with lights for a complete self-contained unit.

Most fiber optic light sources are cooled by fans and use a 150 watt tungsten halogen bulb. Care must be taken when placing this type of light source near the specimen when making long exposures because the cooling fan will introduce vibration. This problem is easily rectified by using an anti-vibration foam pad under the unit to absorb any shaking of the table that might be transferred to the specimen. Using the fiber optic light guides for positioning offers very convenient control for lighting. The light guides can be located to make very flat light or used at a very low angle to get a more raking light that separates the fossil from its matrix background. The fiber optic guides can also be very useful when trying to control highlights that will reflect off the facets in gemstone photography. You can use one guide on the specimen and the other aimed at the edge of the specimen to create a sort of rim-lighting effect to separate the specimen from the background. Fiber optic light guides come with small focusing lenses on each guide. Many have a step wedge neutral density filter that adjusts the intensity of the light without changing the voltage. In this manner the color temperature of the light does not change. The fiber optic guides can also be used to form transillumination used as for a background, add color, or create a vignette bringing more interest to the specimen. See Figure 14.4B and C.

Electronic flash is an excellent choice as a light source. Flash can be more expensive to buy and more difficult to operate. If you are a collector or just beginning to use photography, flash can be frustrating to learn, but with a little experimentation and practice, great results can be achieved. Even when using the latest TTL technology, achieving good exposures can be a challenge until more experiences are gained with the light source. Through-the-lens metering is designed to be used with average scenes that are mostly gray or mid-tones. If specimens are placed on dark backgrounds, getting the correct exposure will not be automatic even when using automatic. Additionally, small on-camera flashes have surface areas that are fairly small. Sometimes called a point light source, it's difficult to control the lighting on specimens that have multifaced reflective surfaces.

One advantage to using flash is that it is cool and produces consistent color. It can also be adjusted to work well with larger specimens. Small hand-held flash models are ideal for working in the field and are very portable. Most are powered by AA batteries. Smaller models of electronic TTL flash are available that can be attached to the lens and produce fairly flat light but can also be removed and radio controlled. Configured this way, they can be fired remotely and give a wide variety of lighting choice options. Small flash units also come in the form of a ring. Ring flashes have limitations, especially in gem photography. Since the flash surface goes directly around the lens it yields a very flat light and specular highlights are very difficult to control. Some ring flashes are designed so that they can be removed from the front of the lens and can be held off to the side to give a more controlled and interesting light. Again this type of flash can be used in the field and has the advantage of working well in close-up applications or be used to illuminate cavities of specimens which are often encountered in the field. Ring lights can make circular reflections in a photograph: see Figure 9.4.

One of the most flexible and portable close-up flash systems on the market today is the Nikon R1 Wireless Close-Up Speedlight system – see Figure 14.3A and B. This system is ideal for use in the field because of its portability for documenting small finds. The light is very easy to control because of its wireless TTL metering capabilities. The small flash units can be used directly connected to the front of the lens or taken off and placed around the subject for more control of the light. Where this unit really shines is in making small studio set-ups. As many as six small flashes can be used at one time giving very precise lighting control. It is recommended to shoot with your camera tethered to a computer. Most camera manufacturers provide software that enables capture to a computer monitor or laptop.

Evaluating the lighting on a large screen seeing how it works for the subject can be extremely valuable.

LED panels have been a true game changer for this type of work. The advantage of LED panels is that they come in many different sizes and run relatively cool. They use a fraction of the energy that regular incandescent or tungsten bulbs use and are cheaper to run (see Figure 14.6C).

LED lights for photography are available in either daylight or tungsten balance, as well as bi-color. You can also buy these lights as kits and they are very economical. Bi-color lights are the most versatile because the Kelvin temperature can be changed right on the back of the light panel. They come in many different sizes and shapes from very large to very small, so are very versatile. Another less expensive option is, if you have sockets for hot lights you can purchase LED bulbs that are balanced for daylight and run much cooler. With the addition of a white transparent background and a set of LED lights, a lightbox can be created for use with transparent or translucent objects.

LED lights convert 75 percent of the energy they use into light, compared with a standard tungsten bulb that is only 25 percent efficient, which is a significant improvement. With

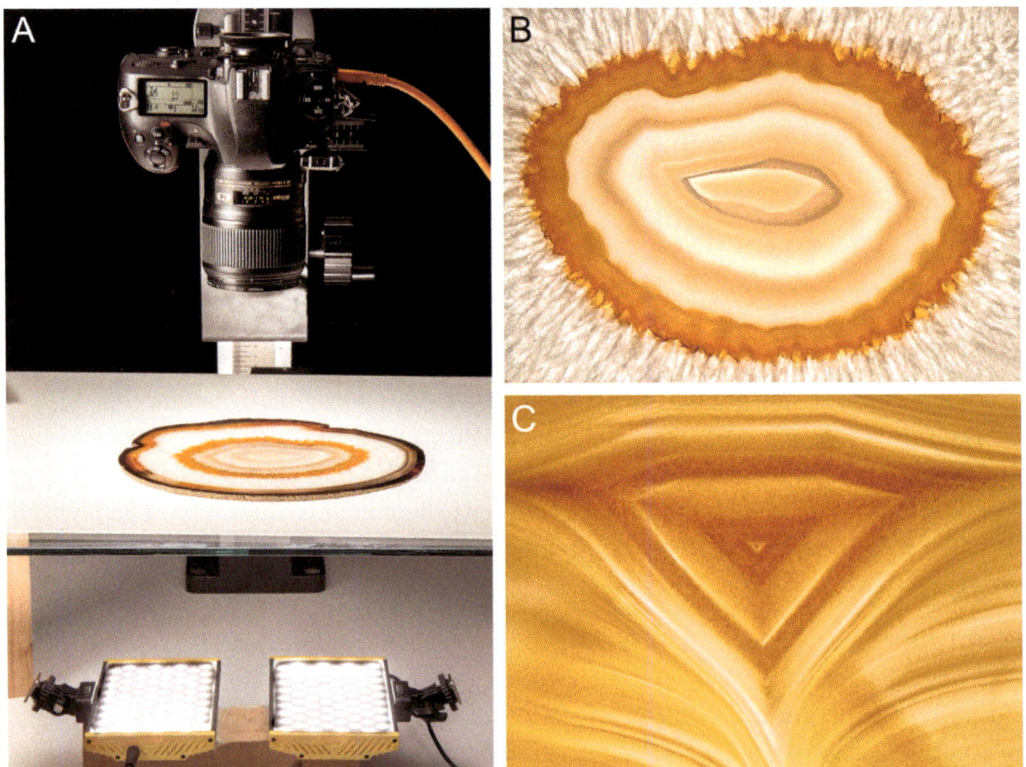

Figure 14.6 (A) Frosted plexiglass used as a light box with LED lights below providing daylight-balanced transillumination for Agate photography. (B) and (C) LED lights can become a powerful tool for several lighting applications using samples of different agates from Mexico. The lights can easily be reversed and used for top lighting (see photographing slabs and polished specimens). Images courtesy of Norman Barker.

many of the newer panels the color temperature can be easily changed from daylight to tungsten and anywhere in between. This is a great advantage if you want to warm up one light while keeping the other balanced for daylight. Another advantage of LED lights is that you can switch from still photography to video seamlessly.

Background Selections and Reflectors

In many ways, background selection is a matter of individual taste, but is very important. Backgrounds should be selected with the idea that they should be neutral and not compete with the specimen, but should compliment it. For example, when photographing gems and some minerals that are transparent and have facets and angles, colored backgrounds can compete with the specimen and even affect the overall color quality of the specimen as well as the individual crystal faces with color casts. If doing close up work, even if the background is not going to be in shot, the use of a black velvet background is recommended. Care must be taken with strictly using black backgrounds because in many cases it can make the specimen look flat and appear to be floating in space with no anchor point (see Figure 14.7). Black velvet absorbs most of the light that falls on it. It's easily cleaned, but most importantly it reduces any extraneous light from bouncing back on the specimen that would affect the color. It also lessens image-flare which can reduce the overall contrast in the final photograph. You can read more considerations for background in Chapter 6.

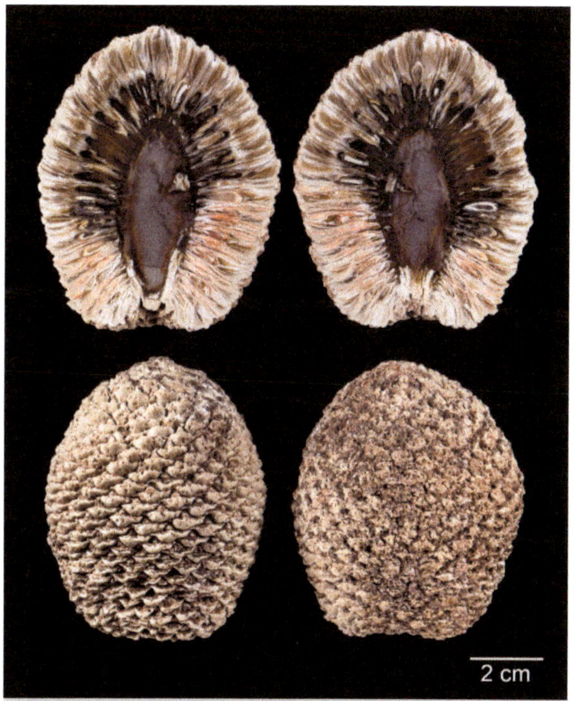

Figure 14.7 A pine cone from *Araucaria mirabilis*, from the Middle Jurassic period, 160–165 million years ago. *Araucaria mirabilis* is an extinct species of conifer from Patagonia, Argentina. This silicified cone has been sectioned and polished to reveal the well preserved seeds within. Each half was photographed individually at the same magnification, on a black velvet background. Then the images were composited together using Adobe CC, 2020. Images courtesy of Norman Barker.

Another background choice can be found in using colored paper. This paper comes in 18 × 24 inch sheets that can be purchased at an art supply store. For larger specimens, rolls of seamless background paper in neutral colors like gray, white and black are good to have on hand. A light gray background is extremely versatile and can be positioned to give a wide variety of shades from very light to almost black. Graduated backgrounds are also commercially available that go from pure white to deep black and all the grays in-between. This type of graduated background makes the specimen stand out and separates the background with the illusion of some depth. As mentioned earlier, Savage Translum® is an excellent choice because the specimen can be elevated, the background can be lit with filters underneath, and any color can be introduced with colored gels. Glass or plexiglass can also be used when the specimen is placed directly on top and elevated above a colored background.

Reflectors, bounce cards, light baffles and diffusion material can be purchased commercially, but often a trip to the art supply store is better value. Different types of 4-ply board are available that come in a variety of colors and surfaces and can be cut to any size needed to fit the specific needs of the specimen at hand. They can be used as bounce cards to add light to certain areas or

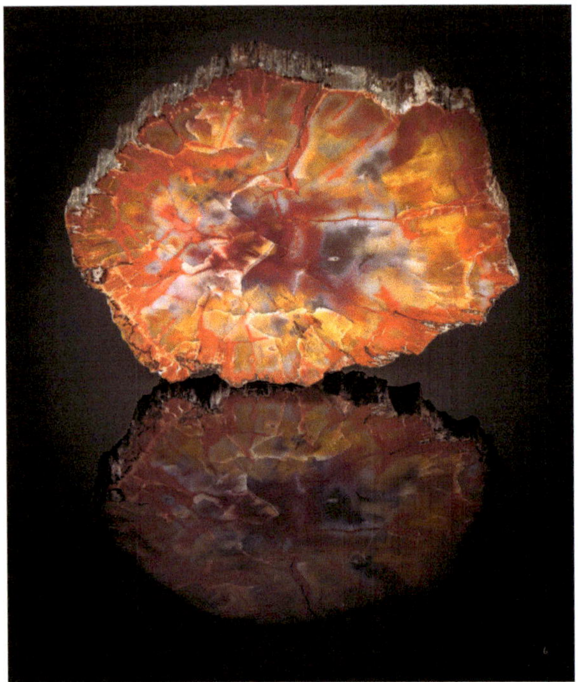

Figure 14.8 An interesting background choice for both gems and fossils is the use of black glass which can be used to add depth to an image by introducing reflections. The subject of this picture is petrified wood from *Araucarioxylon arizonicum* which is the state fossil of Arizona. It was collected in the Petrified Forest, Arizona. Image courtesy of Norman Barker.

as light blocks to hold back light from other areas. I find white, silver and gold the most often used. Whether photographing gems, fossils or minerals, use of diffusion material to control blocked-up highlights is essential. Generally it's always best to use a diffused light source that will control highlights while retaining detail in shadow areas. Keeping in mind that diffused does not mean flat, you still need to light the specimen to have the appearance of three-dimensional contour. Small silver bounce cards (see Figure 14.3C) or even small mirrors can be used to bounce light into small crags or crevices and control the light on the surface of a facet. This is where the gooseneck arms of a fiber optic light come in hand. The light can be focused as a point light source or diffused or even bounced off another surface to control the light falling on the specimen. Don't be afraid to experiment. Cloth backgrounds are seldom used in mineral photography. But, a technique used in archeological and museum photography can be useful. A product called pseudo-suede material can make a great background. This can be purchased at a fabric supply store and is available in many colors. It's great for photographing finds in the field (see Figure 14.3D). The fabric is portable and easily cleaned and brushed and will give a pleasant neutral background.

As previously mentioned, even before the specimen is placed on the background, evaluate the specimen using a magnifying glass and inspect for dust and other particulate matter. I use a small anti-static brush and a manual air blower with a large rubber bulb to dislodge fine particles and grime. Doing this ahead of shooting will save you a lot of wasted time sitting in front of the computer retouching the final image in Adobe Photoshop.

Using Filters

It's amazing when I think back to just a short time ago. I recall the numerous and different types of filters that were used for film photography. Filtration was essential for light balancing, color correction and color conversion. Oddly enough, film is experiencing a resurgence and there is a whole generation that only knows digital imaging using a smartphone. Many young artists think they have discovered a new medium of self-expression. I don't miss the darkroom and all the stained clothes and lengthy exposure to harmful chemicals. So much of what was accomplished using color filtration and film can now be done with computer programs in much less time. In this difficult and challenging type of photography, the conversion to digital has been a welcome transition.

The one type of filtration outcome that can't be performed using image processing is the effect of cross polarization. Using a polarizer to darken blue skies or control reflections on water or glass is very common, but true cross polarization is accomplished when the light source and the camera is outfitted with polarizing filters. This lighting technique is often not only used to control reflection, suppress glare and specular highlights but can often tease out unseen details in fossils or gems. The setup is not complicated to create. The camera lens is fitted with a standard linear polarizing filter and the light source is fitted with a piece of linear polarizing material as well. These sheets of polarized material can be purchased from Edmund Scientific or other vendors. The filter on the lens must be capable of rotating to obtain maximum extinction of reflections from the specimen. Polarization occurs at a 90 degree angle to the other polarizer in the system.

UV Radiation and Geological Minerals

Fluorescence can be observed when certain minerals are exposed to invisible ultraviolet radiation. Once exposed to this high energy, minerals fluoresce (give off light) depending on the particular compounds found in each specimen. Electrons in fluorescent minerals absorb ultraviolet energy and release this energy in the form of visible light or glow. The recommended specialized equipment for this work is a sturdy tripod for long exposures, a short-wave UV lamp, and a Kodak Wratten 2A filter which is pale yellow in color and absorbs ultraviolet radiation. This helps control some of the purple color fringing. There

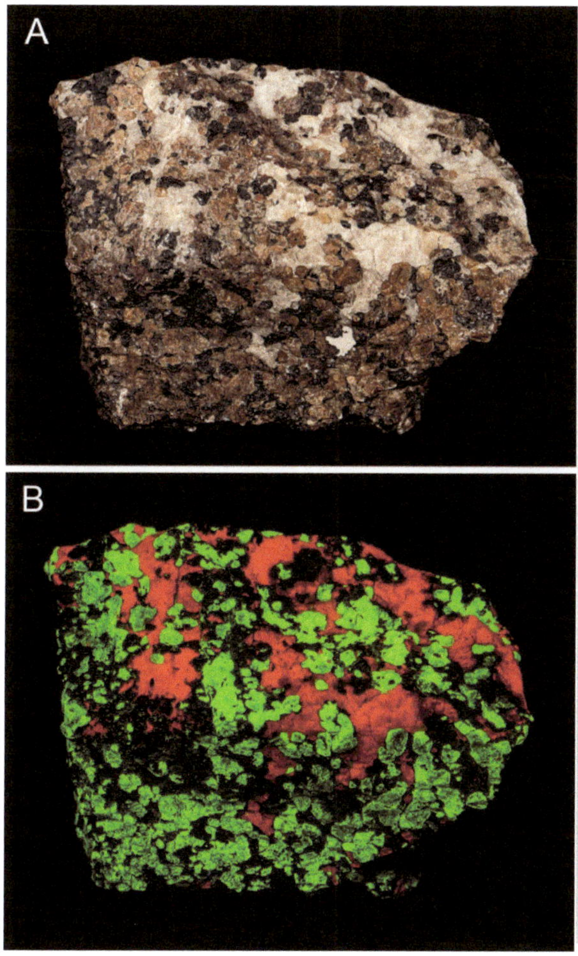

Figure 14.9 In this comparison, a visible light image (A) is compared to an image made with shortwave UV radiation (B). This mineral was collected from the Sterling Hill Mine in Franklin, New Jersey and when exposed to shortwave UV radiation, the specimen glows. The mineral calcite is red, willemite is green and franklinite is black. The mine is now closed and has been turned into a museum. It was once the richest source of zinc ore deposits in the world. For a small fee you go collecting on the rock pile which is full of highly fluorescent minerals, a true Mecca for mineral collectors. Image courtesy of Norman Barker.

is usually no easy way to determine your exposure times because the signal can be weak. The best way to get a good exposure unfortunately is by trial and error. It's not unusual for exposures to run into the minutes depending on the excitation. The advantage of the long exposure especially for larger specimens is that one can use a light painting technique to get even overall lighting of the specimen. You can see more about fluorescence techniques in Chapter 15.

It is believed that only about 15 percent of the more than 5,000 different types of minerals will fluoresce under ultraviolet radiation. Fluorescence also depends on the specific wavelength of the light used for excitation. UV radiation is electromagnetic energy with wavelengths of 160–400 nm. Ultraviolet radiation is used in many scientific applications, ranging from forensics and medical diagnostics to curing substances like inks, coatings, and glues. This type of radiation is usually divided into three types: (1) UV-A is longwave ultraviolet radiation with wavelengths of 315–400 nm, (2) UV-B is mid-wave ultraviolet radiation, 280–315 nm, and (3) UV-C is shortwave ultraviolet radiation, 160–280 nm. Care should be taken when working with UV radiation in a photographic session. Protective eye glasses should be worn to protect your eyes from stray UV radiation. The extremely interesting thing about fluorescent minerals is that the color you observe is emitted from the specimen itself and not reflected light from the surface of the mineral. The colors that are observed during our daily interactions with things are mostly caused by reflected light. The glow that can be observed from fluorescent minerals is unique and can be used to help

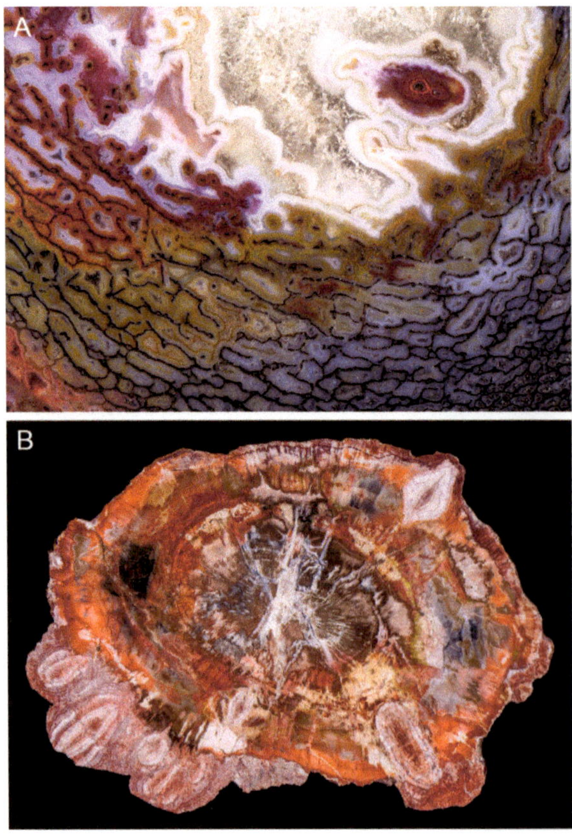

Figure 14.10 Dinosaur bone cut slabs can make wonderful specimens that show off the well preserved cellular anatomy of various cold blooded creatures. (A) A polished slab of dinosaur bone from the Morrison formation, Colorado. This specimen was cut on a large diamond blade saw and highly polished with a buffing wheel. (B) *Araucarioxylon arizonicum*, also called rainbow wood, showing fungus pockets that are clearly visible in the upper right and bottom left. Image courtesy of Norman Barker.

in specimen identification. Some collectors go to great lengths to show off their treasures under very controlled lighting conditions, which can make a magical display showing all the colors of the rainbow.

Slabs and Polished Specimens

Often times fossils can be purchased from rock, mineral and gem shows like the largest showcase in the world, The Tucson Show where people gather to buy, trade and admire some of the most beautiful treasures from around the world. Every February for a couple of weeks the city turns into a Mecca for rock hounds and gem collectors alike. Museum quality specimens can be purchased and if you are willing to haggle a little bit, exceptional deals can be had. Much of this material is cut and polished and ready to be exhibited in collections.

The age of these geological formations corresponds to the middle part of the dinosaur period called the Jurassic period, when there were no other very large animals. Finally, the bone itself is five or six times the thickness of the thickest long bone in an adult male human and, therefore, it had to have come from a very large beast indeed. Contrary to all appearances, the red structure within the quartz-filled medullary cavity is not a nutrient artery. Soft tissue rarely becomes fossilized, and when it does, it is rarely replaced by quartz. The red color comes principally from hematite, an iron oxide. The bone in the lower right is metaphyseal bone, also called spongy bone, with blue chalcedony between the spicules.

Slabs can be photographed using a simple copy stand, LED lights and a specimen leveling stage. Sometimes the actual grinding marks can be seen on the polished specimens, which interferes with making a clean photograph. Figure 14.10B shares a polished slab of *Araucarioxylon arizonicum*, a fossilized wood, showing fungus pockets. This can usually be minimized using a light coating of mineral oil. The oil will fill in cracks and crevices and is easily removed once finished. It's recommended to try coating a small corner as a test to start. If the specimen has a porous sandstone base it could absorb or be stained from the addition of the oil. A well-defined pocket of fungus (Figure 14.10B) appears in the lower left corner and upper right in this cross-section of *Araucarioxylon arizonicum*, which is the most common of the trees found in the Petrified Forest near Holbrook, Arizona. Care must be taken when on Federal lands, in most National Forests it is illegal to remove any specimens of petrified wood from the park and can result in stiff fines and even prison time.

In this example, various and different characteristics of the sample have been revealed. One view is superior. The choice will depend on what was the original goal or intent of the photograph: How will it be used, and for what purpose? I believe both 14.11D

and 14.11F are the most successful. The lighting in 14.11F is the most dramatic and I believe the lighting in 14.11B is too flat. It is important to keep in mind that there are subjective elements to scientific photography, and in the end the matter of personal taste will influence results. Often I will produce several lighting set-ups. Then, when editing, I have some flexibility of making a final image selection that best meets the needs of the original objective.

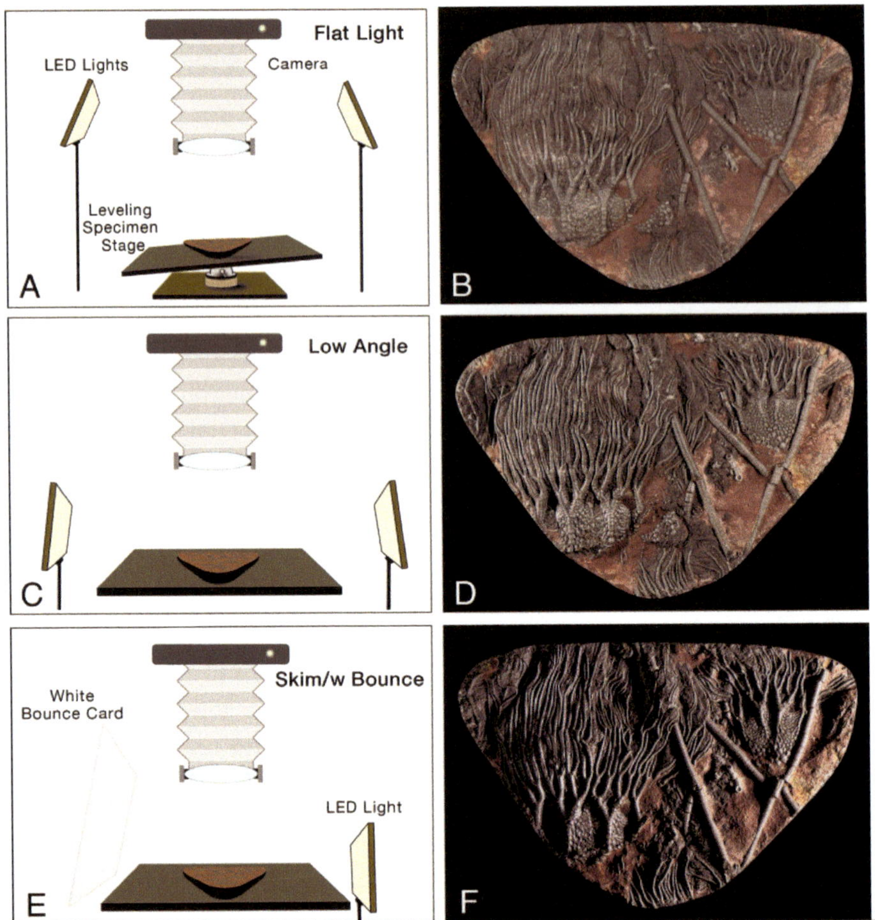

Figure 14.11 Unique specimens of museum quality often come from dealers that have done collecting on private land. This figure was created to show how crinoids look with different lighting. A convenient and easily assembled leveling stage is a great time saver. This device will help to keep the specimen level and maximize best possible focus. Using two sheets of aluminum drilled with ¼-inch tread so a mid-weight tripod ball head can be sandwiched between (see Figure 14.11A). With the ball head the top plate can be easily tilted into position and keep the specimen parallel to the digital sensor. This is a time saver so the specimens do not have to be shimmied into level. This sample in this composite is *Scaphocrinites elegans* which is an early crinoid, a term derived from the Greek krinon, meaning "lily." Though they do resemble lilies, crinoids are not plants but invertebrate animals – stemmed echinoderms, relatives of the sea urchin, starfish, sand dollar, and sea cucumber. During the Paleozoic era, a great many varieties of stemmed echinoderms lived in the sea. Some were free-floating; others were rooted to the bottom of the sea, with a swaying stalk that supported a vase-shaped body (theca). Extending upward from the theca were pinnules or branches, probably carrying finger-like cilia, which swept food into a mouth that faced upward at the apex of the theca. Part of a theca is seen at 12 o'clock in this photograph, with a clump of branched arms matted together in the lower half of the picture. The two columnar structures and the small cylinders at the top left are stems and pieces of stems. Ring-like structures similar to those in the human trachea gave stability to the stems, but because the stems were rigid, they had little flexibility. The figure shows three distinct lighting set ups for different purposes. A flat shadow-less lighting is demonstrated in (A) and (B), a low angle skim lighting to add more three-dimensions in (C) and (D), and one low angle light with a bounce card for fill light, the whole idea being to separate the fossil from its matrix, in (E) and (F). Images courtesy of Norman Barker.

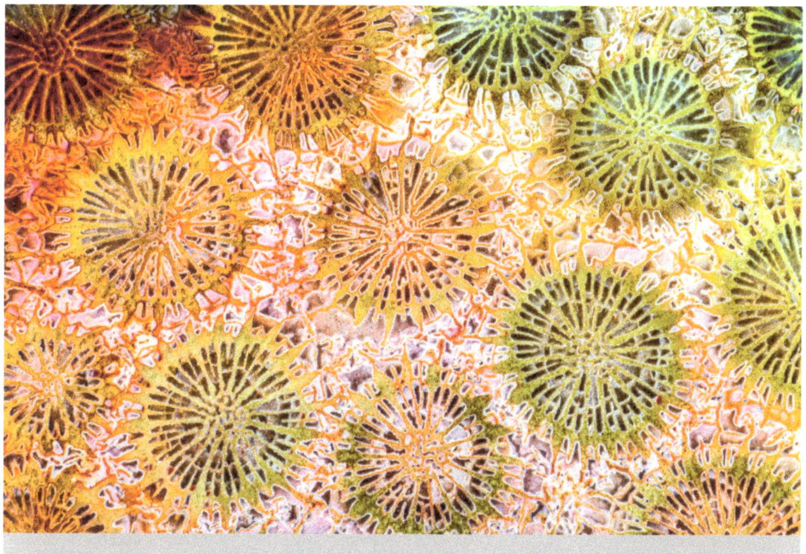

Figure 14.12 Red fossil sponge coral from Indonesia is an amazing sample because of its subtle beauty and is highly prized for fine jewelry. Agatized coral can exhibit a rainbow of natural colors, ranging from white and pink, to brown, gray, black, red and yellow. Depending on how the specimen is prepared, it can be very three-dimensional or a cut and polished slab as in the specimen captured here. During the process of replacement over millions of years this specimen calcium carbonate or coral is replaced by silica. This unusual system of preservation can result in different mineral concentrations depending on the sea conditions. Silica can range from light nebulous opaline to dense forms of chalcedony or quartz. Image courtesy of Norman Barker.

Closing Thoughts

For some, the study and collecting of fossils, minerals and gems can be a fascinating and rewarding pastime. Whether a seasoned rock hound or occasional collector, there is always something new and interesting to marvel at and stand in awe at the wonders of the earth. Collecting can turn into a passion, addiction, and lifetime commitment. There are fossil and mineral clubs all over the country and they often meet monthly. They are great places to learn new things, collect new techniques, and have new experiences from like-minded people who share a similar passion. But most of all it gives you the opportunity to see and share new material for making beautiful photographs! Sometimes that's the only way you can make connections and get the chance to see and handle some beautiful and sometimes very expensive specimens. For the passionate collector, creating high quality images can be a labor of love and a pleasure to illustrate for others, the wonders of nature.

A special thanks to Drs. Giraud Foster and Eric Seifter for their willingness to share specimens and knowledge along with their inspiring enthusiasm for all things geological.

Suggested Reading

Handbook for Scientific Photography 2nd ed., by Alfred A. Blaker. W.H. Freeman and Company, 1989.

Paleobotanical Spendor, by Norman Barker. Man and Lion Press, 2016.

"Suggestions for Photographing Jewelry", by G. Foster and N.J. Barker, *J Gemology*, 22(5): 287–291, 1991.

"The Use of Polarized Light in Photography of Macrofossils", by P. Crabb, *Paleontology*, 44: 659–664, 2001.

Light Science & Magic: An Introduction to Photographic Lighting, by Fil Hunter, Steven Biver, and Paul Fuqua. Routledge, 2015.

Ancient Microworlds, by Giraud Foster and Norman Barker. Custom and Limited Editions, 2000.

Picturing Science and Engineering, by Felice Frankel. MIT Press, 2002.

Envisioning Science: The Design and Craft of The Science Image, by Felice Frankel. MIT Press, 2002.

Handbook of Nature and Scientific Photography, by Charles Laun. Alsace Books, 1988.

Manual of Close-up Photography and Photomacrography, by Lester Lefkowitz. Watson-Guptill, 1979.

The Focal Encyclopedia of Photography and Digital Imaging, Theory and Applications, History, and Science 4th ed., by Michael R. Peres (ed.). Focal Press, 2007.

Laboratory Imaging and Photography, by Michael R. Peres (ed.). Routledge, 2017.

Modern Rockhounding & Prospecting Handbook 2nd ed., by Garret Romaine. Rowan & Littlefield, 2018.

Digital Photography for Science, by Enrico Savazzi. Lulu, 2011.

Photographing Minerals, Fossils and Lapidary Materials, by Jeffrey A. Scovil. Geoscience Press, 1996.

Stoppes' Guide to Photography and Light: What Digital Photographers, Illustrators, and Creative Professionals Must Know, by Brian Stoppe and Janet Stoppe. Taylor & Francis, 2012.

Photomacrography: An Introduction, by William White (ed.). Focal Press, 1987.

Websites

www.webmineral.com – A mineralogy database and excellent resource.

"6 Tips for the Best Gem Photos", by Amanda J. Luke. *Gemological Institute of America*, https://www.gia.edu/gia-news-research-six-tips-best-gem-photos#.XfaWH74rP-0.email, 2014.

www.geology.com

Chapter 15 Imaging with Invisible Spectrums

Adrian Davies

Introduction

There are many subjects where revealing the hidden patterns or features within can be made visible by photographing them using wavelengths outside of the range of human vision. Ultraviolet (UV), infrared (IR) and polarized light photography have been used for more than a century to make the invisible world visible to human eyes. UV and IR photography date back to the early years of the twentieth century. Various techniques utilizing UV and IR have, and continue to be, used extensively in the technical fields of medicine, geology, forensics, security, cultural heritage, and botany. Landscape, portrait, wedding and fine art photographers also use the techniques extensively. UV reflectance photography (UVR) has been used in an advertising campaign for sunscreen, to illustrate its effectiveness, while an IR image of lions in the Serengeti won the title of Wildlife Photographer of the Year in 2014. It has often been said that photography is one of the few subjects where art and science come together, and nowhere within the genre is it more apparent than with the photography of invisible radiation. This chapter will look at the various techniques used for recording UV and IR, including UV reflectance (UVR) and UV fluorescence (UVF), little known IR luminescence, and also the use of polarized light for revealing invisible stress patterns in a variety of materials.

The Electromagnetic Spectrum

The light humans can see represents is just one small portion of radiated energy of the electromagnetic spectrum. Visible light (to humans) is approximately 400 to 700 nanometers (nm), including the colors of violet, blue, green, yellow, orange, and red. Shorter wavelengths, beyond the violet end of the visible spectrum include X-rays, gamma rays and UV, while beyond the red region of the visible spectrum we can find IR, heat, and radio waves. X-rays and gamma rays are used extensively in the field of medicine as diagnostic tools and can produce useful and interesting images using highly specialist equipment. Heat too can be imaged using special thermal imaging cameras.

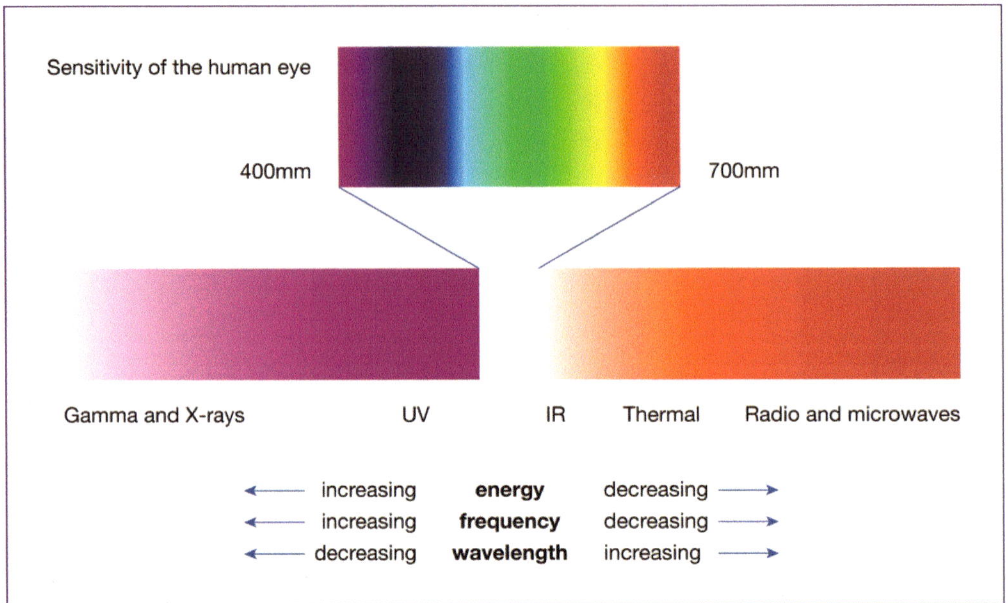

Figure 15.1 The electromagnetic spectrum, arranged according to frequency and wavelength.

UV and IR photographic techniques were used frequently in the days of film photography, for medical, forensic, and other technical applications, as well as landscape photography (in the case of IR). All black and white films were naturally sensitive to UV wavelengths, while various IR emulsions, both black and white and color were available. Today, modern digital cameras can be modified to better record UV and/or IR.

Camera Modification

Modern digital cameras record images using a light sensitive silicon chip, of either a CCD or CMOS type. The sensor contains a grid of light sensitive picture elements (pixels) which usually is in the millions. Light falling on a pixel produces an electrical charge, in proportion to the amount of available light. This charge is converted into a digital signal in the camera. Covering the pixels, in most sensors (Sigma sensors use a different type) is a mosaic of transparent color filters, red, green, and blue. As the light passes through these filters its color is recorded. This filter pattern is known as a Bayer pattern filter. You can read more about sensors in Chapter 8. The sensor in most digital cameras is naturally somewhat sensitive to UV, visible, and IR wavelengths of radiation, but because most photography does not need the UV and IR components of the spectrum, these spectrums are removed using filters. These filters are situated by the manufacturer in front of the sensor. A special hot mirror is an example of an IR cut-off filter. To modify a camera and allow it to record UV and/or IR, the cut-off filter will need to be removed. While it is possible to do this yourself (and there are various online videos showing how to do it), it does require a good deal of technical skill and a dust free environment. The sensor is probably the most expensive element of a digital camera and I recommend the conversion is carried out by qualified technicians. Some of the companies are listed in the appendix. Most camera types can be converted – compact, DSLR, mirrorless, and "action" types alike. Cameras that use interchangeable lenses will be most flexible for a range of subjects.

There are basically three types of conversions possible. UV only, full spectrum only, and IR only. It is very important to choose the correct conversion type, that will allow you to record the images in which you are interested.

UV Only Conversion

A UV only conversion is where the filter is removed from the front of the sensor and replaced with a filter that absorbs all visible and IR wavelengths. This filter will transmit only UV from approximately 330 to 400nm. No other filter is required to be placed in the front of the lens. A UV only is an expensive conversion because a UV transmitting filter is expensive. Multi-spectral images using visible light wavelengths are not possible if you create a UV only conversion. One problem associated with a UV only conversion is that because the Bayer filter overlays pixels, having this new filter combination often produces softer images that reveal strange colors. A couple of companies offer a service for "de-Bayering" a sensor, which involves grinding the Bayer filter away from the surface of the sensor which is an incredibly delicate operation! It is very expensive, but gives images of much higher UV resolution because each pixel records UV and it is very useful in dermatological photography.

Full Spectrum Conversion

In this case, the term "full spectrum" indicates that the camera will be sensitive to UV, visible, and IR wavelengths but it will not record heat. Heat is recordable using a thermal camera. A full spectrum conversion occurs when the hot mirror (or other) filter is removed from the front of the sensor, and replaced with quartz or specialized glass, allowing UV, visible, and IR wavelengths to be recorded. Wavelengths reaching the sensor are determined by filters placed on the front of the lens. In the case of UV and IR photography, these filters are visually opaque. This makes the camera's viewfinder virtually useless on a DSLR camera. A full spectrum conversion is the only type of conversion that can be used for multi-spectral images, simulating insect vision, for example.

IR Only Conversion

An IR only conversion is where the hot mirror (or other) filter is removed from the front of the sensor, and replaced with an IR filter of a specific IR wavelength, usually in the 720 or 760nm range. This will yield excellent monochrome IR images, but does not record visible color wavelengths.

Recommendation

If both UV and IR images are required from the same camera, you will need a full spectrum conversion. If you will be recording only monochrome UV or monochrome IR, then one specific conversion will be more useful, and the camera's viewfinder will remain operable.

	Advantages	*Disadvantages*
Full spectrum	• Highly versatile – one camera can be used for both UV and IR photography. • Can use a wide range of filters over the lens for recording various wavelengths. • Cost effective. • Can be used for "normal" visible light photography with a hot mirror filter over the lens.	• You will need various filters to determine which wavelengths are recorded. • There may be a restriction as to which lenses can be used, depending on filter diameter. • With DSLR cameras, you will need to focus and compose the image before adding the filter (or use the live view facility if your camera has it). • Can only practically be used with a tripod and static subjects. • Focus calibration will be for either UV or IR, so you will need to do some tests to check on the extent and direction of the focus shift for other wavelengths.
UV or IR only conversion	• Can specify a particular internal filter for a specific use. • Can focus and compose the image through the viewfinder, or via the live view facility, so can be used for live subjects such as portraits or weddings for example. • Can use virtually any lens, from the widest fisheye to the longest telephoto (UV will require a UV transmitting lens). • Can shoot video if your camera has the facility, using the camera viewfinder for framing, zooming, and panning. • No need for filters for the camera lens. • A UV only conversion will be relatively expensive due to the high cost of the internal filter.	• The camera can only be used for recording one specific range of wavelengths. • Cannot be used for multi-spectrum photography with filter "stacks", for example. • The camera cannot be used for visible light photography. • UV only is expensive.

When choosing a camera for conversion, it will be a good idea to choose one with a live view feature.

Ultraviolet Photography

There are two types of UV photography – photographing the UV reflected from a subject (UVR), and photographing the fluorescence emitted from a subject when excited using an UV source (UVF). The techniques and equipment required for each are different.

Reflected UV Photography

UV radiation has wavelengths of that are shorter than 400nm down to approximately 100nm. It is usually divided into three regions according to its wavelength: UVA, UVB, and UVC. The shorter the wavelength, the more harmful the radiation to human skin and other biological tissues such as the retina.

Short-wavelength UVC (approximately 100–280nm) is the most damaging type of UV radiation. However, it is completely filtered by the ozone layer in the atmosphere and does not reach the earth's surface.

Medium-wavelength UVB (approximately 280–315nm) is very biologically active but cannot penetrate beyond the superficial skin layers. It is responsible for delayed tanning and burning. In addition to these short-term effects it affects skin ageing and significantly promotes the development of skin cancer. Most solar UVB is filtered by the ozone layer in the atmosphere as well.

The relatively long-wavelength UVA (approximately 315–400nm) accounts for approximately 95 percent of the UV radiation reaching the Earth's surface. It can penetrate into the deeper layers of the skin and is responsible for the immediate tanning effect of the skin. It also contributes to skin ageing and wrinkling. For a long time it was thought that UVA could not cause any lasting damage, but recent studies strongly suggest that it may also accelerate the development of skin cancers. There is also increasing evidence that IR radiation (760–1440 nm) may also be harmful to the skin, and various sunscreens now specify protection against it.

For most UV photographic purposes, the wavelengths from 350–400nm will be most useful. Specialized quartz UV lenses will be needed to record images in UV wavelengths down below 320nm, while normal uncoated glass lenses can record images from 400 to 350nm.

Requirements for Reflected UV Photography

There are four elements are required for making reflected UV photographs. These are:

- A camera that is sensitive to UV.
- A lens that transmits useful amounts of UV.
- A filter that transmits only UV wavelengths (for a full spectrum camera).
- Energy or light source that is rich in UV emission.

Most digital cameras will require modification to enable the recording of UVR. Most lenses contain optical glass and crown glass, which is used in windows. Because of the construction and elements of modern lenses, most have poor UV transmission. They transmit virtually nothing below 350nm. This transmission issue is compounded because of the multiple coatings used on the lens elements of many modern camera lenses. These coatings are there primarily to minimize the transmission of UV. It is not difficult to conclude that this means most modern lenses are virtually useless for UVR photography.

Over the years, various manufacturers have produced UV transmitting lenses made specifically from quartz. They were aimed primarily at medical and forensic photographers. The following lenses were excellent for UV photography, however, sadly all of these lenses were discontinued many years ago. They were rare when made and now can occasionally be found on eBay or other resellers. They will always command very high prices.

- Nikon UV-Nikkor 105mm *f*/4.5
- Hasselblad UV-Sonnar 105mm *f*/4.3
- Pentax Ultra-achromatic Takumar 84mm *f*/4.5

There are several quartz lenses designed for UV photography still manufactured today. A selection is shared below:

- A copy of the Nikon UV-Nikkor 105mm is still available as the Rayfact Nikon UV-105mm *f*/4.5 Multispectral Imaging Lens distributed by Company Seven.
- Coastal Optics, a U.S. company, produces two UV lenses made with quartz elements: the UV-VIS 105mm *f*/4 UV-Micro-Apo CoastalOpt (a direct replacement for the 105mm UV-Nikkor), and the CoastalOpt UV-VIS-IR 60mm Apo Macro. The 105mm has a transmission range of 250–650nm, while the 60mm has a transmission range from 290 to 1500nm. They are both supplied with a Nikon mount.

Accidental UV Lenses

There is a popular misconception that you cannot make UVR images with a normal glass lens. This probably explains why there is a relatively small amount of UVR photography being carried out today. There is no doubt that special purpose quartz lenses are best suited for UVR photography and they will have the advantage of achromatic correction in the UV region. This means that they bring two wavelengths of radiation to the same point of focus. This correction requires little or no focus shifts when used for UV photography. These lenses are capable of photographing using wavelengths deeper down into UV to approximately 320nm. Most non-quartz lenses will restrict effective use to approximately 365nm and above, but making an image is better than having no image. Quartz lenses will also transmit far more UV, and will therefore be effectively several stops faster than a conventional glass lens when used for UVR photography. There are, however, several other types of lens which can be used as alternatives, often very successfully, for UVR photography. A lot of research has been carried out by photographers with various types of lens to see which ones transmit useful amounts of UV. Lenses from fax machines, reprographic cameras, enlargers and various scientific instruments have been tested for use in this spectral region. There are several databases available online hosting various forums listing "accidental" lenses such as those designed originally for other purposes, but found to be suitable for UVR photography. Details of these can be found in the Resources sections of these forums.

One particularly useful type of lens for UVR photography, often available at reasonable prices from used equipment sellers, are lenses specifically designed for photographic enlargers. These lenses, although made of crown glass, generally had no real UV absorbing coatings on the elements. These enlarging lenses came in a range of focal lengths. One of the specifications for a good enlarging lens was the requirement to have good blue light and UV transmission. This transmission was appropriate for use when exposing monochrome photographic papers designed to be sensitive to UV radiation and blue light. These papers were "safe" to use in red light in a darkroom. The wavelength transmitted by such lenses was usually limited to around 350nm.

One of the best enlarging lenses that was found to work extremely well for UVR photography was the Nikon El-Nikkor enlarging lens, available in a range of focal lengths including 50mm, 63mm, 75mm, 80mm, 105mm and 135mm. The oldest ones were the best and had a metal body and scalloped aperture ring on the front of the lens. My "standard" UVR lenses are these elderly El-Nikkor 63mm, 80mm *f*/4 and 105mm *f*/5.6 enlarging lenses, released around 1970. I obtained mine through online auction sites. When used with a cropped sensor camera such as a Nikon D300S, the focal length is magnified by approximately 1.6×. This is due to the crop factor which give the lenses an effective focal length of 88mm (was 63mm), 128mm (was 105mm), and 168mm (was 135mm) respectively. These focal lengths provide long working distances from the camera to the subject, giving enough space to arrange the various light sources and other paraphernalia around the subject. The 105mm, conveniently, is the same focal length as the 105mm micro-Nikkor lens which I use for taking many of the control images for comparison alongside the UV images.

In a Nikon technical brochure (NEX 49) from the 1970s, there is a section that says:

> El-Nikkor lenses are corrected for chromatic aberrations, not only for visible light, but also for near UV rays, so that the image formed of the visible light exactly coincides with that of the UV ray, and focusing can be adjusted perfectly.

It goes on to say:

> another important point is that special glass is not used for making these lenses because such glass is apt to absorb UV rays, resulting in the production of an enlarging lens with deficient brightness. In addition, an anti-reflection coating for the wavelength 400nm is applied to increase the transmission of UV rays. Thus, the lenses perform uniform spectral transmission covering the range from visible light to UV ray.

With a full spectrum camera converted and calibrated for UV, I have found that the focus shift for UV reflected photography with theses lenses is minimal, though I would still recommend carrying out focusing tests for optimal sharpness, as well as stopping the lens down by at least two stops.

When attaching filters onto the front of these lenses, there is a screw thread in the lens barrel. The size of this thread for the 80mm and 105mm lenses is an unusual 34.5mm, requiring the use of a filter stepping ring for attaching most filters.

Most enlarging lenses are attached to the enlarger by a screw thread (e.g. Leica 39mm thread) and have no built-in helical focusing mechanism of their own, usually being focused on the enlarger by means of a bellows system. They will need to be mounted on to the camera via an appropriate adapter (e.g. Leica 39mm screw thread to camera-specific bayonet). Focusing is achieved through the use of an appropriate extension in the form of extension tubes or an extension bellows unit. A particularly useful type of extension tube which can be found on the Internet has a helicoid focusing ring built into it, enabling the lens to be focused normally. Enlarging lenses (and other similar types of lens) have only manual aperture rings, and there is no electronic linkage with the camera, so exposure will need to be adjusted manually on the camera.

35mm *f*/3.5 Lenses

Another range of lenses found to be very good at transmitting useful amounts of UV, and favoured by many UVR photographers are various models of 35mm *f*/3.5 lenses, of which the Novoflex® Noflexar 35mm *f*/3.5 lens is a good example. This type of lens generally has a relatively simple construction, with relatively few glass elements, and can be found in a range of camera fittings. They are particularly useful on APS-C and Micro 4/3 cameras, where, due to the sensor size, the effective focal length becomes approximately 50–70mm due to the crop factor. A large number of these lenses were produced in the 1970s, with manual pre-set apertures, and many have been found to transmit UV down to around 360nm. Again, lists of useful models can be found on the Internet, on some of the websites listed in the resources section at the end of this chapter.

Focus Shift

The focal length of a lens is dependent on the wavelength of light used, amongst other things. Similar wavelengths may be focused at different focal distances when different glasses are used in the lens. Both UV and IR wavelengths are brought to a different point of focus than visible light. Images made using UV may be out of focus for this reason. When having a camera converted to shoot UV (or IR), this shift can be managed by having the camera calibrated at the time of conversion. If you have a full spectrum camera, you need to specify which wavelengths you wish it to be calibrated for, UV or IR.

It is relatively easy to determine the focus shift for a specific lens when used in UV (or IR).

- Place a millimeter ruler scale at a 45° to the lens axis (several proprietary lens calibration systems are shared in the resources section).
- Focus the lens at its mid-point.
- Make two exposures. One in visible light and one in UV. The amount of focus shift can be estimated visually.
- The shift will be greatest when the lens is used for close-up or macro photography.

UV Transmitting Filters

The basic requirement for a UV transmitting filter is that it should absorb all visible light and IR wavelengths, and transmit only UV wavelengths. The traditional filter used for analogue UVR photography was the Kodak Wratten 18A "Wood's glass" filter. These filters were usually supplied in square blocks of glass from 2–5mm thick. Gelatine, commonly used for making photographic filters, could not be used for UV work because gelatine absorbs UV wavelengths. Wood's glass filters have good UV transmission from 320–400nm, peaking at 365nm. However, they also transmitted a small but significant amount of IR radiation. When used with monochrome film, this IR "leakage" was of no consequence, because panchromatic films were not sensitive to IR. However, digital sensors on cameras converted to "full spectrum" are highly sensitive to IR. To make a purely UV image, the IR needs to be filtered out with an IR absorbing filter such as a hot mirror filter.

Baader U Filters

Currently, the gold standard for many UVR photographers is the Baader U filter, designed originally for astronomical use. It is also known as the Venus filter, designed to fit inside a

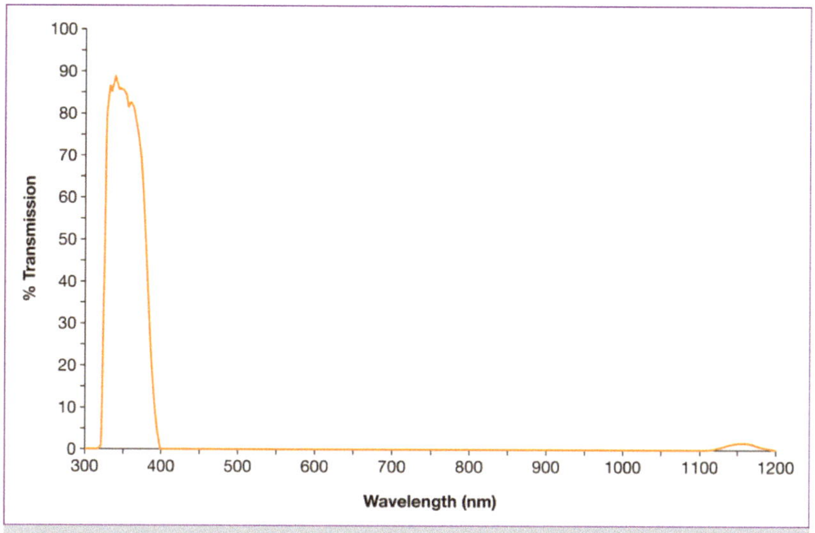

Figure 15.2 The spectral transmission curve for the Baader U UV transmitting filter. Note the complete absence of any IR wavelengths, important when used with a full spectrum camera.

telescope, where it's primary use is for making observations of the strongly UV reflective clouds around the planet Venus. This filter transmits wavelengths from 320 to 390nm, with 85 percent transmission at 350nm. It completely absorbs visible light and IR wavelengths. It is only available in either 1.25 inch (31.75mm) or 2 inch (48mm) diameter sizes, and requires the use of filter stepping rings to fit it onto most photographic and enlarger lenses.

The Baader filter is dichroic, which means it will display a different color when looking at the filter from each side. This could be yellow and magenta for example. Some users recommend reversing the Baader U filter in its mount. Astronomers use it in the opposite direction on telescopes so that the magenta side of the glass faces forward, probably as a precaution against internal flare from the highly reflective surface. I always use a lens hood with this filter to minimize flare.

Drop Down Filter Holders

If you are using a visually opaque filter with a full spectrum converted DSLR camera, you will need to compose and focus the camera before placing the filter in front of the lens. A useful way to speed up this focusing is to place the filter into a gelatine filter holder, such as the Nikon AF-1. This device screws into the front of the lens and has a hinged section which drops down for insertion of the filter. Most of these devices have a screw thread designed to add a lens hood on the opposing side. By using a step-down ring of the appropriate size it is possible to mount the UV filter using the featured threads, so that the filter can easily be placed in front of the lens quickly.

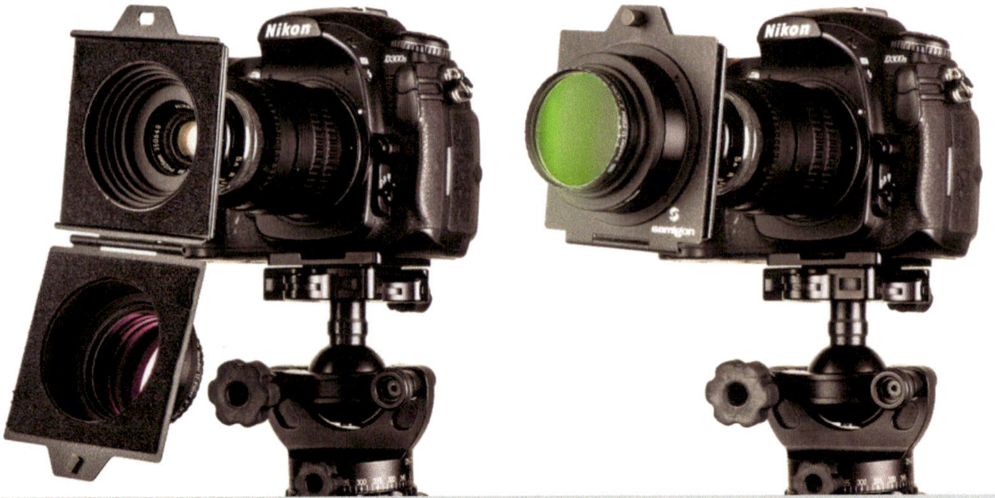

Figure 15.3 This Nikon D300 is a full spectrum-converted camera. It is equipped with a 105mm El-Nikkor enlarging lens and Baader U filter mounted on a gelatin filter holder. The lens is mounted to the camera using an extension tube with a built-in helical focusing thread. Images courtesy of Adrian Davies.

Other UV Transmitting Filters

There are several other astronomical filters which perform very similarly to the Baader U, including a slightly cheaper version from PrimaLuceLab in Italy, called the UV filter. A variant of the Baader U filter is the Andrea "U" Mk II filter available from UVROptics.com, which, according to the manufacturer, has better transmission of wavelengths in the 370–400nm region of the spectrum, creating shorter exposure times. The same company has a Straight Edge U filter, having a peak at 379nm. This particular dichroic filter performs best when the incoming light is aligned closely to zero degrees angle of incidence. Dichroic filters shift their cut-off wavelengths according to the angle of incidence of the incoming light, so they will not perform as well with wide-angle lenses. Several other UV transmitting filters are available, such as the Schott UG1, UG11, Hoya 340 and 360, and B & W 403. Some of these latter filters transmit small but significant quantities of IR ("IR leakage"), and need to be used in conjunction with IR blocking filters (hot mirror) such as the Schott BG38, BG39 or BG40, or the S8612.

A selection of UV transmitting filters and their equivalents

	Wratten 18A	*Hoya U-340*	*Schott UG 1*	*B + W 403*	*Baader U*	*Andrea U*
Approx UV transmission nm	310–400	260–390	290–410	290–400	320–390	346–383
Approx IR transmission nm	670–830	680–750	690–1100	700–1000	>1150	>1100

Note that several of the filters above transmit small but significant amounts of IR, which, if below 1000nm, will be recorded by a full spectrum converted sensor.

UV photographers have experimented with various filters (and other equipment) for UVR over the years, and many of their results can be found on the ultravioletphotography.com forum, where there is a wealth of highly recommended and practical data.

UV Radiation Sources

To start this topic, sources used for UV and IR imaging should be referred to as radiation sources and not light sources, since the UV and IR spectrums are invisible to the human eye which is only sensitive to visible light. Several light sources that do contain some UV emissions include sunlight, LED lighting, electronic flash units, and mercury vapor lamps.

Sunlight or Daylight

Sunlight has a peak transmission of around 480nm, but contains only approximately 3–4 percent UV radiation when the summer sun is directly overhead and there are no clouds. Of the UV in sunlight, the falloff of UV solar irradiance from 400 to 300 nm (or below) is steep and rapid. Atmospheric absorption and scattering limits using sunlight for UV photography, although sunlight can be a good, but variable, source of UV.

UV radiation increases rapidly with altitude, and at 5,000 meters there is 40 percent more UV than at sea level. Snow and ice reflect up to 80 percent of UV radiation, which is why skiers and mountaineers need to wear UV protective goggles to prevent snow blindness.

Given the right sensor and filter combination, it is possible to use daylight as a light source for UVR photography. However, because its emission is variable due to the time of day, the season, or weather conditions, it is not useful when a consistent series of images is required of subjects over a period of time.

Gas Discharge Lamps

Gas discharge lamps, particularly those of the mercury vapor type, emit good quantities of UV. Low pressure versions emit 90 percent at around 254nm, and are used for germicidal purposes. Germicidal lamps are used to sterilize workspaces and are tools used in biology. High pressure versions are also available, emitting mainly 365nm wavelengths, which fit into standard light bulb sockets.

Metal Halide

Metal halide lamps use various gases and a mercury discharge source to produce a regenerative cycle. This is similar to the quartz halogen type of bulb, but has increased and specific spectral outputs. A popular example for UVR and multispectral photography is the EyeColor Arc bulb (M770D). This lamp needs to be operated using a ballast that may require a dedicated electrical service. The wiring is straightforward, but a licensed electrician should perform the work.

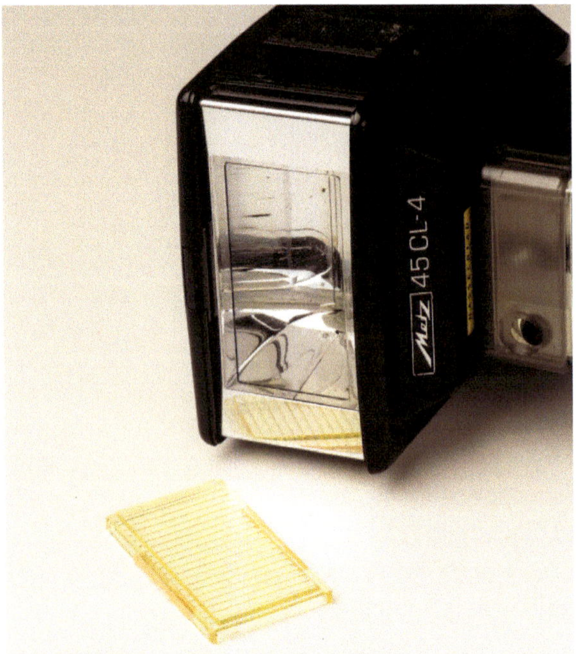

Figure 15.4 An older Mecablitz 45CL-4 flash gun, converted to "full spectrum" applications by removing the UV absorbing filter from the front of the flash tube. Image courtesy of Adrian Davies.

Electronic Flash

Electronic flash units that use Xenon filled tubes offer the most consistent and potentially most powerful light source for reflected UV photography. One or two small modified flash guns, that are placed reasonably close to the subject should give adequate power for an aperture of approximately *f*/11 with an ISO setting of 400. However, like lenses, manufacturers of flash guns sometimes coat the tubes with a UV absorbing filter, or attach a UV absorbing filter over the tube, to prevent it from causing a color cast in conventional photography. Removing the coating from the tube is difficult, so select flash guns that have a UV absorbing filter and remove that. Good examples of this type of flash are the Mecablitz CT45 range, Vivitar 283, and Canon 199A. The Mecablitz flash is particularly good because it is powerful, and it is easy to remove the plastic UV absorbing filter. This converts the flash into a full spectrum unit, emitting good quantities of UV down to around 365nm.

Safety note:

- Some of the older flash guns mentioned above have a high trigger voltage. This can damage electronics in some modern cameras. Another way to trigger the flashes is to use the built-in pop-up flash on the camera and attach a slave to each additional external flash unit. Line of sight for each one is required for consistent triggering
- Also take great care when disassembling electronic flash guns. Their internal capacitor may retain a significant and potentially high amount of dangerous charge. Always discharge a flash unit fully before ever taking it apart. If you are at all unsure, get a qualified technician to perform this work.

LED

Modern LEDs (light emitting diodes) are manufactured with very specific spectral outputs and several emit UV wavelengths. Additionally, several powerful UV hand-held flashlights (torches) are available, such as the MTE U303, and the increasingly popular Convoy S2+ UV model. These products are remarkably inexpensive and can give enough output for successful UVR and UV florescence work. The Convoy S2+ UV model emits wavelengths of approximately 365nm.

Best UVR Photography Practices

In the UVR images of the sunflower (*Helianthus* sp.) shown here, the flower was first placed in a suitable vase full of water, then secured gently using a clamp to prevent it moving during the session. Black velvet was used as the background. A full spectrum-converted Nikon

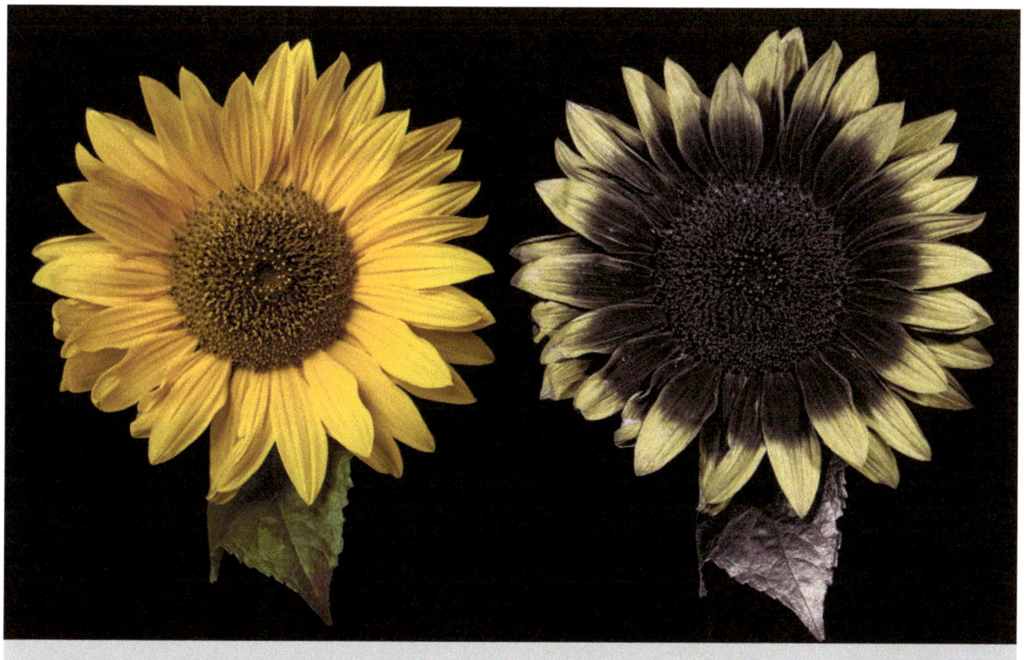

Figure 15.5 A Sunflower (*Helianthus* sp.) was photographed using visible light (left) and UVR (right). For the UVR image a Nikon D300S full spectrum conversion camera with 105mm El-Nikkor lens and Baader U filter was used. Two Mecablitz 45CT1 full spectrum converted flashes produced an exposure of $f/11$, 400 ISO. Images courtesy of Adrian Davies.

D800 camera was fitted with a 105mm El-Nikkor lens, mounted on a variable extension tube. A Baader U filter was placed on a Nikon AF-1 drop-down filter holder.

Lighting for the UVR image was provided by two full spectrum-converted Metz 45CT1 electronic flash guns, using the full power setting which were placed approximately 63.5cm and 20cm from the flower. They were synchronized to fire together using a slave unit. Control images were also shot, using two Elinchrom studio flash guns using a "normal" Nikon D800 camera with a 105mm micro-Nikkor lens, using the same tripod so that direct comparisons can be made. Using this set up, it is possible to achieve an exposure of $f/8$ at 400 ISO. It is important to use the lowest ISO setting as possible to minimize noise. Focus the camera with the filter in the drop-down position, then raise it just before the exposure is made.

White Balancing Reflected UV Images

When recording reflected UV images using a sensor with the Bayer filter in place, you will get images that vary in color characteristics that range from a strong red and up through magenta, depending on the camera. While you can convert these images to black and white at a later time, it may be useful to "white balance" them at capture, particularly if you are posting the images to online forums or working with colleagues in different institutions. Normal white balancing accessories such as an 18 percent gray card, do not work in UV applications. One material that has been found to work well for UVR is gray PTFE tape. Wrap several layers around a small (e.g. 5cm square) tile, and place this in front of the subject. To achieve a correct white balance, shoot in RAW file format, and use RAW conversion software such as PhotoNinja, Adobe Camera Raw, or other software that does not neutralize the image too far from a normal gray at capture.

Making Control or Reference Images

For much "invisible light" photography, you will want to create a visible light control image to compare with the invisible spectrum image. Whenever possible these should be produced with the same camera or a camera with the same size and type of image sensor, and a lens with the same focal length. To achieve the correct color, you can use an 18 percent gray card or other color balancing accessory.

Applications of UVR Photography

As we have seen, UVR plays a major role in botanical study, showing how flowers might appear to insects for example. Several cosmetics companies are using or have used UVR to show how to correctly apply sunscreen, both in stills, but also using real time video cameras.

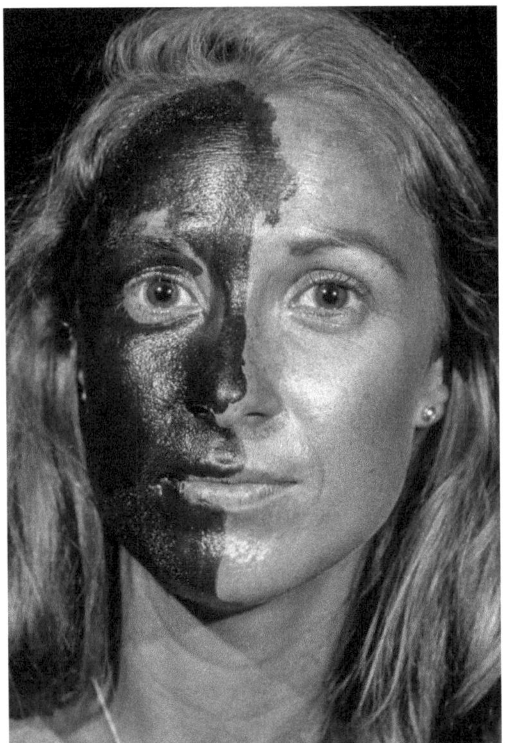

Figure 15.6 A UVR image of a model with sunscreen applied to one half of her face. Note the areas missing from around the eye and nose. A Nikon D300S full spectrum camera that was converted was used equipped with 105mm El-Nikkor lens, and a Baader U filter. Two full spectrum Metz 45 CT4 flash guns, created an exposure of 1/160th second at *f*/11. Image courtesy of Adrian Davies.

A great deal of research has been carried out by medical photographers and clinicians over the years looking at the potential uses of UV (and IR) in medicine. Over time its use has declined greatly. This is primarily due to other, and possibly better, diagnostic techniques. Much of this research concentrated primarily on dermatological conditions and possible use in the diagnosis, for example, of pigmented lesions.

UVR photography is used to show the surface detail of skin and other tissues, often with a great increase in clarity over visible light photography. UV radiation is absorbed strongly by tissue pigments such as melanin, and the technique can be used to show the absence of skin pigment or accentuate changes in pigmentation in comparison with normal skin. Applications include monitoring vitiligo, melanomas, scleroderma, moles, and freckles.

UVR images tend to be sharper than visible light images and fine detail is often enhanced. Blood absorbs UV radiation strongly which, when coupled with its superior definition, will show vascular conditions in greater detail. UVR techniques are also used by forensic photographers, for a variety of applications including the enhancement of surface detail when searching for shoe prints, for example.

Animal Vision

It is well known that many animals including birds and invertebrates such as spiders, butterflies, and bees have sensitivity to UV. Like flowers, many butterflies, particularly yellows and sulphurs such as the brimstone (*Gonepteryx rhamni*) and Cleopatra (*Gonepteryx cleopatra*), appear differently in UV. An increasing amount of research has been carried out over the last few years into spider webs, and how they ensnare their prey. Research has shown that many spider's webs reflect UV. Possibly this is to attract their insect prey to the web in a similar way that flowers attract bees and other insects. A recent study, using the wasp spider (*Argiope bruennichi*), showed that most of the prey species found in their webs had a greater sensitivity to UV than other groups of insects.

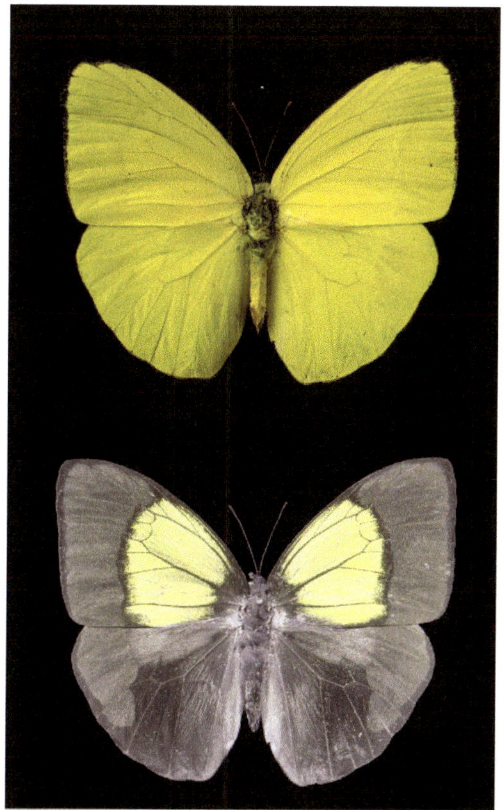

Figure 15.7 Top, a male straight line sulphur (*Phoebis trite*) butterfly from Peru was photographed in visible light. The bottom image was created using UVR. It is thought that the bright areas of the wing in UV are used to signal females when the male is flying. A Nikon D800 full spectrum converted camera with 105mm El-Nikkor lens and Baader U filter was used to make the bottom image. Two full spectrum modified Metz 45CT flash guns provided the UV exposure. UVR image was "white balanced" in Photo Ninja using a home-made gray scale constructed from gray PTFE tape. Images courtesy of Adrian Davies.

Multi-Spectral Imaging

There are many applications where photographing a subject in different wavelengths can reveal useful information. Many satellite images are combinations of images that were made using different wavelengths and multi-spectral images of paintings and artifacts can prove extremely valuable for analysis.

The example shown earlier of a flower photographed in UVR only tells part of the story. Like humans, many insects and other animals have trichromatic color vision, and much research is currently being carried out into animal vision. The human eye has color sensitive cone shaped cells (cones) on the retina responding to blue, green, and red wavelengths, constructing all the other colors in the brain from a combination of these three (trichromatic vision). Many animals have cells sensitive to UV (including many species of bird). One of the most studied insects is the honeybee (*Apis mellifera*), which has trichromatic color vision, with cells sensitive to UV, blue, and green. By using a combination of filters allowing UV, blue, and green to reach a full spectrum sensor (not UV only) then a color image can be recorded, which shows how a honeybee might see a subject. It is difficult, if not impossible, to know the relative amounts of each color perceived by the bee. A good combination to try is a Schott S8612 filter stacked together with a Schott UG5 filter. This combination is often referred to as a "bee vision" filter.

Like the UVR images shown earlier, this image can be white balanced using the gray PTFE target. While we can never be exactly certain what a bee "sees", the flower image provides a very good idea. This approach produces some very interesting images. Of course, there are many thousands of species of bee, and they probably all "see" the world slightly differently!

UV Fluorescence

Of all the techniques outlined in this chapter, photographing fluorescence emitted from various substances is one of the easiest. UV fluorescence (UVF) is the visible light emitted from a subject when lit with UV radiation (light) in a dark room. Subjects include a range of minerals (including fluorspar, named after the phenomenon), some plant materials, including pollen and some woods, some insects, invertebrates, and some inks and pigments.

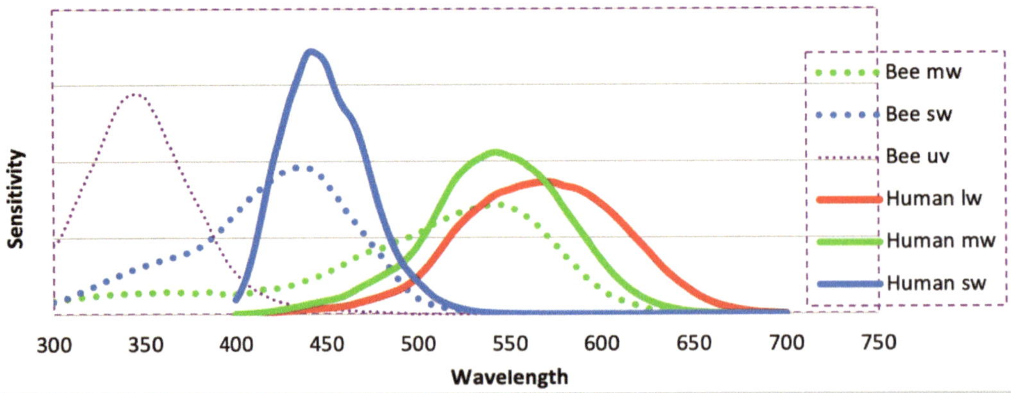

Figure 15.8 This graph shows the relative sensitivities of human color vision (solid lines) and that of the honeybee (*Apis mellifera*) (dotted lines). Note how the honeybee can see wavelengths down to around 300nm but has no sensitivity past around 600nm. A filter combination of a Schott UG5 and Schott S8612 with a full spectrum converted camera mimics the honeybee vision. Image courtesy of Adrian Davies.

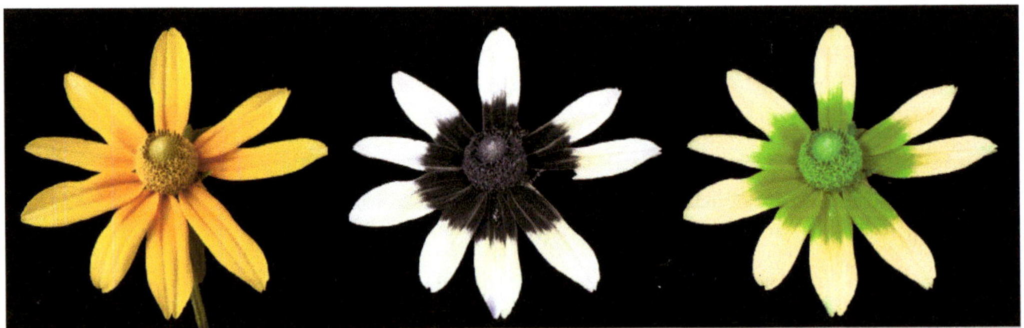

Figure 15.9 A black-eyed Susan (*Rudbeckia* sp.) flower in visible light, UVR, and "bee vision". For the UVR image (center), a Nikon D300S full spectrum conversion camera with 105mm El-Nikkor lens and Baader U filter was used. Two Mecablitz 45CT1 full spectrum converted flashes. *f*/11, 400 ISO. For the bee vision image (right), a Nikon D300S full spectrum conversion camera with 105mm El-Nikkor lens and filter stack consisting of Schott S8612 and UG5 filters was used. Two Mecablitz 45CT1 full spectrum converted flashes. *f*/16, 400 ISO. Images courtesy of Adrian Davies.

Every page on a U.K. passport has an invisible image of a bird or animal which can be made visible with UV. Certain dermatological conditions show fluorescence, such as Tinea capitis, and doctors often keep a UV Wood's lamp to help detect it. With the advent of new cheap UV sources, new subjects are being found all the time, such as a rare woodlouse on the island of St Helena, a new rock type (yooperlites, a form of sodalite), and the fluorescing beak of the Atlantic puffin (*Fratercula arctica*). Engineers routinely use UVF for detecting fine cracks in metal, by using fluorescent penetrating oil. Some glassware, manufactured in the 1920s and 1930s, contains trace elements of uranium, which fluoresces in UV.

UVF is a form of luminescence, where short wave invisible radiation is converted into visible radiation of a longer wavelength through its interaction with the subject technically called "excitation". In the case of UVF, invisible UV radiation is converted into visible light, which can be seen and photographed with a normal cameras and lens. In the case of UVF, radiation of around 365nm is converted into visible light from 400–600nm, for example. Forensic photographers and art conservators will often use very specific wavelengths for exciting certain substances such as blood or paint pigments.

FLUORESCENCE

In order to understand what fluorescence is, it is necessary to discuss briefly the structure of atoms. Atoms consist of a nucleus containing protons, surrounded by electrons. The electrons orbit around the nucleus at their lowest energy state. When the atom absorbs energy in the form of heat or light, the low electrons are pushed into an "excited state". When this energy is lost, the electron returns to its lowest state, and the discharge of energy is transformed into light of a longer wavelength. Not all wavelengths of energy can excite atoms.

From a photographic point of view it is important to realize that atoms can only absorb energy at certain specific values (wavelengths) and not all other values create excitation. Planck's theory states that an electron will only become excited to a higher level if a photon of light contains the appropriate energy to excite it to that level. In other words, molecules are selective in the wavelengths of light that they will absorb. This selectivity is determined by the electron energy levels that each element possesses.

Once excited into the higher state, the electron will quickly fall back to its original level, releasing energy and emit light which is described as fluorescence. To ensure a continuous flow of energy from the atom, a continuous light (radiation) source is required. The fluorescence will disappear immediately on the removal of the excitation source. The difference between the excitation wavelength and emitted wavelength is known as the Stoke's Shift.

Luminescence, Fluorescence, and Phosphorescence

There are several terms related to light emission and fluorescence, which can cause confusion, and it is worth clarifying them.

Luminescence

The visible light emitted from a subject as the result of a stimulus, such as excitation from a UV source, or a chemical reaction, is luminescence. Different types include bioluminescence (as seen in glow worms, fireflies, some marine plankton, and some fungi), chemi-luminescence, the result of a chemical reaction, as seen in glow sticks for example, and electro-luminescence, as exhibited with LED lighting.

Fluorescence

The visible light emitted from a subject when illuminated or "excited" by a shorter wavelength source such as UV is called fluorescence. The emission ceases immediately following the removal of the excitation source. Another interesting form of fluorescence is IR fluorescence, where the subject is lit with visible light of a certain wavelength, usually cyan in the region of 480nm, and IR wavelengths are emitted. Because the wavelengths emitted are invisible and they are IR, it is not known if the effect persists after the removal of the light source, so this phenomenon should strictly be called IR luminescence. It is used primarily nowadays by art conservators looking at pigments.

Phosphorescence

This is a similar phenomenon to fluorescence, but the light emitted continues for a significant time after the radiation is removed.

Practical UV Fluorescence Photography

Any camera and lens can be used for recording UVF, although the camera should perform well when used for long exposure times. You will need to carry out the photography in a completely dark room and use a solid tripod. The main requirement is a powerful UV source, which now are most commonly found as LED devices, such as the Convoy S2+ UV torch, which is relatively inexpensive. Be careful not to use a background which itself might fluoresce, such as paper, and some fabrics. Black velvet makes a good choice, though any speck of dust may fluoresce causing distracting elements which may need to be retouched out of the final image.

Painting with Light

Hand-held lights such as the Convoy S2+, can produce quite a narrow beam, with a distinct hot spot, which may not allow even coverage of the light onto a specimen. A useful technique is to "light paint" the subject. Using a long exposure of perhaps 20–30 seconds, release the shutter (with a cable or remote release), then gradually move the light around the subject, making sure to cover the top, bottom and both sides evenly. Shorter exposure times will make it difficult to evenly cover the subject with the light. Try to use a low ISO setting (e.g. 200 ISO) to minimize the noise in the image, but ultimately the ISO will determine itself based on the availability of brightness or not.

Make sure, if you are photographing delicate subjects such as flowers, that you not to create any movement of the air, which may cause them to move during the exposure, and even hold your breath if necessary for the duration.

Safety note: Never look directly at a UV source and always wear wrap-around style UV absorbing safety goggles when shooting UVF images.

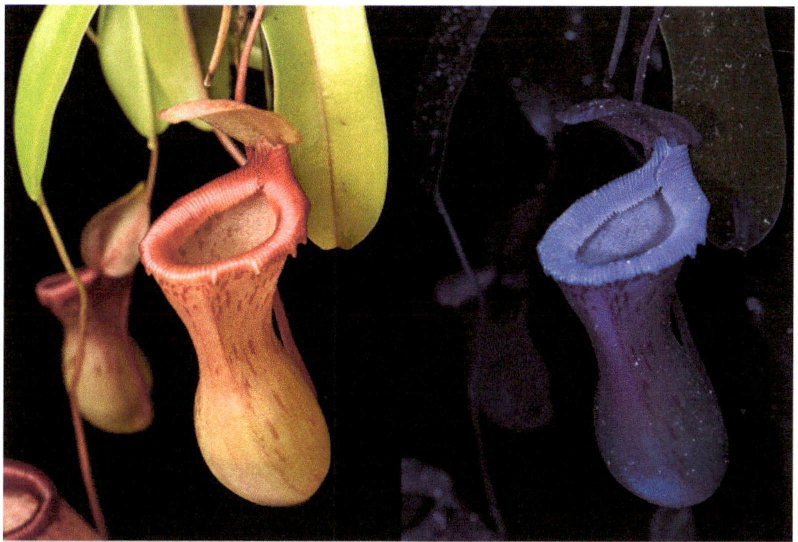

Figure 15.10 A tropical pitcher plant, *Nepenthes × pacifica*, in visible light (left) and fluorescing UV emission (right). Note the numerous dust and lint particles on the leaves and pitcher in the UVF image. The UVF picture was made with a Nikon D810 with 105mm micro-Nikkor lens. The exposure was 15 seconds at $f/16$, "light painted" with a Convoy S2+ UV light. Images courtesy of Adrian Davies.

Note: Even the best UV sources may "leak" some visible light, creating images with a distinct blue cast. It is possible to buy a UV transmitting, visible light absorbing filter specifically for the Convoy light – such as a Hoya U-340. A simple test is to shine the UV source onto a metallic object in a dark room. If you can see the object there is some visible light leakage. Metal does not fluoresce with UV excitation.

Similar to UVR photography, you may wish to produce a control or visible light image alongside the fluorescence image. Use electronic flash or LEDs to light the subject, using the same camera and lens to ensure the same sized image.

Some Applications

Mineralogy

Various minerals fluoresce well in UV, and the fluorescence colors can be an aid to their identification. Some examples include:

Mineral	*Fluorescent color produced with 365nm UV radiation*
Calcite	Red, orange, blue or green depending on source of specimen
Fluorspar	Blue, yellow, green, red depending on source of specimen
Hyalite	Bright green
Sodalite	Bright orange to red
Willemite	Bright green
Zircon	Orange

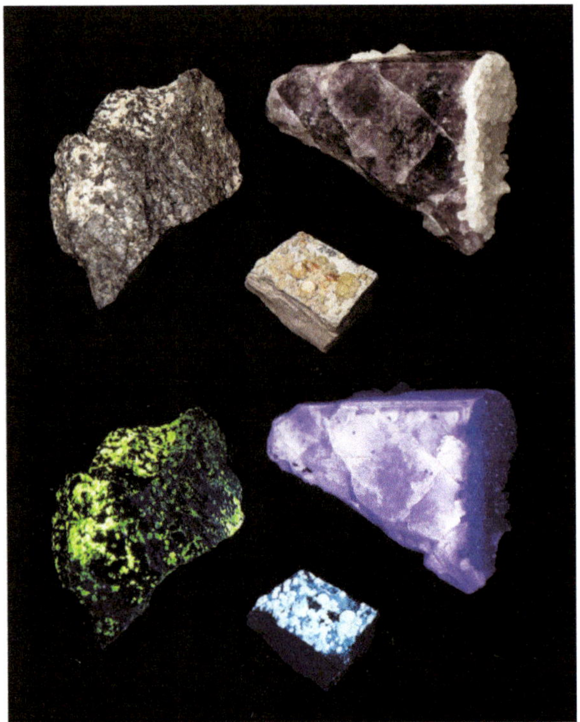

Figure 15.11 A selection of minerals fluorescing in UV radiation. Left: Willemite, Right: Fluorspar, Centre: Wavelite. These pictures were made with a Nikon D810 camera using a 105mm micro-Nikkor lens. An exposure time of 15 seconds at *f*/16, "light painted" with a Convoy S2+ UV torch. Images courtesy of Adrian Davies.

Mold

Some species of mold fluoresce, and UVF is used to find mold on citrus fruit and nuts in packing houses for example.

Invertebrates

Some insects and invertebrates fluoresce in UV. One way of finding scorpions at night in their rainforest or desert habitat is to use a UV source.

Plants

Various plants and plant derivatives fluoresce in UV. These include some species of lichen, the pitchers of some carnivorous plants, various species of wood, the pollen in some flowers, and the quinine found in tonic water. The interior of several fruits fluoresce too when illuminated with UV, often producing attractive images.

IR Luminescence

A far less well-known type of fluorescence is IR, where the subject is lit with visible lighting, usually cyan, although UV is sometimes used and an object will emit longer wavelength fluorescence in the IR region. Because the fluorescence is invisible, it is not known if the fluorescence persists once the light source is removed, so it should strictly be called IR luminescence.

For the photography you need a cyan light source, such as an LED torch emitting 490–520nm, and a full spectrum or IR only converted camera with an IR transmitting filter such as a 720 or 850nm. Use the same light painting technique as for UVF. The photography needs to be carried out in a totally dark room. Suitable subjects include the chlorophyll of many plants, pigments, and various human tissues. The technique is used by art conservators, examining paint pigments which may exhibit the phenomenon (those containing cadmium show particularly strong IR luminescence).

IR Photography

IR photography has a long history, with the first published images dating back to 1911, by Robert Wood of UV "Wood's glass" fame. Black and white IR films first became routinely available in the 1930s. One of the main characteristics of IR imaging is its ability to penetrate haze, which was utilized as early as 1930, when Captain A.W. Stevens of the US Air Force shot an image of Mount Shasta in California shrouded in mist from a plane at an altitude of 23,000 feet and a distance of 331 miles. Because of the mist, the camera had to be orientated by compass. It was captioned the "World Record Long-Distance Snapshot" and

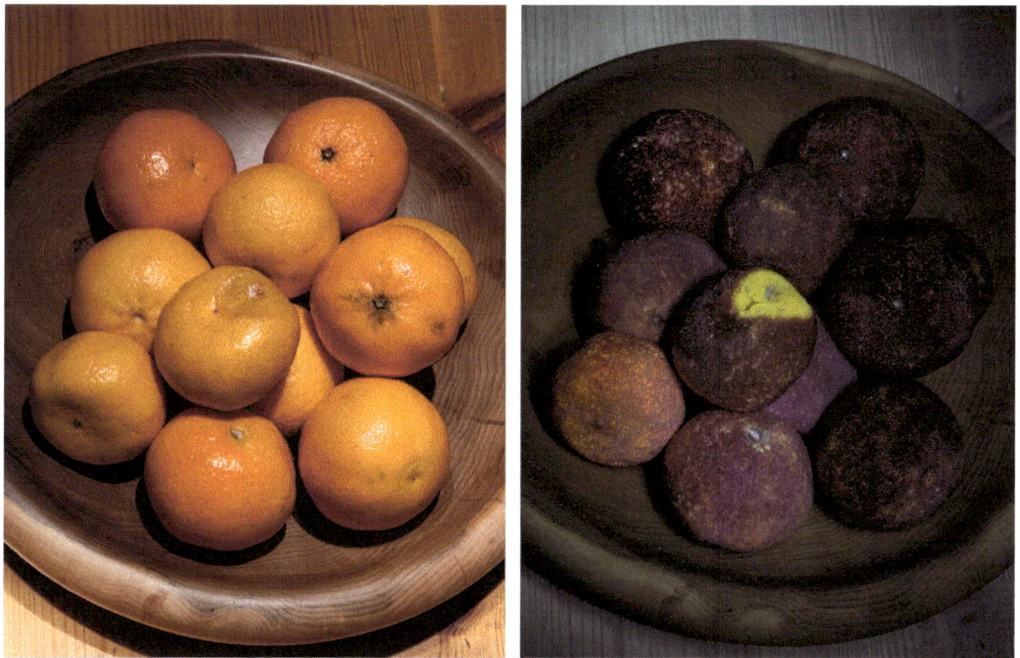

Figure 15.12 Satsumas in visible (left) and UVF (right). Note how the mold found on one of the fruits fluoresces in UV. The UVF picture was made with a Nikon D850 camera fitted with a 105mm micro-Nikkor lens. UVF image "light painted" with Convoy S2+ UV source. The exposure was approximately 15 seconds at $f/16$. Images courtesy of Adrian Davies.

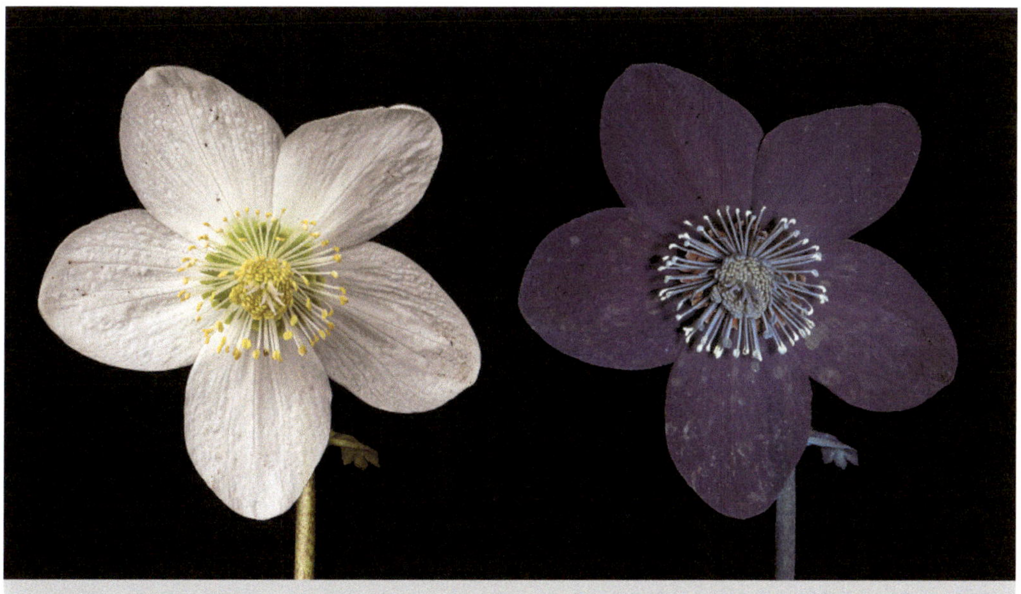

Figure 15.13 Cultivated hellebore (*Helleborus niger*) in visible light (left) and fluorescing in UVF (right). Note how the pollen at the end of the stamens fluoresces. In the UVF image "UV was painted" with Convoy S2+ UV creating an exposure time of approximately 15 seconds at $f/16$. Images courtesy of Adrian Davies.

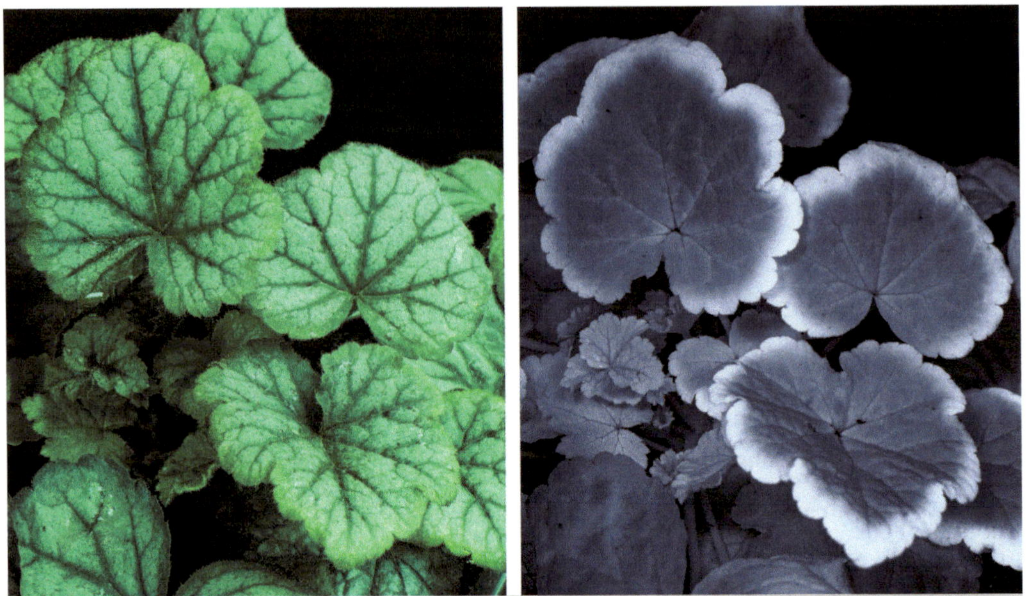

Figure 15.14 An IR luminescence image of garden perennial Heucherella "Twilight" (right image). Chlorophyll luminesces well in IR, and in this case shows a distinct pattern on the leaves, virtually invisible in visible light. The IR image was made with a Nikon D300 full spectrum converted camera, outfitted with a 105mm Series E Nikon lens, with 850nm IR filter. A LightWave LED Cyan light was used with a 20 seconds at $f/11$ exposure. Images courtesy of Adrian Davies.

showed remarkably little haze on the mountain. Another famous image, made in 1935 was shot from a stratospheric balloon from a height of 72,000 feet, and showed the curvature of the earth and the division between the troposphere and stratosphere and the atmosphere. IR photography was later used by military organizations for detecting buildings which had been covered in camouflage netting, for example.

Color IR film became available in the 1940s and was again used for military applications, though it later became used by foresters for detecting diseased trees, and by creative artists for its highly distinctive appearance. Kodak produced "Aerochrome III", primarily designed for aerial, military, or cartography purposes, in 120 film size until fairly recently. The technique became commonly known as False Color IR, due to the way in which colors were shifted along the spectrum. Several images and album covers from the 1960s and 1970s used false color IR imagery, including the iconic *Hot Rats* record sleeve by Frank Zappa.

Today, IR imaging is used by landscape and portrait photographers, medical and forensic photographers, art conservators, and in various remote sensing applications. IR photography was used in the production of Martin Scorsese's 2019 film *The Irishman*, to help with the "de-aging" process of the actors in post-production.

IR photograph records IR wavelengths from 700 to around 1200nm, though for false color images there may be the need to record wavelengths down to perhaps 600nm with a full spectrum converted camera. Wavelengths longer than 1200nm, in particular thermal, cannot be recorded with a full spectrum camera, and a specialist thermal imaging camera will be required.

As previously discussed, IR photography requires a converted camera, either a specific IR conversion, usually, though not always, 720nm, or a full spectrum conversion, allowing the recording of some color as well.

IR Camera Conversions

There are several options available when having a camera converted for IR photography. These include (note that conversion companies may give them different names) in 550nm or 590nm spectrum. Known variously as "Super Color", "Goldie" or "Golden Glow" filter. This filter transmits orange and red wavelengths as well as IR, producing a striking color scheme, which can be further manipulated in post processing to produce other hues. Images shot with this filter can be very attractive without any post processing.

"Enhanced Color": 665nm

This filter transmits red and IR wavelengths, producing more subtle colors than the 590nm conversion.

"Standard" IR: 720nm

This is probably the commonest IR conversion, where the filter transmits IR together with a small amount of visible red color, producing IR images with a subtle hint of color. This can be removed later if necessary in an image processing program. It produces slightly less contrast than an 830nm filter.

Deep Black and White: 830 or 850nm

This filter transmits only monochrome IR and this is the best choice for black and white only IR images.

To review, the 720 and 850nm conversions cannot be used for color IR images.

Lens Filters for IR Photography

If using a full spectrum converted camera, you will need to place an IR transmittance filter over the lens. There are several different types that transmit only IR wavelengths, such as 720, 760, 850, and 950nm. The 720nm does transmit a small amount of visible red wavelengths and it is possible to buy sets of the four filters shared above.

As with the IR camera conversion, you can also use other filters which transmit IR plus some visible wavelengths such as a "Goldie" filter which transmit down to 590nm, for example. Some of the camera conversion companies have galleries of images shot with different filters to show their effect.

NDVI: Normalized Difference Vegetation Index

Another conversion available from some companies is the NDVI. The "Normalized Difference Vegetation Index" is a simple graphical indicator that can be used to analyze remote sensing measurements typically, but not necessarily, from a satellite in space. These images are analyzed to determine whether the area in the image contains live green vegetation or not. It is routinely used by farmers using drones to capture the images and used to assess the health of their crops.

Lenses for IR

Virtually any lens can be used for IR photography – the multi-coatings on modern lenses do not absorb IR and transmit it freely. In the days of film, many lenses had an IR focusing spot engraved on the lens barrel, helping minimize any focus shift resulting from the IR.

One issue though, found with some lenses, is that of "hot spotting" where there is a distinct circular area of brightness in the center of the image. The effect is not limited to one type of lens, or make of camera, and is highly unpredictable. It is not always easy to remove it using an image processing program. It is not clear why some lenses do this, and others not, and, in the end, only practical tests will show whether a particular lens is good for IR photography or not. Some of the conversion companies listed have databases of lenses which they have tested and found suitable for IR photography. As with any form of photography, it is always a good idea to use a lens hood when shooting IR images.

Focusing

IR wavelengths are brought to a different point of focus than visible light which is beyond the plane of the image sensor. If you are using a specific IR converted camera, one where there is an IR filter directly in front of the sensor, it is likely that the conversion company will have calibrated the camera and there will be no focus shift. If you are using a full spectrum camera and the focus shift has been calibrated for UV, then you will need to make an allowance for focusing difference for the IR image.

Older lenses designed for film cameras often had an IR focus spot on the lens barrel. If the lens was focused on a certain distance, this distance was placed alongside the red spot to account for the IR focus shift. This system should still give a good focus correction for IR converted cameras.

If you are using a DSLR camera with Live View capability, most types will display the IR image on the LCD screen on the rear of the camera. It may be very dark and it may be useful to use a specially designed loupe, such as the Hoodman Hoodloupe. This device fits snugly over the screen and offers a 1:1 magnification of the image on the screen. This can really help focus the image precisely. Mirrorless cameras will not suffer from a focus shift since the image is displayed on the LCD as it will be recorded.

Resolution in IR Photography

Just as hot spotting is unpredictable when using different lenses for IR, lenses forming images with high resolution is also a variable. Generally speaking, the resolution (sharpness) of IR images is less than that of UV or visible spectrum images. This may not be readily apparent, and sharp images can certainly be obtained with IR, especially when optimum exposure, low ISO, aperture, and correct focus are used. Image defects may become more pronounced when the image is enlarged, and should be taken into account, particularly when using the images for scientific analysis, for example.

The subject of photographic resolution is highly complex, and relies on a combination of lens resolution, sensor resolution, wavelength, and other factors. However, a simple equation can be used to show how resolution decreases with increasing wavelength:

$$R = x\,(\lambda/NA)$$

Where R is the resolution

x is a constant (usually 0.5 or 0.61)

λ Is the wavelength of light (radiation)

NA is the numerical aperture of the lens.

Thus, generally, with any specific lens, as wavelength increases, resolution decreases.

Some lenses perform better for IR when used wide open, or perhaps one or two stops down from their minimum aperture, with the sharpness declining rapidly as they are stopped down. This is explained by the equation shown above. IR images will be softer than visible light images and this will be affected by the subject being photographed. I have a number of old fixed focal length lenses which used to be used with film cameras. One particular Nikon 28mm *f*/2.8 lens performed very well for IR photography with no hot spotting and formed very sharp images, while an older manual focus 20mm *f*/3.5 lens dating from around 1990, produced superb results with film that had very sharp definition, but makes very soft IR images. This is possibly as a result of this lens consisting of many glass elements. I also tested a range of Nikon 100mm and 105mm lenses, including old and new versions of the micro-Nikkor macro lens, an old 105mm *f*/2.8 dating back to 1970s. The test included an inexpensive Series E 100mm lens. Interestingly, the best performer for IR was the series E lens which was the only one not to exhibit hot spotting. It is always advisable to perform tests with lens/camera combinations and use a range of apertures, particularly before setting out to shoot an important set of images.

Sources for IR

Daylight contains a good amount of IR and is of course essential for landscape photography. The amount of IR is unpredictable though, varying according to time of day, cloud cover and other weather conditions.

Electronic Flash

For smaller, studio-based work, the best, most consistent source of IR is electronic flash. While UV is filtered out by the plastic window in front of the tube which needs to be removed for UV work, this does not affect IR output. So, most electronic flash guns emit good quantities of IR. If you want to use the flash for night-time security surveillance or wildlife photography, with no visible light emittance, then you can cover the flash window with a suitable IR filter such as a Cokin Infrared 720 (89B) A007 67mm square filter. Other acrylic 67mm IR filters are available through online auction sites. Make sure that you seal the edges with black tape to ensure that there is no visible light leakage around the edges. It may be possible to cut acrylic filters to the right size for your flash gun.

LED

Another light source that might be considered for small subjects are LEDs, found in various flashlights similar to the one suggested for UV work. LEDs generally emit a single specific wavelength, for example 850nm, so you will need to check that the wavelength of the LED is higher than the IR filter being used on the camera. For example, a 780nm IR LED will work with a 720nm filter, but will not be visible using an 850nm filter. Advanced Camera Services in the U.K. sells a 36 LED array, which can be custom modified to either 850nm or 940nm for use with IR and full spectrum converted cameras. 850nm LEDs give off a dull red visible glow which does not seem to disturb wildlife, whilst the 940nm version has no visible output, which makes them an ideal light source for security surveillance work. Several of these units can be joined together to provide a larger, more powerful light source.

Characteristics of IR Images

IR wavelengths penetrate human skin and other substances to a depth of 1–2mm and do not record surface detail in the same way that UVR will. Portraits will tend to have a smooth, alabaster like appearance, removing wrinkles and other blemishes from the surface, though subcutaneous blood vessels may be recorded and show up as darker lines. Vegetation too loses much surface detail, giving leaves a very smooth appearance. This surface penetration property is used by art conservators for recording hidden layers of paint or under-drawing in artwork.

Black and White IR Photography

Black and white IR photography often produces some unexpected results, where foliage of a completely different color appears as the same tone in an IR image, showing different colors that reflect with the same amount of IR. While many IR landscape photographers look for bright sun and distinct clouds, it is worth trying IR in all lighting conditions to see which will suit your needs. As previously stated, portraits take on an alabaster like appearance, removing surface blemishes, though sometimes revealing sub-cutaneous blood vessels.

Exposure measurement may be tricky, as the TTL metering in the camera may not give accurate results. Tests may be required for the various filters that you are using. As a general guide, with a full spectrum camera, a 720nm filter may require one extra stop over the daylight exposure, whilst an 850nm filter may require around two stops extra.

Haze Penetration

IR does a very good job in cutting through haze. One of the main reasons for haze becoming apparent in images is light scattering. When light hits an object it is scattered, as it is reflected from the surface of objects, but there are several different forms of scattering. When sunlight shines through the smoke from a bonfire the result is often several hazy rays of light. The sunlight is being scattered by the smoke particles in the air. How light is scattered depends greatly on the size of the object causing the scattering.

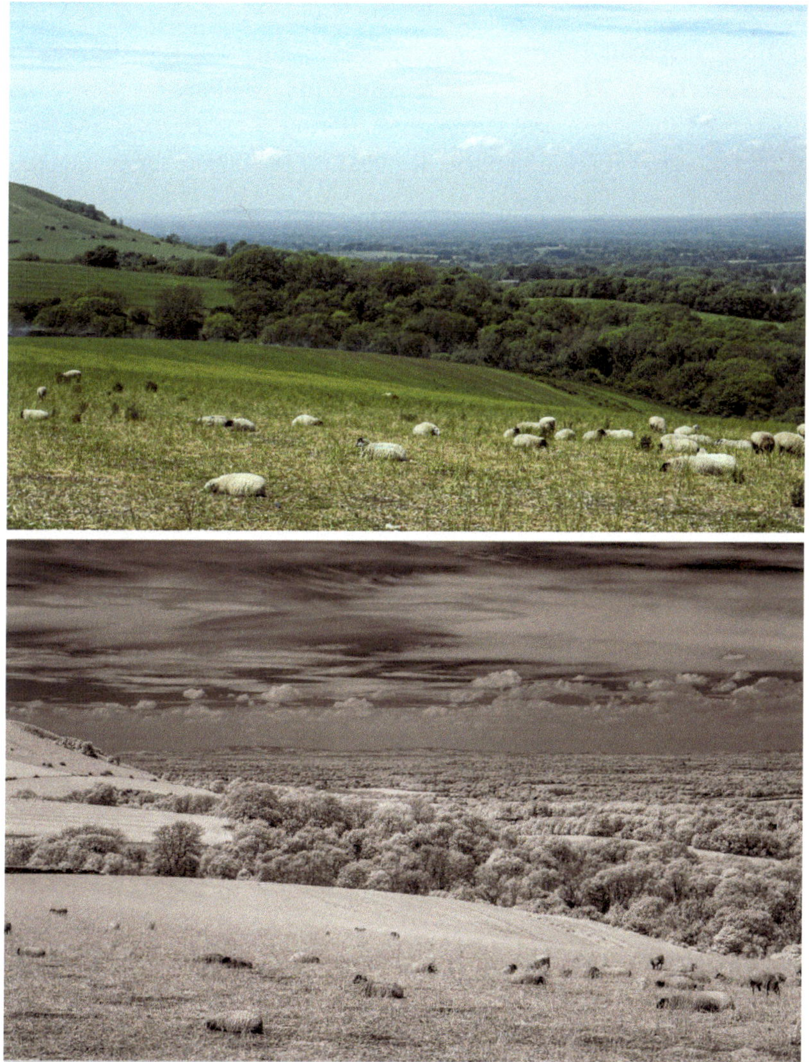

Figure 15.15 IR will minimize the effects of certain types of atmospheric haze. The extra detail visible in the clouds can be quite remarkable. The IR image was made with a Nikon D300 full spectrum camera using a Nikon Series E 50mm lens outfitted with a 850nm IR filter. An exposure of 1/125th second at *f*/11 was used. Images courtesy of Adrian Davies.

"Non-selective scattering" is a mechanism that occurs when particles much larger than the wavelength of the light are scattered. This occurs mainly with larger water droplets, ice crystals, snow and similarly sized atmospheric particles. This scattering occurs equally for all wavelengths. Shooting in IR doesn't offer any benefits in these situations.

"Mie scattering" is a scattering that occurs with atmospheric particles that are approximately the same size as the wavelength being scattered. These particles are typically spherical in nature and include dust, pollen, and water vapor (droplets). Generally, all colors are scattered equally. Clouds appear white since the water vapor is being scattered equally across all colors. Images shot in heavy fog are not usually enhanced by IR photography, due to the relatively large size of the water particles in the atmosphere.

"Rayleigh scattering" occurs mainly on the molecular level, when the particles are much smaller than the wavelength of light being scattered. It is this form that IR photography has the most effect upon. This type of scattering in the atmosphere is primarily caused by oxygen and nitrogen molecules which absorb the light and re-emit it in a random direction, thus scattering the light. However, the amount of Rayleigh scattering that occurs is inversely proportional to the fourth power of the wavelength. IR radiation at 800nm, for example, is thus scattered $1/16^{th}$ as much as blue light found at 400nm for example.

At ground level, all three types of scattering can influence the production and amount of haze, and are highly wavelength dependent. As a result, the haze-penetrating benefits of IR photography are not as strong. Images shot from planes can often be greatly enhanced by using IR photography, as there is generally less dust at higher altitudes. The majority of the scattering at high altitudes is Rayleigh scattering.

Color IR Photography

With a full spectrum converted camera and appropriate filters, colored images can be obtained containing both IR and some visible light wavelengths. Using a 720nm filter such as the Hoya R72 will produce a monochrome image with a hint of color. The images can be subtle or garish depending on your preference. When shooting with IR and color combined, with the 550nm "Goldie filter" for example, the image out of the camera will be predominantly an orange/red color. While this may be quite attractive, it is also possible to apply a whole range of adjustments to the image in programs like Photoshop and Lightroom producing a variety of different color schemes. One commonly used technique, particularly with landscape images, is to "Swap Channels" where the red and blue channels of the image are swapped so that the sky will end up as blue in the final image, with warm toned colored foliage.

Most digital images are composed of three color channels usually (but not always) red, green, and blue, relating to the colors in the original scene. These channels can be seen in the Channels palette in Adobe Photoshop. You will need to check the Show Channels in Color box in Preferences > Interface, to see them in color. The red channel is colored red, the blue channel blue, and the green channel green. It is possible however to re-color the channels, making the red channel blue for example.

Channel Swapping Technique

You will need Adobe Photoshop or a program with a channels facility which enables you to manipulate the different channels of the image, using a feature called the Channel Mixer.

Open the required image in Adobe Photoshop, and open the Channel Mixer dialog box (Image > Adjustments > Channel Mixer).

With the output channel set to red, make the red channel 0 percent, and the blue channel 100 percent. Then switch to the output channel for blue, and make the red channel 100 percent, and the blue channel 0 percent. This should result in blue skies in a landscape image, with warm toned yellowish foliage. The results can be adjusted further using various other adjustment controls.

Photoshop Actions

The above sequence of operations can be recorded and saved as a Photoshop "Action", which can be replayed over subsequent images.

Open the Actions palette (Window > Actions) and in the menu at top right go to New Action. Give the action a name (e.g. IR Channel Swap) then press the Record button. Now everything that you do to the image will be recorded, until you press the Stop button (the square button at the bottom of the Actions palette). All of the operations in the Actions are listed, and can, to some extent, be edited. When you open another image, the Action can be replayed by selecting the Action and pressing Play.

Several websites have various Photoshop actions specifically for IR images, which can be downloaded. One example is the "False Color Action Suite for Photoshop" from Kolari Vision (see resources list). This suite is a set of five actions including: the standard channel swap, a channel swap that only keeps the blue color, a 665 to 720nm action, a 590nm to 665nm action, and a 590 to 720nm action.

False Color IR

As previously discussed, analog color IR photography has been used for a long time, although very few films are still being made. A normal color transparency film used for conventional photography consisted of three light sensitive emulsion layers. One is sensitive to blue, one to green, and one to red wavelengths of light respectively. During the processing, dyes of a complementary color to the emulsion layer are formed in proportion to the density of the latent image. Yellow dye is produced in the blue layer, magenta dye in the green layer, and cyan dye in the red layer. When the transparent film "sandwich" is projected, or viewed against a bright light such as a lightbox, "true" color is seen.

With color IR film, the three light sensitive layers were green, red, and IR blue light was absorbed by a yellow filter incorporated into the film emulsion structure. During processing, yellow dye was produced in the green layer, magenta dye was produced in the red layer, and cyan dye was produced in the IR layer, thus shifting the colors towards the IR end of the spectrum, giving a "false color" appearance.

Green grass and other healthy foliage became red or magenta in the final image, but may appear as blue or cyan in diseased foliage if the amount of chlorophyll was affected, altering the amount of IR reflected. Red flowers appeared as yellow, for example. Thus, false color IR photography could be used to detect plant disease by foresters. Medical photographers used it to accentuate venous patterns underneath the surface of the skin. False color IR images from film had a very characteristic "look" which can be simulated with digital cameras together with post capture processing in programs such as Adobe Photoshop.

Digital False Color IR

Many photographers have experimented with various filters and full spectrum converted cameras to emulate the IR color film. Kolari Vision has produced an IR Chrome filter specifically to emulate the look of Kodak Aerochrome film. It transmits a select combination of IR and visible wavelengths, and needs to be used with a full spectrum converted camera.

There are several ways of simulating the effect with a digital camera, and details for one method are given below.

Simulating False Color Analog IR Images

You will need to use two precisely matched, pixel perfect images, an RGB image, and an IR image of the subject. Ideally you need to use the same, full spectrum converted camera and lens for both images, mounted on a sturdy tripod so that there is no movement between the two exposures. If you use an IR only converted camera, you will need to use a separate camera to record the RGB image, which would make pixel perfect registration of the image virtually impossible. For the IR image I generally find either a 760nm or 850nm filter gives a good result.

You will need to use Adobe Photoshop because Lightroom, Photoshop Elements or similar programs do not have the necessary color channel and layering features to accomplish the technique.

Simulating False Color Film Technique using Adobe Photoshop

The image chosen for this example was shot on a rather dull day, as seen in the control RGB image.

1. For the RGB image I use the same full spectrum converted camera as for the IR image, and place a hot mirror filter (e.g. Kolari Hot Mirror) over the lens to achieve the correct daylight color. I have found that this generally gives images with a strong cyan cast. I shoot two images, one including a gray card or other color balance device, the other without. The two RGB images are then synchronized in Adobe Camera Raw, so that the settings from the gray card image are applied to the image without the card.
2. Open both images (RGB and IR) in Photoshop, and open the Channels and Layers dialog boxes (e.g. Window > Layers).

3. The IR image needs to be colored cyan. Double click on the foreground color palette at the bottom of the toolbox, which opens the color picker. In the R, G, B boxes type a value of 0 in the Red box, and 255 in the Blue and Green boxes. This will give a pure cyan color. Click OK.
4. With the IR image active, go to Edit > Fill > Foreground Color with the blending mode set to Screen. The image should now be a pale cyan color.
5. Make the RGB image active by clicking on it. In the Channels dialog box go to the sub menu, and click on Split Channels. This will open three grayscale images, labeled as red, green, and blue.
6. Make the red channel active, and in the foreground color palette type in the figures: R: 255 and B: 255, similar to step 3. Click OK. Follow step 6 with the red image – it should appear magenta.
7. Repeat again with the green image, typing in the figures: R: 255 and G: 255 so that the image appears yellow.
8. Make the red image active, go to Select > All, then Edit > Copy.
9. Make the cyan colored IR image active, and paste the red image on top of it. It will appear as layer one in the Layers palette. Select the option Multiply in the blend mode drop down box.
10. Make the green image active, go to Select > All, then Edit > Copy, and paste it into the image containing the IR and red images. Again, select the Multiply option in the blend mode drop down box.
11. You should now have a three color image simulating how IR color film would have recorded the scene. The overall brightness, saturation, and vibrancy of the colors can be adjusted in Photoshop afterwards.

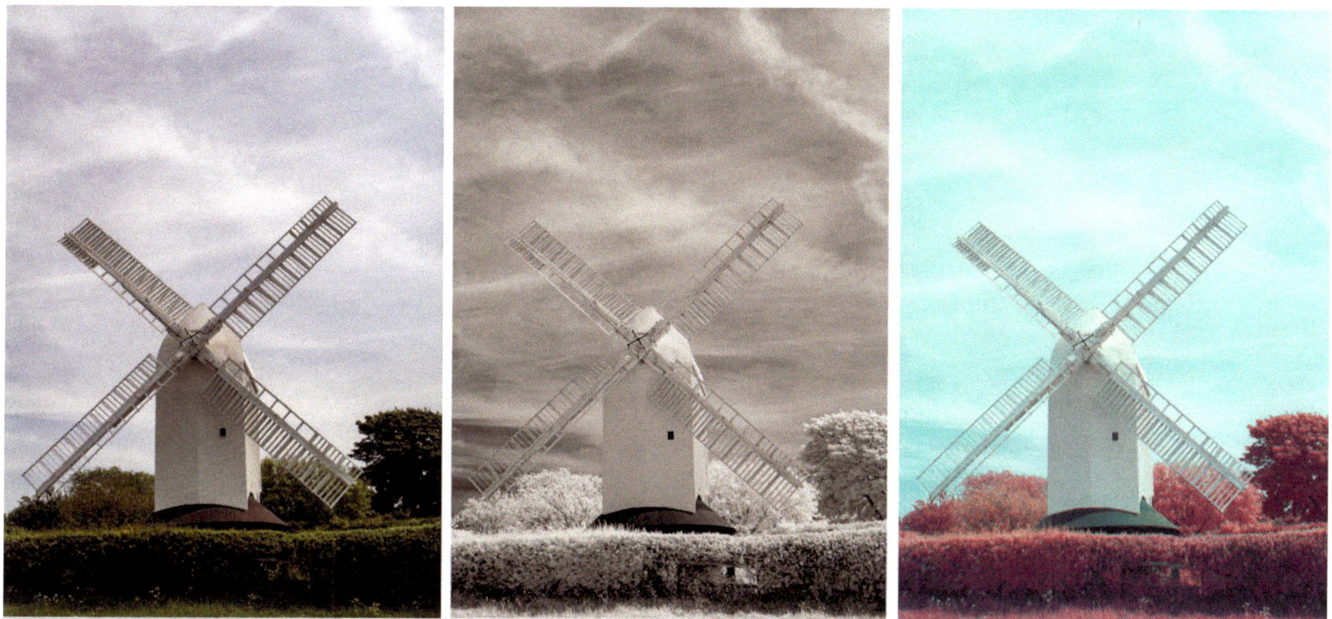

Figure 15.16 Simulation of IR false color using digital technique as outlined in the text. The IR image was photographed with a 850nm filter, final image is composite of IR plus two channels from the RGB image. Images courtesy of Adrian Davies.

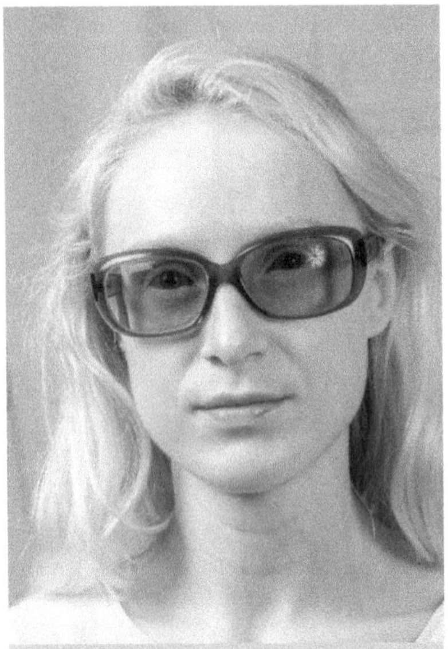

Figure 15.17 An IR portrait of model wearing visually opaque sunglasses. Note how the IR has produced an alabaster like skin texture and made the glasses transparent. Note also the sub-cutaneous blood vessels showing on the neck and forehead. A Nikon D300 full spectrum conversion camera with 100mm Series E Nikon lens, 760nm IR filter with an exposure of 1/125th at *f*/11 was used. Two Elinchrom studio flash heads produced the exposure. Images courtesy of Adrian Davies.

Applications of IR Photography

IR photography has many uses in the field of natural sciences. Because of its ability to penetrate the surface layer of skin, subcutaneous blood vessels can often be seen.

Several devices, including the VeinViewer® are used by medical professionals to accurately position catheters and other devices into veins.

Much wildlife monitoring is carried out with camera traps – either conventional cameras remotely triggered by the subject, or a purpose-built camera trap sometimes called a trail camera with built in triggering capabilities. These can generally shoot visible light or IR images, in both stills and video. Often resembling a device from a science fiction film, these cameras can be left out for considerable periods of time, recording the wildlife passing through a particular spot.

Polarized Light

Most photographers will be familiar with polarizing filters, used by landscape photographers for darkening blue skies and minimizing reflections from non-metallic subjects. Polarizing filters will have no effect on metallic objects. They are very good at minimizing reflections from water surfaces such as lakes or rock pools. Anglers often wear polarizing glasses to help them see underneath the water surface. The effect will depend on the angle of the camera to the water surface, working best at around 37°, which is the Brewster angle.

Polarizing filters are generally available in rotating mounts enabling the effect of the filter to be seen through the viewfinder. They work best at 90° to the sun, and care should be taken to not overdo the effect, as there may be a significant fall off in light level from one side of the frame to the other, particularly with wide-angle lenses, or producing rather garish blue skies.

Note, in the example shown, how the visibility through the water increases dramatically, and the blue sky darkens.

Reducing reflections from shiny subjects such as leaves has the effect of increasing color saturation, enhancing many landscape images.

Polarization of Light

Normal visible light is unpolarized – light rays from the sun vibrate in all directions at right angles to the direction of travel. When unpolarized light passes through a polarizing filter it is restricted to vibrating in one plane. Polarizing filters have a special molecular structure, which restricts the lights rays to one plane, as though passing through parallel railings. The human eye cannot see polarized light (though interestingly some animals can, including many insects and octopuses).

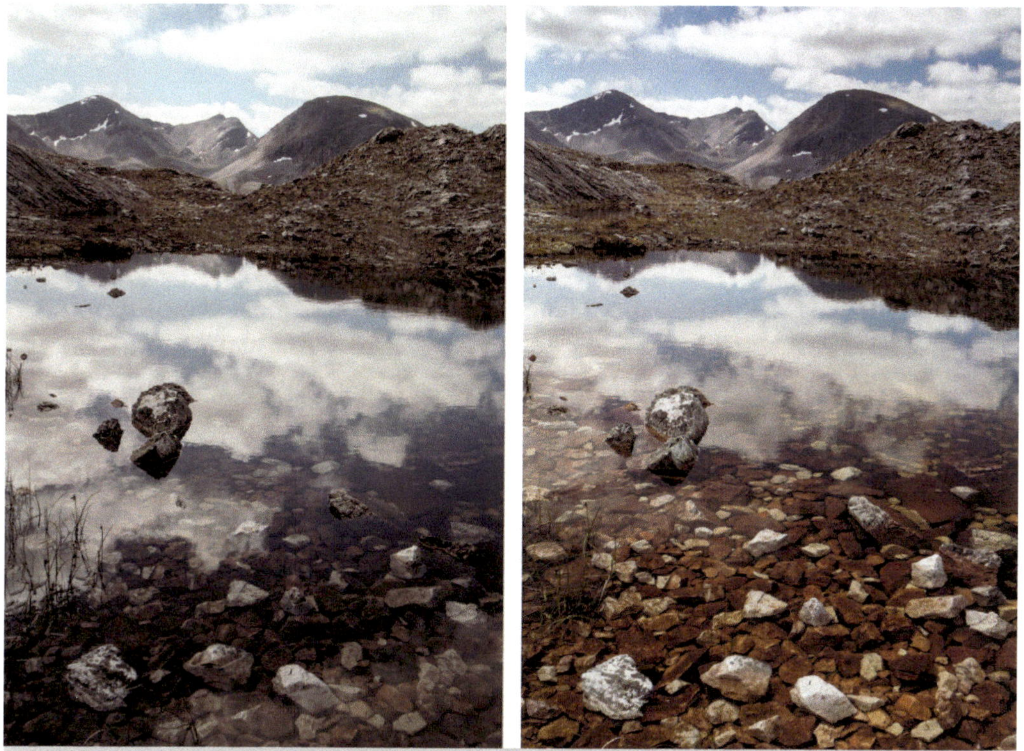

Figure 15.18 A view towards Meall a' Ghiubhais, Beinn Eighe National Nature Reserve, Wester Ross, Scotland. An unpolarized view (left) and polarized light view (right). Note the improved vision through the water surface, generally increased color saturation, and darker sky. The polarizing filter absorbs approximately two stops of light. A Nikon D800 camera with 24–85mm lens was used. Images courtesy of Adrian Davies.

Polarizing Filter Types

There are basically two types of polarizing filters. Older style linear polarizing filters can cause problems with beam splitters used in the through the lens (TTL) metering systems in modern cameras, so most polarizers nowadays are of the circular type. Visually, they look exactly the same. Circular polarizing filters are linear filters coated with a thin layer of suitable material called a quarter wave plate, which twists the polarized light as it is transmitted by the filter.

Other Applications of Polarized Light

Through the use of polarizing material, cross polarized lighting techniques can be used to reveal previously invisible stress patterns in some birefringent materials, such as some plastics, minerals, crystals, and ice for example. The basic idea is to position the transparent or translucent specimen between two polarizing filters and rotate one of the filters in relation to the other.

On a light panel or light box, a sheet of polarizing material is placed underneath the subject, and then another polarizing filter is placed on the front of the lens. Rotating the filter on the lens will reveal a rainbow of colored patterns with a suitable subject.

The technique has been applied for a long time by architects when looking at possible stresses on buildings. A scale model of a building or bridge was built out of a suitable transparent substance such as an Epoxy resin or perspex. Weights would be hung from the model creating stress, which would be revealed in cross polarized light.

In the natural sciences, thin sections of minerals, usually viewed under a microscope with cross polarizers, can reveal a wealth of colors, which in some cases can be used to identify specific mineral types. Similarly, some vitamin crystals, sugars, and salts reveal complex patterns and colors when viewed in cross polarized light. While much of this work will be carried out with the use of a microscope, macro photography can also be used with larger specimens.

When photographing specimens on light panels, make sure to mask off all extraneous light around the subject with black card (or cut a suitable sized window in a sheet of black card) to minimize the effects of flare.

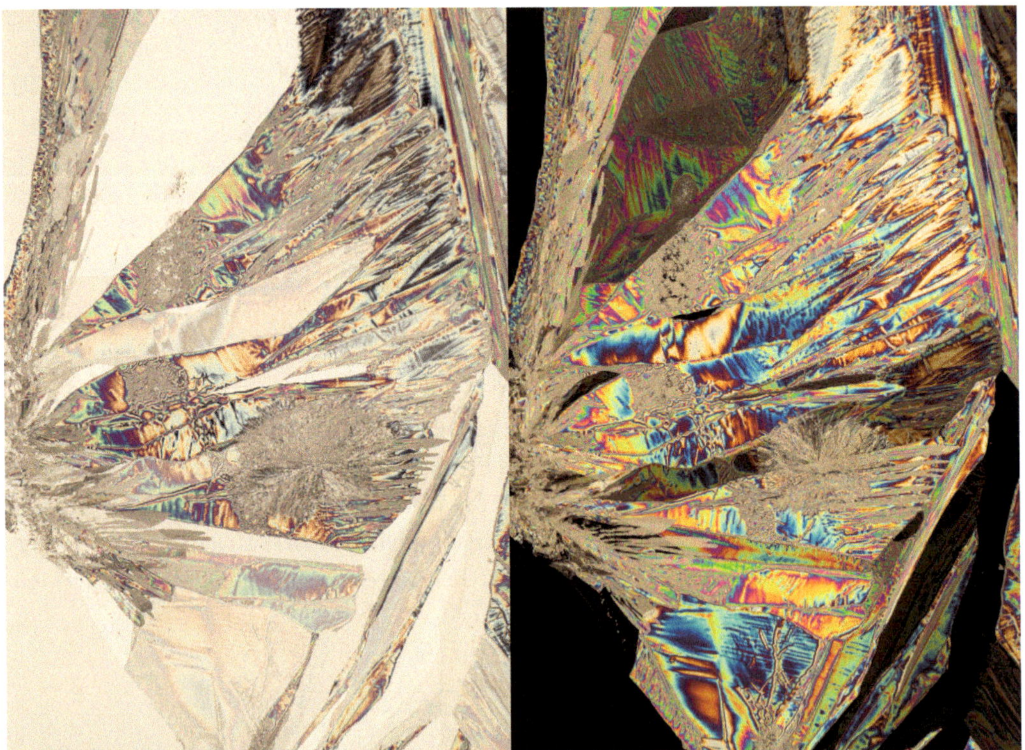

Figure 15.19 Crystals of Epsom salts (magnesium sulphate, $MgSO_4$) in brightfield illumination and cross polarized light. The photograph was made using a light panel with polarizing sheets underneath, and a circular polarizing filter on the lens. A Nikon D810 camera with 60mm micro-Nikkor lens and 150mm extension tube was used. The magnification was approximately 1.5× in camera. The polarized image exposure was 15 seconds at *f*/8, 200 ISO. Images courtesy of Adrian Davies.

Case Study: Preparing a Vitamin Specimen for Polarized Light Photography

Dissolve tablets or powder of a vitamin such as ascorbic acid (vitamin C) in a small amount of water. You will need a saturated solution. Use warm water and keep adding powder until a small amount is left, then add just enough water to completely dissolve it. Place a large drop of the solution onto a glass slide (75 × 50mm works well), place in a warm room and allow to evaporate. The crystal structure will vary according to the rate of evaporation and whether the liquid was smeared over the slide or not.

Using a Laptop or Tablet PC

An alternative to the light panel and polarizing sheet, is to use a laptop or tablet PC screen. Set the screen to give a bright pure white tone (either generate an image in Photoshop for example, or use an app such as Color Screen). The specimen can be placed directly on the screen. Take great care not to damage the screen and the polarizer on the lens used in the normal way. It is always worth shooting both a polarized and unpolarized image to enable comparison.

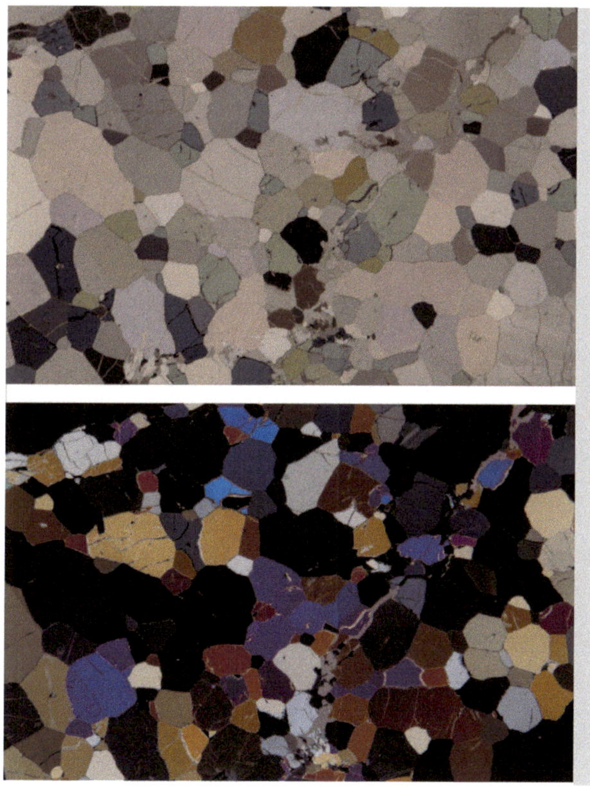

Figure 15.20 A thin section of the igneous rock peridotite, in unpolarized and cross polarized light. The picture was made on a light panel with a polarizing sheet underneath and a circular polarizing filter on the lens. A Nikon D850 camera with 60mm micro-Nikkor lens and 150mm extension tube was used. A magnification of approx. 1.5× in camera was created. The polarized image was 15 seconds at *f*/8 at 200 ISO. Images courtesy of Adrian Davies.

Suggested Reading

UVIRoptics stock a huge range of filters for UV and IR photography including the Hoya U-340 for Convoy torches, and a BUG filter for simulating insect vision: www.uviroptics.com.

www.ultravioletphotography.com: a forum for UV (and IR) photographers, with a huge amount of data on lenses, filters, and other equipment as well as a large section on botanical subjects shot in UV and IR.

A few examples of camera conversion companies:

Advanced Camera Services (UK): https://advancedcameraservices.co.uk.

Life Pixel: www.lifepixel.com.

Kolari Vision: https://kolarivision.com (includes IR lens database).

MaxMax: https://maxmax.com.

Sheets of polarizing material can be obtained from Greenweld Optics: www.greenweld.co.uk.

Chapter 16 Contemporary Light Microscopy Practices

Introduction

When things are too small to see with the naked eye, a magnifying lens will be needed. Light microscopes can magnify subjects from as low as ×5 and up to more than ×1000. Microscopes have been important tools for examining invisible subjects since their invention. There is some debate about who was the inventor of the microscope. Zacharias Janssen, a Dutch spectacle-maker, invented the first truly compound microscope in 1590s and Antonie van Leeuwenhoek, also Dutch, built a simple microscope in 1666. His invention allowed him to observe small organisms. Today, the capabilities of microscopes are amazing. Advances in computing, digital technology, optics, lighting, and software allows microscopes to be used at the frontiers of what it is possible to see.

Pictures made using microscopes are called photomicrographs. While the making of photomicrographs has become easier than it was when film was used, it remains common for new users of photomicroscopes to be intimidated by the instrument, with all of its various knobs and controls. Frequently, this hesitation comes from a lack of knowledge and experience.

Some Challenges in Seeing and Forming Well-Defined Images using a Light Microscope

- Handling, preparing and displaying 2D, but frequently 3D, subjects for examination.
- Working with VERY small working distances.
- Managing the physical and optical behaviors that affect light and influence image formation.
- Optimizing the use of lenses with very small, well corrected, and finite ranges of focus.
- Managing the presence of optical aberrations.
- Managing the isolation of dirt/artifacts.
- Optimal use of sophisticated imaging equipment and related software.

Foundations of Magnified Images

Forming highly resolved images of infinitely small structures requires a working knowledge of applied optics, physics, and software. It also requires being aware the role the sample will play in image formation. This chapter will focus on the core principles and other fundamental approaches needed to successfully form and capture images produced by a light microscope. The chapter will investigate both transmitted and reflected light techniques using compound light microscopes and contemporary digital microscopes.

Users of microscopes rely upon direct observation, whether using eyepieces or evaluation of images displayed on screens. Microscopists are frequently reminded about the influences of human vision and how it directly contributes to what is observable. Often magnified images are dim or poorly formed. Awareness of the influences of human vision and anticipating how to maximize seeing in difficult situations will provide important advantages when looking at nearly invisible subjects and having to determine what is real versus what is thought to be have been observed.

There are a variety of types of microscopes. These include biological, transmitted light, and industrial. They can be upright or inverted, compound or simple, confocal, or digital. This book will not investigate scanning or transmission electron microscopy. These techniques are very important methods used for observation and preserving scientific information, but space limitations precluded our ability to include it in the book.

A compound microscope is a complex optical instrument that has been designed to magnify very small objects. It forms images and delineates fine structures from a part of the larger sample. The ability to see fine structures within a mostly invisible object requires that a microscope create three conditions. These conditions are: create magnification; discriminate fine details within the sample described as resolution; and create contrast or visibility within the sample.

When forming images, a compound microscope produces several types of images. These include aerial or real, and virtual. An aerial image is formed by a positive lens and is an image that floats in space before being displayed on a screen. It can be focused onto a screen or eye that is placed into its optical path at a location where the image is focused. Virtual images are formed by negative lenses and cannot be focused onto a screen directly by themselves. They are mostly used in combination with positive lenses. A compound microscope forms an aerial image that is relayed onto the retina or sent to a camera sensor by a lens called an eyepiece. It is important to keep in mind that the human lens and its refractive properties play a role in seeing sometimes difficult to see magnified images. Sometimes, aerial images create illusions, or the perception that something was seen. This happens in much in the same way people think stars twinkle at night, even though they do not blink, or how some individuals see mirages when in challenging environments such as the desert. The human visual system frequently imparts its influences on seeing when using a microscope. These influences can play a role in image definition, color, and contrast.

Magnification

Microscopes form images that are larger in size than the subject. A compound microscope uses two lenses to create magnification. The initial stage of magnification is produced by the primary lens, or objective. Once formed, the primary image is further magnified by a secondary lens, the eyepiece, which relays the more magnified image to the eye or camera. This is true for both transmitted light and reflected light microscopes. The first lens produces a real image while the second lens by itself produces either a real or virtual image, depending on its optical elements. A real image can be focused onto a screen and a virtual image is one that can be seen, but cannot be focused – without additional lenses – onto a screen. Together they form the microscope's primary image, which is real, inverted, and larger than the subject.

If a 10× objective is used with a 5× eyepiece, the combined image will be magnified 50 times or ×50. When referring to an objective's magnification, typically the number goes before the ×, e.g. 10×, but when referring to the image's magnification, the "×" precedes the number, e.g. ×100. Therefore, ×50 is the correct way to indicate image magnification, and 50× is the correct way to indicate an objective's magnification.

Estimating the magnification of a system can easily be determined by multiplying the magnification of the objective and eyepiece. This calculation provides an approximate magnification. To be precise, a stage micrometer – a microscope ruler – must be used to determine the exact magnification of a microscope's imaging system. Stage micrometers come in different calibrations. Typically, a 1mm scale is common and subdivided into units using microns (μm).

Imaging systems often create image magnifications that are different than those observed in the eyepieces. This is due to variances in the photo/imaging system or the viewing pathway. This plays a role in creating different magnifications of the seen, or image magnification. To manage this difference, many research microscopes have software that enables highly precise measurements once the microscope's imaging software is calibrated. Calibration is accomplished by placing a stage micrometer into the microscope and then programming the imaging to calculate how many pixels are included in a specific reference distance that is established using the stage micrometer. It is also possible to calculate a system's magnification by referencing the image's projection distance from the eyepoint to the sensor.

$$\text{System magnification} = \text{eyepiece (mag)} \times \text{objective (mag)} \times \text{projection distance}/250\text{mm}$$

This equation describes how the magnification can be calculated by measuring the projected image distance from the exit pupil of the eyepiece to the sensor. The projection distance is divided by 250mm. In some literature, the eyepoint, called the Ramsden disk, is where to measure. In complicated research microscopes, this is very difficult to access. The optics are deep into the instrument.

The Objective as a Lens

The microscope's objective lens produces the system's primary image. The quality of the image is determined by the objective's degree of correction. Objective characteristics include magnification, color correction, and flatness of field, amongst others. The primary image is formed in the microscope's body tube. The body tube is part of the microscope where the image travels from the objective to the eyepieces. The intermediate image and the primary image are formed at the intermediate image plane, a location also situated within the body tube. Another important optical location visible within the body tube is the exit pupil of an objective. The objective's exit pupil can be observed in the microscope's body

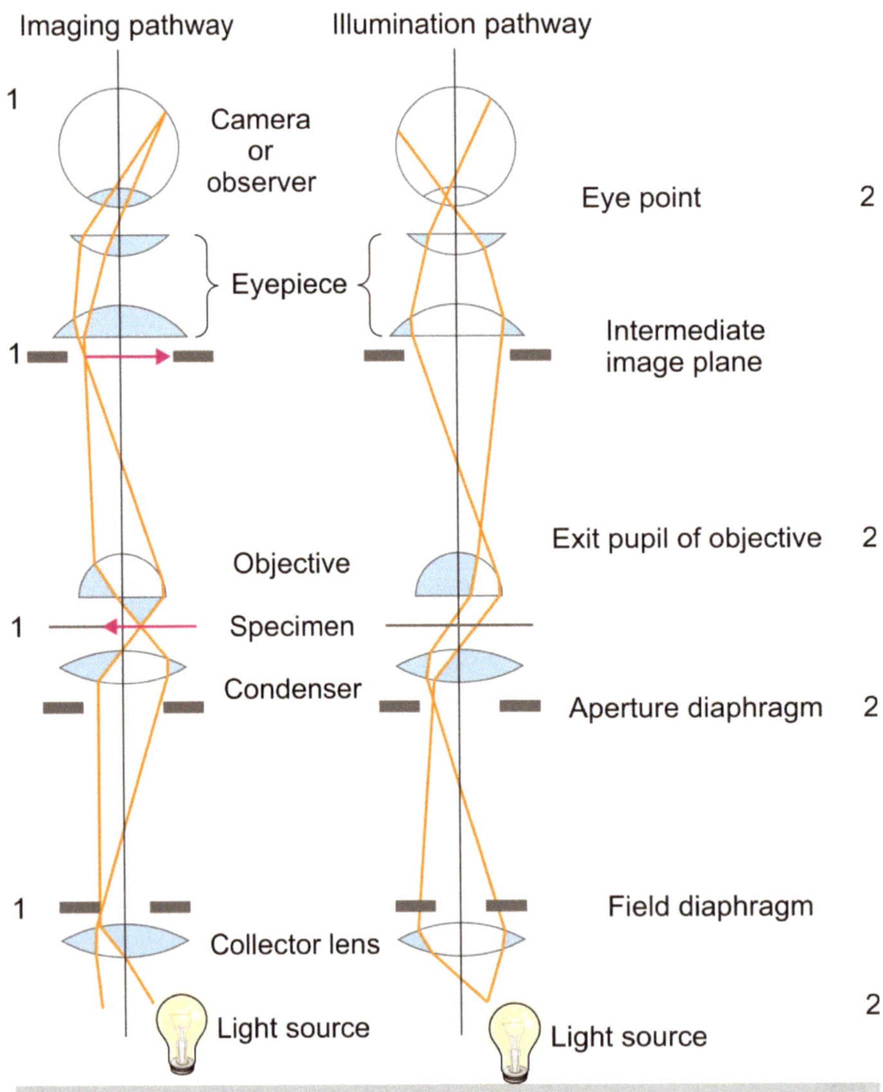

Figure 16.1 The illumination and imaging pathways required for a brightfield compound microscope to create excellent images and illumination.

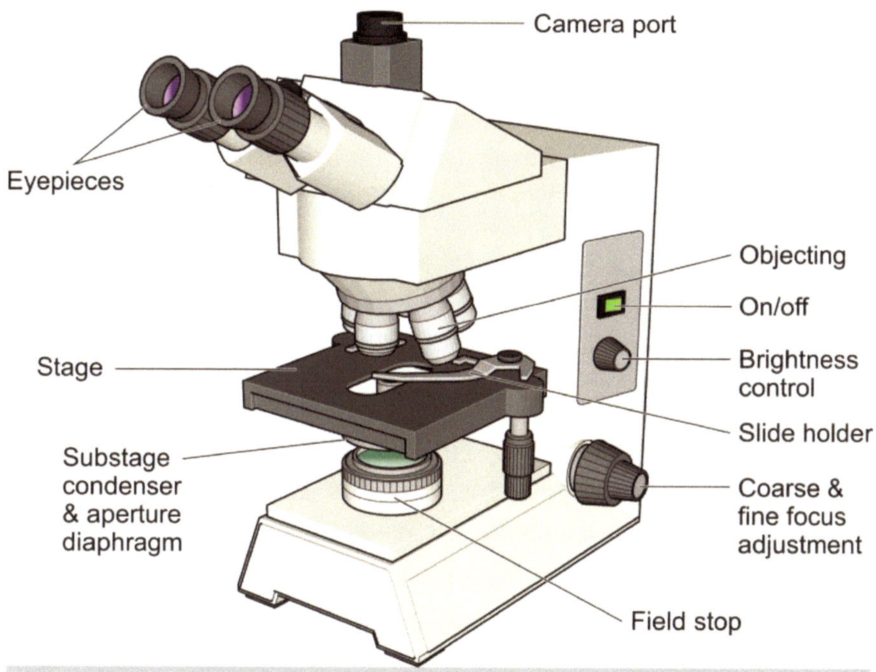

Figure 16.2 The main parts of a light microscope. Knowing where each control is situated on the microscope will lead to improved imaging outcomes. This approach is similar to how drivers must know where all the controls needed to operate an automobile are located without looking for them. This is called "knobology".

tube when an eyepiece has been removed from the instrument. The primary image comes into focus at the intermediate image plane. The intermediate image is also located at the front focal point of the secondary magnifier, the eyepiece. Being aware of these locations can be useful when trouble shooting image quality issues. More than three decades ago, microscopes were built to have a specific tube length, e.g. 160mm. Today contemporary instruments are designed to use infinity (∞) corrected objectives and have no specific tube length requirement.

Once the primary image has been formed and relayed through the eyepieces, the image will exit at a location called an eyepoint. The eyepoint can be observed by holding a piece of paper near to the eyepiece lens. It will be visible as a tiny dot of light situated very near to the eyepiece itself. The small dot will be visible on the paper but may be out of focus. Moving the paper closer or further from the eyepiece will allow exact determination of this location. The eyepoint can be a useful location for imaging. For example, a smartphone's lens can be placed at the eyepoint and used to make photomicrographs. When photographing using a fixed lens camera, the front of the camera lens must also be located at the eyepoint. Hand holding a camera for this purpose is very difficult and a camera stand or tripod is required for best results. With improved sensors, reasonably good photomicrographs using smartphones or fixed lens cameras can be produced with care and patience.

Optical Components of a Light Microscope

Eyepieces

Eyepieces, sometimes called oculars, allow users to view the images produced by a microscope. Many modern and digital microscopes no longer use eyepieces and display images directly onto a computer display. There are eyepieces used for imaging systems and typically they are different magnifications than those used for viewing. Frequently the imaging system's field of view will be less than what is observed in the eyepieces directly. Although you may see two images in the binoculars, a microscope's image is not stereo. The primary image is formed from a single objective having a single point of view. A microscope with only one eyepiece is called a monocular microscope and a microscope with two eyepieces is called a binocular microscope. A microscope with three ports (locations for eyepieces) is referred to as a trinocular microscope and is sometimes called a photomicroscope. Trinocular microscopes are typically designed for imaging and there may be other imaging ports.

Often viewing oculars come equipped with adjustable diopter correction(s) allowing all users to see crisply defined images. There is a neutral position setting (0) as well as (+) and (–) settings. For most users, each eye will require different diopter settings to create the best focus for each eye. You should know which eye is your dominant eye when establishing the correct diopter settings. If the eyepieces are not properly adjusted, you can expect to have eyestrain, or headaches, or both with prolonged use.

There are two types of viewing eyepieces used in microscopes, Ramsden and Huygens. There have been small improvements to their designs over the years, however, they are basically the same as when they were first invented. Depending on the design of these lenses, and the budget of the operator, Huygens negative lenses are most frequently used with achromatic objectives and positive highly corrected eyepieces, while Ramsden lenses are recommended for use with the finest apochromatic objectives.

Photo or Imaging System Lenses

When using a trinocular microscope, the third tube contain an eyepiece lens called the photo or projection lens (eyepiece). It is so named because it relays the magnified image from the microscope to an imaging system. The eyepiece's attributes will be inscribed on the lens barrel. Besides magnification, other inscriptions found on the lens will include the degree of correction and the field of view. The diameter of the eyepiece aperture (in mm) is called the field of view number, or field number (FN).

With a known FN, it is possible to calculate the diameter of the sample's imaging field. An eyepiece with a larger field of view is useful and practical and allows for more ease in scanning a sample to locate an area of interest.

$$\text{Field size} = \text{field number/objective magnification (Mo)}$$

All eyepieces do the same thing, they magnify the primary image and then relay it for viewing or for imaging. The photo eyepiece will usually be of higher optical quality than the viewing eyepieces and in many cases, the photo eyepiece may be of lower magnification than the viewing eyepieces. In fact, the photo/optical relay system may have a magnification that is adjustable when different specific magnifications are required. Some photomicroscopes may also have an additional optical zoom lens that is highly corrected and can be used to achieve just the right magnification based on object size. Many older imaging lenses will have the letter P or PHOTO inscribed on the lens. This information means that the lens has been specifically designed for imaging applications and has been corrected for flat field applications. Photo relay lenses are made and manufactured to the highest degree of correction which minimizes the presence of aberrations. Contemporary imaging eyepieces have been designed to work with small sensors and may have magnifications of .5× or .7× for example. Conversely, the viewing eyepieces on that same microscope could be 5× or 10×.

Beam Splitters

All microscopes use some type of prism or beam splitter to direct light to different pathways within the microscope. Once an image has been formed by the objective, the image must be relayed to the viewing eyepieces or to the imaging system. Some instruments allow the image to be viewed in both places simultaneously. At the prism, the primary image is directed to the viewing eyepieces or to the imaging system by changing the pathway. Changing the light pathway can be accomplished either manually or automatically in research-grade microscopes. When photographing, the prism should direct the light to the camera's imaging lens and away from the viewing eyepieces.

Adjustments to the prism will cause the light to be re-directed either partially or completely. Prisms can influence the light re-direction in various percentages based on design. For example, some instruments send 20 percent of the light to the viewing eyepieces and 80 percent to the imaging system. There are 50:50 prisms or 0:100 types. If the instrument's prism is of low quality, optical degradation can occur to the image as a consequence of poor optical glass, or lack of correction to the prism. Some flare or dispersion might be introduced by a prism. When performing fluorescence microscopy, a 100 percent beam splitter is the ideal choice because a fluorescence image has very low brightness and using a 50:50 prism would be detrimental to an imaging outcome. Dirt, unfortunately, is a common subject that finds its way into a microscope and may be on many surfaces. The dirt will be recorded as poorly defined dark spots.

Condensers

Managing light travel in a microscope is integral for good imaging outcomes. Located beneath the stage of a transmitted light microscope is a lens group and diaphragm assembly called the substage condenser. Inverted microscopes have this condenser, however, it is above the objective. The substage condenser is responsible for creating a cone of illumination required to form a highly resolved image of a specimen containing both contrast and resolution. The substage might be the least understood optical element in a microscope. Often this lens assembly is completely neglected or used incorrectly by new operators. Similar to an objective, a substage condenser has a numerical aperture (N.A.), which describes its resolution potential as well as its degree of correction. More about numerical apertures will be explored later in this chapter. The substage condenser contains a diaphragm called the aperture diaphragm. This diaphragm influences the resolution and contrast of an image that is produced by the objective. The condenser's N.A. should be at least equal to that of the objective for effective influences on an instrument's performance.

There are several types of substage condensers used in brightfield or transmitted light microscopy. There are also condensers used for contrast producing techniques, such as phase contrast technique. In a reflected light microscope, the objective will behave as the microscope's condenser. Three common types of brightfield condensers are the Abbe condenser, a flip-top or swing-out condenser, and achromatic-aplanatic condenser. There is also a type of condenser called a universal, that is capable of creating various techniques including brightfield (bf), darkfield (df) and Phase (ph).

The most common condenser is an Abbe condenser. The Abbe condenser, like any other condenser, focuses light that will interact with the specimen prior to entering the objective. Condensers have adjustable controls needed to operate them correctly. One control adjusts the distance of the condenser to the sample stage, and the other controls characteristics of illumination when using the aperture diaphragm. Adjusting the height of the condenser changes the size of the diameter of the beam of light. The aperture diaphragm controls the brightness, contrast, resolution, and DOF of the image. Abbe condensers typically have an N.A. of 1.25 and are useful for all types of microscopy and all magnifications. They serve as a sort of general-purpose condenser and are relatively inexpensive. They can operate with a range of objectives including the 4× and up through 100× objectives. They are not fully corrected for aberrations and will have comparatively high dispersion potentials. They will produce very satisfactory results.

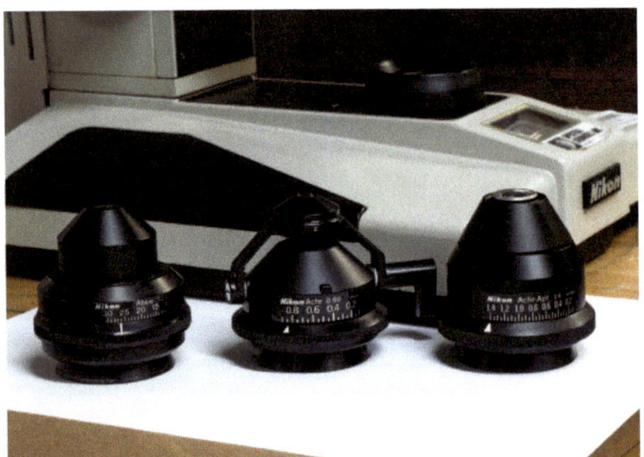

Figure 16.3 Left to right is an Abbe condenser N.A. 125, a swing-out or sometime called flip-top type with an N.A. of .90, and an achromatic-aplanatic condenser with an N.A. of 1.40. Many contemporary microscopes can be outfitted with what is called a universal condenser. See Figure 16.11 to see the angle of cones of illumination from these condensers. Images courtesy of Michael R. Peres.

Swing-out, sometimes called flip-out or flip-top condensers, are also very common and are suggested for use when low magnification objectives such as a 2× or 4× objectives are routinely used. They have numerical apertures that are in the range of .65 to .95 because of they have been designed for use in air. They are versatile and their corrections are on par with achromatic lenses.

An achromatic-aplanatic condenser is the most highly corrected condenser and is corrected for spherical aberration and chromatic aberration. It is expensive and typically has an N.A. of 1.35 or 1.4. There is also an aplanatic condenser which is not as highly corrected as the achromatic-aplanatic. It was frequently used for black and white imaging applications.

A Deep Dive into Objective Lenses

The single most important lens in a microscope by far is the objective. The objective lens produces the primary image and

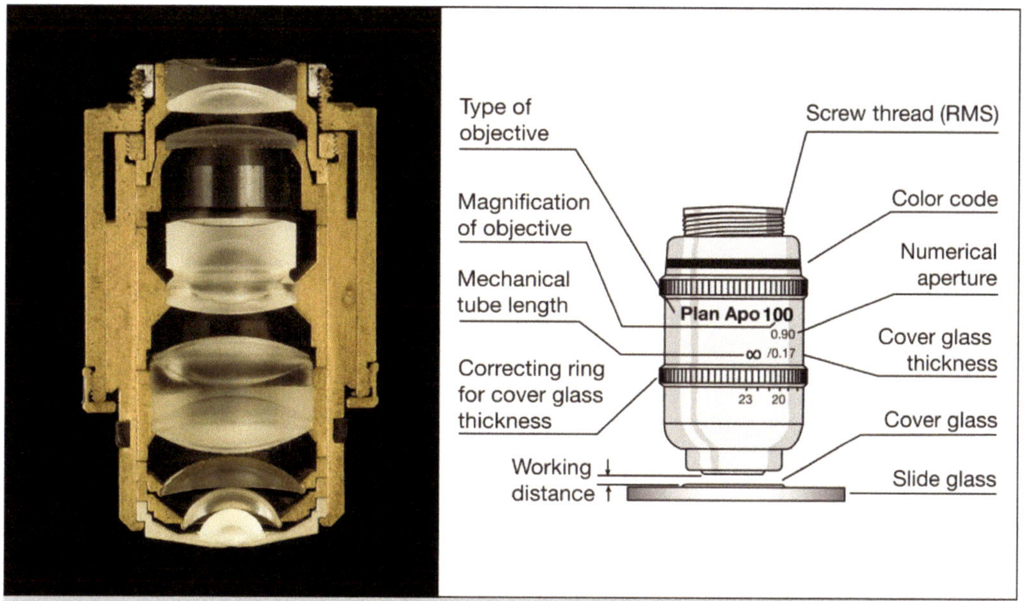

Figure 16.4 View inside of a 100× APO Zeiss objective. On the right is an illustration of a 100× objective displaying the nomenclature inscribed on the lens barrel. Images courtesy of Michael R. Peres.

creates the first stage of magnification in the microscope. The objective lens creates the image attributes including contrast, resolution, and magnification as well as depth of field. There are inscriptions engraved on an objective that show its corrections and capabilities. It can be useful to familiarize yourself with all of the objectives on your microscope. Objectives are complicated lenses with many internal lens elements that create the image's attributes and control aberrations.

An important correction for an objective is flat field or PLAN. PLAN refers to the objective's correction for flat field use. Frequently biological microscopes are used to examine subjects that have been cut and prepared. Often these samples are thinly cut sections of materials placed onto glass slides for ease of handling. Because of small differences in working distances, variances are amplified by magnification. Consequently, small differences in sample working distances make focusing across the entire field of view challenging optically. If a lens is not corrected for this aberration, curvature of the field will be observed. A lens not corrected for flat field performance will reveal good focus in the center of the image or in the outside regions of the sample, but not in both regions at the same time. For visual work, flatness of field is a smaller concern since focusing can be continually adjusted but when imaging, a flatness of field aberration is a big problem.

Another inscription describes the objective's magnifying power. The magnifying power is indicated this way, 10×. Each objective on the nosepiece or turret will display a magnifying power. An objective lens, in addition to its magnifying capability, also has an equivalent focal length similar to traditional photographic lenses. The 10× objective has an equivalent focal length of 16mm. This can be calculated by dividing the objective's magnification into the mechanical tube length. Samples are typically located approximately one focal length from

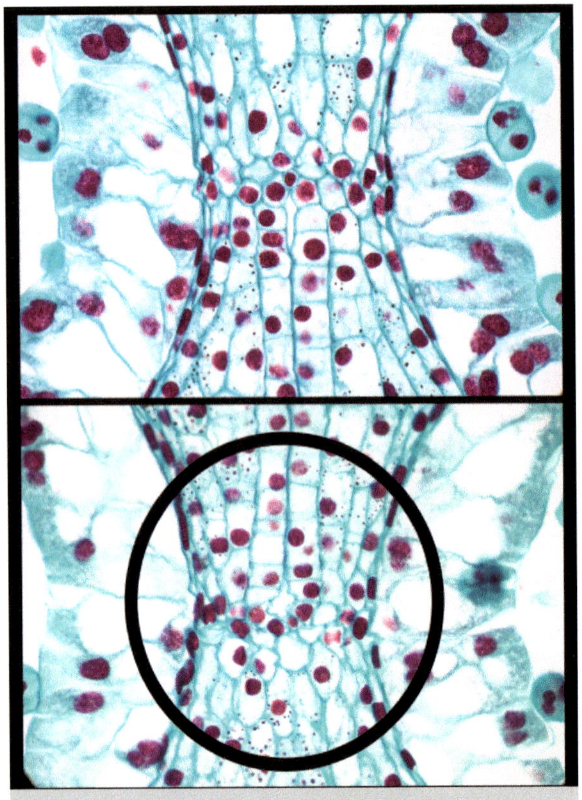

Figure 16.5 The effect of two different objective lenses. One corrected for flat field and the other not. The top image was produced using a PLAN objective and the bottom image used a non-PLAN objective. Notice the image area outside of the circle. Images courtesy of Michael R. Peres.

the front of the objective (when the image is in focus). A typical range of objective magnifications found on a transmitted light microscope includes: 2×, 4×, 10×, 20×, 40×, 60×, and 100×. These represent common magnifications, but are not the only choices for both transmitted and reflected light microscopes. There are hundreds of types of objectives designed for many unique applications.

Explaining Numerical Aperture

The quality of an objective lens is predicted by the objective's numerical aperture (N.A.). Often N.A. is shared as a decimal such as .45 or it can be higher than 1.00 for higher magnification objectives that are used with immersion oil. When the objective is used in air, the N.A. will be less than 1.00. When it is used with immersion oils, the N.A. will be greater than 1.00. The N.A. shows by way of a number, a ratio of the light available for image formation measured against what is actually gathered up by the lens. The higher the N.A., the better the image definition or optical resolution. The higher the N.A., the more the objective will cost. N.A. is also related to magnification. Typically, lower magnification lenses will have lower numerical apertures and higher magnification objectives will have higher N.A.s. Higher N.A.s will create more optical resolution but will contain less DOF.

Forming Highly Resolved Images: The Role of Diffraction and Resolution

It is useful to understand how an objective forms an image. Diffraction, a complex subject, plays a critical role in the process. In microscopy in particular, many aspects of image formation are influenced both positively and negatively by diffraction. This idea was initially proposed by Ernst Abbe in the late 1880s. There are other light behaviors that contribute to image formation including refraction. Refraction is the bending of light as a consequence of a speed change when the light enters a new medium. Diffraction plays an even greater role than refraction in image formation because it influences the objective's optical resolution. It is an unfortunate reality that as the magnification of an image increases, the image often becomes less defined. This outcome is caused by the influence of diffraction. When radiated energy encounters an opening (aperture), many interactions of the traveling light and other elements begin to occur. The sample and its inherent physical structures will determine how small, or large, the effects of diffraction are.

An aperture in an optical pathway where light is traveling will act like a light source itself. The edges of small openings cause some traveling energy to be re-directed into new directions. These new directions will contain less energy than the original light, but will contain critically important structural information about the sample. The size and direction of this behavior will be dependent on the diameter of the aperture. If the aperture is wide, most of the energy will travel through the aperture unaffected. In this situation, most of the peripheral energy will be bent only slightly outward away from the original travel direction. In this condition, there is minimal effect on an image. If the aperture is small, the bending of light away from the original travel direction will be greater. This becomes important because the sample itself is often a porous grating and will behave like many small and separate apertures. Objects that are porous are sometimes referred to as diffraction gradients. The smaller the fine structural details, the greater the light will be affected upon exiting the sample. The ability to delineate small details in a sample is described as resolution. The smaller the structural details, the more difficult it becomes for the objective to resolve these details. For a microscope objective to resolve detail, it must be able to collect all the radiated energy that interacts with the sample. This of course is impossible. The challenge arises when the energy leaves the sample after going through the aperture diaphragm in the microscope. As the energy interacts with various light rays from across the sample, interference occurs. Interference can either lead to more visibility by creating contrast or lead to data loss and poor resolution. This outcome is the result of the objective being unable to gather all the diffracted rays required to form highly resolved images.

Demystifying Concepts about Numerical Apertures

When characterizing a traditional photographic lens, a useful indicator is the maximum aperture or *f*-stop of that lens. *F*-stops can serve as indicators of light transmission. A lens's maximum aperture can be determined by dividing the focal length of the lens by the maximum clear diameter of that same lens. Microscope lenses are also described by their light gathering potentials. The factors which influence their performance would be the medium

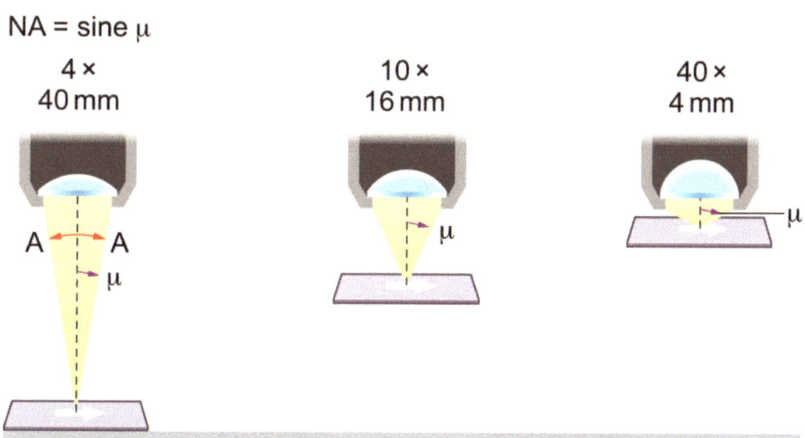

Figure 16.6 The relationships between working distance, magnification, and numerical aperture (N.A.).

in which they are operated, and the acceptance angle of the diffracted energy into the objective.

The objective's N.A. can be mathematically calculated using the following equation. Information about an N.A. is provided to users and calculated by the manufacturer.

$$\text{N.A.} = \text{n Sine } \mu$$

Where n is the refractive index of the medium that the objective has been designed to be used in, and Sine μ is the acceptance angle of that specific objective lens. Air has a refractive index of 1.00, water 1.33, and immersion oil 1.53.

The N.A. of an objective is influenced by the focal length of the objective. Shorter focal length lenses will have shorter working distances and possibly a greater angle of acceptance, which leads to more resolution. As the magnification goes up, so will the corresponding numerical aperture. The objective's N.A. will dictate the smallest of structural details that can be resolved. This distance can be determined using the following formula:

$$D = \frac{\quad}{\text{N.A.}_{\text{obj}} + \text{N.A.}_{\text{cond}}}$$

Where D = the minimum resolvable distance between two points, and λ = the wavelength of light used. 550nm is often used in this equation to represent average white light and is derived by averaging the wavelengths of the visible spectrum.

Airy Disk

An instrument's optical performance can be predicted by considering the production of an airy disk and the diffraction potential of the system. An airy disk is a predicter of how effective the best-focused spot of light from a point of light formed by a perfect lens with a circular aperture will appear as an image. An airy disk is only limited by the system's diffraction. The diffraction pattern resulting from a uniformly illuminated circular aperture is characterized as having a bright region in the center called the airy disk, and the disk is surrounded by concentric rings of diminishing brightness. The airy disk plays a central role in indicating a microscope's resolution by playing a role in diffraction. When the airy disk spreads, the outcome is called a point spread function and plays a significant role in image clarity and predicts whether independent points will remain delineated from one another. Many confocal or other high-end research photomicroscopes have controls to manage and influence the point spread function of a lens.

A sample's diffraction potential plays a significant role in a system's performance when operated in conjunction with the objective's numerical aperture. This effect can be better understood when considering how light interacts with a diffraction grating. Light that travels through a grating – an optical component with repeating structures – the light is split into new beams that travel in different directions. Some light will travel straight through the gradient and is characterized as the light's zero order. To the outside of the zero order would be the first order of diffracted rays. The angle and departure of this deviation is a function

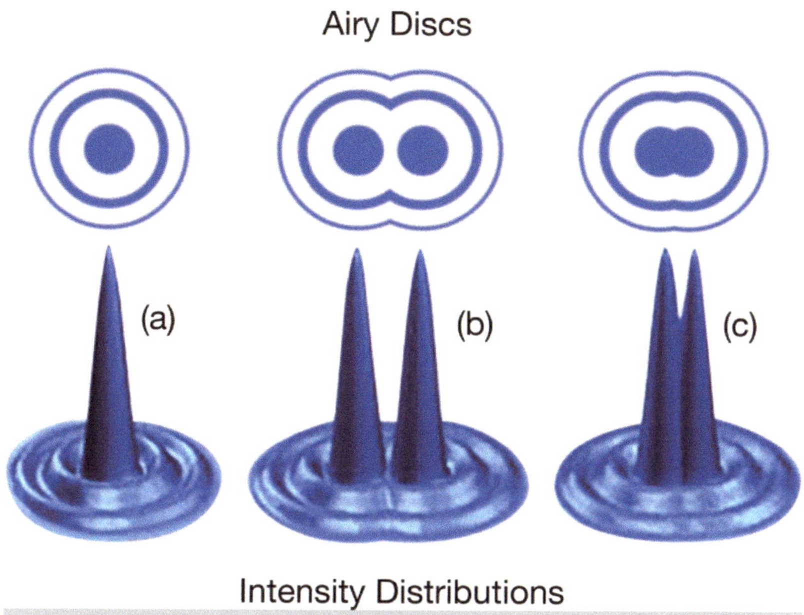

Figure 16.7 An airy disk is a useful measure for predicting how a location (point) will be modified when diffraction affects sharpness. When two airy disks are adjacent to one another, sometimes diffraction will cause the two points to become one. Higher numerical apertures produce better resolution or better image separation of the points seen as distinct locations.

of the size of the aperture. Adjacent to the first order rays, and further from the zero order would be the second order of diffracted rays. The zero order principally is responsible for providing the illumination used for the background in brightfield applications. The diffracted rays carry the information about the sample's structure. It is ideal to direct as many of the diffracted rays into the objective as possible. The greater the acceptance angle of the lens, the easier it becomes to collect the diffracted rays. It is possible to see the diffracted rays at the exit pupil of the objective in the body tube.

N.A. defines the highest image magnification that an objective can produce before the image begins to lose visual crispness. This is evidenced as a condition where the magnification reveals only space and not structure. The magnification where detail starts to diminish is called empty magnification. Predicting where empty magnification begins is accomplished by multiplying the N.A. of the objective by 1000. Magnification is a multiplicative process. Empty magnification needs to take into account the magnification of the image size in print or screen display.

$$\text{Empty magnification } (\times) = \text{N.A.} \times 1000$$

Objective Corrections

Another inscription found on an objective will be the degree of color and spherical aberration correction that the objective has been optimized for. Common or general-purpose objectives would be characterized as achromatic or ACHRO objectives. A higher quality objective would be a fluorite (FL) objective, while apochromatic (APO) objectives would be considered the highest quality objectives available. Achromatic objectives are by far the most common. ACHRO objectives will not have ACHRO inscribed on the lens barrel. ACHRO lenses are spherically corrected for green, and chromatically for blue and red. Additional improvements can be made to achromatic lenses by adding fluorite to the glass. A fluorite objective will optically perform better than an achromatic lens but is not as good as an apochromatic objective. APO objectives will have a noticeably higher N.A. The cost of a fluorite or an APO objective will be higher than an ACHRO objective. Having an objective with a higher N.A. will lead to more optical resolution displaying a higher degree of color accuracy. Apochromatic objectives are spherically corrected for two colors and chromatically for all three colors. Another color correction that might be found on an objective is VC or violet correction. The following list shows what the NA of three objectives of varying correction might be.

Achro 10× objective – N.A. = .25

Fluorite 10× objective – N.A. = .30

APO 10× objective – N.A. = .45

Other inscriptions refer to the instrument's optical body tube length requirement and whether the objective requires a cover glass between the objective and sample. Most contemporary instruments use infinity (∞) corrected objectives. However, in the past, 160mm was the standard tube length. The Royal Microscopical Society proposed a tube length standard of 160mm that lasted for more than 100 years. Infinity (∞) objectives have advantages because the same objective can be used for many applications such as phase contrast, brightfield, and DIC for example. Infinity corrected objectives focus at infinity. They use longer working distances and allow the insertion of optical components into the optical pathway at different locations during use.

Using an objective where it has not been optimized will affect the quality of the characteristics of the image it produces. When incompatibility exists, it often leads to the creation of aberrations or other diminished quality. If the objective has been designed to examine subjects with a cover slip in the optical pathway, it will have .17 on the barrel. If the lens works best without a cover slip, it will have (–) or 0 inscribed on it.

Spherical aberrations are where light is refracted differently in the middle of the lens than outside areas, resulting in low contrast and hazy images. At lower magnifications, spherical aberrations can be managed by creating more contrast using the aperture diaphragm, or during image processing, where contrast can be added. At higher magnifications, the problem is more severe and may not produce a usable image. Photographing with incorrect coverslips will cause spherical aberration. It is possible to cheat when using lower magnifications, where image contrast can be restored using image processing software. Coverslips come in many thicknesses and materials. Plastic cover slips will absolutely degrade an objective's performance and should be avoided.

Environmental Influences

There are environmental conditions that can lead to inferior images from microscopes especially when the subject's illumination is dim or displays poor contrast. These conditions make it generally difficult to see. Rooms without windows are recommended or at the least, adjustable lighting is recommended when imaging. Gas lift chairs are also recommended for extended shifts at a microscope and should be re-adjusted for each session. It is important for an instrument's eyepieces to be situated at the height of the user's eyes. The best practice when sitting at a microscope requires an operator to raise or lower the chair to locate their eyes at the height of the eyepieces. This will minimize body strain from poor posture. Not managing these conditions, including using poorly adjusted eyepiece diopters, will lead to headaches and the production of inferior results. Creating better conditions for seeing will lead to the formation of better images that possess excellent focus and resolution.

Fundamentals of Operating a Microscope

Operating a microscope is not overly complicated but does require using controls in specific ways and in a specific order. This is needed to produce the best possible image from a wide range of samples. There are practical approaches that can improve efficiency and image quality. When first turning on a microscope, select a brightness that is comfortable for viewing. It is easy to change the brightness of the lamp and this will not affect the optical performance of the microscope. More brightness does not lead to better pictures. Adjusting brightness is an acceptable practice for viewing, however, changing brightness is not recommended when imaging. Brightness can be reduced using the microscope's voltage, by using neutral density, or with a polarizing filter. Some photomicroscopes may have a symbol for use as the brightness setting for imaging. This setting was useful for film photography but now can play a role in standardizing a bulb's color temperature useful for best imaging practices. Once the brightness has been established, it remains important not to change the lamp brightness (using tungsten halogen lamps) during an imaging session. Consistent and easy-to-manage digital image files are produced when the color temperature is NOT changed. Using N.D. filters is the ideal mechanism to change a microscope's brightness for imaging related activities. If a microscope has LED bulbs, brightness can be changed without creating large variances in color temperatures. Bulb brightness changes create only small variances in color with LED technology. LED sources can be of two types. One is continuous and the other pulsed. If the bulb is a pulsed type, its brightness can be changed with no color change. If they are of the continuous type, when they are dimmed their color will change slightly by such a small amount that it is almost negligible.

Using Eyepieces

When looking into eyepieces, it is ideal to focus the microscope using your dominant eye. To start, position all diopters to the eyepieces to their neutral positions. Next, using the dominant eye, focus the microscope into sharp focus using the course or fine focusing knob. Once focused for the dominant eye, focus the image for the non-dominant eye using the eyepiece's diopter lens only. By focusing in this manner, both eyes will have the same focus based on an operator's unique visual requirements. Viewing eyepieces are typically not highly corrected and should not be used for imaging unless there is no other option.

Focusing an Image

A microscope's focusing mechanism moves the stage up and down. The stage is the large rectangular plate where samples are placed. There are typically two knobs on the instrument used for focusing, however, many research microscopes can be focused using a computer or external control panel. Frequently one knob controls the coarse focus and the other, the fine focus. By moving the stage, the distance between the front of the objective and the subject is changed. This distance is called the working distance. The fine focus is capable of moving the stage in single micron increments (μ) and can be useful in determining the thickness of a sample. By focusing on the top of the sample and recording this number (from the fine focus setting) and then focusing to the bottom of the sample and referencing this number, the difference can be found by subtracting one value from the other. This number will provide an approximate thickness in μ. To be precise, the approximate thickness can be multiplied by the refractive index of the material that surrounds the subject. A common biological mounting media has a refractive index of 1.53.

Operating with Very Small Working Distances

It can be helpful to know the stage's travel direction when making focus adjustments without looking at the stage. This can easily be determined by watching from the side of the microscope when rotating the focus adjustment knob clockwise or counterclockwise. Once the travel direction is known, operators should bring the front of the objective as close to the subject as possible by adjusting the coarse focus. Next, rotate the focusing knob to make the subject move away from the objective. It can be practical to perform this procedure a few times using various objectives. Knowing the stage's travel direction will allow focusing to be managed without concern for driving the objective into a sample! When increasing the working distance, the subject will always move further from the lens.

Interpupillary Distance

The distance between an operator's eyes is unique. Many microscopes allow for this distance to be adjusted for individual users. This distance is referred to as the interpupillary distance. It is easy to measure this distance using a ruler, but many microscopes display this number when the oculars are adjusted. Typically, the tubes where the eyepieces are located can be moved inwards or outwards. To make the proper adjustment, look into the microscope and move the tubes until both images are seen as one. This may require a few attempts. A normal adult's interpupillary distance is in the range of 59mm to 69mm.

Using the Microscope's Body Tube

It is possible to improve an image's quality by examining optical behaviors in the body tube. To evaluate optical behaviors, remove an eyepiece. With the eyepiece removed it is possible to see into the body tube when the microscope's light is on. The body tube is the pathway where light travels from the objective lens to the eyepieces. The microscope's body tube's primary function is block ambient light from entering the microscope. The aperture diaphragm is visible in the tube at the exit pupil of the objective, and the lamp filament in true Kohler microscopes can also be observed there. This location will be visible in what appears to be the bottom of the body tube. By engaging the diaphragm, it is possible to see it come into, or leave, the tube's diameter. By watching the AD to the diameter of the tube, an operator can establish a system's best chance for optical resolution.

Nosepiece

The part where the objective lenses are joined to a microscope is called a nosepiece or turret. Turrets are designed for four, five, or six objectives. The nosepiece rotates in a circular motion underneath the body tube and chassis. An objective that is properly engaged should click or lock into place. Microscopes can use both rear or front-loading nosepieces. Front loading nosepieces allow users to see the objective at the front of the instrument rather than in the rear. A few companies still produce front-loading nosepieces, however, rear loading nosepieces are more common. They can be rotated manually or by computer when using advanced research photomicroscopes.

Substage Condenser and Adjustments

The substage condenser must be adjusted for optimal image formation. Located near to the condenser will be a knob used for making height adjustments to the condenser. Often the knob is used to move the condenser up and down. It is important to position the condenser at the right height for the objective and sample requirements which maximizes image characteristics when properly adjusted. Each objective will have a different condenser setting requirement. Establishing the correct distance will control the size and quality of the cone of illumination that the condenser produces.

Using the Field Diaphragm or Field Stop

There are two critically important diaphragms in a microscope. One is the field stop or field diaphragm. The other is the aperture diaphragm. The field stop is situated in the base of a compound transmitted light microscope and found directly below the substage condenser. For reflected light microscopes it will be located near to the illuminator. The field stop controls the diameter of the illumination beam that enters the microscope's substage condenser. There is a mechanism to changes the field stop's diameter. The proper diameter is determined by the objective's requirements and corresponding magnification. The field stop controls the diameter of the illumination and does not change brightness. Opening or closing the diaphragm causes the diameter of the illumination to change size. If the field stop is too small, it will encroach into the image. If the field stop is larger than the diameter of the objective, flare light will be formed. Flare light will reduce image contrast.

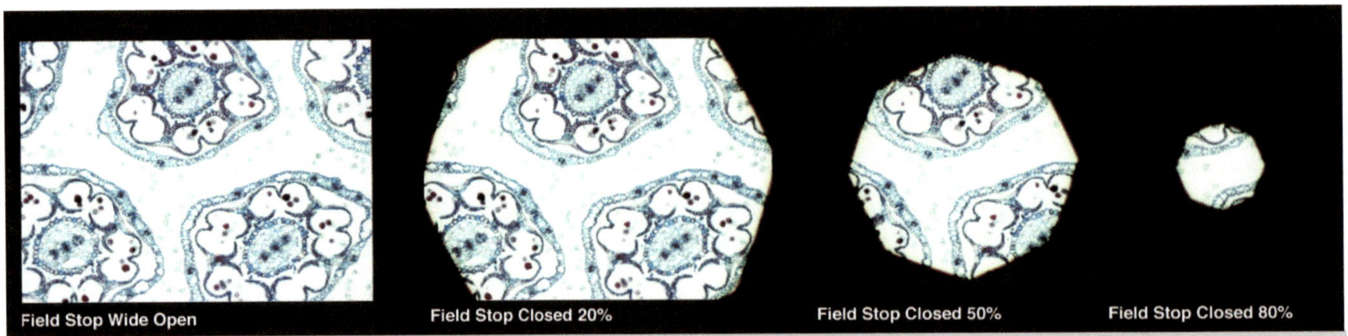

Figure 16.8 The field stop controls the diameter of illumination but not its brightness. In the left image, the field stop was completely opened and exhibits the same brightness as the image on the far right where the field stop was completely closed to its smallest opening. Images courtesy of Michael R. Peres.

Light and Lamps

Typically located in the rear of a light microscope will be the light source. Most microscopes still use tungsten halogen lamps, although LED bulbs are becoming increasingly common. One advantage of tungsten halogen bulbs compared with tungsten bulbs is that a halogen bulb maintains color temperature for life and tungsten bulbs change color as they age. A halogen bulb has a life expectancy of several hundred hours. LED lamps can last for 20 years, however, they are more expensive. An LED bulb's color temperature is more consistent than a tungsten halogen lamp.

You should learn how to access a microscope's bulb and determine how to change it. Many bulbs have long life expectancies but eventually will need replacement. A bulb may be characterized by its brightness and the spectrum that it emits. Bulbs will have a variable color temperature and the color temperature is shared as Kelvin degrees. 3200K was the recommended color temperature for film because film was formulated for one specific color temperature. Color temperature does not play as critical a role in digital photography because an image's white balance can be easily adjusted. Changes in bulb brightness will cause color temperature variances. Tungsten halogen bulbs should not be directly handled because doing so will leave finger oils on the bulb. It is useful to wear gloves or use tissue when handling bulbs.

Fluorescent illuminators use various bulb types including xenon, sodium, or mercury vapor lamps. These bulbs are very expensive and have a recommended length of usage. LED light sources are also making inroads into the fluorescence lamp market. They are, in some ways, very practical because they can produce customized excitation spectrums that can have real advantages for exciting specific fluorophores.

Aperture Diaphragm

The aperture diaphragm (AD) is the most important control on a light microscope. Located in the substage condenser on a transmitted light microscope, this diaphragm controls image contrast, resolution, depth of field, and brightness. Operating this diaphragm is crucial to forming the best possible images. Leaving the diaphragm in a more open position creates images with high resolution, but shallow DOF, and less contrast. Closing the AD down produces images that possess more contrast, and exhibit more DOF, however, they will have low resolution. The diaphragm's setting for each sample will be different. The aperture diaphragm will also affect image brightness but should NEVER be used to adjust image brightness.

Establishing Proper Brightfield Illumination when using Transmitted Light Microscopes

Creating illumination that is uniform and produces images with maximum detail was first proposed by August Köhler in 1893. Creating the brightfield illumination for a transmitted light microscope is not overly complicated. Variances in illumination can easily be avoided, and using the aperture diaphragm correctly will lead to images that exhibit full resolution and structural visibility. Achieving Köhler illumination when using low power magnifications is more challenging than when using higher magnification objections. Some contemporary imaging software can correct for illumination errors and this feature is called shading correction.

Establishing Proper Köhler Illumination

1. Turn on the microscope and set the interpupillary distance.
2. Adjust each eyepiece's diopter as needed. Start with the dominant eye. Focus the image by changing the working distance using the dominant eye. Next, using the diopter correction collar, focus the non-dominant eye diopter. The focused image should be achieved in both eyepieces.
3. Close the field stop fully.
4. Using the substage condenser, focus the image of the field stop until the blades are well defined and in the same plane as the focused image.
5. Center the image of the field stop using the substage condenser centration screws. The centration should be visible in the field of view.
6. Open the field stop until it just leaves the field of view.
7. Set the aperture diaphragm based in the sample's requirements.

When changing objectives, it is important to re-establish Köhler. Typically, once established, changing of magnifications will require only minor corrections.

Setting Up Köhler Illumination and More

1. *Prepare the microscope for use*: It is important to establish a comfortable viewing brightness. This will make using the microscope comfortable until final adjustments are made for imaging.
   - Set the interpupillary distance.

- Adjust the height of the chair to locate your eyes directly across from the eyepieces without effort.
- Adjust the eyepiece diopters as needed. Sometimes you will find leaving them in their neutral position is best.

2. *Focus the image*: Move the stage to the highest level using the focusing knob and bring the image into rough focus by increasing the working distance. Once the rough focus has been achieved, using the fine focus control, critically focus the image. Once you have established these conditions, it may be necessary to re-adjust the eyepiece diopter settings and refocus the image. Sometimes samples are easier to see in different circumstances. It may be difficult see much of anything when first starting to work.
3. *Close the field stop (FS)*: Close the field stop (FS) to its smallest size. This should be done while looking into the microscope, however, looking into the microscope is not necessary. Depending on the state of the microscope's alignment, you may not be able to see anything, or the image of the FS will be clear and visible in the center of the field of view. In either situation, close the diaphragm slowly and continue to make minor adjustments as needed. You may need to use the substage condenser at this stage of the alignment. It may be necessary to keep the image of the field stop visible at all times when establishing Kohler illumination.
4. *Focus the image of the field stop*: Move the substage condenser up and down until an image of the field stop is sharply defined in the oculars when the FS is in its closed position. The image will be focused when the edges look very crisp and appear black when focused by the condenser. The sample's image MUST also remain sharply focused when the field stop image is sharp. You may notice color fringes (red or blue) at the edges of the FS. This indicates the presence of some chromatic aberration. Chromatic aberration is a lens deficiency, where the various spectral components RGB are all brought to focus in different locations. Depending on how out of align the field stop is, this step of focusing the field stop image could be easy or challenging.

   Achieving parfocality of the specimen and the image of the field stop may be a bit more difficult in some situations, depending on the state of the alignment or the sample thickness. Parfocal describes a condition where all the images are in the same plane of focus. You may need to move very slowly, making small adjustments, e.g. closing the FS, or making minor height adjustments to the substage condenser. Be patient and study the field of view to determine the location of the field stop relative to the edges of the field of view. Once the image has been successfully located, it may be advisable to begin to center the substage condenser, which moves the image of the field stop. Move the substage condenser using the centration tools typically located just below, or on a bracket that holds the substage condenser under the stage. Move the adjust screws gradually and observe what happens to the image of the field stop. Move the image of the field stop roughly into the center of the field of view.
5. *Center the image of the field stop*: Use the centration adjustment screws on the condenser assembly and look into the microscope. They are typically two of them. Adjust both screws, either individually or together, to move the image of the field stop into the center of the field of view. If the image of the diaphragm is crisp, the correct condition has been established, if not, you may have to go back through each of the prior steps to improve

the result. The goal is to view a focused and centered image of the field stop that is superimposed over the focused image of the specimen when looking into the eyepieces.

6. *Open the field stop until it just leaves the field view*: Once the image of the field stop is focused and centered, the final adjustment for the FS can be made. Looking into the microscope and grasping the field stop, begin to open it. Bring the field stop to the edges of the field of view and, at this location, assess the centration of the image of the field stop to the edge of the field of view. If additional centering is needed, using the substage condenser centration knobs, make final adjustments. Once truly centered, open the field stop until it is just outside the field of view and is not encroaching into it. Sometimes this setting will be different for photography than for viewing. It is best to adjust the field stop for the requirements of the imaging system. You may also notice the image of the field stop will get less defined as it is opened. Focus should be assessed based on the central region of the field of view and not at the periphery.
7. *Adjust the aperture diaphragm*: Establishing the correct aperture diaphragm setting is probably the most important adjustment you can make. The next few paragraphs will describe how to create the ideal settings for any sample. The aperture diaphragm is located in the substage condenser but is visible in the body tube when an eyepiece has been removed. When adjusting the aperture diaphragm (opening and closing), it can be observed as it moves into the tube with an eyepiece removed. It can be very useful to move the diaphragm through its entire range and observe the brightness and location changes.

   It is important to have the diaphragm partially interject itself into the body tube. How far is the question. When the diaphragm is fully open, the diaphragm will not be visible in the body tube and when the diaphragm is completely closed, it is may cover more than 50 percent of the tube's diameter. A general starting point for the diaphragm setting is when 80 percent of the tube is open and 20 percent is covered. If the AD is outside of the tube, it will have NO effect on image formation and if the AD is closed to its smallest size, it will create significant contrast, increase the image's DOF, and SIGNIFCANTLY diminish image resolution.

   The aperture diaphragm controls image DOF, intensity, resolution, and contrast. The proper use of the aperture diaphragm will determine the characteristics of an image and will depend on the sample.

   It is possible to see how the aperture diaphragm influences resolution, contrast, and DOF. First establish proper Köhler illumination using a 10× objective and use a microscope slide of a subject that has easy to see structure. It is important NOT to use a weakly stained slide. Remove an eyepiece and observe the aperture diaphragm moving in the tube at the exit pupil of the objective once the sample has been critically focused. The image of AD should appear as a bright disk of light, and an image of the dark aperture diaphragm at the periphery of the tube should be visible when moving the AD though its range. Grasp the aperture diaphragm, look into the body tube, rotate the diaphragm, and observe the operation of the diaphragm within the microscope.

It bears repeating that that substage condenser also has a numerical aperture. When the aperture diaphragm is moved into the pathway, the numerical aperture of the condenser is reduced. It can be useful starting point to match the N.A. of the objective to that of the condenser. The diaphragm will have minimal if any effect on image formation when its N.A. matches the N.A. of the objective. Only when the diaphragm interjects itself into the illumination pathway will image enhancement or degradation occur. Put the eyepiece back into the microscope.

The aperture diaphragm plays a critical role in the characteristics of an image. The four image attributes controlled by the aperture diaphragm are.

- Image contrast
- Resolution
- Depth of field
- Brightness

The aperture controls these four outcomes by controlling the microscope's diffraction. The degree of diffraction required to create an optimized image will be influenced by a subject's characteristics. Blood cells, for example, do not exhibit any changes from use of the AD, however, other samples may be noticeably degraded.

The proper setting for the aperture diaphragm is dependent on the needs of the sample and based on visual judgments. A person just learning about microscopy will need to take the time necessary to learn how to see the effects of diffraction and then become proficient in mastering the power of observation. It is very easy to see contrast, but it is not so easy to see resolution.

The final establishment of the AD setting should be done when looking into the microscope. By rotating the diaphragm through its range and refocusing as necessary, details can change and should be noted. By studying the effects of the diaphragm situated at the edges of samples, a good balance of contrast and resolution can be found. Once established, critically focus the microscope. It may be necessary to re-focus throughout the process.

As the diaphragm is closed, the image will get darker, possess more contrast, and demonstrate more depth of field. As the diaphragm is more open, the image will be brighter, exhibit less contrast, and possess less depth of field, but it will have more fine detail. Somewhere in this range will be the ideal setting for the specimen being examined. Every subject will have a different requirement for the creation of optimal results. Be careful not to perceive contrast as resolution and detail. It is quite easy to add contrast to an image during image processing and it is much more difficult to form optical resolution.

Reflected Light and Industrial Microscopes

Very frequently, subjects are opaque and require the use of reflected lighting. Reflected light can be produced from a fiber optic light that is shone onto an opaque subject placed on the stage of transmitted light microscope. While not ideal, it is possible to examine opaque subjects in this way.

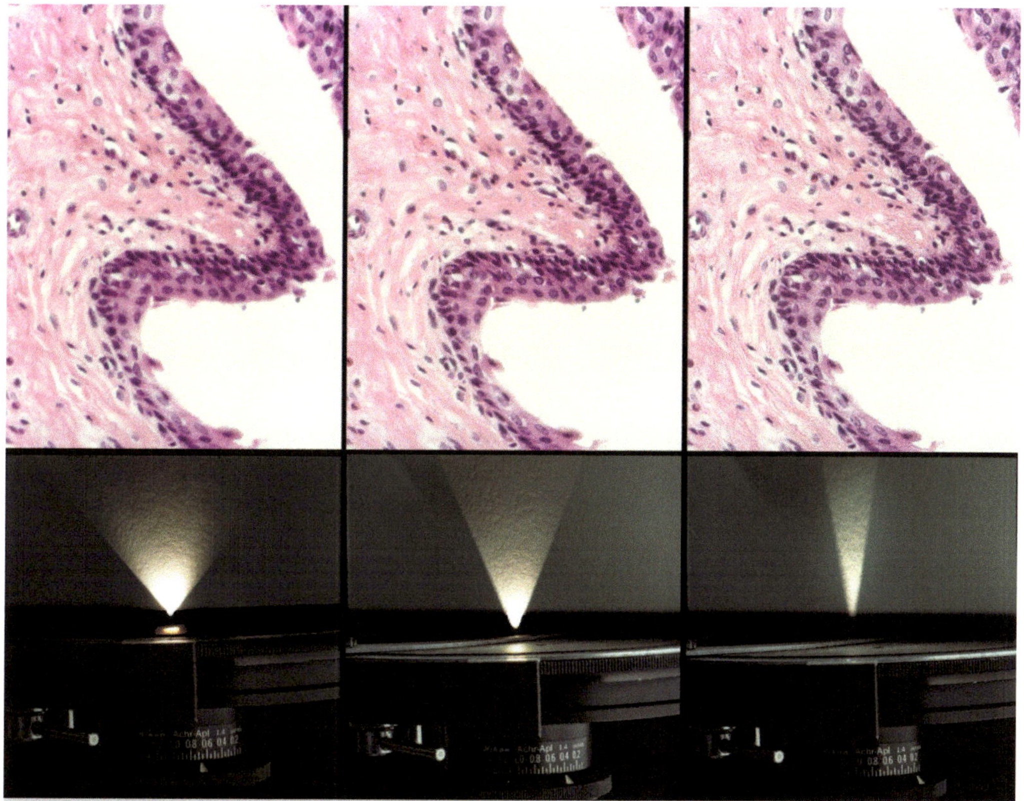

Figure 16.9 In these photographs, the aperture diaphragm has been set to three different positions. Each photograph was made using a 10× APO – N.A. = .45 objective, and an achro aplanatic substage condenser of N.A. = 1.40. The photograph on the left was made with the aperture diaphragm in its maximum opening for the 10× lens (N.A. = .45) creating the maximum resolution and least amount of contrast. Notice the angle of light leaving the condenser. The middle image was made with the N.A. set to approximately .3, which created the optimum balance of contrast and resolution for the objective. Notice the angle of the light becomes smaller when the aperture diaphragm is closed. The image on the right has the most contrast and least amount of resolution as a consequence of the aperture being nearly closed to its smallest setting, an N.A. of .15. The angle of light is also the smallest when operated in this position. Images courtesy of Michael R. Peres.

There are specialized reflected light microscopes sometime called industrial or metallurgical microscopes. These microscopes are used for inspection of metals, ceramics, and so many other materials.

Lighting that comes from above is called episcopic. Many reflected light microscopes produce both axial (specular) and darkfield (diffuse) illumination. These types of illuminations will reveal various features of the sample that otherwise might be suppressed. Episcopic illumination can also be polarized.

Microscopes that are used in industrial environments are required to be tough, and can take rough handling. Industrial users might examine textiles and fibers, packaging and related materials, electrical components such as capacitors, parts of medical devices, parts of

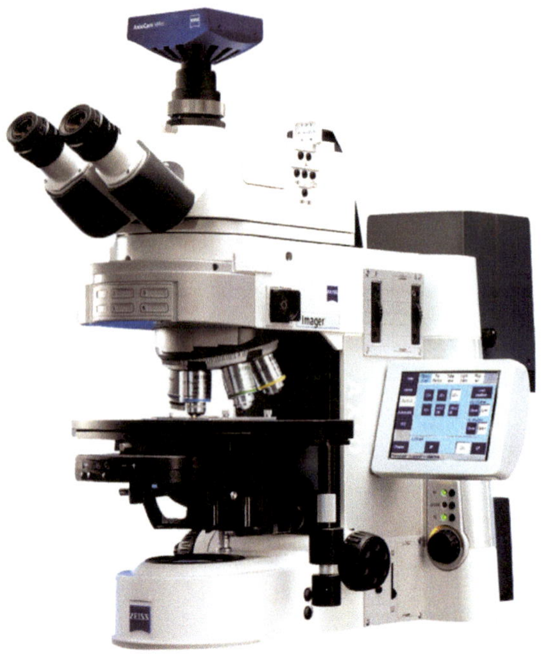

Figure 16.10 A reflected light photomicroscope. Notice the illumination box is located above the stage to the rear of the microscope. Image courtesy of Carl Zeiss Microscopy, USA.

motors such as turbines, or pieces of automobile parts. Sometimes using strobe lighting, industrial microscopes can find their way into machine vision systems.

Reflected light microscopes have all the same parts as transmitted light microscopes. These parts operate in the same way, however, there are additional components in the microscope to achieve specialized top lighting.

Darkfield Transmitted Light Illumination

Darkfield illumination produces images with high contrast, that appear white against a field of black. Darkfield microscopy requires the use of a special condenser but requires no special objectives. Creating a dark background has some challenges and good results are dependent on the right sample and its preparation.

A darkfield condenser illuminates samples using oblique angle lighting. Light that normally would illuminate the background (zero order) is subtracted from the system using an opaque disk located where the aperture diaphragm would normally be found. The diameter of the background illumination should be equal to or greater than the N.A. of the objective. Anything in the system that is transparent will refract light and will be illuminated, especially DIRT. Cleanliness in sample preparation is an absolute must.

There is noticeable light loss in darkfield illumination. The zero order of illumination is the brightest and strongest part of the illumination. The disk stop, if properly sized, removes the zero order illumination and results in low light. Exposure times may be longer than expected when compared to brightfield methods. Contrast is generated by this technique. Live and sometimes motile or unstained materials are a challenge to photograph because of subject movement and subsequent long exposures.

Establishing Darkfield Illumination

Köhler illumination is not used in darkfield. The brightfield condenser will need to be replaced with a darkfield condenser and the field stop must be opened to its maximum diameter. Diffused light is also required. The condenser must be centered and adjusted to its correct height in the system. This can be achieved by placing a small piece of white paper where the sample normally would be located. Adjust the condenser's height to the position to where the "circle" or "donut" of light produced by the condenser becomes a dot/point directly under the objective lens. When looking into the binoculars, center the illumination to cover the entire field of view evenly with light. A dry darkfield condenser works with 10× and 20× objectives. For lower magnification work, a typical darkfield condenser may

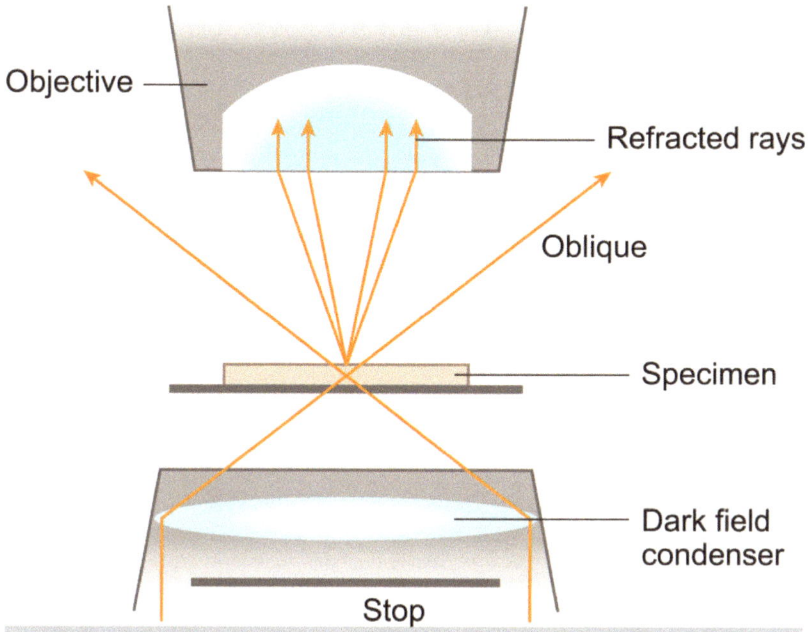

Figure 16.11 The basic lighting pathways used in a darkfield condenser. Oblique angle lighting is produced and shone through the sample. From oblique angles, illumination becomes trapped within the subject and light is refracted into the objective.

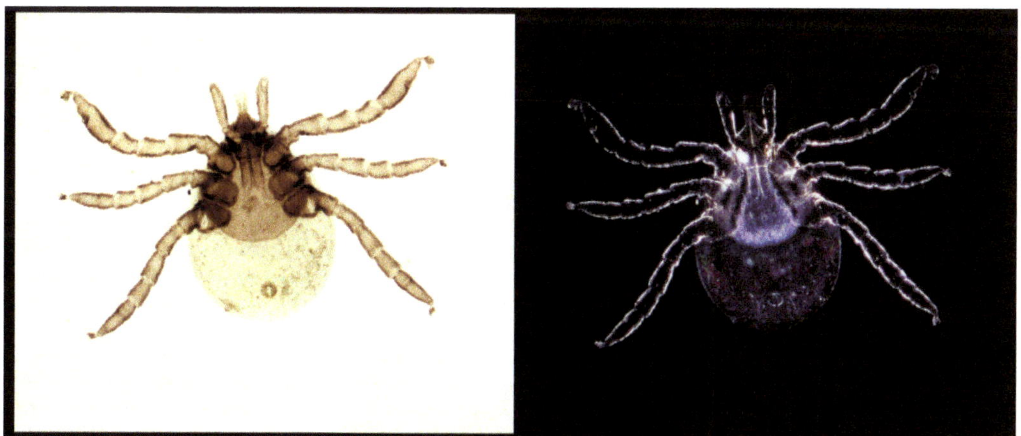

Figure 16.12 A brightfield illumination picture (BF) on the left and darkfield (DF) on the right. The image features an *Ixodes dammini* nymph. A 4× objective was used to photograph the samples. Images courtesy of Michael R. Peres.

not be able to cover the field of view of the objective with light. For applications requiring more magnification, e.g. a 40× objective, an oil darkfield condenser will be required. A darkfield condenser's numerical aperture MUST be higher than the numerical aperture of the objective lens. This is the exact opposite of brightfield methods.

Modifying a Brightfield Condenser for Darkfield Illumination

If a darkfield condenser is not available, it is possible to create oblique angle illumination using an Abbe condenser. Open the aperture diaphragm fully and insert your smallest finger into an Abbe condenser. This will create a sort of quasi-darkfield which is strikingly easy but requires you to hold your finger in the microscope. The means you have only one hand to use the microscope. An achro-aplanatic or flip-top condenser cannot be modified for DIY darkfield. The lenses in the condenser prevent access to the correct location within it.

It is also possible to create a central disk stop from a circle of black construction paper. Mount the stop into the condenser using a piece of plastic, or by using a non-permanent adhesive. The improvised disk stop needs to be located in the center of the condenser where the aperture diaphragm (AD) would be located, and perpendicular to the objective's optical axis. It is possible to determine precisely how large the disk stop should be by pulling out an eyepiece and establishing the aperture diaphragm setting that is equal to the body tube's diameter. Then remove the substage condenser and measure the size of the AD's diameter. A disk stop can then be cut and located where the AD would be located. Be certain that the aperture is fully opened when inserting the disk.

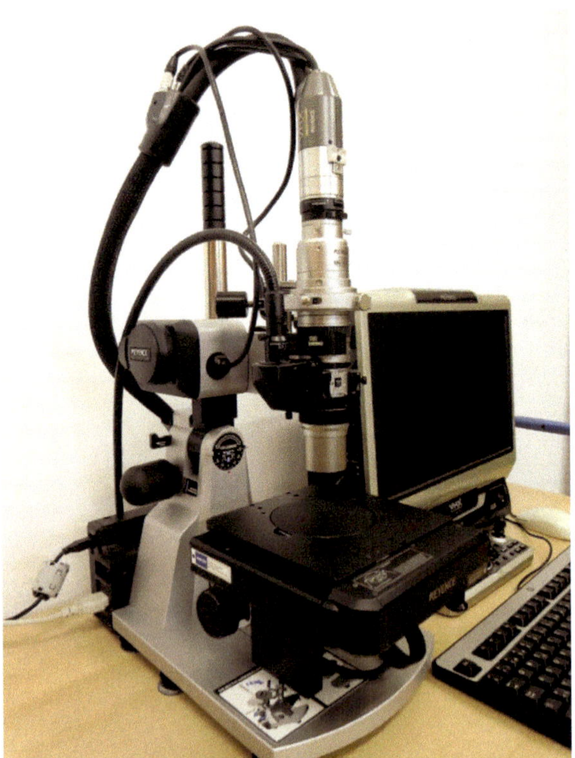

Figure 16.13 Keyence digital microscope. Digital microscopes have no eyepieces and use a single objective. The image is displayed on the computer monitor. Image courtesy of Keyence Corporation of America.

Challenges in Darkfield Microscopy

Darkfield microscopy creates high contrast images by using oblique angle lighting and the refractive properties of the sample. This method can become a challenge because, not only does the specimen refract light, but internal particles or air bubbles in the preparation will also be illuminated by the lighting. Time and attention should be paid during sample preparation to maximize cleanliness. Glass slides will need to be carefully cleaned, paying attention to dust and particulate materials. When photographing live aquatic materials using darkfield, it may be necessary to use deionized water and remove dissolved air bubbles before imaging. Samples should be prepared as thin as possible because too much sample will produce a challenge in DOF issues, because the object behaves like a light itself. When the entire sample is out of focus delineating a focused portion of the sample may be impossible.

Digital Microscopes

A digital microscope is not a traditional optical microscope. The image is not viewed through eyepieces, but rather a digital microscope displays its images on a computer screen and requires software to operate. Digital microscopes typically have in-built illuminators and their images are formed by digital devices for observation on a tablet, laptop, or large monitor. There is no way for direct visual examination of objects because there are no eyepieces. Contemporary digital microscopes are available both as a turnkey self-contained systems including all parts, or a microscope and controller that couples with a PC workstation supplied by the customer. While primarily reflected light microscopes, they can be used for transmitted light applications as well.

Digital microscopes are available as inexpensive USB models, and can range up through advanced industrial models that can cost from \$50K to more than \$100K. Many low-priced microscopes are similar to webcams in quality, but can be useful for field work and operated using a laptop computer which provides its power and software.

One of the earliest digital microscopes was made by a Japanese company in 1986. That company is now named Hirox. Its initial design required the use of a controller box. It was equipped with a zoom lens that was connected to a CPU. The product at the time used S-Video technology. Around 2005, Hirox introduced an all-in-one unit that came ready to use with a monitor and computer as a single item. Over the years, digital microscopes have become more compact and versatile, and they provide excellent portability for users who take their microscopes into factories.

The invention of USB communications technology has allowed a proliferation of these microscopes that vary in quality, performance, and cost, and they create relatively high-resolution images directly to a computer. The majority of these powerhouses are equipped with built-in LED light sources.

Digital microscopes allow for the inspection, image capture, and measuring of a sample's characteristics using one interface. Digital microscopes are very easy to use and are advancing material sciences and engineering. This is accomplished because of how sample properties can be investigated using visual and analytical processes. Digital microscopes allow for real-time 2D/3D measurements and other advanced image and sample analysis. The various measurements can be made directly to a screen, saved as image data, or used to generate scientific reports.

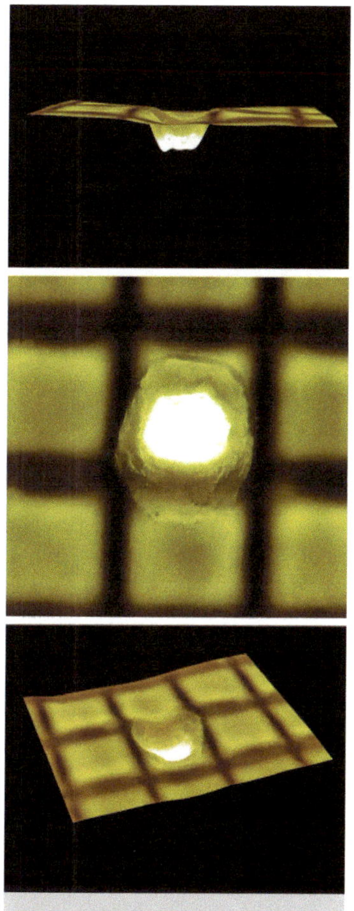

Figure 16.14 Three 3D images showing a common band-aid and a punched ventilation port through the plastic and gauze. Image courtesy of Ashley Crichton.

A digital microscope's magnification cannot be calculated using traditional methods. Instead the magnification is calculated based on magnification of the monitor's display. There are many monitor sizes and image magnification will vary depending on devices.

Stereo Microscopes

Stereo microscopes have replaced almost all photomacroscopes. It is rare to find someone still using a true photomacroscope for low magnification photography. Stereomicroscopes are faster to use and are certainly more flexible than a macro system. They provide a range of magnifications and they can quickly be adapted for use with both translucent and opaque objects.

A stereomicroscope is different from compound microscopes. While they have a primary objective and eyepieces, that is where similarities end. Stereomicroscopes do some things very well, however, may be limited for other reasons. One of the greatest advantages of a stereomicroscope is its ease of operation, and the image is oriented correctly to the way it is seen in the eyepieces. This means the left side of the object is represented by the left side of the image. When moving or dissecting an object, this orientation alignment can greatly simplify experiments. This makes this instrument ideal for dissection, cleaning, or manufacturing where operators can focus on the work and not worry about orientation. In all other magnified imaging systems, the image and the object are in the opposite orientation to one another. An operator must constantly remember this, especially when working with delicate materials that can be damaged in handling or when moving them incorrectly.

The objective in a stereomicroscope forms two optical images because there are two optical pathways in a stereomicroscope and the optical pathways forms two different points of view. This allows the image from a stereomicroscope to produce true stereopsis or stereovision for viewers. An image formed by a stereo microscope is said to appear in 3D. The microscope itself is located on a vertical column or post that allows its height to be adjusted. This adjustment allows the microscope to be used with objects of various thicknesses. The instrument can be quickly raised or lowered on the column and the objective's working distance can easily be changed.

There are two basic adjustments on a stereomicroscope. One adjustment is used for focusing or changing the working distance of the microscope, and the other is for changing the microscope's magnification. The focusing knob moves the instrument up or down and, unlike compound microscopes where the sample is moved, the stereomicroscope itself is moved and the sample is stationary. There is only one focus control used on a stereomicroscope, unlike a compound microscope where there is a fine and course focus adjustment. A stereomicroscope sometimes also uses a Barlow lens. This lens increases working distances and decreases magnification.

There are two ways that magnifications can be adjusted in a stereomicroscope. The initial magnification is dependent on the microscope's primary objective. Many stereomicroscopes have various objectives such 1×, 1.5× or 2× as the primary objective lens. In some microscopes, these objective lenses are interchangeable. This objective lens forms the initial stage of magnification. Many stereomicroscopes also have a zoom range of approximately 7–12 times. This adjustment is not fixed and allows an operator to have an entire range of fine adjustments available to optimize the amount of magnification. The viewing eyepieces also have magnification. These can range from 2.5× to 10×. When all the various components of a stereomicroscope are factored as possible elements, these instruments can create images with a magnification of ×2 up to ×250 at capture.

Stereo is achieved as a result of viewing the images from two optical pathways, however, the imaging system utilizes only one pathway. An image produced from a stereomicroscope is not stereo. In fact, the lenses by design are not installed at a 90 degree axis to the sample and this can lead to some minor challenges when managing imaging. Each objective is offset approximately 6 degrees from 90 degrees and produces an off axis view from the left and right side of the sample. When imaging, the image is formed from a slightly offset point of view. This can be easily addressed by positioning a sample at a similar angle of approximately 6 degrees away from perpendicular when the sample is underneath the objective lens.

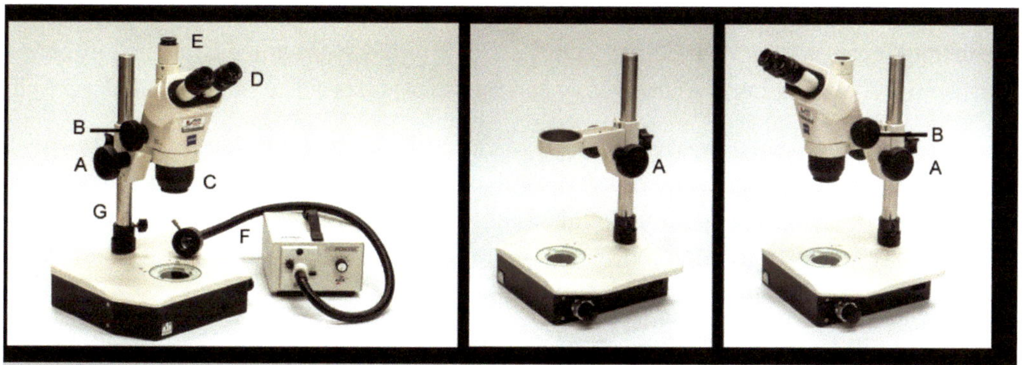

Figure 16.15 A stereomicroscope has different parts to a compound microscope. In this illustration the basic parts of the stereomicroscope are revealed in the three views. Part (A) is the focus controller which changes the working distance is shown here. Part (B) features the magnification changer. Typically, this control will have a 10× zoom range. Part (C) shows the primary objective. Part (D) features the viewing eyepieces and part (E) is the imaging eyepiece. Part (F) is the instrument column attached to the base that has a built-in illumination port. Images courtesy of Michael R. Peres.

The offsetting of the imaging pathway can create minor image defects since by design the lens is not optically centered over the sample. Because of advances in lens design, corrections and higher numerical apertures are more common in stereomicroscopes than in the past.

Photographing

It is possible to use a variety of cameras on a microscope. Both a DSLR and specialized microscope cameras with various capabilities have both advantages and disadvantages.

Specialized Microscopy Cameras

Instrument cameras have been designed to use on microscopes or telescopes. They require camera controller software such as Zen® from Carl Zeiss Microscopy or Elements® from Nikon. A camera and its corresponding software can be a powerful tool used to record, in some cases, more data than you can see. Almost any type of camera can be used for microscopy. At a basic level, all cameras are capable of collecting photons, recording them as variances of brightness, and then creating image files. Various features in the camera and its software create color images from the pixel information collected at capture. All cameras have shutters and a basic method for measuring the amount of light required to create a proper exposure.

Instrument cameras offer advantages when properly matched to imaging needs. They are relatively compact and offer various features across models ranging from very basic or those capable of working in low-light, have very high-resolution, or monochromatic cameras. The basic features of these cameras include preview or Live View, selection of exposure modes, exposure adjustment controls, sensor sensitivity adjustments, gain, binning, flexible digital resolution pixel (megapixels), white balance controls, image contrast capture modes, and adjustable dynamic ranges. Some cameras will have other unique spectral response capabilities, color adjustment features, bit depth settings, and other features such as image sharpening. High-end cameras frequently can measure and perform image analysis as well.

Once an image has been properly formed, it can easily be captured. Given the wide range of products and operating systems, it is NOT possible to share particulars for any given camera. Operating an instrument camera should be relatively straightforward. The basic software requires no special preparation for use, however, advanced software products can be initially overwhelming with their many features.

Once an image has been formed and directed to the camera, the camera controller software needs to be initialized. The image should first be focused on the display of the imaging system and not in the eyepieces. Instrument cameras must be operated in the Live View mode. The camera can be operated in fast or slow focus modes. When operating in the fast focus mode, typically half the camera's pixels are used for focusing. The capture image resolution is established separately. Some samples make it impossible to reshoot, and establishing the proper capture pixel resolution is an important initial decision. This determination must be made at the beginning of an imaging session. When time is short, using fast focusing makes sense. When time is not a factor, slow focusing allows for assessing better images that are more optimized and display more defined features of the sample. Microscope cameras offer various resolutions and bits per channel across the models and types. Some cameras

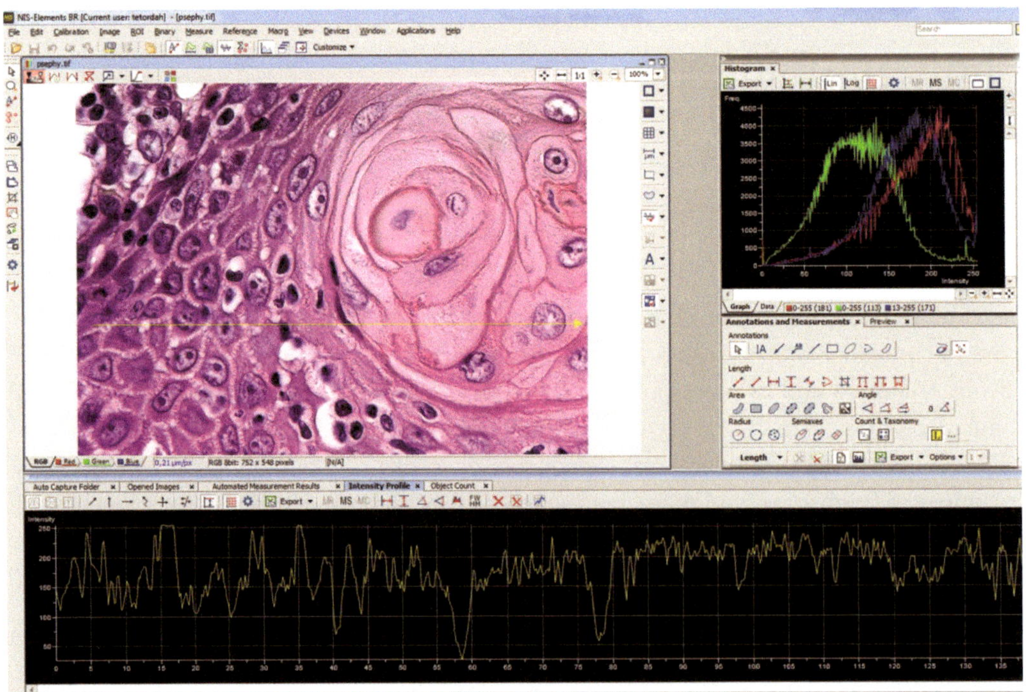

Figure 16.16 A photomicroscope's camera software is different to the menus found in a DSLR camera. This illustration reveals the appearance of a screen using Nikon Elements software. Included in the dialogue box is the Live View, color LUT feature and other important image data. Image courtesy of Torgeir Dahlen/BergmanLabora.

are equipped with piezo motors, allowing them to make multiple recordings that produce more highly resolved digital files.

Once critical focus has been achieved, and the optimized image attributes confirmed, the exposure can be finalized. There are two modes of operation for instrument cameras – auto and manual. Selections of one versus the other is based on preferences, or other contributing factors. In manual mode, the operator needs to select the proper time increment needed to create a correct exposure. The selected time will depend on a sample's brightness and other factors that are affected by the microscope and sample separately and together. Seeing the displayed image and corresponding histogram provides all the required information for the careful selection of an exposure time and confirmation of other image attributes. When operating in the manual mode, shorter times lead to darker images and longer times lead to brighter images. Manual mode can be very effective when working in fluorescence applications and offers tighter control of exposure.

Exposure

When a camera is operated in automatic mode, the exposure a camera produces will create images that exhibit an 18 percent gray tone. The average exposure setting is indicated by (N) normal, or (0). For brightfield microscopy it is suggested to increase the auto exposure using the software's exposure compensation control. A normal adjustment would be +1 or +1.5 E.V. (exposure value) when examining a sample that possesses average thickness and contrast. When working in darkfield illumination, the exposure compensation would be –2.0 E.V. Observing the image in preview mode is a useful tool for determining the best exposure. It is important to make exposure judgments from the brightest region of the sample. Images will be more acceptable when there are a few regions that are dark, rather than regions that are too bright and overexposed. An ideal exposure has detail in both the bright and dark tones. Sometimes there are controls to critically adjust the dynamic range of the sensor, allowing image optimization using the camera's software to achieve both.

A sample's contrast influences exposure. If the sample is very thin or has poor definition, a good exposure will require different considerations than if the sample is opaque and possesses dark tones, even when a microscope brightness has not changed. The sample's internal visibility or contrast is a due to the sample's thickness, the stain darkness/concentration, and the type of stain, for example. To make better images it is helpful to profile the capture contrast or modify the gamma of the sensor in the camera to match the sample characteristics. Some cameras will allow contrast to be changed in different ways. Samples that are thin or low contrast will benefit from a sensor setting having an increased contrast or gamma. For samples that are dark and thick, it is useful to select a low contrast setting. This feature will be located in the camera's software, and will be described in different ways. Modifying the shape of the image's histogram in the LUT (look up table) is one common method to optimize image characteristics. Other software controls refer to scenes and choices, which provide another way to optimize recording. These scenes might include fluorescence, brightfield, darkfield, or high, medium, or low. Carefully selecting exposure and contrast settings is an enormously important step needed to record the maximum data from a sample. RAW file formats are not typically found as choices in instrument cameras.

When using a camera for the first time, it is suggested to create exposure tests and evaluate how the various software features function. When evaluating new software, a good exercise is to create an exposure series going from −2, −1, 0, +1, and +2 using a typical sample and changing nothing else. It can also be useful to make a series of exposures containing various contrast settings as well. As experience is gained using the camera and its related software for various samples, it will become easier to record more and more information in future sessions. Like anything, practice makes perfect.

Color

A camera's color reproduction is controlled separately from a camera's white balance. Both can be changed. White balance describes the process of equalizing the data from a white target in a scene that has equal units of RGB information. When considering a transmitted light microscope, the clearfield or background is an excellent and uniform target for use in white balancing. Measuring and calibrating the sensor to this value can be useful for the creation of standardized images overtime. It is possible to apply white balance by averaging the whole scene or measuring a specific number of pixels in a region using the software's eyedropper. For brightfield images, a white point brightness of 240 in the RGB pixels is ideal. When averaging, no part of the stained sample can be located in the illumination pathway. The sample must be removed to establish the best white balance. Whatever method is selected, all will provide effective results consistently as long as the process is managed. Adopting best and consistent practices will create effective photomicrographs time and time again.

The amount, and the purity, of a color that is reproduced is often associated with image saturation. Sensors and software create color in slightly different ways, based in part on the characteristics of the filters used on pixels. For this reason, there will be variances across brands and models. Certain colors will reproduce with increased saturation or purity and other colors will be more correctly reproduced. When a sensor is more responsive to specific regions of a spectrum it is described as not being equalized. Pixels that are adjacent to regions where this situation is present, can be affected at capture. Decreasing the color saturation by a small amount can be an important way to manage all of the useful and precise image data that will be recorded. Some cameras will display saturation as color contrast. Proper management of this capture attribute can assist in the clarity of the image attributes. Sometimes adjacent pixels are affected by minor exposure errors as well. This image defect is called blooming. It is an adjacency effect. Blooming is characterized where pixels adjacent to one another contain exposure information gained from a neighboring pixel. De-saturating or diminishing exposure at capture can manage this affect and control how the volume of color is recorded. When managed, more sample structures will be visible.

Changing Sensitivity

In certain situations, there will not be adequate brightness to make a good exposure. To increase a sensor's sensitivity, gain is used. Increasing gain will cause the sensor to become more sensitive to light. Using too much gain, however, will cause the production of digital noise in an image. Because creating image artifacts is a negative outcome, increasing a camera's sensitivity should be chosen when there are NO other ways to make a correct exposure. For example, sometimes there are neutral density (ND) filters, or internal polarizing filters in the optical path, and these can be removed to make more brightness. If gain is required, select the least amount of gain to start and increase the increment as needed.

Another way to increase a camera's sensitivity is binning. Binning is the grouping of pixels together so they will act as one. By grouping pixels together, the individual pixels behave as one large image site. Larger pixels collect more light and create shorter exposures. The disadvantage of binning is that the number of pixels that create an image file will be lessened by the factor of the binning.

Many instrument cameras offer features such as image annotation, embedding magnification bar scales, or basic image processing. There are also many filters that can be selected for use at capture such as sharpening. Annotating can be very useful for later use and so can be the addition of the bar scales placed into image files. It is important to keep in mind that any image processing work done on a proprietary file format will not be compatible with image processing software such as Adobe Photoshop or GIMP (GNU Image Manipulation Program). Image J software frequently can open proprietary file formats but not all. Capturing and saving in a proprietary file format has advantages but also saving as TIFF can be practical. JPEG, while a universal format, compresses file data and should never be selected for capture in critically important work.

Using a DSLR on a Microscope

Digital single reflex cameras and many mirrorless cameras can easily be adapted for photomicrography. They will require an adapter for use on a microscope. There are advantages and disadvantages for their use on a microscope. Mirrorless cameras function in much the same way as a DSLR but do not use a reflex viewing system and are more compact. For this reason, they have become very popular for travel and applications where the size of the camera matters. Many of the newest models are also excellent for use with video. If the camera has a movable LCD screen it is advantageous, since it may be challenging to stand above the camera and observe the display.

DSLR Camera Advantages

There are many camera settings that can be used effectively for optimizing the recording of image data. These include:

- Variable ISO settings and noise reduction management.
- High resolution.
- Portability and ability to use cameras for other imaging applications.
- Color reproduction and color management choices.
- Various file format choices including RAW.
- Live View and video capabilities of high-end cameras.
- Bit depth options.
- Operated using battery or AC power.

DSLR Camera Disadvantages

- Source of vibration when used in magnified imaging applications.
- Coupling to a microscope.
- Frequent set-up and tear down of camera from microscope.
- Direct Focus into the viewfinder.
- Changing menus settings on a small screen.
- Battery consumption when not using AC power.
- Delayed shutter activation when operating in mirror lock-up mode.

Attaching and Operating the Camera

Coupling a DSLR camera to a light microscope is another process. It is possible to suspend a DSLR camera over the microscope's photo eyepiece using a vertical copy stand or tripod. The lens should be removed. When a camera is suspended, it is practical to use extension tubes or a bellows to act as a light baffle required to block ambient light. Ambient light will create flare and other image defects when used in this way. Flare will lower image contrast and create poor definition or visibility. If an extension tube or a bellows is not available, it is possible to make a tube using black construction paper. The sensor will need to be carefully aligned over the eyepiece and the sensor will need to be perpendicular to the optical axis. Using a spirit level or smartphone leveling app can be useful as an aid in making a good alignment. The sensor needs to be parallel with the microscope's stage. The camera should be at the distance from the sensor where the circle of illumination produced by the eyepiece is large enough to fully cover the sensor and not produce a circular image. The formation of circular images is not desirable for research, but can be fun for artistic work. Moving the camera further from the eyepoint will increase the circle's size and image magnification. When moving the camera further from the eyepiece, the image will become dimmer and the circle of good definition will become larger.

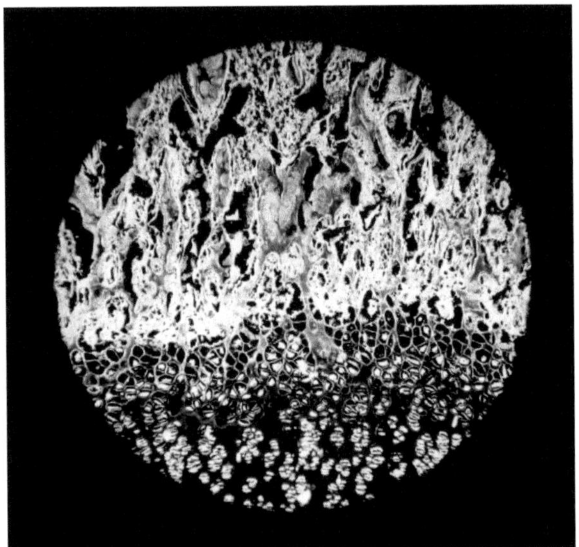

Figure 16.17 When an image is relayed from the eyepoint to the camera, the projection distance must be large enough to produce a circle of good definition that is larger than the sensor. If the distance is too short, the circle of good definition will not fill the rectangle of the sensor, leading to circular images. This circular image features a cross-section of human developing bone using darkfield illumination. Image courtesy of Michael R. Peres.

If a more structured system is desired, coupling the camera to the microscope will require an optical and/or mechanical couple. Many manufacturers of microscopes such as Nikon, Olympus, or Zeiss sell adapters for DSLR cameras. There are many configurations as you might imagine, and it is necessary to correctly select a coupler for a specific type of camera as well as the type of microscope. C-mounts, bayonet mounts, or other particular mounts can be suggested by a vendor. Coupling might be considered ideal, however, ensuring the optical elements in the optical coupler are of the proper quality should also be evaluated before purchasing. "You get what you pay for" will define the characteristics of many optical elements. Inferior optical elements will diminish image quality significantly. A full sensor digital camera will require a different magnification than a cropped sensor camera and some optical couplers will have lenses, and others will be without glass elements.

Focusing the image in a DSLR camera can be accomplished in several ways. It is possible to project the microscope's image directly into the DSLR without a lens and focus in the camera's viewfinder. Seeing a well-defined image can be challenging. The characteristics of the DSLR viewfinder make seeing fine detail of the magnified image difficult. The image may never appear really sharp or crisp in the viewfinder. It is also important to set the camera's viewfinder reticle setting for your eye. There will be an eyepiece diopter adjustment control, which establishes best near to far focus setting for each operator.

It might be possible to focus an image using a DSLR camera's Live View mode and for mirrorless cameras, this is expected. Live View displays the image from the sensor directly onto the computer or LCD screen. This allows precise viewing, focusing, and composing. In a dimly lit room, this can be very helpful for precise control and location of critical definition within the sample. A small disadvantage of Live View might be that the preview window for live viewing might be smaller than desirable when photographing dynamic events. And more importantly, there will be a delay in triggering the shutter when changing from Live View to opening the shutter. Triggering a camera can be a source of vibration and needs to be carefully managed.

Vibration is a very real problem in photomicrography. Vibration can be produced by many sources within a camera, as well as from environmental sources. Vibration will make images appear not sharp. These images maybe focused but appear blurry because of image movement during exposure. Computers and internal fans used in light sources can create vibration if they are situated on the microscopy table. Fiber optic lights can also have fans used for cooling a hot bulb and these can be a source of vibration. There can also be environmental sources of vibration including elevators within buildings, subways that go under buildings in major cities, as well as sometimes the powerful HVAC systems located on the roof of a building. Anti-vibration tables are ideal solutions to isolate vibration, but are very expensive. There are vibration reduction settings in some cameras and VR lenses, which are more common. Image processing can be somewhat effective in removing some of the artifact of image shake but is not an ideal solution.

DSLR cameras themselves can be a source of vibration. Simply depressing the shutter can jar the camera and create image shake. A DSLR camera uses a mirror that allows for the viewing of the image in the reflex viewer. When the shutter is depressed, the mirror moves out of the way and the focal plane shutter opens and closes, allowing the light to strike the sensor. All of these mechanical processes can have significant influences on image stability and can produce vibration. The mirror will create camera shake and the movement of the focal plane shutter also can make vibration. Minimizing the effects of vibration in the image can be controlled through the use of slower shutter speeds rather than by using shorter times.

There are few things that can minimize the production of camera vibration. Triggering the camera from the computer and not using the shutter depress button directly is a good place to start. If tethered, the shutter can be activated using Adobe Lightroom or other camera

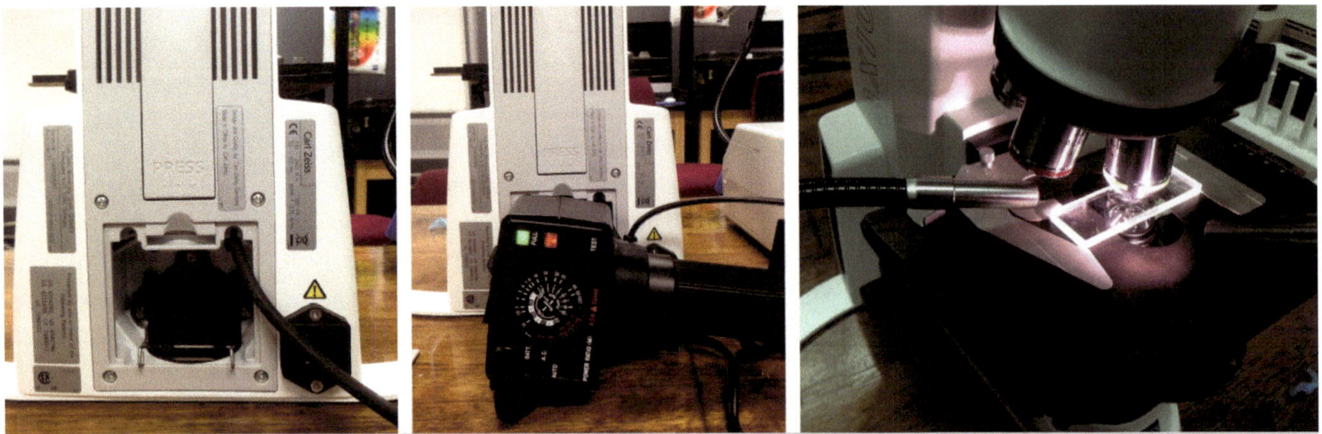

Figure 16.18 The microscope's built-in illuminator was removed (left) and a hand-held electronic flash placed where the microscope's light would normally be located. On the right, the stage and sample are illuminated with a fiber optic light, used for focusing. Images courtesy of Michael R. Peres.

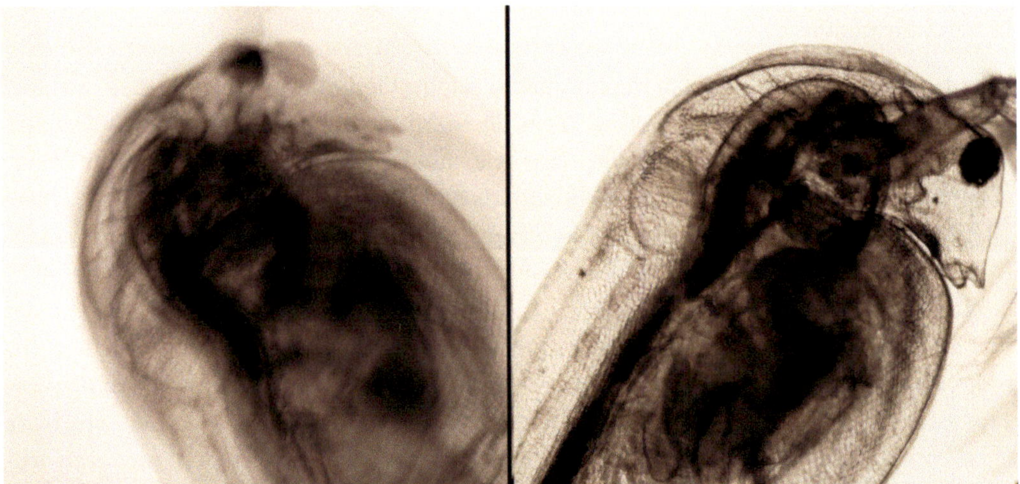

Figure 16.19 This picture shows the improvement that can be expected when changing from a continuous light source to a short duration light source for photographing highly active aquatic invertebrates. Images courtesy of Michael R. Peres.

controlling software. You could also consider using the camera's self-timer. This will also dampen any vibration caused by the operator. It can be practical to use an electric remote shutter release, too. Any or all of these three approaches will be helpful in controlling vibration that is produced when firing the camera.

Probably the most useful tool for vibration control using a DSLR camera is by selecting the camera's mirror lock up feature. For mirror lock up, focusing is accomplished first and then the mirror is locked up. Once the camera has settled down, the exposure can be made. An image that is vibration free is a great accomplishment using a DSLR camera. Sometimes it is suggested to make multiple photographs of a sample using the same settings, hoping that one of the files will be slightly sharper than the others.

Case Study: Photographing Live Material using Electronic Flash on a Microscope

Photographing live materials using a microscope presents numerous challenges. Since magnification is a multiplicative process, object movement is amplified by the microscope's magnification. It is possible to slow an organism's motility and/or activity, but they are still moving. A far better way to stop motion is to use an electronic flash as the light source.

In this example, *Daphnia* sp. was photographed. The immature organisms were placed into a welled microscope slide using several drops of de-ionized water. No cover slip was added to the wet mount. The continuous light source must be removed and replaced using a hand-held or small studio electronic flash unit. In certain circumstances there will be no modeling or focusing light. Fiber optic lights can be used to provide focusing illumination.

Suggested Reading

Practical Digital Photomicrography: Photography Through the Microscope for the Life Sciences, by Brian Matsumoto. Rocky Nook, 2010.

An Introduction to Digital Photomicrography, by Brian Matsumoto and Carol Roullard. Crowood Press, 2017.

Photography with a Microscope, by Frederick William Dickes and Fred Rost. Cambridge University Press, 2000.

Chapter 17 Photographing Small Animals in Captivity

Charles Hedgcock

Introduction

Although photographing small live animals in situ is considered "best", and certainly more fun, it may be required to sometimes photograph them under captive conditions. Whether photography takes place in the laboratory, controlled field conditions, or in a mobile field studio, keep in mind that the well-being of these subjects is critically important and can be challenging. Keeping the specimens alive and healthy should be your first objective. The photographer and assistants must take great care handling the animals properly while in captivity. Wild subjects should be retained only as long as necessary. The photography should be accomplished efficiently and then the animal returned back to its place of capture and released. This will minimize the risk of injury or death of the subject.

Many studio techniques described elsewhere in this book can be adapted to small live subjects as well. The difference between lighting a Volkswagen® Beetle and lighting a scarab beetle is mostly a matter of scale. You should think of the subject first as a surface to be properly recorded. Once that assessment has been made, you can design and build the necessary enclosure with its containment and create lighting that is specific to the subject's particular needs.

When photographing small, highly mobile, and often unpredictable subjects, two key ingredients for successful imaging outcomes are being knowledge about the subject, and having a lot of patience. You should have an understanding of the animal's behavior, or work with someone who does. In addition, knowledge of the animal's environmental preferences will allow for proper containment choices, and the positioning of props for more biologically accurate and naturalistic settings. Understanding animal behavior and preferred conditions also allows for the anticipation of the subject's movements that will create more satisfying images.

The fact that these live specimens have been photographed in controlled conditions should be made clear in all notes and metadata associated with the images. An image of a contained animal should never infer the animal was photographed in the wild, if it was not. You can read more about this in Chapters 4 and 13.

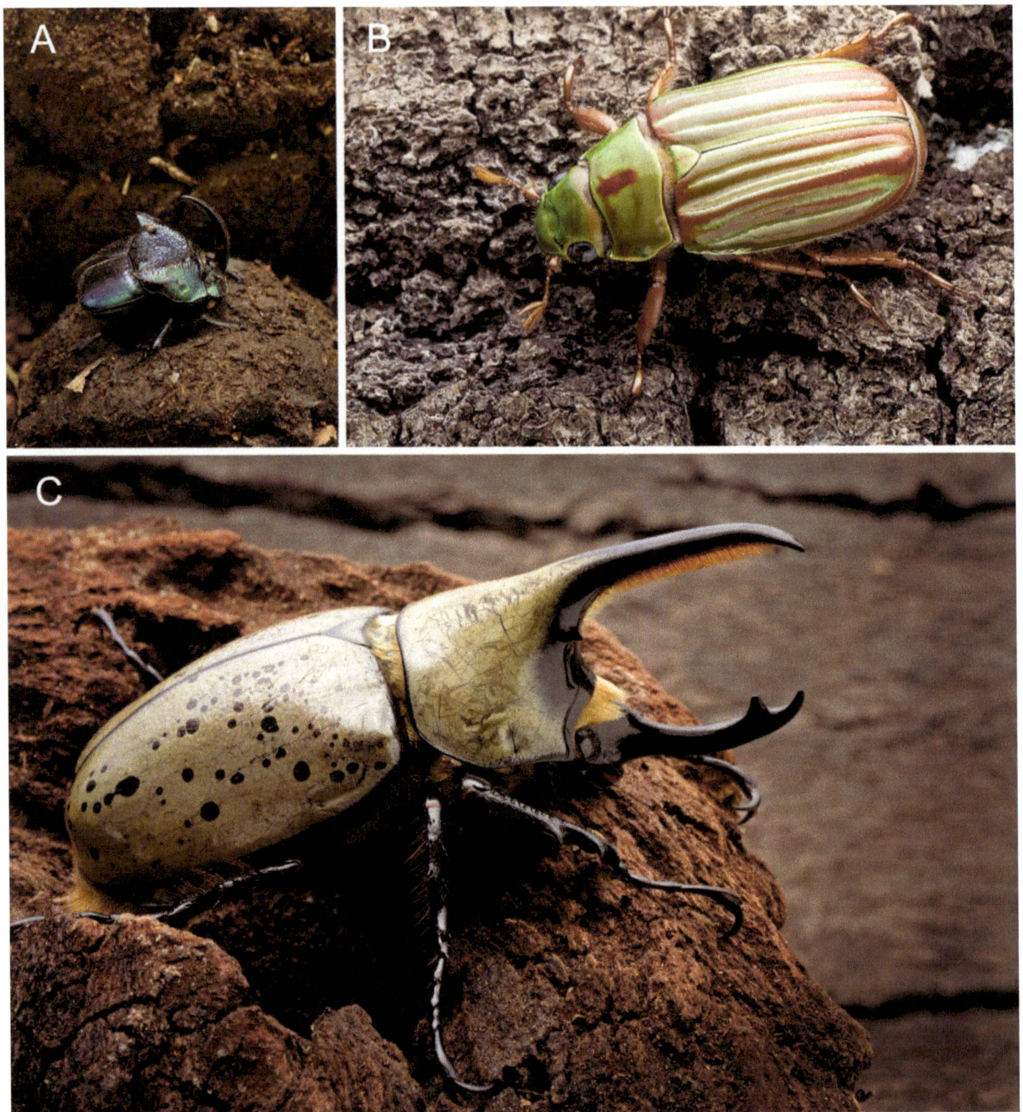

Figure 17.1 Scarab beetles, in the family Scarabaeidae, a very diverse and often beautiful group of insects. The difference between lighting a scarab beetle and lighting a Volkswagen Beetle is mostly a matter of scale. Both will benefit from a large diffused light source. Image (A): Male dung beetle, *Phanaeus quadridens*, Mesa Tres Rios, Sonora, Mexico. Image (B): Jeweled scarab, *Chrysina adelaida*, Mesa Tres Rios, Sonora, Mexico. Image (C): Male Western Hercules beetle, *Dynastes granti*, Rincon Mountains, Arizona, USA. Images courtesy of Charles Hedgcock.

Careful Handling of Living Organisms Maximizing Their Safety and Yours

Amphibians such as frogs, toads, and salamanders have very sensitive skin that must be kept moist to prevent rapid desiccation. Having clean hands that are free from chemicals, perfumes, insect repellant, or other compounds will prevent absorption of these potentially harmful compounds through the amphibian's skin. Additionally, some amphibians may exude toxins themselves, so you should wash hands after any handling of the animals, and avoid touching sensitive areas on your face including your eyes or mouth. For some species, powder free surgical gloves may be necessary for handling. Paying careful attention to handling the animal ONLY when necessary and keeping your hands clean will protect both the subject and the photographer.

Working with a knowledgeable assistant is helpful for managing the animal, will allow you to follow the animal, and have the camera ever-ready for responding to the perfect pose. It will be a random event.

Though most small subjects are not venomous, many subjects can, and will, bite. A painful bite may cause you, or others on your team, to react in a manner that endangers the subject and/or allows it to escape. More than once I have seen inexperienced handlers bitten by large, non-venomous lizards. The surprised handler shrieks and vigorously shakes the finger with the lizard still attached. As you can imagine, this sends the animal flying through the air. In many settings this outcome can injure the subject or allow it to escape. It may disappear into nearby shrubs or rodent holes, under laboratory cabinets, or worse yet, out the door into your facility.

A variety of tools are helpful in managing, wrangling, and photographing small subjects when in captivity. Forceps of various types and sizes are useful for reaching into aquaria and other containers to position props, clean out debris, or for placement and removal of the subject. A stout piece of wire approximately 30cm long – a bent coat hanger wire works quite well – with a small hook on the end, allows the easy maneuvering of small invertebrates, arranging lizards so that the tails will be in the camera frame, and managing small snakes. The end of this hook must be absolutely smooth to prevent any possible injury to the subject. Place your enclosure on a "Lazy Susan" turntable allowing the enclosure and the specimen to be rotated, without needing to move about excessively or rearrange lighting. Medium sized spring clamps, available at hardware stores, are a simple way to hold up fill cards but can serve many other useful purposes in the studio.

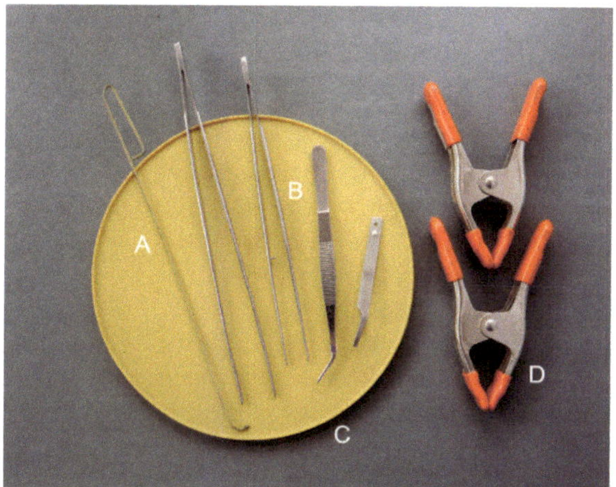

Figure 17.2 A variety of tools will be helpful in managing/wrangling/photographing small subjects in captivity. (A): a small hook made from a stout piece of wire approximately 30cm long, for maneuvering invertebrates, managing small snakes, and arranging lizards so that the tails will be in the camera frame. (B): various types of forceps and sizes useful for positioning props, cleaning out debris, and for placement and removal of the subject. (C) a "Lazy Susan" allowing the enclosure to be rotated and minimize the need to re-arrange lighting. (D) a variety of medium sized spring clamps that serve many useful purposes in the studio. Images courtesy of Charles Hedgcock.

Venomous arthropods such as scorpions and centipedes should be carefully contained and handled or maneuvered only with tools to avoid being stung or bitten. For oblique, low-angle views of subjects such as scorpions, a system of double containment may be helpful. A small shallow tray of soil and other environmental props allows the placement of the camera at a low, near ground-level angle. To discourage escape from this shallow tray, it may be positioned atop glass jars or lab glassware – use your imagination – inside a larger, deeper, and empty tray. Should the subject flee the shallow tray of substrate, it will fall down into the deeper tray, allowing for easier re-capture.

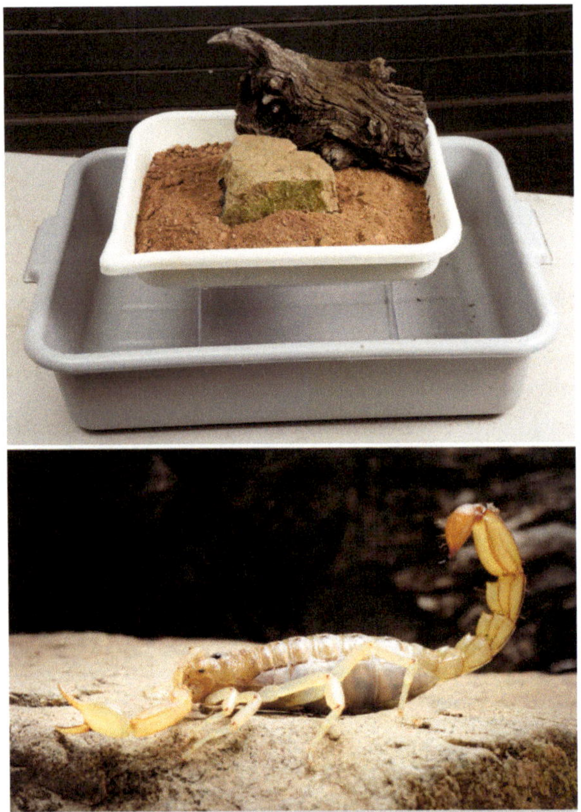

Figure 17.3 With oblique, low-angle views of subjects such as scorpions, a system of double containment may be helpful. Here an empty plastic aquarium is used to hold a shallow tray of soil and props atop a larger, deeper tray. Should the subject flee the shallow tray of substrate, it will fall down into the deeper tray below, allowing for much easier re-capturing. Arizona striped-tail scorpion, *Paravaejovis spinigerus*. Images courtesy of Charles Hedgcock.

Venomous snakes and lizards must be carefully contained. It is recommended to have at least one highly experienced wrangler to manage the animal and make sure that you do not get too close to the subject. There should be no spectators during a snake photo session. This will minimize bite risks and additional confusion or stress.

Fast moving specimens may be quite difficult to photograph. This will be a test of your patience. For subjects that are uncomfortable out in open spaces, or that become agitated during photography sessions, it may be useful to create a temporary covering to help them calm down and feel more secure. Creating a small hiding place is easy to accomplish by placing a small piece of cardboard or similar material with bent corners in the chamber. This will create a low, protective space for a nervous subject to retreat under. After a few moments and when the subject has had time to adjust, the cover can be lifted carefully with forceps and, for a short time, the subject may be photographed in its more relaxed and natural state. You will need to work quickly because the subject will likely not remain relaxed for long! This approach may need to be repeated several times.

On very rare occasions, subjects must be cooled down prior to photographing them. It is best to avoid anesthetizing animals by placing them on ice or into a refrigerator, since this approach will put them at risk and lead to unnatural behaviors, posture, or coloration. It is better to rely on knowledge of animal's behavior and implore your inner patience.

Selecting a Background

Natural backgrounds can be aesthetically pleasing and provide additional information about a subject. This might include plants and the types of soil associated with the subject's natural habitat. This approach provides visual hints about habitat types and elevation. Background materials might include just a few leaves, a bit of tree bark, or a few stones. These items should be collected at the time of capture and are useful in providing the most accurate information possible about habitat. If you frequently photograph subjects from particular habitats, it is practical to keep materials on hand in the laboratory or studio for re-use when needed.

Solid or simple backgrounds can provide a less cluttered view of the subject. A white or light gray background leads to a simplified representation of the subject, and is often preferred for scientific journals or posters produced for scientific conferences, leading to simplified representations of the subject. Various materials can be effective as a solid background, but avoid most paper products because they often will have a textured surface. Smooth, blemish free surfaces are generally preferred. One-quarter inch acrylic panels work well and, if

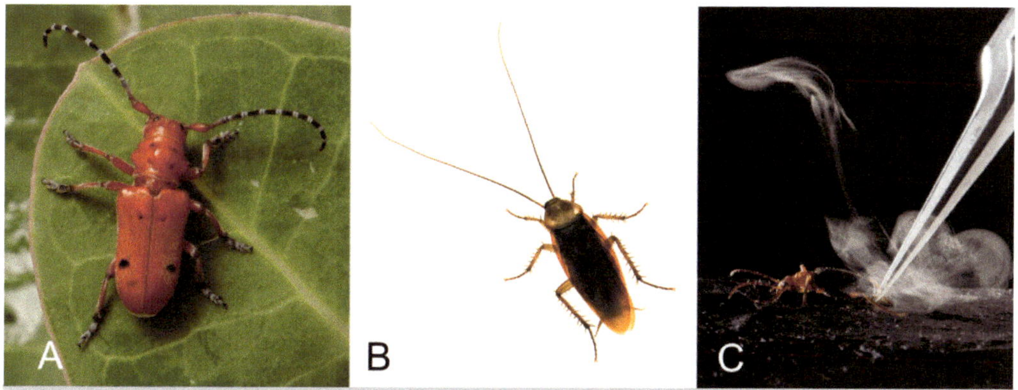

Figure 17.4 You should select a background based on the expected use of the image. (A): Red-femured milkweed borer, *Tetraopes femoratus*, photographed on its host plant. (B): American cockroach (*Periplaneta americana*), photographed on a white background for a scientific paper and poster. Careful exposure prevented the loss of the fine detail in structures such as antennae and the spines on the legs. (C): Bombardier beetle, *Brachinus elongatulus*, photographed on a black background with rim lighting to show the toxic spray of the beetle's chemical defense. Images courtesy of Charles Hedgcock.

protected from scratching, can travel well. This material will last for many field or studio sessions. Other materials such as paint samples from hardware stores, a piece of erasable white board, or scraps of laminated counter-top material have also been used successfully. Small blemishes or other imperfections in the surface may be removed during digital post-processing without imparting false information to the science. Careful exposure control will prevent a bright white background from becoming over-exposed and diminishing the fine details of the subject such as antennae, legs, and feet.

Black backgrounds can be quite dramatic but need to be handled carefully because they may cause dark subjects to blend into them and visually merge with the background. Taking extra care with lighting and positioning will help prevent this. It may be helpful to light the subject from above and behind, providing a rim light that separates subject from background.

Lighting Small Living Subjects

Lighting of small living subjects is best accomplished with electronic flash units. Small, relatively inexpensive battery powered flash units are easy to position, will freeze most animal motions, and provide a great deal of light brightness, allowing you to choose small apertures that produce adequate depth of field. White reflector cards may be used to bounce light into shadow areas. This allows for the use of a single light source and eliminates cord clutter in what often is a small working environment. While the textured surfaces of some subjects are better revealed with a direct, undiffused flash, many subjects are better captured using diffused light. A collapsible 20 inch × 24 inch diffuser can be easily made by using PVC pipe sections and a sheet of white rip-stop nylon with elastic corners. By collecting an assortment of white foam-core reflectors and small stands, a system can be adapted to work for most small subjects. These can be set up quickly in the laboratory, the field studio, or on the tailgate of a pickup truck. More on lighting can be read in Chapters 9 and 14.

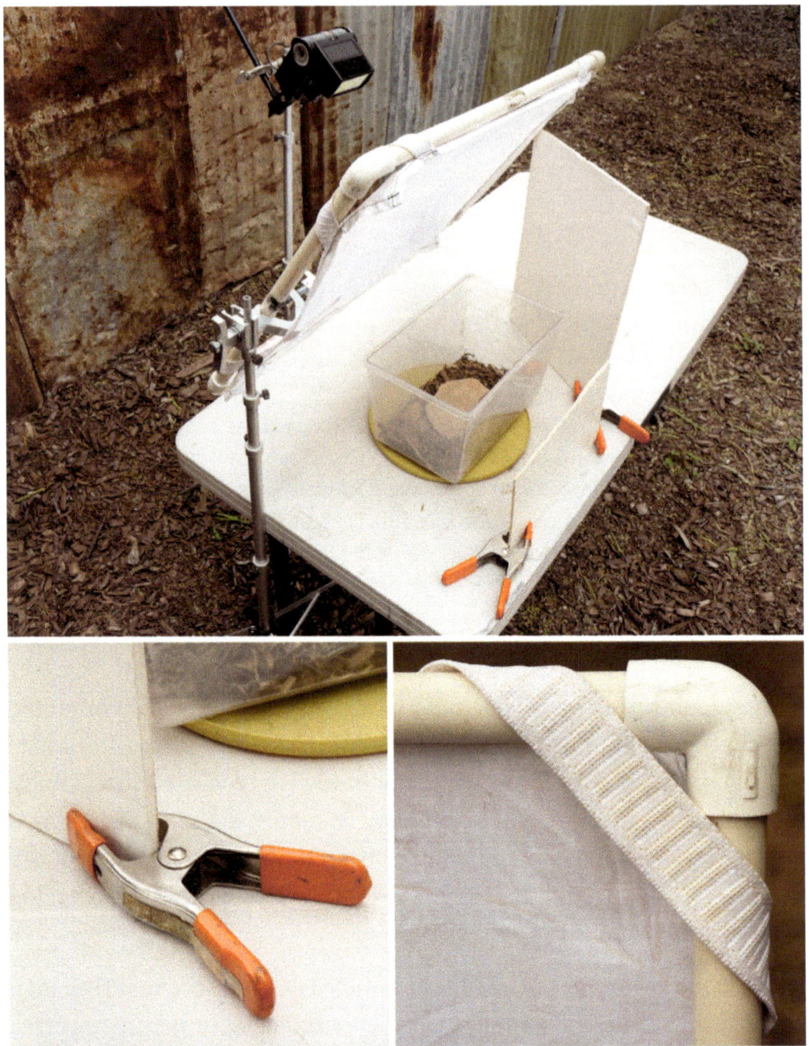

Figure 17.5 A simple portable studio that can be set up quickly in the laboratory, the field studio, or on the tailgate of a pickup truck. This system uses a single battery powered flash unit and a rip-stop nylon diffuser with elastic corners stretched onto a 20 inch × 24 inch PVC pipe frame. The diffuser breaks down easily for transportation and storage. White reflector cards, supported by spring clamps, are placed around an aquarium or other container to bounce light back into the shadows. The container is placed on a Lazy Susan to allow the photographer to rotate the enclosure, and the specimen within. Images courtesy of Charles Hedgcock.

Scale

Including a scale when photographing biological subjects is often expected. Do not include coins, pocket-knives, or other random objects with unknown dimensions to show scale. Rather than cluttering the image with foreign objects, carry a small thin metric scale in your camera bag for this purpose. The scale should be placed on an axis with the subject, and will provide genuinely useful information. If located near the edge of the frame, it may be cropped out at a later time for aesthetic purposes. Photographing gentle and safe-to-handle

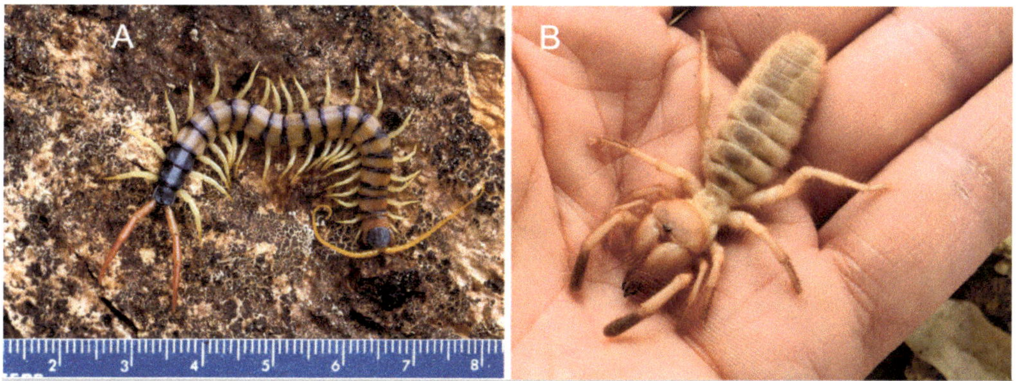

Figure 17.6 A scale should be included when photographing biological subjects. Photographing gentle and safe-to-handle animals in your hand can also illustrate scale. (A): Tiger centipede, *Scolopendra polymorpha*, Tucson Mountains, Arizona. (B) Solpugid, *Eremorhax pimanus*, Sierra Huérfana, Sonora, Mexico. Images courtesy of Charles Hedgcock.

animals in your hand can illustrate approximate scale and creates a less intimidating portrait of subjects that some viewers may find frightful. As always, it is important to consider the photographer's and animal's safety first and foremost.

Controlled but not Contained

Some larger and/or slow-moving subjects, while photographed in controlled conditions, may not require much containment at all. Larger beetles, for example, may be reluctant flyers and background props can simply be placed on a lab bench or other convenient surface. The beetle is then placed on the background and followed as it walks about. An assistant may be helpful to be ready with a quick hand or insect net to capture those that decide to fly after all! Large stick insects and caterpillars will generally cling to a stick or other prop and can be photographed quite easily in the studio. Here it may be useful to work one handed, turning a piece of bark or stone where the subject may be roaming, with your other. Terrestrial snails, for example, may be photographed as they crawl slowly about on a stone held in one hand while operating the camera with the other.

Photographing in Aquaria

Inexpensive and readily available commercial glass aquaria make excellent enclosures to photograph a variety of small subjects – both terrestrial and aquatic. Lizards, small snakes, insects, and a variety of aquatic organisms can all be readily documented in these clear vessels available in most pet supply stores. The clear glass of the aquarium allows for easy lighting from outside the enclosure and as a benefit, the smooth glass will make it difficult (though not impossible!) for most subjects to escape.

Plastic aquaria are easily transported safely for use in a field studio and have a lesser chance of breaking than glass. They can be purchased in an assortment of sizes that can be nested together for transporting. They are only suitable for photographing using a downward view

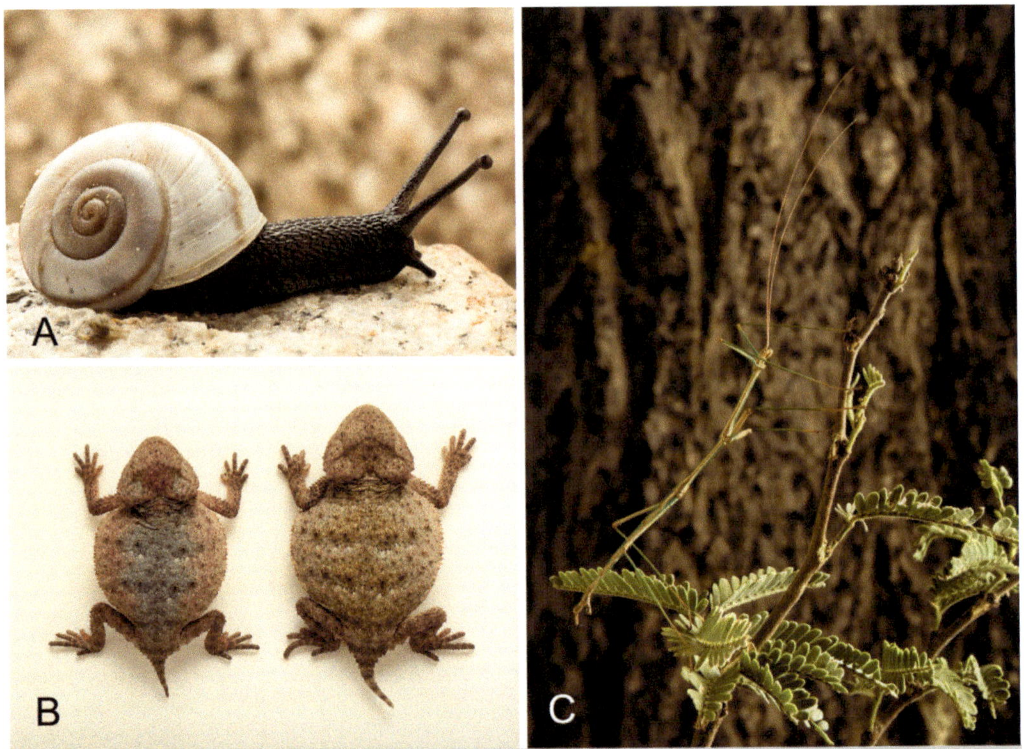

Figure 17.7 Some subjects, when photographed in controlled conditions, may not require containment. (A): Eastern desert snail, *Eremarionta rowelli*, photographed while the snail crawled on a stone held by the photographer in the field studio. Tinajas Altas Mountains, Barry M. Goldwater Air Force Range, Arizona. (B): Rock horned lizards, *Phrynosoma ditmarsi*, photographed loose on a table top in the field studio. Sierra Manzanal, Sonora, Mexico. (C): Arizona walking stick, *Diapheromera arizonensis*, photographed uncontained while clinging to a mesquite branch in the field studio. Rancho Puerta Blanca, Sonora, Mexico. Images courtesy of Charles Hedgcock.

through the open top because the plastic sides are prone to scratches and create optical distortion. Used in a field studio, they are practical for photographing many subjects, and can be used to contain animals temporarily while preparing to photograph them.

A word of caution: mammals, and even some insects, may chew their way through plastic screened tops often used on aquaria. Snakes, if kept too long in such containment, may also rub their rostrum raw trying to find a way out. If snakes must be held overnight, it may be best for them to be kept in a securely knotted snake bag that is placed in the aquarium for safe keeping, rather than allowing them to be unrestrained in the aquarium where they may injure themselves trying to find a way out. When in the field, care must be taken when storing animals overnight so that they don't fall prey to nighttime camp raiders such as raccoons and skunks.

Glass aquaria are breakable which makes them a delicate tool for use in a field studio. They are however relatively inexpensive, readily available, and come in an assortment of sizes. Custom aquaria can be made from sheets of glass cut to size and glued together with silicone

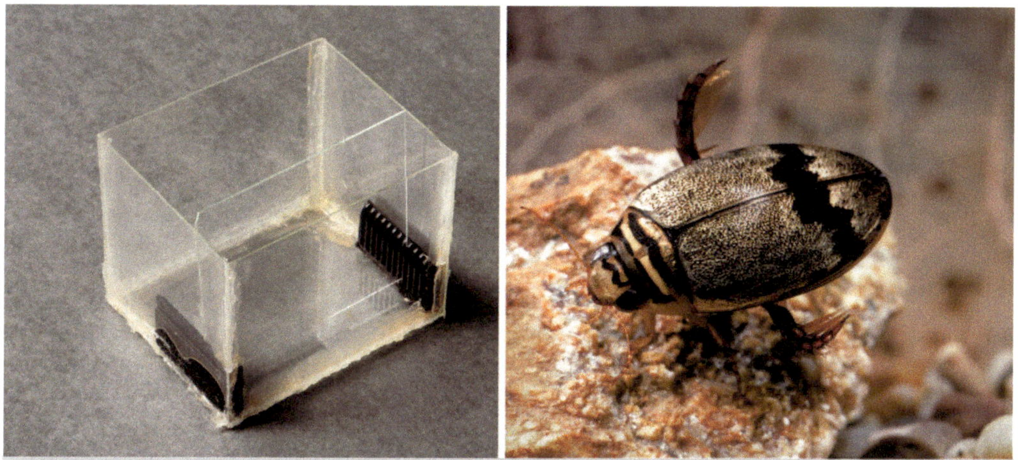

Figure 17.8 A predacious diving beetle, *Thermonectus nigrofasciatus*. Small custom aquaria for aquatic specimens can be assembled using silicone or other appropriate waterproof adhesives. A separate piece of glass can be used to constrain movement of the subject and keep it closer to the front of the aquarium. A large leaf, out of focus in the background, prevents reflections in the glass at the back of the aquarium. Images courtesy of Charles Hedgcock.

or appropriate waterproof adhesives. Small aquaria can also be built from pre-cleaned microscope slides, large microscope cover slips, or products such as Kodak's Projector Slide Cover Glass (#B 324) that may be available online.

If working with aquatic animals in the field, set up your photography field studio close to the aquatic "home" of the subject. This will minimize excessive time out of its native environment. If possible, setting up in the shade will help prevent overheating and make the studio more comfortable for both the photographer and subject.

Aquatic specimens that need to be photographed in water require a clean water supply without particulate matter that will mar the image. Native water for use in aquaria should be collected from deeper, clear portions of the pond or stream, if possible. Native water can be filtered in a variety of ways. Perhaps the simplest method is to use two layers of white material, such as a T-shirt. This will remove all but the tiniest of the particles in the water, making for a cleaner image, free from floating debris. A small amount of native water may be pumped, using a common back-packing water filter to give the photographer perfectly clean water. This method may be too tedious for large quantities of water, however. Bottled water, purchased from the store provides very clear water, but you may require quite a large amount for larger aquaria. The water should be the same temperature as the native water, so float the bottle of store-bought water in the native source for a time to allow the temperatures to equalize. The specimen itself may also need to be rinsed off before placing in the clean water of the enclosure.

When adding water to the aquarium, air bubbles may become attached to the inside glass surfaces. Remove them using a long cotton swab or a soft brush. Forceps and a small pipette can be used to remove debris. Be sure that the outside glass surfaces are clean and dry as well. Allow the water-filled enclosure and any subjects within a chance to settle down before starting to photograph. Be mindful that there will be enough dissolved oxygen in the water for most organisms, but it is not an infinite supply. Gill breathing animals may use up the available oxygen if left too long in the enclosure. When you are finished photographing, drain, clean and dry the aquarium immediately. Water that is allowed to sit in the aquarium over time may evaporate and create water lines on the glass.

Positioning the camera perpendicular to the glass/water surface will help avoid possible refraction errors caused by oblique angles of view. Lighting from above, rather than from the sides, provides a more natural looking lighting position. Positioning the flash units at an oblique angle to aquaria will minimize reflections from the lights as well. When photographing down through the open top of an aquarium, lighting can be easily managed from outside the tank. Fill cards and reflectors may be placed outside – even on occasion inside – the enclosed space. White fill cards may be positioned on either side of the tank to bounce some light into the shadows. A separate piece of glass placed within the aquatic environment can help limit the subject's range of movement, keep it closer to the front of the aquarium, and make it easier for you to follow the subject.

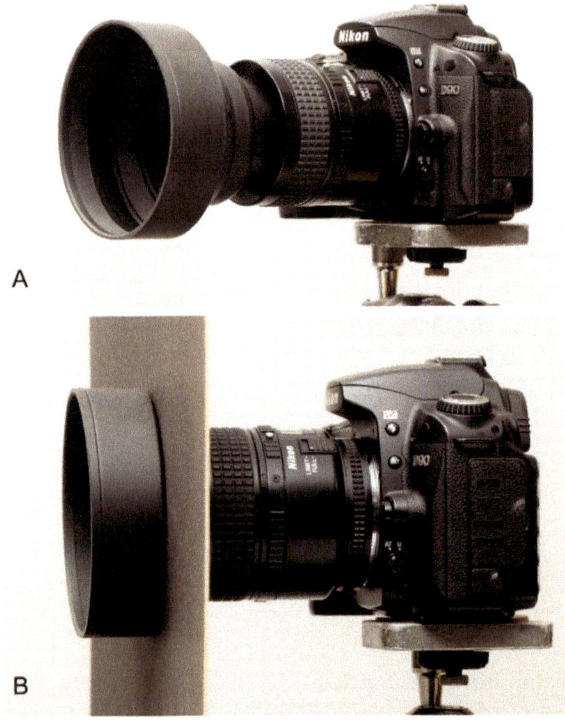

Figure 17.9 (A): A flexible rubber lens hood that is placed directly against the glass of an aquarium wall will block any possible reflections while simultaneously protecting the glass from being scratched or damaged. (B): By photographing from behind a large black card with an opening in the center for the lens and held in place with a rubber lens hood, most reflections will also be eliminated. Images courtesy of Charles Hedgcock.

Avoiding reflections in the glass of an aquarium, or the water surface, if photographing down through the open top, will be a concern. The smooth surface of the water and the aquarium walls may act as a mirror that will reflect back to the camera lens any bright objects. You may be surprised to find camera model numbers, hands, or even your shiny forehead reflected in the smooth, mirror-like surfaces. See Figure 5.6. Because of this, you need to be mindful of what the glass wall or water surface is "seeing." When photographing down into an aquarium filled with water, the photographer's body can usually be positioned to shade the surface, thus blocking bright objects such as the sky, ceiling fixtures, or tiles. The glass surface at the rear of the aquarium may also reflect back into the camera. Reflections here might include the back of the subject itself, creating a double image. Props can be positioned to block this surface, or the flash balanced with the ambient light on background material outside the tank, to prevent a dark background that is more likely to show reflections.

Another simple remedy for avoiding reflections when photographing through the glass sides of aquaria is to outfit the lens with a flexible rubber lens hood. The hood and lens should be placed directly against the glass aquarium wall, blocking any possible reflections, while simultaneously protecting the glass from being scratched or damaged. If the photographer needs to be some distance away from the glass, a simple trick is to use a large black card. Cut an opening in the center of the card the size of the lens and hold the card in place with the rubber lens hood. By photographing from behind the card, most reflections will be eliminated.

It can be useful to work using one hand when photographing living subjects. For example, looking through and operating the camera can be managed with one hand while gently prodding the subject into position, or manipulating the aquarium on a "Lazy Susan" as the animal moves about the enclosure.

Heavy clear plastic sandwich bags (Ziploc) may sometimes be used to contain very small and highly mobile subjects. Leaves or other simple background material can be inserted, then the bag held around the lens with one hand while operating the camera shutter with the other. Lighting through the bag is easily accomplished. This technique prevents the subject from escaping and allows light to easily pass through the clear bag.

Use your imagination when devising containment for your subjects. Look around your studio, laboratory, or home for other possible choices. As well as the aforementioned containers, I have used Petri dishes, darkroom processing trays, glass baking pans, and plastic shoe boxes. Since they don't appear in the photograph, the possibilities for enclosures are many.

Chapter 18 Time Based Imaging

Ted Kinsman

Introduction to Time-Lapse Imaging Best Practices

The science, the art, and the interest in time-lapse photography comes from the need to speed-up time processes using imaging. This type of imaging – when practiced correctly – can manipulate time and display events in an accelerated, yet precise way. Time-lapse photography creates moving pictures that are displayed at rates slower than the standard 30 frames per second used as an industry standard.

Sometimes time-lapse photography is portrayed as a novelty or used for comedic reasons. It is, however, an excellent way to share characteristics of activities that take too long to observe in real time. The true advantage of time-lapse photography shines when motion picture analysis is required. Objects such as clouds that are moving, plants that are growing, or chemical reactions that require days or months to run to completion, can be filmed, and then sped up by a significant factor for viewing. A good example might be found in the growth of a seed. When planted in the right environment, it will sprout and then appear to grow into a small plant in the matter of a few seconds. Time-lapse films have become jewels in the scientific and educational world because they share new processes and observations required for new discovery. The time-lapse photographer enters new territory with each assignment. Patience and access to the required specialized equipment are absolutes to produce work in this field.

Theory

To record time-lapse sequences, the camera takes a picture when activated and then waits to make the next exposure using a pre-determined duration in between frames. This time presents one of the most difficult considerations required for the production of an effective time-lapse video. Choosing how long the camera will be off between frames is foundational to creating an effective time-lapse film. This time is called the wait time and is defined as

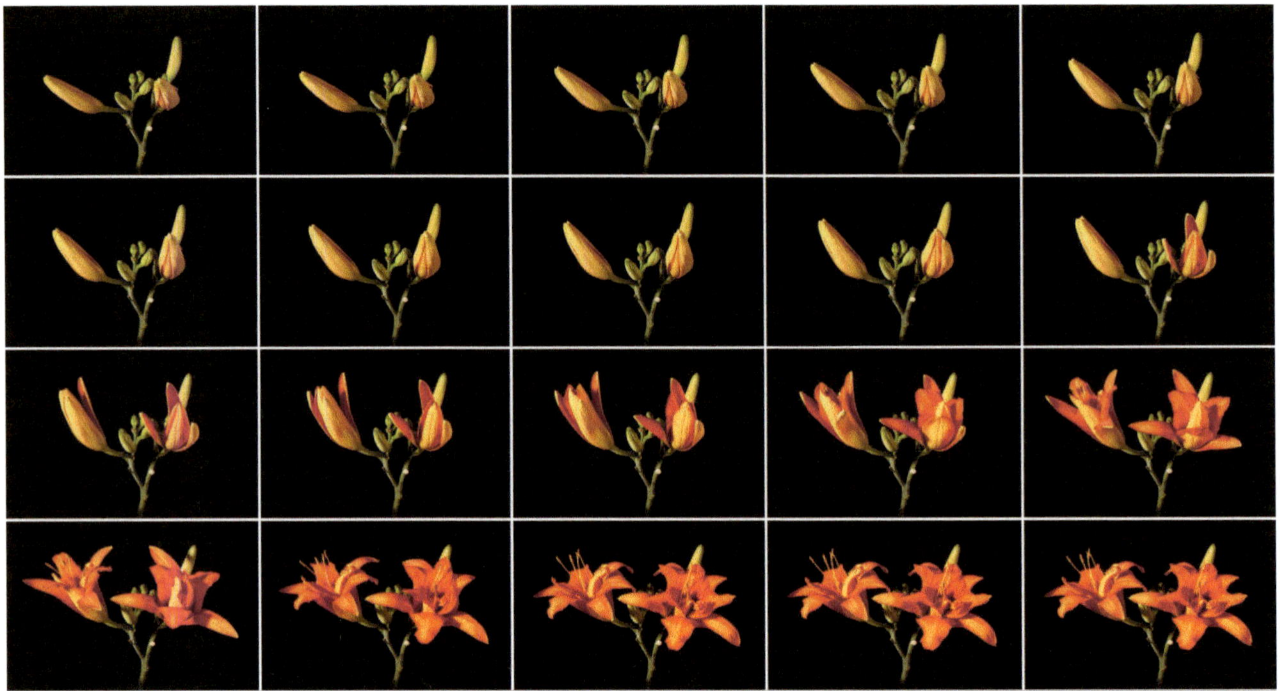

Figure 18.1 Composite picture featuring Hemerocallis day lilies opening. The time between each exposure was 5 minutes. Image courtesy of Ted Kinsman.

the time between frames. The wait time can range from a very short increment such as 1 second between exposures, which creates a highly sped-up display, or possibly exposing one frame every 3 hours, that would be used for photographing a very slow moving event such as an acorn growing into a seedling. In each situation, there will be a specific wait time that gives the best result. The techniques described in this chapter should be viewed as general recommendations and used to help you focus on the criteria needed to select the optimal wait time.

The longer the time between exposures, the more sped up the event will be when the final movie is displayed. When longer times are chosen between exposures, it becomes more difficult to record the event. Events that last for long times create challenges in managing all of the variables that might be present during an event's life cycle.

To determine the time between exposures in a typical time-lapse event, you will need to determine how long the event lasts from start to finish. You will also need to know how long the finished sequence should be projected in display. With these times determined, the total run time for the event is divided by the total number of frames which is equal to the wait time between each exposure.

The following calculations are used as an example, demonstrating how to compress a 2-day event into a 15 second video.

Event duration: 2 days = 48 hours = 2,880 minutes = 172,800 seconds

Required video run time: 15 seconds projected at 30 frames per second

Number of exposures: $30 \times 15 = 450$ images

Wait time: 172,800 sec/450 images = 384 second wait time = 6.4 minutes between images

The above example is based on ideal circumstances. When filming a sequence, it is practical to over-shoot the event. That is, expose more frames than are needed. The extra exposures will ensure that you get the shot, even if the subject's movements occur quicker than expected. Working this way will allow the video editor to adjust the speed of the finished video by dropping frames during video editing if needed.

It is challenging to determine the correct wait time when the exact duration of an event is not known. Take the example of filming a panoramic cityscape that would include a night to day sequence. It is difficult to know exactly how the shadows will fall and the amount of light that will be available if clouds move in or out of the scene. There are also variables that can never be fully accounted for, and for some scenes, parts of the sequence will need to be filmed during more than one day.

For fluid motion in a video created from individual frames in the time-lapse sequence, the subject only needs to move a small distance between each frame. How small of a distance change between frames is a bit of a judgment call and subjective. If you can detect just a small change of position between individual frames, then your timing is right. The resulting shot can always be sped up during editing in the playback rate selection. When filming long duration events, remember that it is quite common for scenes to change at variable rates. An example would be a construction site where cement might be poured in a day, steel goes up in a week, and then the finished brick work takes a month. All of these processes will require their own display rates in the finished video, although they might be all filmed at the rate of the fastest portion of the vent.

Long Exposure Time-Lapse Events

Another way to record a time-lapse event is by using a long exposure time in-camera. In this technique, the shutter remains open for long periods of time and is only closed to advance to a new frame on the media. Since the shutter is open much more than it is closed, there is a high probability that a random event will be photographed. When using this technique, the camera's shutter uses a timed exposure, using prescribed times before advancing to the next frame.

Since the shutter is open more than it is closed, there is a high probability of capturing some unique images. This technique can turn stormy seas into a magical mist and cars moving in traffic at night into tracers of only red lights. This is also useful for capturing meteor displays, the Aurora Borealis movements, and lightning storms.

One aspect of long exposure time-lapse filming is the duty cycle of the sequence. Using the duty cycle is important for capturing random events, for example a lightning storm. In more traditional time-lapse events, a frame is exposed for a short duration, the shutter closed, and then the camera waits before exposing the next frame. For long exposure time-lapse videos,

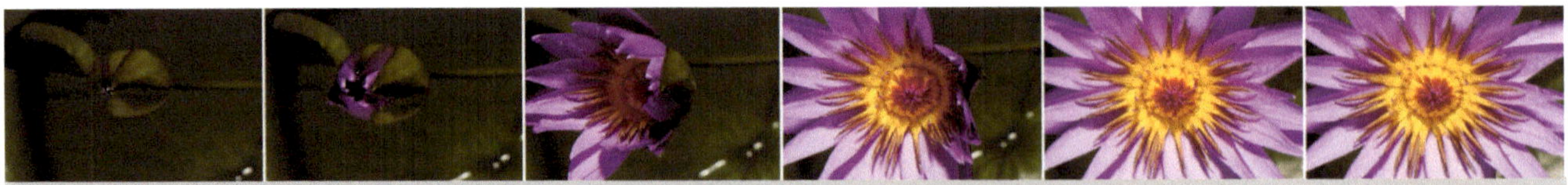

Figure 18.2 A blue Nymphaeaceae water lily opening. The length of time from beginning to end was 40 minutes. Image courtesy of Ted Kinsman.

the shutter will remain open for long durations and then quickly move to the next frame. This increases the probability of capturing the event.

An example of this might be helpful in gaining a better understanding of a duty cycle. In this example, a camera is programmed to make an exposure that lasts 1/2 a second and the camera is also programed to have 300 seconds between exposures.

- The first step is to convert the fraction to a decimal 1/2 = .5 second.
- Step two is place the exposure time in the numerator, and the wait time into the denominator .5sec/300sec = .00166.
- In this example, the shutter is open less than .16 percent of the time.

To capture sporadic events such as lightning, a duty cycle of 90 percent or greater is desired.

So, in this example and if trying to photograph lightning, the shutter should remain open for 30 second increments, and then the camera should be programmed to advance to the next frame in less than 1 second. This timing creates a duty cycle of 30 seconds that spans 31 seconds and produces to a 96.7 percent chance of success.

All time-lapse photographers suffer from "second guess" syndrome. It occurs when starting to create a time-lapse and the camera is being set-up and taking pictures. At that time, the photographer wonders if the correct wait time has been chosen. It is very tempting to change the wait time, but doing so will wreck the sequence. A sudden change in motion displayed by having a few percent change will be quickly noticed by a viewer of the video. If you are filming using 1 second wait times, and then switch to a 1.2 second wait time, this will be noticed when viewing the final film. Remember, it is always better to expose more frames than you will need. You can always speed up a sequence, but it is much harder to slow down recording an event in mid-production. There is no real substitute for having a real subject timing test. A timing test allows you to observe and measure how long the duration of an event actually lasts. A timing test can use several exposures a few minutes apart. The motion between the frames can be determined after capture by placing the two different images into two different layers in image processing software. Using the differences in location can be effectively used to detect the changes that occurred over time. Only a slight difference in location is desired to produce a smooth time-lapse sequence.

When producing critical video recording of unique events, creating a finalized timing test will ensure that the director of photography or the scientist will be satisfied with the timing sequence and will not require the creation of more footage, and the event is smoothly chronicled.

More on Wait-Time

Time-lapse videos create the compression of time which allows high-speed events to be viewed using an expansion of time. To make videos that are smooth and less choppy, more frames are required that have longer time increments between frames. In the end, keep in mind that the final determination of time increments between photos is subjective. It really boils down to personal taste and expectations.

Photographic Considerations

Similar to other applications of photography – such as increased DOF imaging, or wide field high-resolution applications – nothing should change to the camera's capture settings or to the artificial lighting (if used) during imaging. If exposures change during capture, the images when projected will exhibit "flicker" or brightness variances while being played throughout the video. This can be very distracting for viewers when watching time-lapse videos. When time-lapse work is being performed using ambient or found light, this flicker from the lighting may become a significant part of the video if not equalized. When documenting changes in dividing cells using a microscope or growing crystals on a piece of glass, the only changes that occur should be changes to the subject and not the background, exposure, or camera operations. If the lighting brightness and quality can be maintained during the recording of the event, it is recommended to use a camera's manual exposure controls to minimize any variances produced by the automatic exposure control of the camera. If the lighting does change during the event, using the camera's automatic exposure mode will be more practical to better manage the numerous and different exposure times required for the various brightness changes during the filming.

The stability of the camera, lights, and lenses should all be tested before a time-lapse sequence is started. Cameras can change sensitivity with temperature and if this is an issue, it will need to be pre-tested. Some inexpensive flashes will also change exposure discharges with temperature. This is the result of internal fans running and changing the amount of light produced by the flash.

The two most important sources of changeable exposures or flicker in time-lapse work is lens aperture and camera shutter inconsistencies. To best manage aperture variances from automatic aperture controls, it is recommended to use a lens with a manual aperture. There are numerous older lenses readily available on eBay and other resellers. It is also possible to convert a lens mount which allows these older lenses, such as a Pentax lens, to be mounted onto a modern Nikon or Canon camera body. A 42mm Pentax mount converter would be required for this conversion.

The second source of flicker in time-lapse videos is from the shutter. Longer shutter times will display less flicker than shorter ones during the making of many exposures. If this is a concern, choosing longer exposure times can be created by using neutral density filters to darken the light entering the camera. Dense neutral density filters can lead to exposures that are longer, such as half a second or slower. Be advised that all cameras have shutters, and each will behave differently. All equipment should be pre-tested before use on important work. Never simply arrive to make a time-lapse production with system components that have not been fully tested or use equipment that does not create results that are 100 percent satisfactory. Having all things optimized is critical to mission success.

Most of the time it is impossible to absolutely eliminate all the flicker produced by a system. Once all of the images have been recorded, the final sequence can be run through de-flicker software that will amortize the variations visible across the frames.

Digital Formats and Equipment Considerations

For maximum control, image capture settings should be selected by considering how much post-processing will be required. "Less will be more". Since the vast majority of contemporary time-lapse work is shown as videos, the images will, in all probability, be shot as a RAW file or as a JPEG file if space on memory cards needs to be managed. The work surrounding making hundreds or thousands of TIFF or RAW files that subsequently will require image processing and then importing into video editing software cannot be ignored when constructing your Best Practices strategy. Image processing takes time, and unless

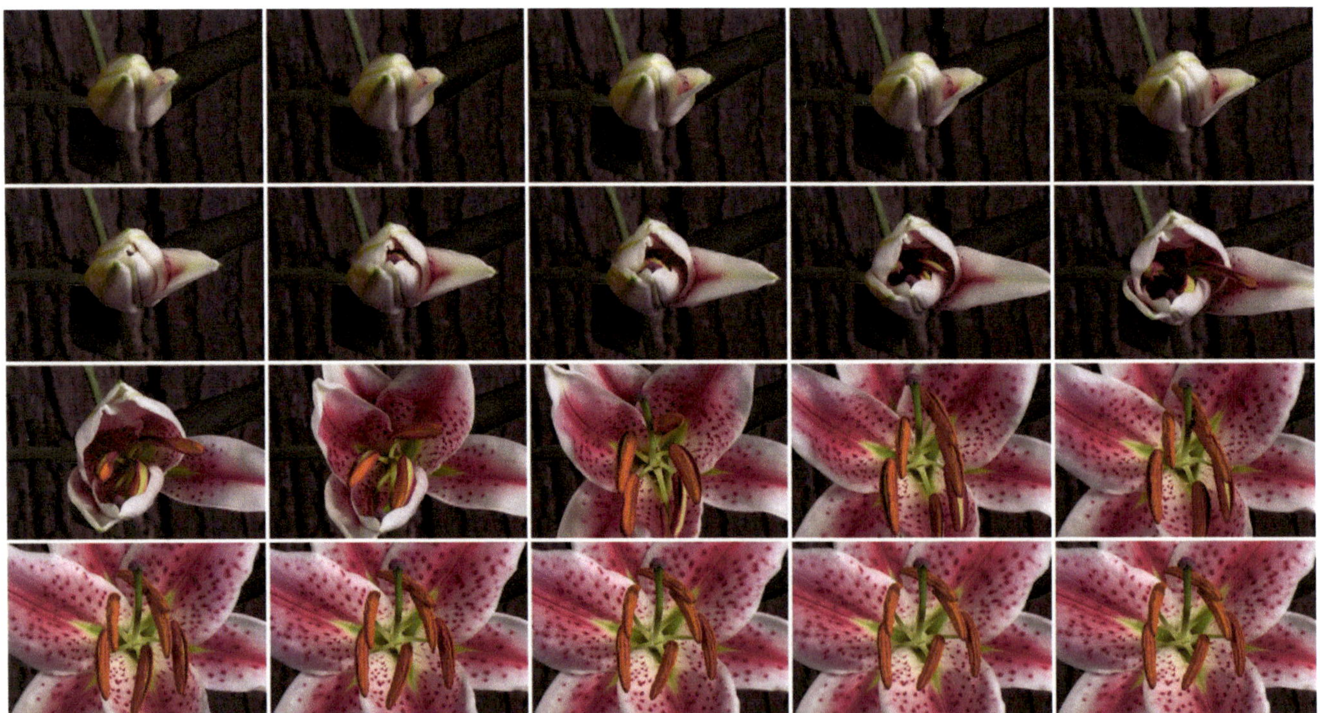

Figure 18.3 The opening of a Lilium orientalis "Stargazer". This series of pictures took 5 hours from start to finish. Image courtesy of Ted Kinsman.

a scientist photographer aspires to spending a considerable amount of time processing images, efficiencies should always be sought out. There are many trade-offs with preserving image integrity when working with JPEG files. This will be mostly evident in data compression or other image artifacts, but the time savings can be enormous.

Every camera and software program will have differences in how they operate for this type of work. It will be impossible to offer specifics for all products used in various situations. For discussion, this chapter will cover global considerations and settings for image acquisition and image processing. These suggestions should be adapted as necessary to your own equipment and applications. Using a DSLR camera is a very good choice for time-lapse photography because of its features, the high quality of its output, and excellent and precise controls. Instrument cameras offer advantages as well, but have a downside because of their need to have a stable power supply. Effective time-lapse work mandates that the camera and the optics cannot change any positions during imaging. Even the smallest changes can be disruptive to the sequence.

Intervalometer

Once the camera position has been identified and the various exposure attributes set in the camera's software, it is important to have a practice run once before attempting to produce important work. Making a time-lapse movie is the product of many individual exposures. Producing this type of work without automation can be accomplished, but it will be fraught with challenges. It is more practical to program an electronic controller to make the exposures using an internal timing device called an intervalometer. This device allows you to program a camera to take pictures using specific time increments between exposures as required. Intervalometers are valuable tools and allow precise and autonomous operation of the camera over a long period of time without an operator being present. An intervalometer can be embedded in camera software or operated as a remote trigger.

Many high-end Canon cameras use a Canon TC80N3 trigger, and newer Nikon cameras have a time-lapse mode built into them. Older Nikon cameras will have an interval-timer shooting mode. Regardless of the camera and control methods, it is very important to make a test sequence of images ensuring that all functions work correctly and image attributes are correct.

On average, the majority of time-lapse videos have a duration of approximately 12 seconds. Many of mine have been edited down to only a few seconds. Many times though, the length of filming will be much longer than the targeted 12 seconds. This is the result of waiting for the event to begin and then exposing useable frames. An example is found when waiting for a flower bud to open and one frame is exposed every 3 minutes for 4 days. The event being logarithmic and not linear creates a lot of images that will not be included in the final footage.

Regarding the timing of the exposures, it is always safe to err on the side of taking frames too quickly. Sequences can always be sped up in the digital editing world. This often happens. Another suggestion is that if there is magnification in the imaging system, use shorter increments of time between frames.

The Best Way to Determine the Timing of a Sequence

It is often very difficult to determine exactly what the wait time should be between images in a time-lapse sequence. One technique that works well with plants is to expose one image every minute for 20 minutes. Pull these images into Photoshop and then use the layers attribute to look at the differences between the images. You can then pick a time delay that generates a small difference between files. You can then use this time as the wait time between future images. This will allow you to guesstimate if the correct delay should be 1 minute or 15 minutes for a particular plant.

Photographing a Time-Lapse Sequence

1. Set the time-lapse shooting function in the computer software or in the camera's menu.
2. Reformat the card before exposing a time-lapse sequence. Losing a few critical frames can render the project unusable. Fast cards are also suggested.
3. Set the time when to start the shooting sequence.
4. Set time increment between exposures.
5. Set the time to establish how long the session experiment will last.
6. Consider the size and capacity of the memory card. Photographing for prolonged exposures over time will create significant amounts of data. It would be a catastrophe if the event could not be exposed from start to finish as a consequence of lack of storage on a media card.
7. Disable and power off all settings such as: LCD preview, auto focus, and image stabilization on lens.
8. Consider using an un-interruptible power supply for the studio equipment if your community is subject to power outages.
9. Select the quality of the file format. There are advantages to using RAW vs JPEG. RAW allows the files to be effectively managed, but will create larger files which will require more time for image processing. JPEG files have great advantages because of speed in handling, but data will be compressed and artifacts can be added.
10. Do not use auto white balance settings. Custom white balance settings will provide the cleanest files.
11. Set the camera's optimal exposure attributes including ISO, aperture, and shutter speed choices. It is vital to expose for the whites using a test exposure frame. Shadows can be adjusted for exposure in image processing, but white details can never be added back to a file. Select the aperture based on the DOF requirements.
12. Keep in mind that slower shutter speeds have more repeatable exposures than short exposures. Exposures should be ¼ second or longer. Use N.D. filters to get the exposures in this range.

Once considerations for all the above aspects of the process have been finalized, the individual frames can be exposed. Some cameras will produce a series of exposures that may require a small amount of image optimizing before they are inserted into time-lapse assembly software such as Adobe After Effects. Some cameras are capable of creating the time-lapse movie as one of its features when attributes from the video mode within the camera itself is selected. It is good practice to record and keep the time-lapse sequence as individual frames. If the camera is responsible for making the finished movie, and the power runs out, all of the content for the sequence will be lost.

Image Editing Time-Lapse Photos

After the shooting phase has been completed, there could be hundreds or more image files. The number of frames is a function of the settings that were selected before starting. In some situations, there can be advantages to improving the image attributes before merging them into the time-lapse video. These changes to image attributes might include brightness, cropping, white balance, and so on. Editing each photo one at a time would be time prohibitive. There are ways to batch process the individual files using Lightroom or Photoshop software.

To create a time-lapse movie directly in Lightroom, a presets module for time lapse will need to be added to your software. These presets can be found on the Adobe website and downloaded. Once a module has been added to the presets folder, the choices are quite intuitive. There will be a frame rate choice, and other export mode choices will also be located in the module. It is a relatively easy activity to work-through.

Final Thoughts About Time-Lapse Videos

As a closing thought, remember that mechanical shutters are limited to 150,000 to 250,000 cycles for modern cameras. An image is not free, and each cycle of the shutter uses up a finite amount of the cost of the camera. For instance, a $3,000-dollar camera that is limited to 200,000 shutter cycles uses 1.5 cents of the cost of the camera on each frame. Keep this price in mind if you have an assignment that requires shooting 200,000 images. The life of the shutter is different for every make and model of camera. If you produce time lapse videos, it is recommended to always have a spare camera body available.

Case Study: Making a Time-Lapse of a Plant Growing

Plants have several conditions that are required to make them grow and bloom. These conditions include temperature, humidity, light quality, and light duration.

Of these variables, temperature seems to be the most important. Controlling an average temperature can be achieved using a heating pad. There are flat waterproof heaters sold for sprouting seeds. These heaters are perfect for supplying heat to a plant or an enclosure. The heating pad is placed directly under a pot to supply bottom heat to the plants.

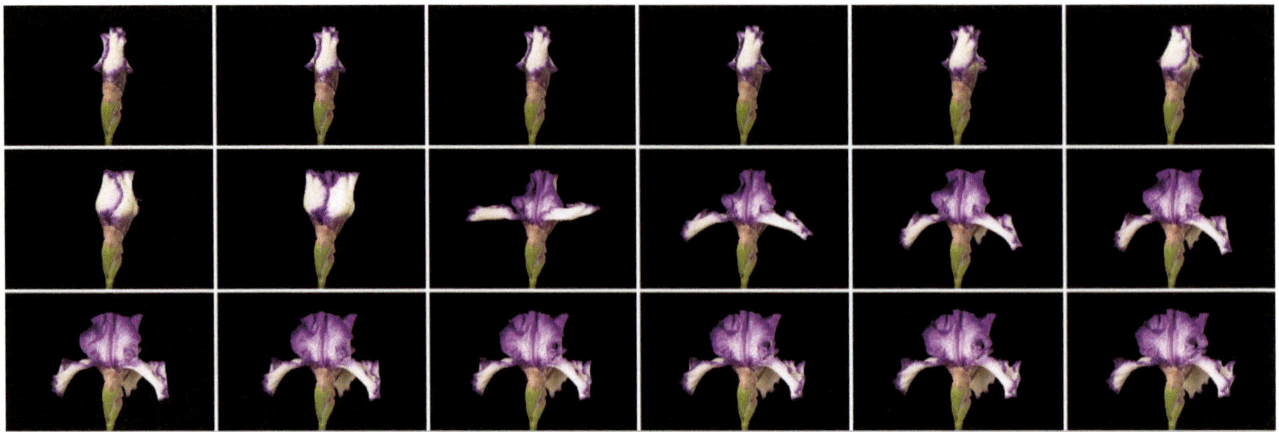

Figure 18.4 An iris germanica or bearded iris opening. This flower took 90 minutes to open. There is a challenge to determine where the 90 minutes will be. Image courtesy of Ted Kinsman. Image courtesy of Ted Kinsman.

Light Duration

The amount of light and its duration can be either computer controlled, or managed using house light timers used in series with the desired lights. Plants can become stressed if lights are left on for 24 hours a day and some will not bloom, or will bloom very slowly, in this condition. The duration of light should be consistent with what the plant's normal exposure to light would be. If plants are collected from the wild, try to keep the grow lights in phase with the current sunrise and sunset times when they were collected.

Light Quality and Spectrum

The spectral output of grow lights can be very important to some varieties of plants. Morning glories require a spectrum close to the natural solar spectrum to ensure blooming. The duration of the light is also very important. The simple act of turning on photo lights every few minutes can mess up the photo cycle of the plant. Poinsettias are very particular about having an uninterrupted dark or "rest" period.

When a plant is ready to bloom it should be moved into the photo studio at the appropriate time to photograph the desired behavior. Many plants will still grow and bloom if a part is cut off and moved into the photo studio. Flowers can be cut, placed in water, then photographed. As long as the action takes less than a few days, the photography can be completed before the plant wilts or begins to die.

Many flowers move a lot as they bloom and should be wired in place. Very fine piano wire can be used to make a support that is hidden from the camera, and keeps the flower from moving during the several days required for a flower opening series. Dandelion buds can be picked before they bloom and, since the stem is hollow, the supporting wire can be easily inserted into the stem.

Many plants require special techniques. The processes created often lead to new discoveries about the motion of plants. Most plants have never been photographed using time-lapse imaging, so new discoveries are found often.

Many, if not all, plants exhibit phototropism. That is, they follow the sun as it travels across the sky. Leaves will rise to meet the light and then bend down at night. Plants like dandelions will do this on a continuous basis and their flowers will follow the sun. This can lead to problems if the grow light is in a different orientation to the plant than the camera's lens. What happens is the plant bends towards the grow light and away from the camera. A way to control this problem is to have the lights placed near to the camera. This main grow light keeps the flower or leaf pointing in the correct direction. The wire trick mentioned above can also be used to limit the plants motion, as long as the wire support can be hidden from the camera.

Timing of Exposures

The best timing sequence for plant exposures is to have one image exposed every 3 minutes, or a bit longer if the plant is particularly slow growing. Flowers that open with the sun such as morning lilies and morning glories will completely open in 3 hours. Plan your duration in response to the length of the final footage. It is often best to test timing and technique for the specimen before filming.

Grow Lights

There are many discussion groups online about grow lights that emit large regions of the solar spectrum. Currently there are a number of LED models that are designed specifically for plant growth. The light level or brightness is very important. A minimum 200-watt light is recommended for a small space. The majority of time-lapse plant work requires bringing a cut flower into a studio. The goal for the light is to keep the lighting cycle in the studio in synchronization with the outdoor natural lighting. A cut flower placed in the dark will often not bloom. Keeping the flower in a bright and warm environment will force the cut flower to bloom.

Managing Both Photo and Growing Lighting

There are several ways to manage artificial lighting for plants in the studio. The easiest lighting solution for plants is to use white LED panels and leave the light on for the duration of the photo shoot. If cut flowers are photographed and the total duration of the photo shoot is only a few days, this is an excellent solution.

For longer time-lapse plant sequences that span several weeks, special spectrum LED grow lights should be selected. These lights are turned off for each exposure using a custom-built switching circuit. In this type of circuit, the grow light is turned off and the LED photo light is turned on so an exposure can be made. After the exposure, the LED photo light is turned off and the LED grow light is turned back on. Systems like this are relatively easy to build and can use an Arduino microprocessor and solid-state relay.

There are a number of problems that need to be managed if you are using flash photography over several weeks. It is difficult to find a flash unit that creates rock solid exposure consistency over a span of several weeks because of possible temperature changes in the studio. The easiest thing to use is a continuous light source with an exposure time longer than a quarter of a second.

Before starting the imaging sequence, ensure the set-up is stable, free of insects, and plants are freshly watered. Watering a plant during a sequence can change the color of the soil and flesh up the leaves. Keeping insects out of the photo studio can be challenging. Insect tape can be used and always pre-sterilize the soil before adding the plant. Various forms of insecticides can also be used, but these are not recommended indoors in a studio since many cause health issues when used indoors.

Never underestimate the value of testing. It is very important to test your system before using it. It is important to eliminate all the bugs – electrical problems and real bugs too – before you start creating a real sequence of images.

Photographing High Speed Events

High-speed photography is a technique that creates images of short duration events. This is accomplished by making one or several very short duration exposures. High speed techniques typically are defined as having exposure times that are shorter than 1/1000 of a second. Modern focal plane shutters produce exposure times of 1/8000 of a second. For events that happen at shorter times, special approaches are required.

Determining the correct exposure time is influenced by identifying the acceptable blur of the subject. This determination is based on subject travel speeds and image magnification. Determining the acceptable blur is subjective.

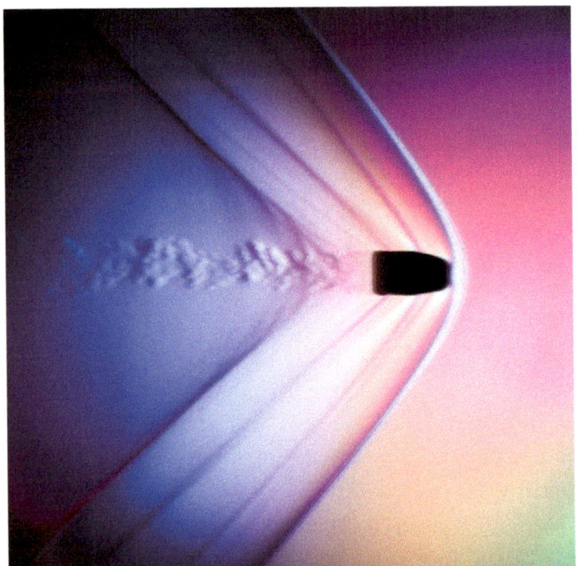

Figure 18.5 This picture is the result of using a high-speed photograph system to capture the air turbulence caused by a speeding .22 caliber bullet. In front of the bullet is a bow shock wave, and in the rear, tail turbulence. This caliber of bullet travels between 1,400 and 1,800 feet per second. Image courtesy of Ted Kinsman.

There are basic approaches for taking high-speed still photographs. One approach uses a high-speed shutter system, and the other uses a short duration flash when the shutter is open. Typical exposure times can be as short as 1ms in duration. To learn more about this subject, it might be useful to examine photographing birds or other flying animals

Case Study: Photographing Birds and Insects in Flight with a Capping Shutter

Background

The objective of high-speed photography is to stop the action of a very short duration event. I use a high-speed flash to create sharp photographs of fast-moving activities. In this manner, the high-speed flash will act like a very fast shutter and reduce image blur caused by subject movement. The use of high-speed flash becomes critical with fast moving animals such as birds, as movement is emphasized when working at high magnifications.

The camera shutter/flash is a successful combination for relatively slow-moving animals. When photographing very fast-moving events such as a jumping cat or a flying bird, special and more advanced techniques are required to stop the action.

The use of the high-speed flash reduces blur, however, the triggering of the flash and camera are difficult to synchronize because these types of events will occur faster than you can respond. Making high-speed photographs is easily accomplished using a camera trigger.

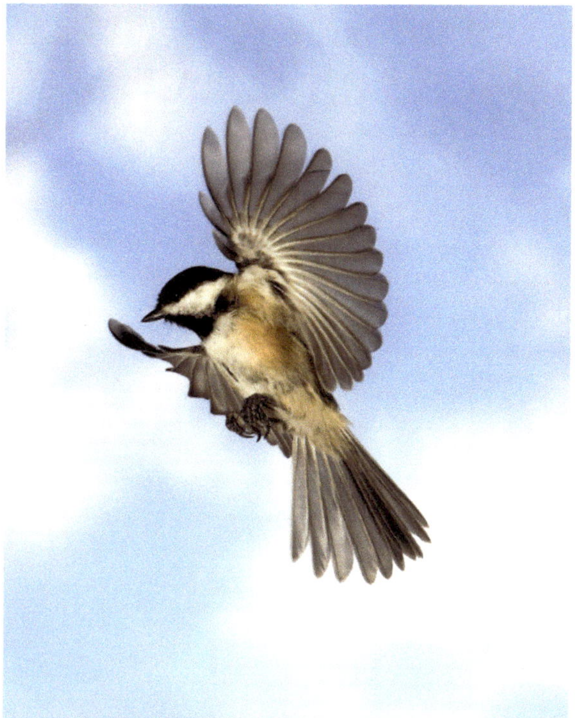

Figure 18.6 A chestnut-backed chickadee slowing down before landing on a bird feeder. Image courtesy of Ted Kinsman.

Required Equipment

A camera trigger is a microprocessor that controls the camera, a sensor, and a flash. In the simplest configuration the camera controller will detect a subject that breaks a light beam which then triggers the camera to set off a high-speed flash. Using this type of system will allow a bird to be photographed in mid-flight as it approaches a feeder. There will be a lag time from when the camera controller sends a trigger command to the camera and when the camera actually takes the picture. This time is called the camera delay and it is critical that you are aware of this delay and plan for how long it is. For a bird flying to the feeder, the light sensor should be placed so that the bird has time to fly into the camera field of view exactly when the camera delay time ends. This solution anticipates that all birds fly at the same speed. After several trials, the correct placement of the camera control trigger can be identified, and numerous photographs of birds can be taken as they come in to land at the feeder.

This setup becomes a camera trap, or a system to photograph animals without the requirement of an operator needing to be present to take the picture. Such systems are highly desirable because the presence of a human will scare your subjects. More importantly, it is difficult to sit in an animal blind for weeks just to get the right shot.

Using this setup, there is an assumption that a portable battery, or access to a standard electrical outlet is available. The use of remote cameras running on batteries typically requires special modifications to the camera to enable it to consume lower levels of current while in the sleep mode. The creation of these systems is beyond the scope of this text.

For this example, I will discuss using one of the better designed portable flashes on the market, the Einstein 640 WS flash made by Paul C. Buff, Inc. These flashes are some of the fastest mono-light heads on the market.

There are specific procedures for setting up a high-speed nature shoot used to photograph birds going to a feeder. Situations like this provide a lot of control over the flight path of the birds. The placement of light diffusers and reflectors can be taken into consideration.

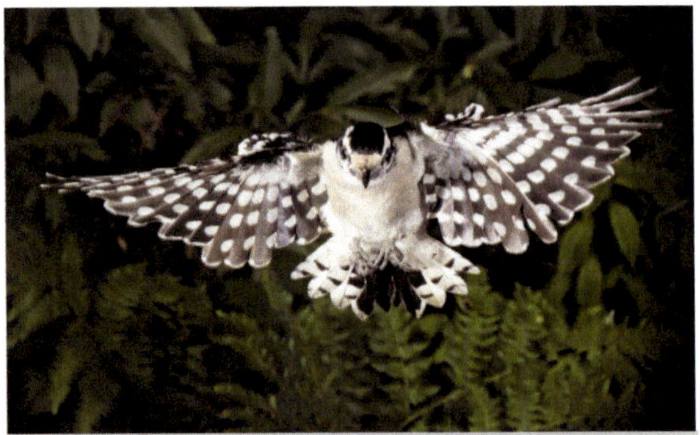

Figure 18.7 A downy woodpecker, still in flight, coming into a birdfeeder. Image courtesy of Ted Kinsman.

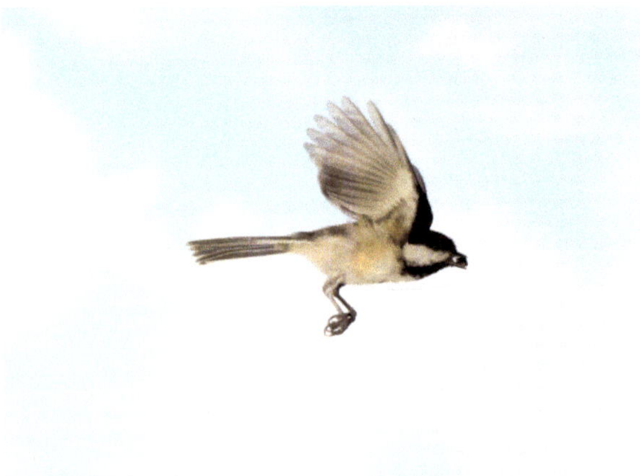

Figure 18.8 A chestnut-backed chickadee in flight after visiting the feeder with a sunflower seed still in its beak. Image courtesy of Ted Kinsman.

The skills developed setting up this simple controlled photoshoot will be invaluable, and will provide guidance when working in conditions that are less than ideal.

In this example, birds should first be attracted to a well-stocked feeder. An infrared beam is used to determine if the bird will be in the correct position relative to the trigger used for the camera system. The camera shutter must be synched to the flash. For best results, determine what the trigger lag of your camera will be. Digital cameras all have different shutter lags with the average time being approximately 80ms. Of course, the smaller the time lag, the better the results. A bird flying at 9m/s will travel .72 meters during the 80ms of the shutter lag. Your results will be improved if the bird slows down while landing on the feeder.

The specialized techniques used for photographing hummingbirds in flight has grown in popularity in the last few years and now there is dedicated equipment specifically designed for this. Hummingbirds move their wings at one of the fastest rates of any bird. With frequencies of wing movements in the range of 12 to 80Hz or flaps per second, the birds are a challenge to photograph. In these instances, the triggering of a normal shutter is too slow to respond to a bird located in the correct location. To decrease the response time of the camera, an external shutter is used. This external shutter can open and close in less than 5ms. This is several times faster than even the best response of the fastest camera on the market. This external shutter is called a capping shutter and is placed on the front of the lens of the camera. The capping shutter is synchronized with a high-speed flash to capture the bird in the correct location and remove all blur from the motion of the animal while the actual camera shutter remains open.

Creating a System

Capping shutter systems can be built at home with optical shutter systems, however, the price of the parts is almost equal to buying a commercial system from Cognisys. The operate a system such as this, follows these steps.

1) The camera controller will open a camera shutter (but not the capping shutter) for typically 30 second increments. An exposure of 30 seconds creates heat and sensor noise and so the timed exposure should be limited to 30 seconds.
2) When the bird flies through the infrared trigger beam during this 30 second window, the capping shutter will open.
3) This will fire the high-speed flash and close the capping shutter.

4) Following this exposure, the internal camera shutter will close.
5) The camera will wait a few seconds to save the image and then the camera cycle starts again.
6) Using this technique with an IR beam set-up next to a hummingbird feeder, you can make countless pictures every day.

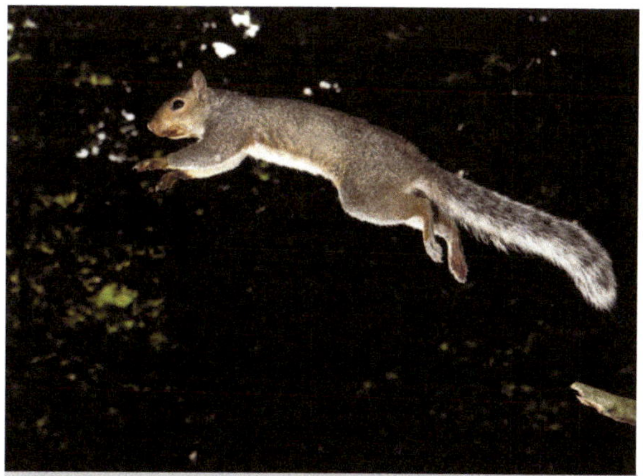

Figure 18.9 A *Sciurus carolinensis* or eastern gray squirrel was photographed in mid-jump after leaving a branch. Image courtesy of Ted Kinsman.

A bird or other subject photographed with just a high-speed flash will create a black background and the results make it appear as though the bird was flying at night. Since birds do not like to fly at night, a background is used – oftentimes a painted background suggesting clouds. Some photographers print out images of the habitat the bird lives in as a background. The use of a background requires a separate high-speed flash, but the resulting image will be quite nice. To modify a black background in image processing is a tremendous amount of work and time and since an average setup can collect several hundred images every day, it is out of the question to digitally add backgrounds.

The techniques illustrated in this chapter can be applied to many animals and insects. It is important to become proficient in the techniques before venturing into the field. You should also create an all-inclusive checklist of all material required before attempting one of these elaborate setups. Should you be ambitious, be sure to carry spares of all vital components.

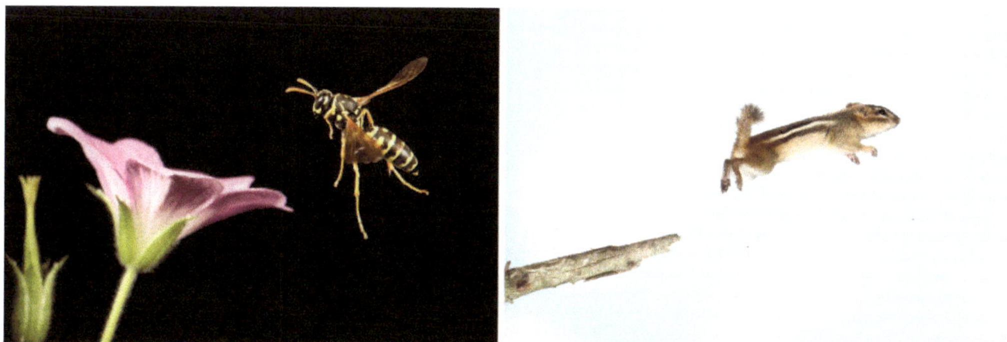

Figure 18.10 On the left is a vespid (paper) wasp. It was photographed using the methods described above. In the photo, the background drops out and the picture appears to have been taken at night. By placing a painted background into the scene, the photographs appear more natural, as shown on the right, where a marmotini or chipmunk was photographed jumping off a branch. Images courtesy of Ted Kinsman.

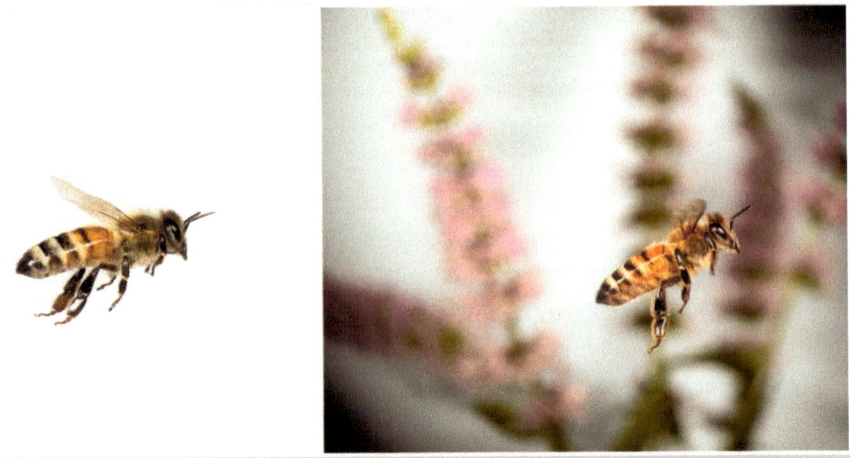

Figure 18.11 An *Apis* or Western honey bee was photographed against a white background and against flowers. It is evident how the inclusion of flowers or other props can change the pictorial result. A Canon 65mm macro lens was used at life size or 1:1. Image courtesy of Ted Kinsman.

Suggested Reading

Time-lapse Photography: A Complete Introduction to Shooting, Processing and Rendering Time-Lapse Movies with a DSLR Camera (Volume 1), by Ryan A. Chylinski. Cedar Wings Creative Press, 2012.

My Ivory Cellar: The Story of Time-Lapse Photography, by John Nash Ott. Twentieth Century Press, 1958.

Part IV After Capture Post Production

Chapter 19

Image Processing in Science

Staffan Larsson

Introduction

Science pictures are more than simply pictures, they are visual representations of scientific data. Science pictures contain visual information about an object or event that may need to be evaluated at another time. To make the data in a digital file more visible, the file needs to be "processed" or clarified after capture. This allows for improved viewing or analysis of the recorded information. Chapter 4 gives a comprehensive overview about the complex issues surrounding the image processing of scientific images. Changing data comes with responsibilities that must be considered when evaluating how much a file should be "modified" before it misrepresents the event or object itself. When images are being optimized or processed, data is being changed. Image processing can be destructive or non-destructive to native data. Best and scientifically accepted practices are required when image processing files that are to be used in scientific publications.

It is accepted by readers and journals that some level of image processing is required for improved image reproduction in printed articles or when images are featured on websites. Technical cameras actually begin processing image data at the moment the image recording attributes are selected. The process of improving sample visibility begins even before making an exposure. Making things visible without embellishment is a fundamental expectation in scientific imaging. Consider that from the start of an imaging experiment, increasing an object's visibility through optical means, lighting approaches, or other scientifically accepted practices makes the "nearly invisible, visible".

There is a distinction between image processing and image analysis. Image analysis is a process that extracts data/information about an image. For example, an image might be analyzed for the frequency of a certain characteristic or color. Images are analyzed in everyday life. Bar codes are an example of a common application of image analysis. This chapter will not discuss image analysis further.

Manipulation, Enhancement, or Clarification

Digital image processing software allows scientists to adjust data in ways that was never possible with film, and the capabilities continue to improve. The precision with which images are processed makes detection of image processing to be almost impossible.

Consider the current conversations about fake images and the narrative that surrounds them. As a consequence, many contemporary journals require image processing histories to be submitted with images that will accompany an article. Unfortunately, while this protocol seems like extra work and a hassle, it serves to confirm the integrity of images by describing the step-by-step image processing procedures that were used. Including the image processing history can validate an image's truthfulness. Unfortunately, this requirement will be an expectation in scientific publishing.

A fundamental decision about processing an image resides with the intent or objective of the creator and their considerations about expectations of an intended audience. A number of issues surrounding approaches and intent were covered in detail in Chapter 1 and re-examined in Chapter 4. An image processing workflow (protocol) in natural science needs to be carefully considered and include procedures that do NOT create image credibility issues.

Another aspect surrounding image processing can be found in the words that are used to describe image processing itself. Image processing is frequently described as image manipulation and this is partly true. Image clarification also describes image processing and so does the term image enhancement. All descriptions are correct but lead to different conclusions about the intent of the maker of the image. In the arts, creating illusions, emotions, or fantasies can be the primary objective for the photographic artist. Because the world of applied photography is large, the term manipulation was part of the original vocabulary used in early conversations about digital photographic practices. Manipulation was adopted as a way to describe image editing. Truthfully, there is no doubt that image content is being manipulated, but in science, manipulation leads to conclusions that there is a misrepresentation of something. Image enhancement is another term used to describe a process intended to change and sometimes improve the visual appearances of an image's data. In science, recorded data is processed to help visualize sample attributes which makes these structures, or features, more visible.

The process of improving what is able to be seen in an image in science is more aptly described as image clarification. An analogy might be found with wearing glasses. Glasses allow a person to see things more clearly. Looking at unprocessed files may be similar to how the world would appear to a person who wears glasses but is not wearing them. Glasses and image processing clarify what it is possible to see.

In the early stages of the adoption of digital photographic practices in 1997, the FBI (Federal Bureau of Investigation) formed the SWGIT group, designed to establish digital guidelines and protocols. SWGIT stood for Scientific Working Group – Imaging Technology and was formed in response to the very dynamic and constantly evolving digital tools and the very new technology. One of the approaches recommended by SWGIT was the adoption of the term clarification for use when describing image processing when providing testimony. It is now easy to understand that when testifying in a court proceeding, the use of the word manipulation could evoke negative connotations from jurors or other officers of the court. The word clarification suggests a more noble outcome and implies less deceit. Language, whether visual or written, has influences that contribute to the perception of a maker's intent.

Image Editing Software

Software is required to see digital images and process them. When ingesting images from a capture device to a computer, the computer needs an operating system, a graphics card, and image editing software to see, process, and view files. This type of software displays image pixels in the manner in which the file was created. Software can be basic or advanced. The more features, typically the higher the cost, but not always. Many shareware, or open source products, are particularly powerful and useful. Image J and GIMP (General Image Manipulation Program) are examples of very useful and powerful open source software. Adobe Photoshop software is arguably the most common and well-known image processing software but there are others. Many industrial and technical cameras from companies such as Nikon or Zeiss come with software that can perform image editing in addition to operating cameras.

Image processing software accomplishes two tasks. It allows pixels to be selected and it allows for the modification of the selected pixels. Each of these tasks is separate and fundamental to using an image-editing program. Pixels can be selected using different tools, either as individual pixels or as a group of pixels. Pixels can be selected by choosing pixel brightness, location, or other characteristics that are similar in large groupings.

Once pixels have been selected, adjustments, or replacement of those pixels – sometimes called painting – can be accomplished. Depending on the software, various selection and/or processing tools are available. The more powerful the software, the more capabilities it offers, and in some cases, image analysis and video editing capabilities are also available. The evolution and improvements to these products is astonishing when evaluated using the lens of time.

Image J Software

Image J is a powerful 64-bit image-processing program. Launched in 1987 as NIH Image, scientists at the National Institutes of Health created this important tool for use in the earliest days of digital imaging. Image J is capable of performing many complex functions including image analysis. It is an open source image-processing program that has been optimized for use with scientific and multidimensional images. Image J is easily expanded and has numerous plug-ins and macros useful for performing a wide variety of image processing tasks More about the functions of Image J can be found online: https://imagej.nih.gov/ij/.

There are countless useful websites dedicated to sharing methods for using Image J. This software is prevalent in science laboratories worldwide and can work on all the common file types such as TIFF, JPEG as well as many other proprietary file types that cannot be opened in other commercial image editing programs.

GIMP Software

The General Image Manipulation Program (GIMP) is an open source bitmap and raster image editor launched in 1995 by Spencer Kimball and Peter Mattis. It allows for easy pixel selection methods and pixel retouching including drawing, re-sizing, cropping and many other important tasks. GIMP is also a 64-bit image processor. GIMP, which began as a college project, has become a highly respected and capable image editing software. Similar to Image J, it is easily expanded with plug-ins and macros. There is a relatively large GIMP following and a number of books have been published about using this software.

Some of GIMP's main features include how it describes color, how it creates selections and paths, its image editing, layers, masks and channels, as well as scripting and automation to name a few. It is also free. Because GIMP is open source software, its features are frequently expanded and changed. Because all of these image processing programs can run on various operating systems such as Windows, Unix, or Linux, it has led to variant products such as CinePaint, GIMP Classis or GIMP Portable. Similar to Image J, a number of resources and user groups are available online: https://www.gimp.org.

Adobe Photoshop Software

Photoshop is arguably the most well-known image processing software in the world. It has an excellent user interface and it is the de-facto program for image processing in the photography, graphic arts, and publishing industries. Photoshop version 1.0 was launched in 1988 and has grown to now include video editing capabilities and so much more. As an image editor software Photoshop uses a raster mode, layers, and supports RGB, CMYK, lab color and spot color as a few of its countless features. There are many excellent books and web resources that discuss every nuance of Photoshop. Many terrific resources such as LinkedIn Learning®, a subscription service, and other tutorial websites, are easily found.

The book was written to focus on the making of scientific photographs, however, image processing is a fundamental step in the image chain required to make things visible. This chapter was written to share practical methods and considerations for using Adobe Photoshop software in the science. Another Adobe product, Lightroom, can also be useful for image processing but will not be covered in detail.

Basic Color Theory

Almost all digital cameras produce RGB image files – except high-end monochromatic grayscale cameras. Since color is fundamental to digital systems, color theory must be a part of any conversation about image processing. How a DSLR camera forms an RGB image was covered in Chapters 6 and 8. Primary colors are defined as colors that can be added together to make other unique colors and are the foundation of a digital image. Red, green, and blue are primary colors. When added together in equal amounts they produce white. Adding red and green makes yellow; adding green and blue makes cyan; and adding red and blue makes magenta. Primary colors are part of human vision. RGB color is used in video projectors, all computer and tablet displays, TV displays, and in inkjet printing. RGB color is also used for web applications, in MS Office, and in imaging editing programs such as Photoshop software.

Subtractive colors conversely are described as secondary colors and include cyan, yellow, and magenta. When put together, these colors subtract light and lead to the creation of primary colors in varying degrees depending on the purity of the colors combined. Equal amounts of cyan, yellow, and magenta will make black, described as the absence of light. Magenta and cyan creates blue. Yellow and magenta creates red, and yellow and cyan leads to the creation of green. The CMYK model is the basis of the offset printing process. Very few, if any, contemporary cameras produce CMYK files because of the smaller color gamut they produce. RGB images can easily be converted to CMYK images using an image editing software but the conversion will change the color gamut of the image.

Complimentary colors are colors that are opposite from one another as shown in Figure 19.1. Combining complimentary colors leads to the production of gray. Yellow is the compliment of blue, cyan is the compliment of red, and magenta is the compliment of green. When adjusting a color cast in a digital file, using the complimentary color is one of the first strategies that should be chosen to moderate the error in color reproduction. Changing a file's color bias can be accomplished by working in the highlights, the midtones, or in the shadows of the image. When done correctly, it will produce a more neutral file. Setting a camera's white balance correctly should always be used to manage the RGB variances.

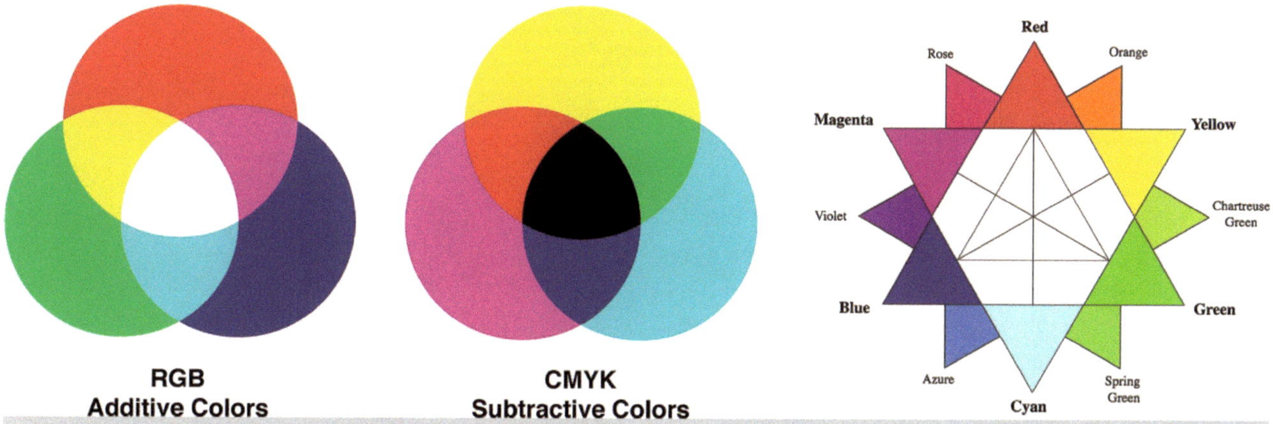

Figure 19.1 The RGB primary colors, sometimes called additives or primary colors (left), the subtractive or secondary colors (middle), and color star (right). Additive colors when combined produce white as seen on the left, and subtractive colors when combined form black as evidenced in the center of the image. On the star, a color's complimentary color is situated 180° across from that color in the triangle. Primary or additive colors and secondary or subtractive colors are located on the star's points. The RGB color model uses red, green and blue. The subtractive or secondary colors include cyan, yellow and magenta and when blended together create the absence of light, or black. Graphic courtesy of Staffan Larsson.

Fundamental Digital Color Models

Digital color is described by an imaging processing program using mode. The most common modes are RGB, CMYK, grayscale, Index color, and LAB color. There are others. Not all software programs can open and process all of these file mode types. RGB color is the largest mode and offers the most capabilities for color imaging. It can record and display more than 16 million variances of color. The number 16 million is the multiplicative of three channels (RGB), each being 8-bit, but it can now be 16-bit as a RAW image file and describe up to 256 tones per bit. CMYK is most commonly used for offset printing and creates a smaller color gamut than RGB files. Grayscale images are 8-bit files and have data in only one channel, the monochromatic channel. Index color is very efficient and uses only 256 individual colors. Converting RGB to index files produces smaller files with the same pixel resolution. Index color is excellent for web applications where color is not a critical component of the subject's features.

Files can be Saved As and converted back and forth across the various modes but may experience small or large amounts of data loss in the conversion process.

Channels

Channels display color attributes of an image as separate red, green, and blue components (RGB). Since pixels respond to brightness differences, the colors displayed on a monitor are mathematical models of the individual RGB values. When using the channel display window, each of the primary color components can be evaluated separately. The red

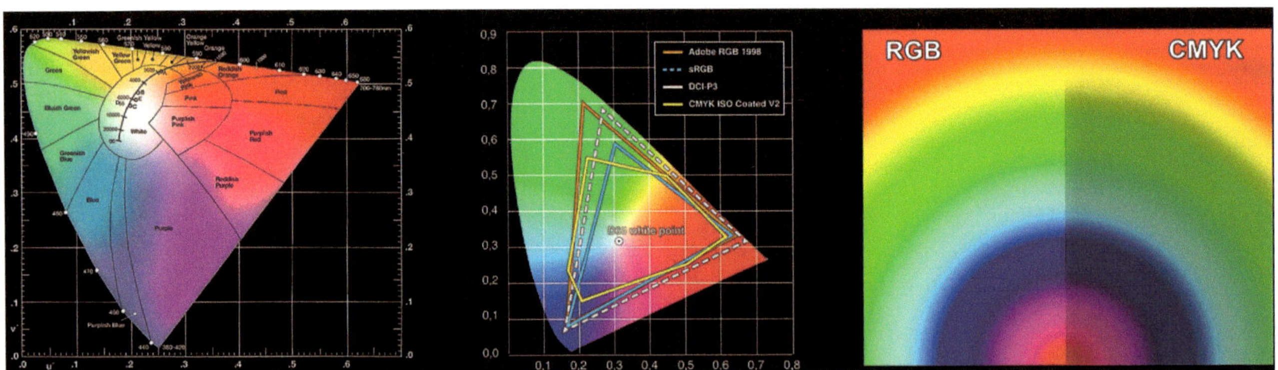

Figure 19.2 In this figure, the Comission Internationale de l'Eclairage or International Commission on Illumination (CIE) diagram is reproduced on the left. The commission was founded in 1913 to provide a group whose mission was to set standards for all things related to light and lighting in 1931. CIE formed a technical committee for Vision and Colour that has been a leading force in colorimetry. The CIE color model was developed to be completely independent of any device or other means of emission or reproduction and is based as closely as possible on how humans perceive color. The key elements of the CIE model are the definitions of standard sources and the specifications for a standard observer. The middle image shares what color gamut is possible for specific file types including CMYK, sRGB, Adobe 1998 and DCI-P3 (in the near future, DCI-P3 will become the new standard for many devices, websites, and software and replace sRGB as the prominent color space), when compared to the CIE diagram's potentials. In this drawing, Adobe 1998 can reproduce the most colors and has become a benchmark for RGB specification standards in color management. The image on the right is a graphical representation of the loss of color data when an RGB image is converted to CMYK. You can see more on color space in Chapters 6 and 8. Graphic courtesy of Staffan Larsson.

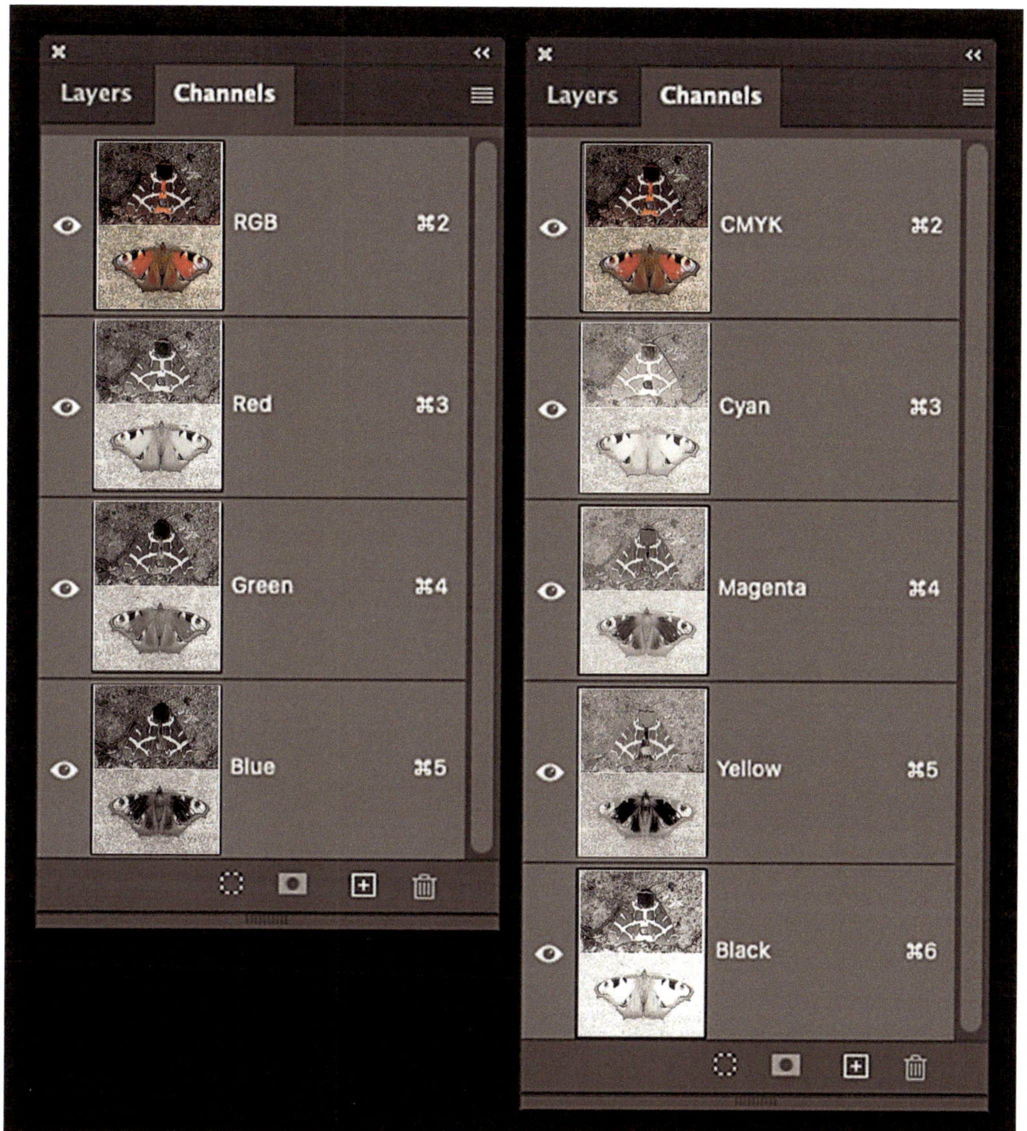

Figure 19.3 Two versions of the channel display. On the left is an RGB image of a moth photograph displayed in its R, G, and B components. On the right, the CMYK version of the same file displayed in C, M, Y, and K components. Adobe product screenshot(s) reprinted with permission from Adobe.

component of an image can be seen by selecting the red channel, the green information can be seen by making the green channel active, and the blue channel would be handled in the same way. Older digital camera sensors were not as uniform in their sensitivities across the three spectrums as modern cameras. Older cameras recorded more data in the red channel than in the blue channel. Using channels can be useful for creating a black and white image from an RGB file, pseudo-coloring an image, or for making stereo anaglyphs.

When using adjustment layers or creating masks in another channel, an Alpha channel might be created. An Alpha channel in Photoshop is a feature that determines the transparency settings for certain colors or selections. In addition to common red, green, and blue channels, you can create a separate alpha channel to control the opacity of an object or isolate it from the rest of your image.

Layers

Layers might be considered to be like sheets of clear plastic containing images that are stacked on top of one another. When image processing, working in layers is not destructive to the native file. The original recorded data remains unchanged in what is called the background layer. Each layer can be seen through the transparent areas of the layers located on top of, or down into the layers below. Visibility of each layer is based on how the layers are merged or blended and then displayed. Layers can be positioned in a specific order, they can be moved, or they can be deleted without compromising the native data. It is possible to change a layer's opacity, which can make some of or all of the content transparent or visible through the layers. It is possible to use layers for compositing multiple images, adding text to an image, or adding vector graphic shapes. The manner in which layers provide transparency throughout the file is controlled by layer blending.

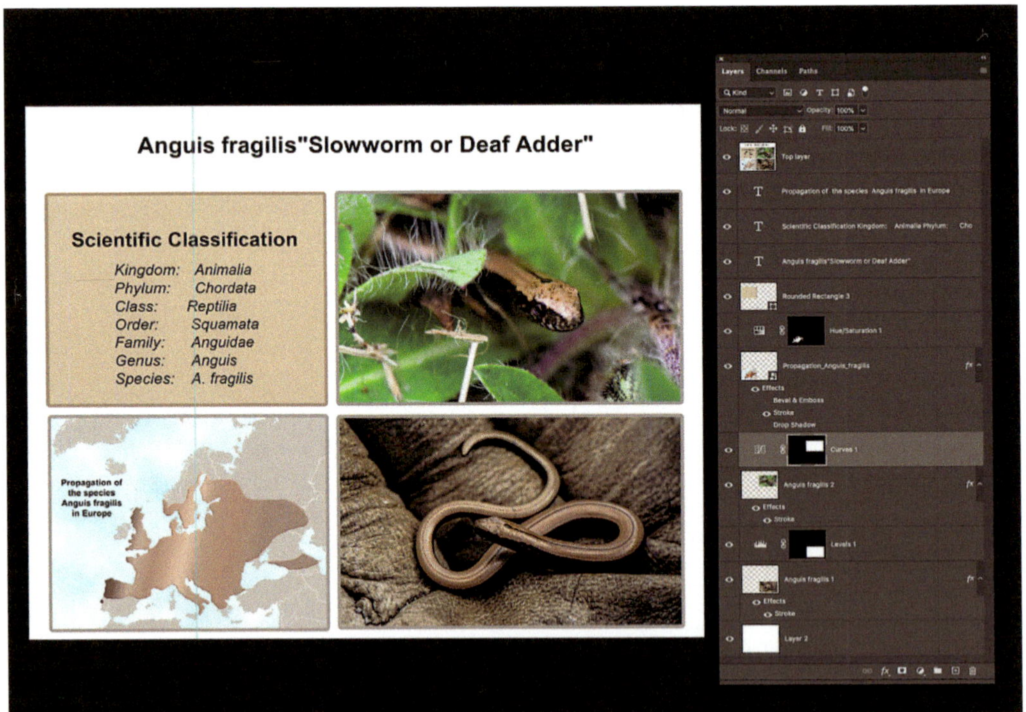

Figure 19.4 This composite image shares (right) the individual layers that were used to make the image featured on the left. In the composite file are two separate image files, one type layer, two image adjustment layers, and one frame layer. Once satisfied that the composite image is complete and no further work will be necessary, the layers can be flattened for publication. Images courtesy Staffan Larsson. Adobe product screenshot(s) reprinted with permission from Adobe

Chapter 12 explores computational photography. Computational photography is an approach that creates images composed of numerous and separately recorded individual image files. Wide-field high-resolution imaging, increased DOF imaging, and HDR are a few of the topics that are covered in the chapter. Additionally, and useful in science, is batch processing used to create standard file outcomes and efficiencies when working with large volumes of files with similar attributes. Chapter 12 shares why and how to batch process files, and how to create macros or actions that enable a scientist photographer to record repeatable sequences of steps – this is useful when automating a lot of similar tasks.

Global Image Editing

Editing images using global methods are techniques that had direct counterparts to methods used in film darkrooms that remain universal image processing techniques, are accepted in science and are still used today. These methods include changing color, changing color balance, changing gray balance, retouching dirt, changing hue and saturation, adjusting levels and curves for contrast, and brightness management to all pixels equally. RGB to gray-scale conversions and RGB to CMYK conversions are also common. Image sharpening using unsharp masking and high band pass filters are also accepted practices in science.

There are also methods for retouching called cloning or pixel replacement. These might be considered similar to retouching a print using a spotting brush and dyes. All of the above methods are considered to be traditional and acceptable image processes for science, except for pixel replacement. Pixel replacement remains an area of debate and there is great uncertainty about its appropriateness since it modifies some pixels, but not all.

Chapters 6, 7, 8 and 9 suggested methods used to create files that will require LESS image processing. These methods included the establishing of optimal image capture settings, preparing a subject for imaging, optimizing the imaging equipment, establishing proper lighting and establishing proper color temperature, the use of a good specimen, controlling the exposure, and establishing a good white balance. Accomplishing these conditions will improve imaging outcomes.

Monitor Calibration

Best Practices require the use of a calibrated monitor for image processing. This is fundamental to the control and precision of image processing outcomes. It will help to ensure that tones and colors are represented accurately on a monitor and that they match the expected outcomes for a particular sample.

The following are recommended steps for the calibration of a monitor:

1) Set the color temperature D50 or D55 (5000–5500K) and specify the color mode as Adobe RGB 1998.
2) Set the gamma to 2.2.
3) Calibrate the monitor when and as needed by the monitor's software prompt.
4) Manage ambient lighting in the image processing room.
5) Situate the monitor in a place without background light/flare and away from windows and not near brightly colored walls.

To get the best quality of display of the images, use a monitor with a wide spectrum (Adobe 1998) gamut that is equipped with a spectrometer software which is important for re-calibrating a monitor as needed. Many contemporary monitors are capable of automatic self-calibration when desired.

Image Processing Tools Overview

When processing digital images, all or some of the pixels will be adjusted. There are specific tools used to select all pixels in a file, or a portion of a file, all the way down to just one pixel if desired. Each of the various selection tools has various degrees of precision and ease of use. I have also learned that I can get very good results by using only a few tools by gaining experience with them. This has helped me produce better work more quickly. It is not important to master every tool at once. If you want to get good, I would suggest you take a basic image-processing workshop. While the content in this course will not be specific to science, the general knowledge is transferrable to any application.

One important first step when image processing is to set up an effective workspace in your image editor needed to display the panels that will be used most often during the editing process. To set up the Adobe Photoshop workspace, go to Window, where you will find the following panels: Navigator, Info, Histogram and History.

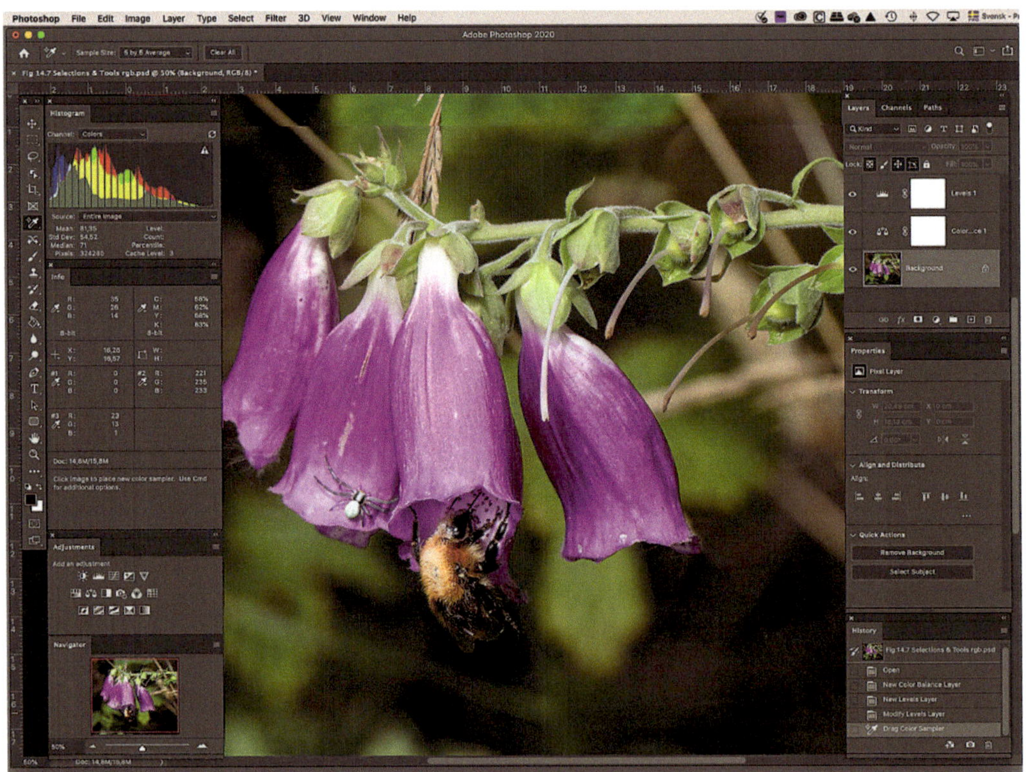

Figure 19.5 The workspace overview windows and panels. It is possible to make your own custom workspace, and once saved this will open for each new session. Adobe product screenshot(s) reprinted with permission from Adobe.

The Navigator panel provides an overview of where in the image you are working or can establish a position for working on an image when zooming in and out of an image.

The Info panel provides the color values of pixels while you are making color corrections, or the information provided from measurement tools such as the eyedropper or color sampler tool. When you work with the adjustments panel or properties panel, the Info panel displays two sets of color values for the pixels under the pointer. The value in the left column is the original color value and the value in the right column is that of the color after the adjustments have been made. The Histogram tool displays the quality and the tonal range of the image. You can preview the effects of image processing when the image data is displayed in the histogram.

The History panel displays the individual steps that were used to process a file during a session and allows an operator the opportunity to go back in time and deselect a step or steps. Each time a new task is used on an image, it is added to the panel. It is also possible to create a document from the History as a screen shot or snapshot. Once the workspace setup is correct, it is recommended to save that setup of the workspace.

Image Size and Display Resolution

One of the most important and first determinations about working on a file is to learn how many pixels are in the file, defining its digital resolution. An image's pixel resolution can be easily determined by accessing the image size function found in the "pull down" Image menu. Located in this menu are a number of commands including image size and canvas. By launching the image size function, a dialogue box appears on screen, sharing the file's size, number of pixels and other attributes. The image size can be changed in this window. Before modifying any image and its size, be certain what each image attribute in the control panel can accomplish. The image size window shares a file's size, the number of pixels in the file, both height and width, as well as the file's display resolution or DPI. Of great importance when changing an image's file size is the proper use of the resample function. A file will have a specific number of pixels that can be displayed in any number of resolutions. LCD or screen displays use 72 dpi, however 96 dpi may be used on 4k monitors for display resolution, and printers require use of 200–300 dpi depending on the printer. Whenever resampling, pay close attention to the total file size. Sacrificing pixel resolution for a particular application is not desirable unless mandated.

Image Editing Tools Overview

- Clone tool: creates a sampling of image data taken from one region in a file and then uses this to replace targeted pixels in another region. This tool is frequently used for dirt removal.
- Crop tool: trims the outside part of a picture using a free or fixed aspect ratio.
- Dodge and Burn tools: the Dodge tool is used to lighten an area in an image. The Burn tool is used to darken an area in an image. The size of the tool and the amount of

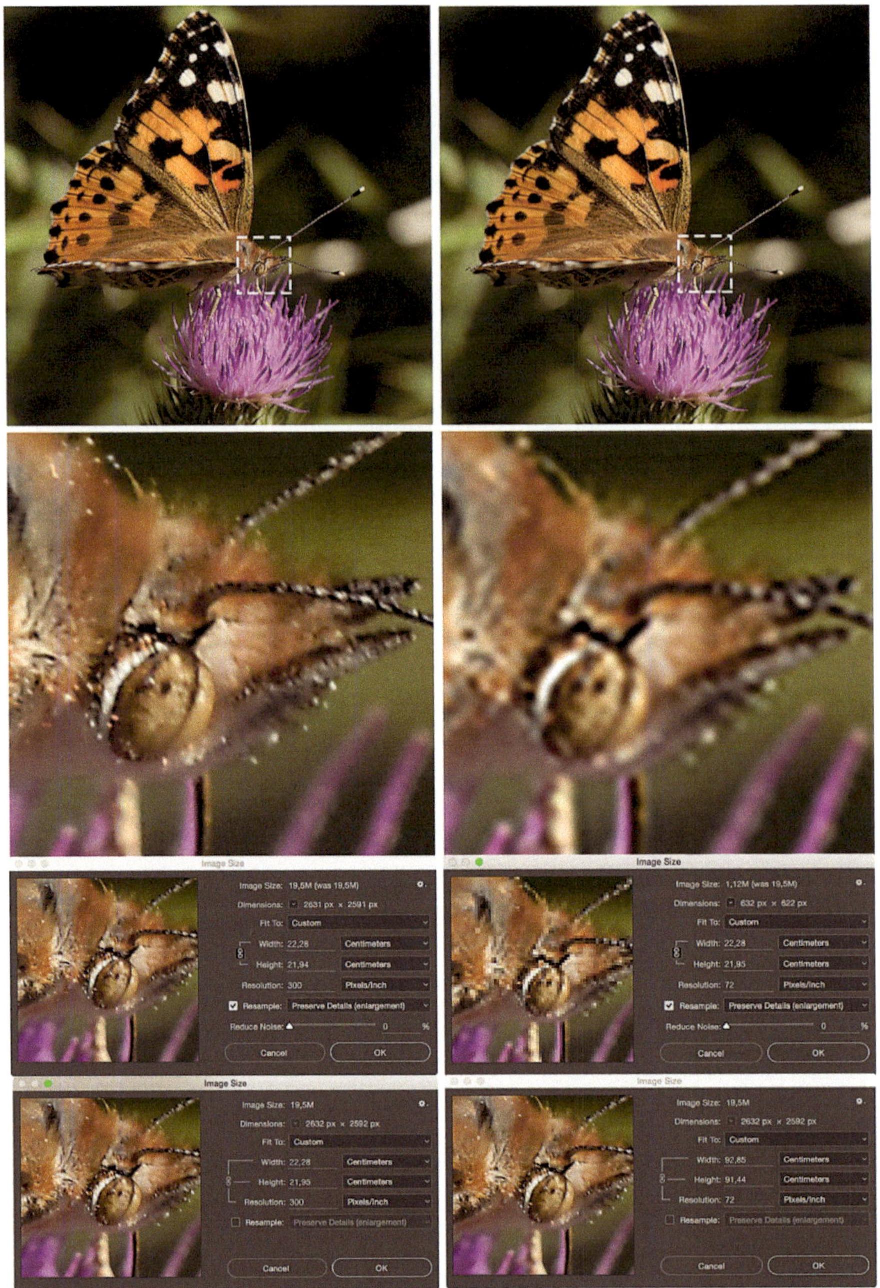

Figure 19.6 This illustration includes important information about an image's display resolution. Image attributes are shared in the dialogue boxes included below each image. Images on the left have a resolution of 300 dpi that was maintained when making changes by cropping only. On the right, all images were created to be 72 dpi. When cropped, the image falls apart because there are not enough pixels to make this adjustment. When changing image size and resolution, it is important to evaluate image size, pixel width and height, resolution and then resample the data for use. Adobe product screenshot(s) reprinted with permission from Adobe

lightening or darkening can be adjusted as a percent. Slower changes are ideal targeted to the highlights, midtones or shadows regions.

- Eyedropper or Color Sampler tools: these measure pixel data in the active layer. They can go down to the pixel level or to a grid of pixels such as 3 × 3.
- Healing Brush tool: this paints/blends pixels using a preselected sample or pattern to repair imperfections in an image. This is an excellent tool for blending background imperfections.
- Lasso tool: selects pixels by drawing around a region to be selected. This selection can be freehand or magnetic.
- Magic Wand tool: selects pixels with pre-determined similar brightness. May also be called the Color Wand tool, which selects pixels based on a pre-determined color.
- Zoom tool: changes the scale of the image display. Allows for individual pixels to be observed greatly enlarged.
- Marquee, Circle, or Rectangle tool: selects pixels in pre-defined shapes such as a circle, rectangle, square, or line.
- Ruler tool: measures distances in an image by counting the number of pixels in a specific distance. Calibrating the Ruler tool can be helpful when calculating known distances.
- Move tool: moves the region of a file containing pre-selected pixels.
- Perspective tool: allows an image's shape to be changed. Might also be called skew or transform.
- Rotate tool: allows an image to be rotated clockwise or counter clockwise in various degree increments. There are pre-set selections for 90°, −90°,180° and custom.
- Select All: a tool for selecting all the pixels in the entire file.
- Text tool: allows text to be created in a new layer on top of the image or background layer. Fonts, colors, and size can all be modified. Adding type creates a new layer in a file and frequently will require the image be flattened or collapsed before saving.

Image Processing Fundamentals

Improving an image's readability or visibility is a two-step process. First, the entire file or parts of the file must be selected prior to processing. Once selected, the pixels can then be modified. Pixels can be modified or changed in an image in two different ways. One method does not modify the pixels but creates a new image layer on top of the original file. This method uses an adjustment layer and preserves the original data so it can always be re-evaluated as long as the layers are not collapsed. The down side to using adjustment layers is that files become larger. This strategy would be characterized as non-destructive image processing.

A more common method is to work directly on the native pixels themselves. In this type of image processing, data is changed. Going backwards is near impossible but most image processing is accomplished using this approach. This method creates files that often remain approximately the same in megabytes as when they were created but pixels in the files are forever changed. If for any reason, when working on an image, there is a need to revisit an older step in the image processing sequence, the scientist photographer cannot go back to the early steps once a file has been saved. If that native file was not properly archived, the

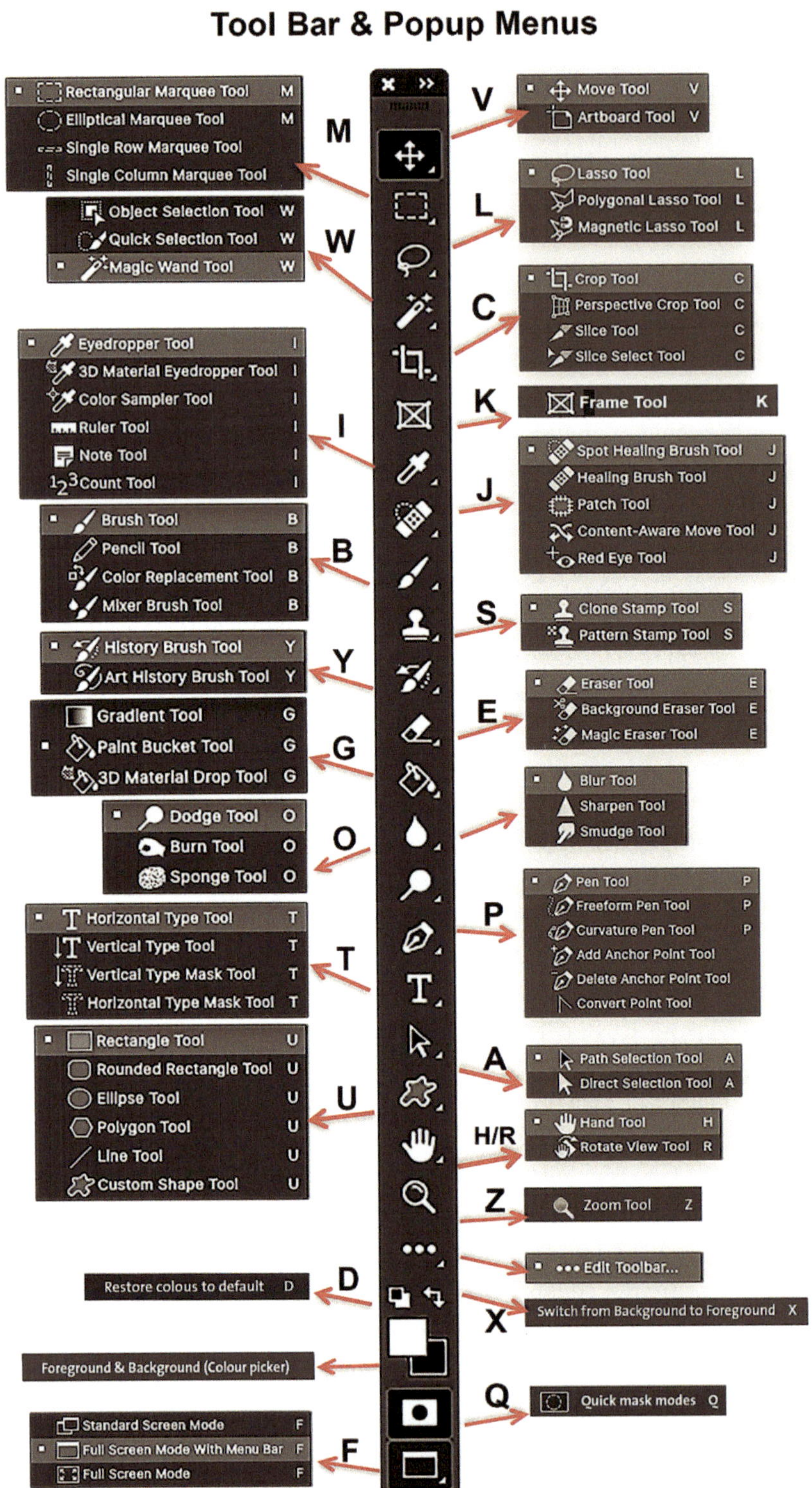

Figure 19.7 The Photoshop 2020 tool bar which has been updated with new tools and features. Capital letters indicate keyboard short cuts that are activated when typing that specific letter into the program. The H (hand) and Z (zoom) tools are particularly useful. Adobe product screenshot(s) reprinted with permission from Adobe

problem can be significant. In either of these two scenarios, the edited file should always be Saved As. This will keep all the individual steps/improvements in one file and the native file is preserved.

File Formats

Before discussing image processing further, it might be useful to review all of the various file formats.

- **TIFF** or TIF: Tagged Image File Format. A lossless and universal file format. It is possible to use LWZ compression but not affect the integrity of image data.
- **JPEG** (JPG): Joint Photographic Experts Group. This file format has a high amount of compression and over time can introduce a loss of quality.
- **PSD**: Photoshop Document. A practical and stable file format specific to Photoshop. It allows repeatable access to editable layers.
- **BMP**: sometimes called bitmap or DIB file format. It is not very common today but was a decade ago.
- **PNG**: Portable Network Graphics format. It is good for Microsoft Powerpoint and Word applications. When a screen grab is made, frequently it is in the PNG format.
- **RAW**: image format from a DSLR camera. RAW is one of the cleanest digital file formats directly from the camera sensor.

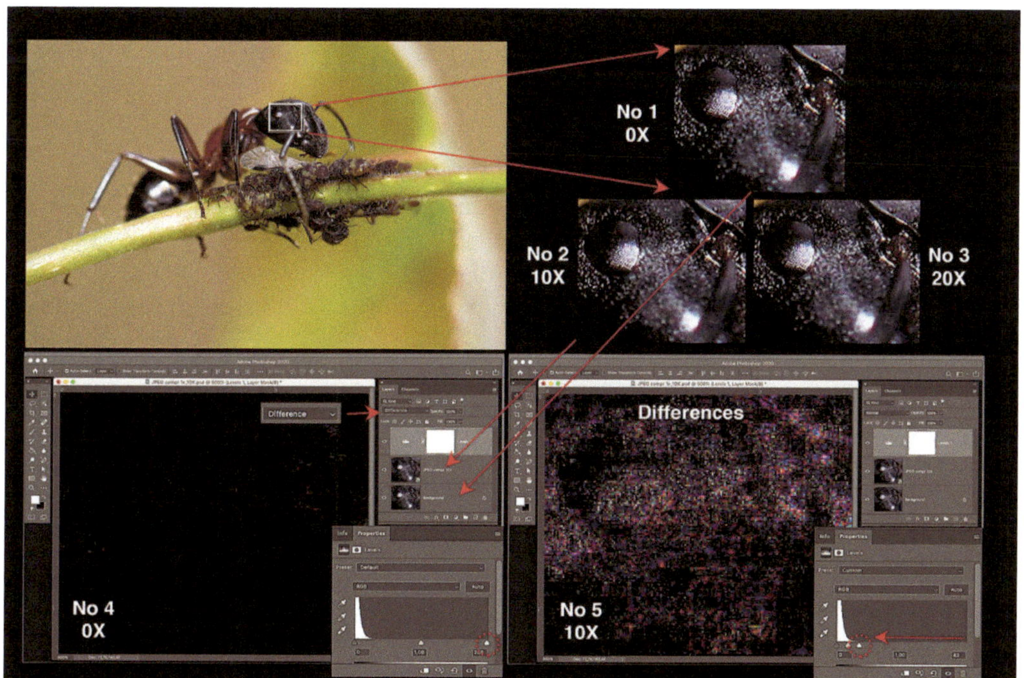

Figure 19.8 It is possible to evaluate the effects of using various JPEG compressions before making a final commitment to how much compression is needed or desired. Adobe product screenshot(s) reprinted with permission from Adobe. Image courtesy of Staffan Larsson.

- **EPS**: Encapsulated Post Script. Often vector format and used in graphic arts applications.
- **GIF**: Graphics Interchange Format which includes fewer colors. It is frequently used for web applications.

JPEG is perhaps the most common file format and has a couple of advantages but also some disadvantages. It is sometimes written JPG. JPEG compression "dehydrates" or removes image data. In some instances it causes loss of important information which becomes amplified if the file is saved several times using low and medium compression. In these situations, fine details and colors will become smoothed out. If a file is saved only once, it is acceptable to use JPEG maximum quality (12–10). To be sure there is no data loss, save the file in the lossless format of TIFF or PSD.

Steps Used to Evaluate JPEG Compression

Start with a native file.

1) Crop and save a part of an image at max jpeg quality (Example 1).
2) Next save a copy of the image at a low jpeg quality (Example 2).
3) Save the max jpeg quality image as a PSD file and copy the low quality image and paste it into new layer 2.
4) Stay in layer 2. Go to the window to the left at the layers panel and click "Normal" and you find different options. Choose "Different". The image turns to black.
5) Add a levels adjustment layer to the file.
6) Adjust levels. Push the small highlight and the mid-tone triangles to the left at the beginning of the right side of the slope. Continue until the noise difference between the images becomes visible.

Improving Contrast and Appearance

There are many image processing strategies that can be chosen to improve an image's contrast and appearance. The image processing can be easy and fast if the image has a good exposure from the beginning. Image processing can be more difficult and time consuming if you start with files that have deficiencies.

For new users of image processing, these are suggested steps for making minor adjustments.

Making Minor Adjustments to Contrast or Color

From the menu bar, select Image > Adjustments > Levels, or Curves, or Color Balance. You can then make the desired changes and save the image file.

It is more complicated when advanced image processing is required. These are the suggested steps for images that require major adjustments. If an image requires more advanced adjustments, you can use the steps shared above but instead of working directly on the file, use the Adjustments layers. You can see more specific information in Figure 19.14(A). Adjustment layers are effective for changing Color Balance, and Color, and Contrast Correction using channels or levels.

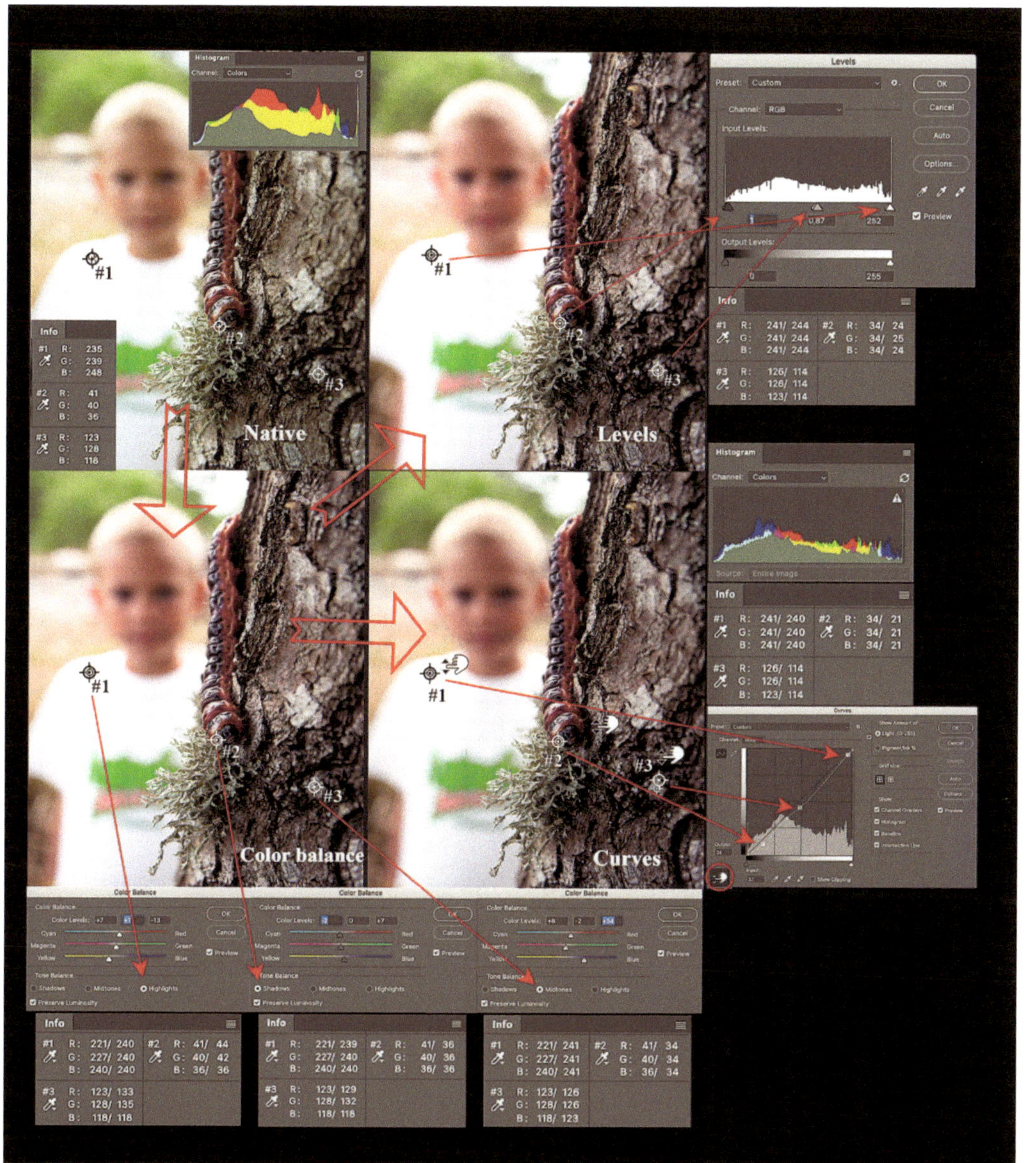

Figure 19.9 This series of pictures reveals how an image's color and contrast can be adjusted, creating a more acceptable appearance. The top left photograph reveals the histogram and the color of the file directly out of the camera. Three color makers have been added to the native file and can be seen as #1 at the highlight white area, #2 at the black area and #3 at the middle gray area in the image. In the dialogue box also at top left, the RGB values are visible. Adobe product screenshot(s) reprinted with permission from Adobe.

The image on the bottom left shows how the RGB levels can be adjusted by using Color Balance. The top left image shows how the RGB levels can be adjusted to produce a histogram that is more blended and without large color spikes. This data can also be seen in the RGB values in the dialogue box. The bottom right image demonstrates how to optimize the image's contrast using an adjustment level. This file's white point has been set to 241, black to 34 and midtones to 126. Adobe product screenshot(s) reprinted with permission from Adobe. Images courtesy of Staffan Larsson.

Setting an Image's White Point Using the Color Balance Function

An image will have a potential luminance ratio of 0 for a black or shadow and 256 for a white or highlight. It is impossible to print a pure black or pure white and so it is important to keep a touch of tone in the shadows and the highlights. A useful starting point is to set a shadow level to a brightness of 5 and the highlight brightness to a value of 240. The ultimate goal is to have the same brightness level and color values in each of the RGB channels for all files. By creating a targeted brightness value, consistency can be achieved file-to-file.

1) Use the color sampler tool on at least one neutral spot on the image but preferably on a highlight (white) region, one on a midtones (middle gray) region and one on a shadows (black) region. The RGB values will be displayed in the Info panel. Set a marker in a region that is useful to monitor. This tool can be found with the eyedropper tool.
2) Next go to Image > Adjustments > Color Balance.
3) In the Properties panel, select Shadows, Midtones, or Highlights to select the tonal range where you want to focus your changes.
4) A second panel is now added into the Info panel. In this window, the changes for the Color Balance panel will be highlighted. You can work on the highlights in the Color Balance panel.
5) Drag a slider toward a color that you want to improve in the image; drag a slider away from a color that you want to decrease in the image.
6) The values above the color bars display the color changes for the red, green, and blue channels. For LAB images, the values are displayed for the A and B channels. These values can range from −100 up to +100.
7) Using the complimentary controller slider, adjust the small triangle in a direction that makes the sample appear more neutral. Keep an eye on the Info panel and try to reach the value of 240 in a white region in the Blue, Green, and Red channels.
8) You can now set the white point and gray balance.

Changing Contrast

There are two simple ways to change contrast. If you need minor adjustments, choose Levels. For major adjustments and more precise control, choose Curves.

Using Levels to Increase the Contrast

1) Select Image > Adjustments > Levels.
2) Continue to use the information second window from the three samples in the Info panel you made with the Color Sampler tool. All changes in the Levels panel will be highlighted and visible there.
3) Move the small white triangle located on the bottom of the histogram on the right side, to the left. This will manage important information in the highlight region of the histogram. Next, move the small black triangle located on the bottom of the histogram on the right side, to the right. This will manage important information in the shadows. Finish the editing by adjusting the midtones, moving the gray triangle located in the middle to the right or to the left.

4) Go to the RGB window. Click the left mouse button on the window and all the RGB channels will appear.
5) Choose the Blue channel. Push the small white triangle on the right hand side of the histogram (highlight) to the left. Stop where the slope starts at the histogram.
6) Continue to adjust the Red. Go back to the RGB window again and change from the Blue to the Red channel.
7) You have now set the white point and some more contrast (gray balance).

Using Curves to Increase Contrast

Continue to use color samples and information from the Info panel that was made in the Color Balance.

1) Go to Image > Adjustments > Curves.
2) Activate the hand in the left upper corner with a mouse click.
3) Move the mouse pointer into the area where you want to change contrast.
4) The pointer becomes an eyedropper (Color Sampler tool). Hold down the left mouse button and move the mouse up and down. The contrast changes will be visible in real time. In the Curves display window, small, added makers for the adjustments and movements will be visible. Stop when you are satisfied with the result. Go to the next location you want to adjust and repeat the procedure.
5) You can also use the same procedure in channels to adjust colors (Figure 19.13).

Setting the White and Black Point with Levels in a Simple Way

1) Begin by making a Levels adjustment layer.
2) Select the white Color Sampler (eyedropper) at the Layer panel. Double click the eyedropper and it will open the Color panel. Locate the small circle in the upper left corner. This location represents R 255 – G 255 – B 255 = total white.
3) Set a new standard default white point (gray point value) with the value of R 240 – G 240 – B 240. You can also set the luminescence L from default L 100 to a new default number of 95 and it will lead to the same result. Locate the small circle in the upper left corner and notice that it has moved. This new location in the white corner represents a new setting in the color space. Save by clicking OK.
4) a. Go to the black eyedropper in the Levels panel. Double click the eyedropper and it will open the Color panel.
   b. Locate the small circle down in the left corner. This location represents R 0 – G 0 – B 0 = total black.
   c. Set a new standard default black point (gray point value) with the value of R 5 – G 5 – B 5. You can also set the luminescence L from default L 0 to a new default L 5 and it will lead to the same result. Locate the small circle down in the left corner and notice that it has moved. This new place in the black corner represents the new value in the color space for black (gray point value). Save by clicking OK.
5) This will become the new default setting unless changed at another time. This setting is ready to be used as a fast and useful way to establish a white point setting on future files.

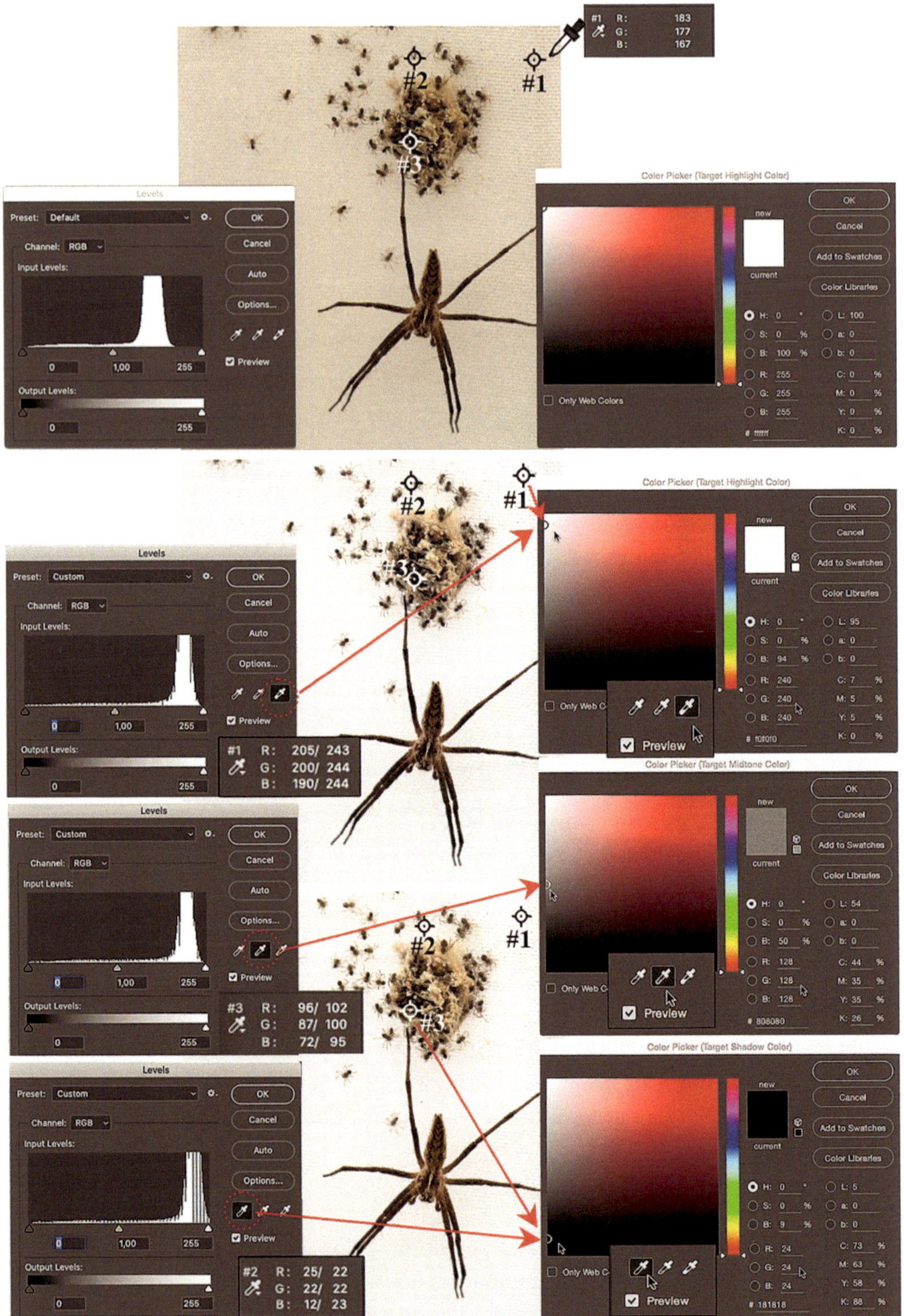

Figure 19.10 This composite file shows how image processing can be used on a file (upper image) when setting the best black point and white point settings. The images in the middle reveal how to adjust the black triangle and white triangle (point) on the image's histogram. The image on the bottom reveals how these corrections would appear having a white point value of 243 in each channel. Adobe product screenshot(s) reprinted with permission from Adobe. Images courtesy of Staffan Larsson.

6) To use the new default setting adjusting white balance, go to the white eyedropper in the Levels panel. Position the eyedropper on a neutral place on the specimen. Click the left button and take a sample. Now you have a white point. Observe the new histogram.

Changing Contrast by Setting White and Black Points using Levels

This is a method that can be used to reduce unnecessary information in the luminance ratio of the gray scale.

Setting the White Point

1) Open a file and select the Levels control feature. To evaluate the highlight, hold down the Alt key and grab the small white triangle on the right hand side of the histogram (highlight) to the left.
2) Move the control while holding down the Alt key. Stop when you see the first white spots in the black parts of the image. This is the start of highlight information in the image. You have now set the white point.

Setting the Black Point

1) This is a way to reduce unnecessary information in the luminance ratio of the gray scale.
2) To evaluate the shadow, hold down the Alt key and grab the small black triangle on the left hand side of the histogram (shadows) to the left.
3) Move the control while holding down the Alt key. Stop when you see the first black spots in the white parts of the image. This is the start of highlight information in the image. You have now set the black point.

Color and Contrast Corrections

Color and Contrast Corrections using Channels and Levels

1) Set three sample markers made with the Color Sampler tool, one on a highlight, one on a midtone spot, and one in a shadow. Notice the Info panel. All changes in the Levels panel will be highlighted and visible there.
2) There are three ways to find the Adjustment Layer menus:
   a. Go to Layers > Adjustments Layers > Levels.
   b. Click on the Adjustment Layer icon at the bottom of the Layer panel and create a Levels Adjustment layer.
   c. Click on the Levels tool icon 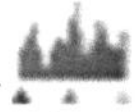 in the Adjustments panel which is directly above the Layers panel. To adjust the shadows and highlights manually at the Levels panel, drag the black and white Input Levels slider to the edge of the first group of pixels at either end of the histogram.
3) Next, move the small white triangle located on the bottom of the histogram on the right side to the left. This will reduce the presence of irrelevant information in the highlight region of the histogram. Next, move the small black triangle located on the bottom of

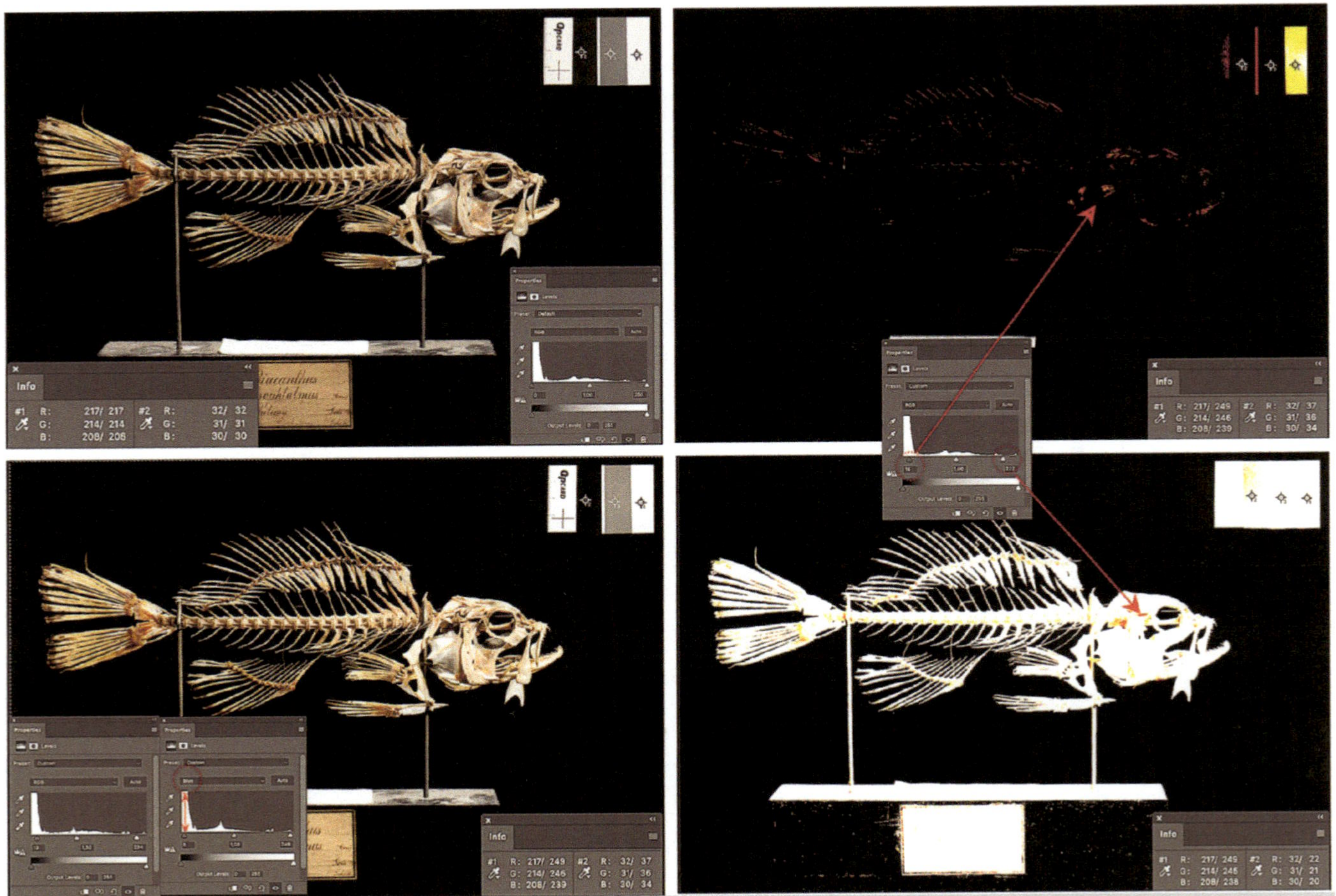

Figure 19.11 The screens that would be visible when optimizing a file's brightness attributes leading to color correction. On the top left is how the file would appear as opened. Set a color marker/picker into a region of primarily white. On the top right, once the Levels tools have been activated, the image's black point is set based on the histogram to 31 and the white point to 232. On the bottom left, selecting the Red channel, the image's brightness is also set to 232. To complete the improvement, in the bottom right image, the Blue channel is then set to 232. Adobe product screenshot(s) reprinted with permission from Adobe. Images courtesy of Staffan Larsson.

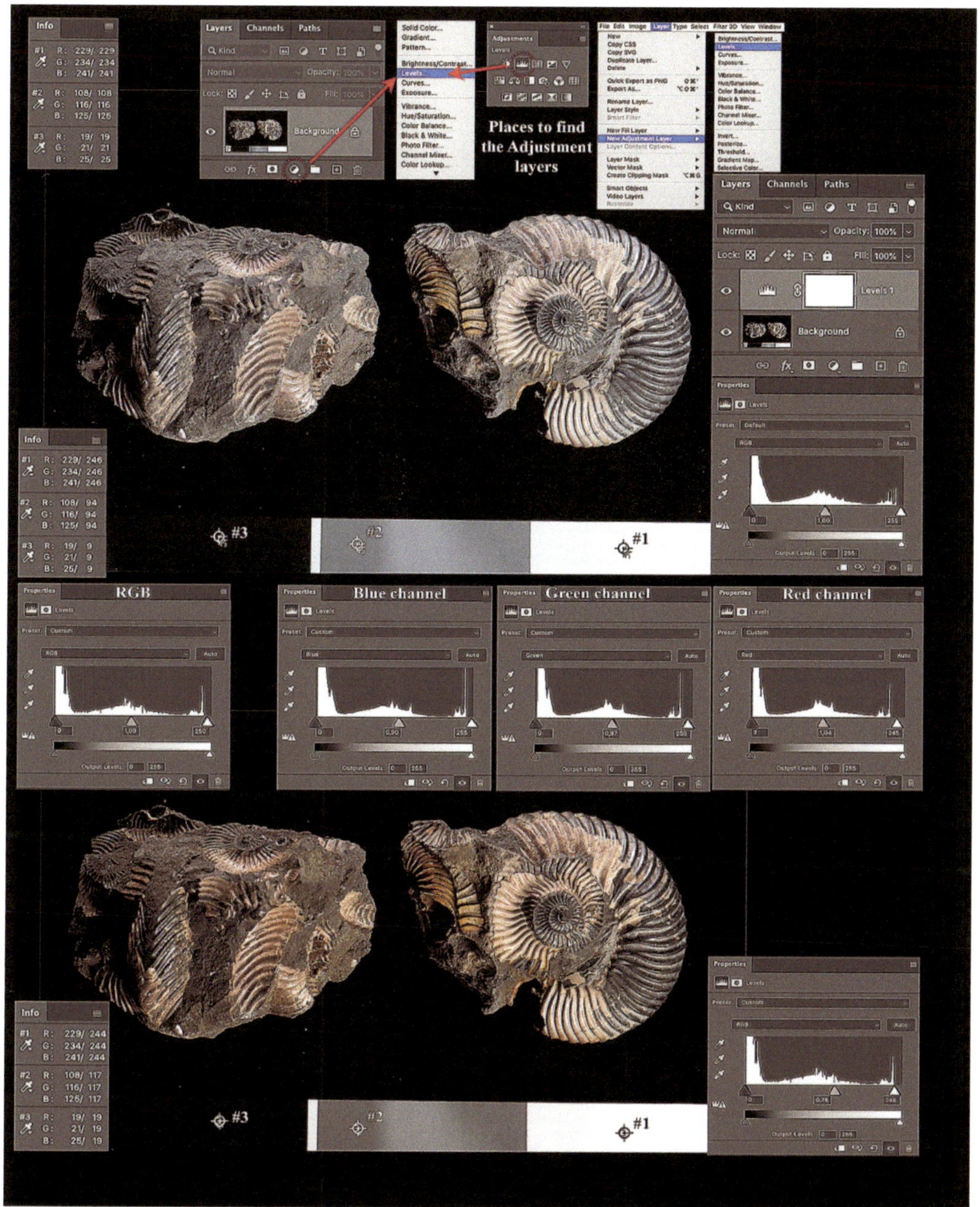

Figure 19.12 The screens that would be visible when optimizing the file's brightness attributes when used for color correction. At the top is how the file would appear when opened. On the bottom of this illustration is an image from a three-part calibration target that can be purchased from a photographic dealer. It should be included to facilitate good color correction if there is no distinct black, white or gray spots in the subject. Set a color marker/picker into a region of primarily white (#1), middle gray (#2) and dark gray/black (#3) surface visible in the calibration card. Between the images to the left and once the levels tool has been activated, the image's black point can be set based on the histogram to 9 and the white point to 246. To complete the improvement, adjust each channel, Blue, Red, and Green, by selecting and moving the black, gray and white triangles to reach an equal result in each channel for brightness, shadows and midtones. I used 244, 19 and 117. Adobe product screenshot(s) reprinted with permission from Adobe. Images courtesy of Staffan Larsson.

the histogram on the right side to the right. This will reduce unimportant information in the shadows. Finally, adjust the midtones by moving the gray triangle located in the middle to the right or to the left.

4) Go to the RGB window at the Levels panel to adjust tones for a specific color channel.
5) Select an option from the Channels menu. Press the left mouse button on the window and all the RGB channels will show up. To begin adjusting the color balance, it is usually accepted to use the middle value of one of the RGB channels as a starting point. In this case, choose the Blue channel. Push the small white triangle at right side of the histogram (highlight) to the left. Stop where the slope starts at the histogram.
6) Continue to adjust the Red. Go back to the RGB window again and change from Blue to Red channel. You have now set the white point and some more contrast (Gray Balance).

Color and Contrast Correction using Channels in Curves and the Directed Adjustment Tool

1) Open the file and access the Color Sampler tool located with the eyedropper tool.
2) Set a color measurement marker in a highlight area (#1) and, if it's needed, also in the midtones (#2) and shadows (#3) with the Color Sampler tool.
3) Make a new Curves Adjustment layer.
   There are three ways to find Adjustment Layer menus.

   a. Go to Layers > Adjustments Layers > Curves.
   b. Click on the Adjustment Layer icon at the bottom of the Layer panel and create a Curve Adjustment layer.
   c. Click on the Curves tool icon in the Adjustments panel.

4) Activate the hand in the left upper corner with a mouse click.
5) Move the mouse pointer into an area in the image you want to change contrast.
6) The pointer becomes an eyedropper (Color Sampler tool). Hold down the left mouse button and move the move the mouse up and down. The contrast changes will be visible in real time. In the curves display window small, added markers for the adjustments will be visible. Stop when you are satisfied with the result. Go to the next location you want to adjust and repeat the procedure.
7) Use the same procedure in each of the RGB channels to adjust colors.
8) Check the Info panel while you are making the color balance adjustments. Try to achieve balance by separately adjusting the curve in each RGB channel by using the Directed Adjustment tool and moving the markers on the curve, touching the regions for the highlights, shadows, and midtones.

Adjustment Layers for Improving Color Balance and Contrast Correction using Channels in Levels

1) Open the file and access the Color Sampler tool located with the Eyedropper tool. Can you reference Fig 19.14A as well. Set a color measurement marker with the Color Sampler tool in a highlight in the window area (#1). Continue with one in the midtones (#2) to the upper left of the hand, and finally one dark neutral area (#3) to the right.

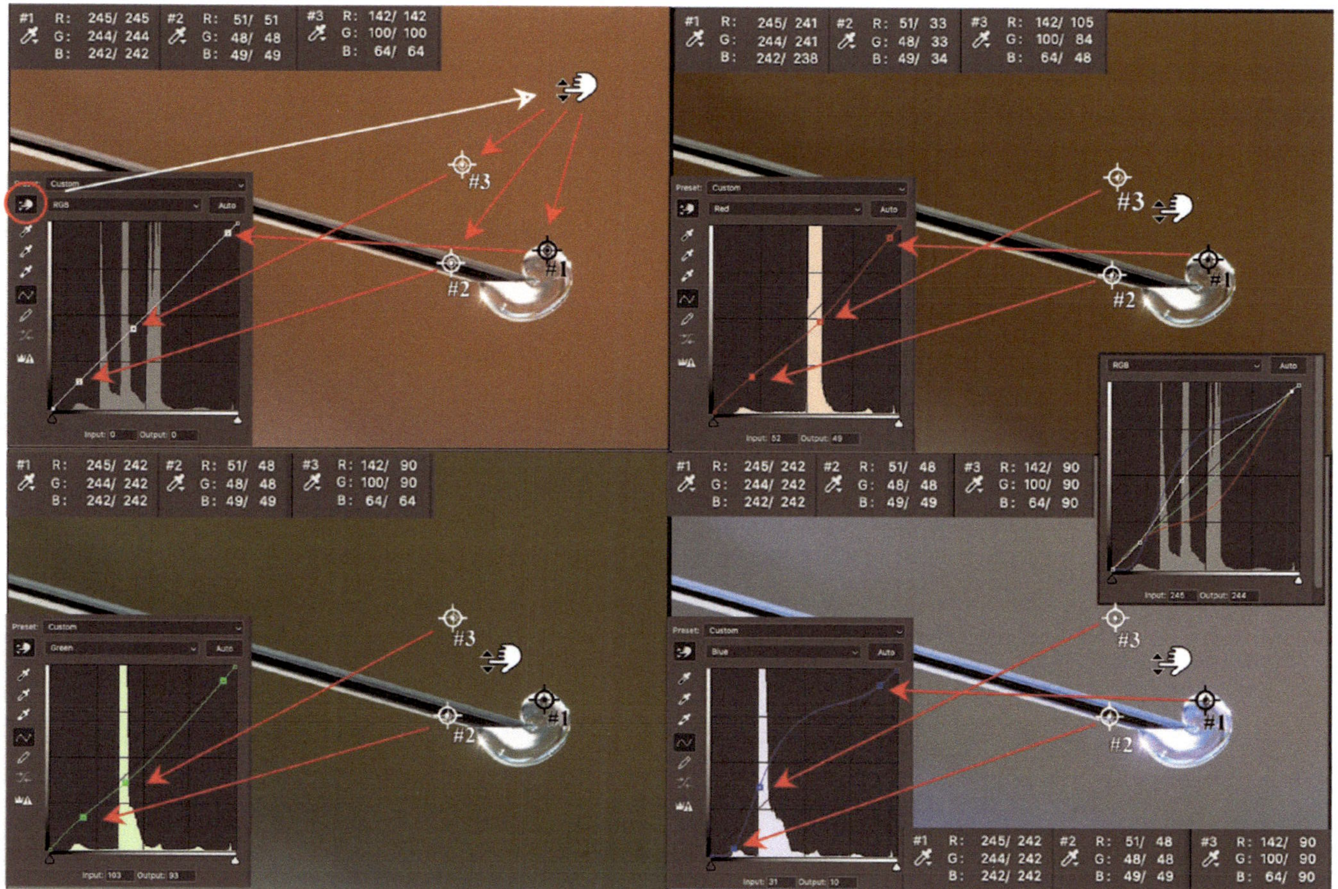

Figure 19.13 A tiny drop of chemotherapeutic gel on an injection needle. It has been photographed in ambient tungsten light and looks yellowish. To achieve the original colors, image processing was required to produce accurate and neutral color. This figure demonstrates how to use Curves for optimizing an image. In the photograph in the top left, the file is opened, and Color Picker is added to a primarily white area. In the photo on the top right, using Curves, the image is adjusted to create a value of 235, matching the brightness of the Red channel when the file was initially opened. In the photo on the lower left, the Blue channel is next set to 235 by adjusting the shape of the curve. The lower right image has been finally set to a 240 RGB value and a small refining of the black point setting. Adobe product screenshots reprinted with permission from Adobe. Images courtesy of Staffan Larsson.

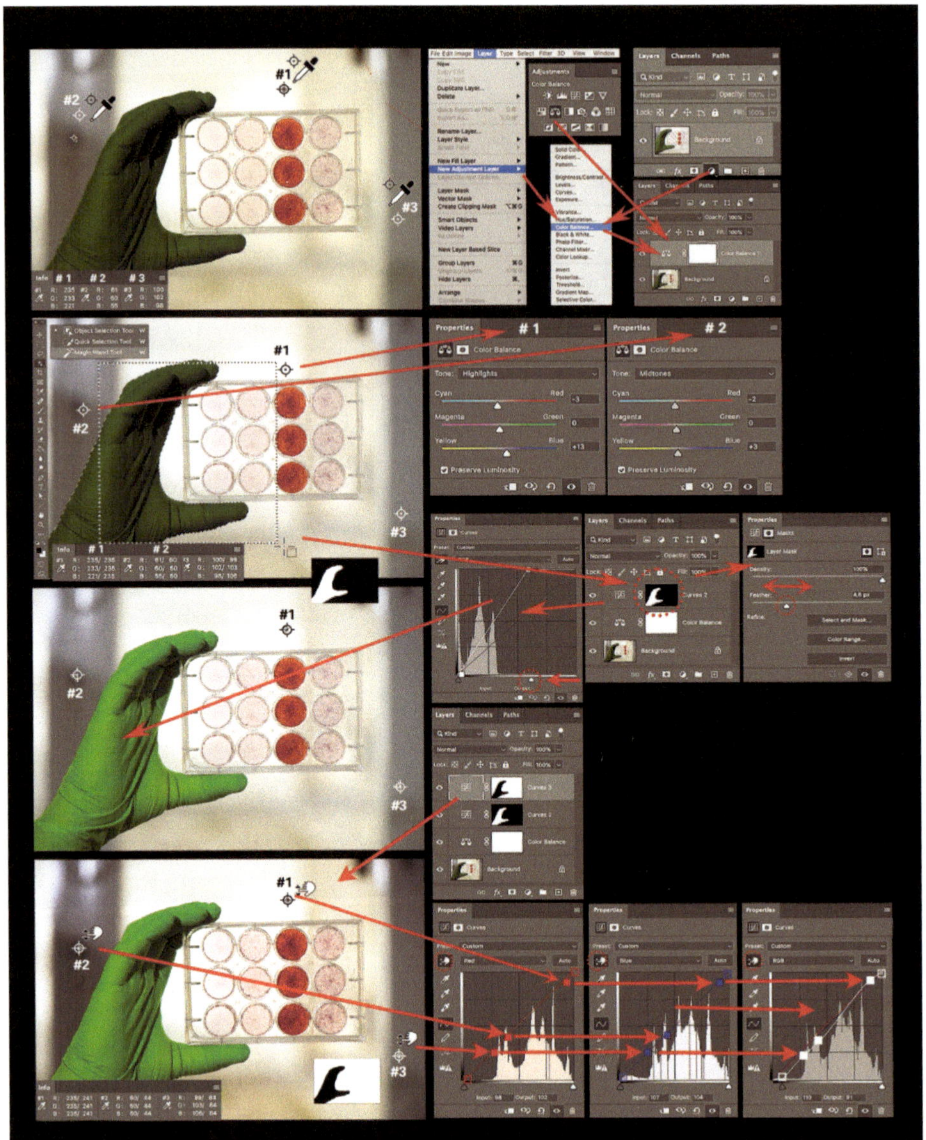

Figure 19.14A The advantage of using Adjustment layers in image processing, is that you can always return to any place in the image processing sequence after making an adjustment you do not like. The native file remains as the bottom layer and is unaltered. This figure of specimens in a welled plate was exposed using backlight from a window, creating flare and lowering contrast. The hand became dark and a shadow. My goal was to achieve balance between shadows and contrast and manage the backlight flare. This series of pictures reveals how an image's color and contrast can be adjusted leading to a more acceptable appearance. The photograph on the top reveals the exposure and the color directly out of the camera. Three color markers have been added to the file and can be seen as #1 at the highlight white area, #2 at middle gray area and #3 at the dark gray area as a black was absent in the image. In the dialogue box also on the top left, the RGB values are visible. The middle image shares how the RGB levels can be adjusted to produce a histogram that is more blended and without large color spikes. This data can also be seen in the RGB values in the dialogue box. The image on the bottom demonstrates how to optimize the image's contrast using a level adjustment level. This file's white point has been set to 221. Adobe product screenshots reprinted with permission from Adobe. Image courtesy of Staffan Larsson.

2) Make a new Color Balance adjustment layer.
3) Start to adjust the color balance in the highlights (#1) and the midtones (#2).
4) Next step, in the second image, is to adjust the unexposed green hand. To make a separate adjustment of the hand, you have to make a selection that will end up as a mask of a Curve layer.
5) To select the hand, you can choose between the tools behind the Lasso and Magic Wand tools. One of the options behind the Magic Wand tool is the new Object Selection tool.
6) Drag and cover the hand with the Object Selection tool. When finished, the "marching ants" will appear. Save them or go to choose Curves Adjustment layer. The "marching ants" will create a mask in Curves layer. Only the highlighted white area is transparent and will let the Curves make the adjustments within it. The black parts will cover the image and the curves will not be adjusted. If you click on the mask you can soften up the mask edge.
7) Go to Curves layer. To exclude the unnecessary shadow, move the black triangle in the left corner to the beginning of the slope. Remove the unexposed information.
8) Create a new Curves layer to make the final contrast adjustment.
9) Activate the hand in the left upper corner with a mouse click.
10) Move the mouse pointer into the area where in an image you want to change contrast. The pointer becomes an eyedropper (Color Sampling tool). Hold down the left mouse button and move the move the mouse up and down. The contrast changes will be visible and live in real time. In the Curves display window, small added markers for the adjustments and movements will be visible. Stop when you are satisfied with the result. Go to the next location you want to adjust and repeat the procedure.
11) You can also use the same procedure in Channels to adjust colors.
12) The last step is to make final adjustments in contrast and color balance, excluding the hand.
13) You can see the steps shared graphically in Figure 19.14B.
14) Create a new Curve Adjustment layer.
15) To exclude the hand from the new actions you have to make a mask. Copy the mask from before by following these steps. Press the mouse button on the mask of the hand. Hold down the Alt key and move the mask to the Curve layer above. "Replace the layer mask" will appear. Release your finger from the mouse button. You will see a copy of the mask.
    a. You will need to invert your layer mask, just hold Command + I on Mac (or Control + I on Windows).
    b. The white area in the mask is transparent and will let through the actions from the curves.
16) To make the final Curve adjustment, see the previous section about Curves (Figure 19.13).
    a. A good recommendation, if your file contains many layers, is to make a top layer that includes everything important. You also can switch off alternative layers and save them for later.
    b. To create the top layer, use Command + Alt + Shift + E on a Mac (Ctrl + Alt + Shift + E in Windows).

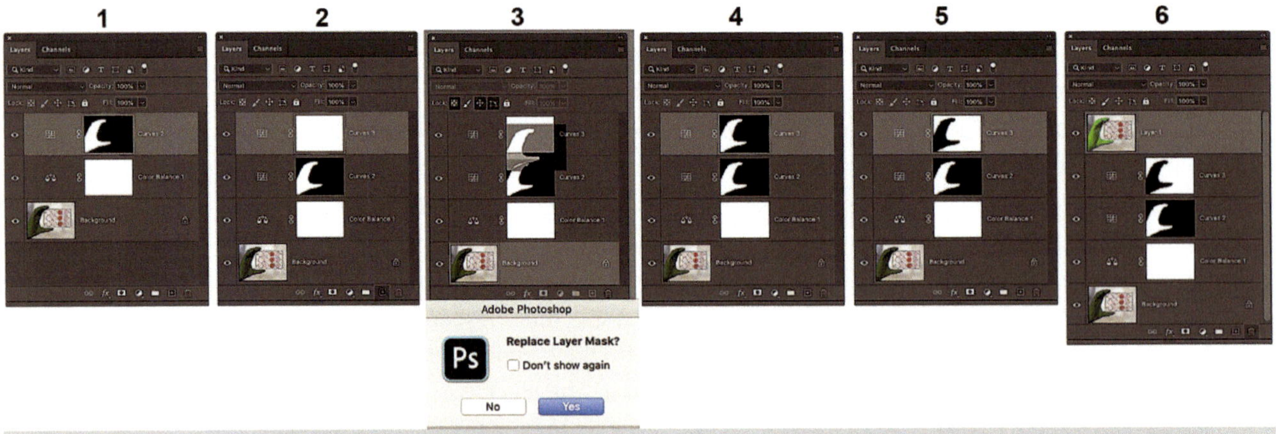

Figure 19.14B This figure reveals the steps used in .14A and is reproduced in larger detail. Adobe product screenshot(s) reprinted with permission from Adobe.

Converting RGB Files to Black and White (Grayscale) – Converting Directly to Grayscale

1) Go to Image > Mode > Grayscale. Select Discard when prompted.
2) Next, adjust the image contrast. Select Image > Adjustment > Levels or Curves.
3) In the Levels histogram, move the small black triangle to the left (shadow region) and the small white triangle to the right (highlight).
4) This reduces the information that is not important in the image.

Converting RGB Files to Black and White (Grayscale) – Using Split Channels

1) Click on the Channels panel.
2) Select the channel with the most black and white information.
3) Click on the icon in the top right corner to open the menu. Select Split Channels.
4) Now each RGB channel will become a separate black and white image. Compare and delete the weakest channels and maintain the channel with the best tonal data. Save using a new name.
5) Adjust the contrast if needed: select Image > Adjustment > Levels curves
6) Go to Image > Mode > Grayscale and save

Converting RGB files to Black and White (grayscale) - Channel Mixer

1) Go to Image > Mode > Adjustments > Channel mixer.
2) Activate the Monochrome option box at the bottom left of the Channel Mixer panel by checking that box.
3) The image will become grayscale but remains an RGB image. In the panel adjust the black and white information for each channel as needed. Adjust the small triangles until you reach the correct contrast/grayscale representation. Compare the RGB image

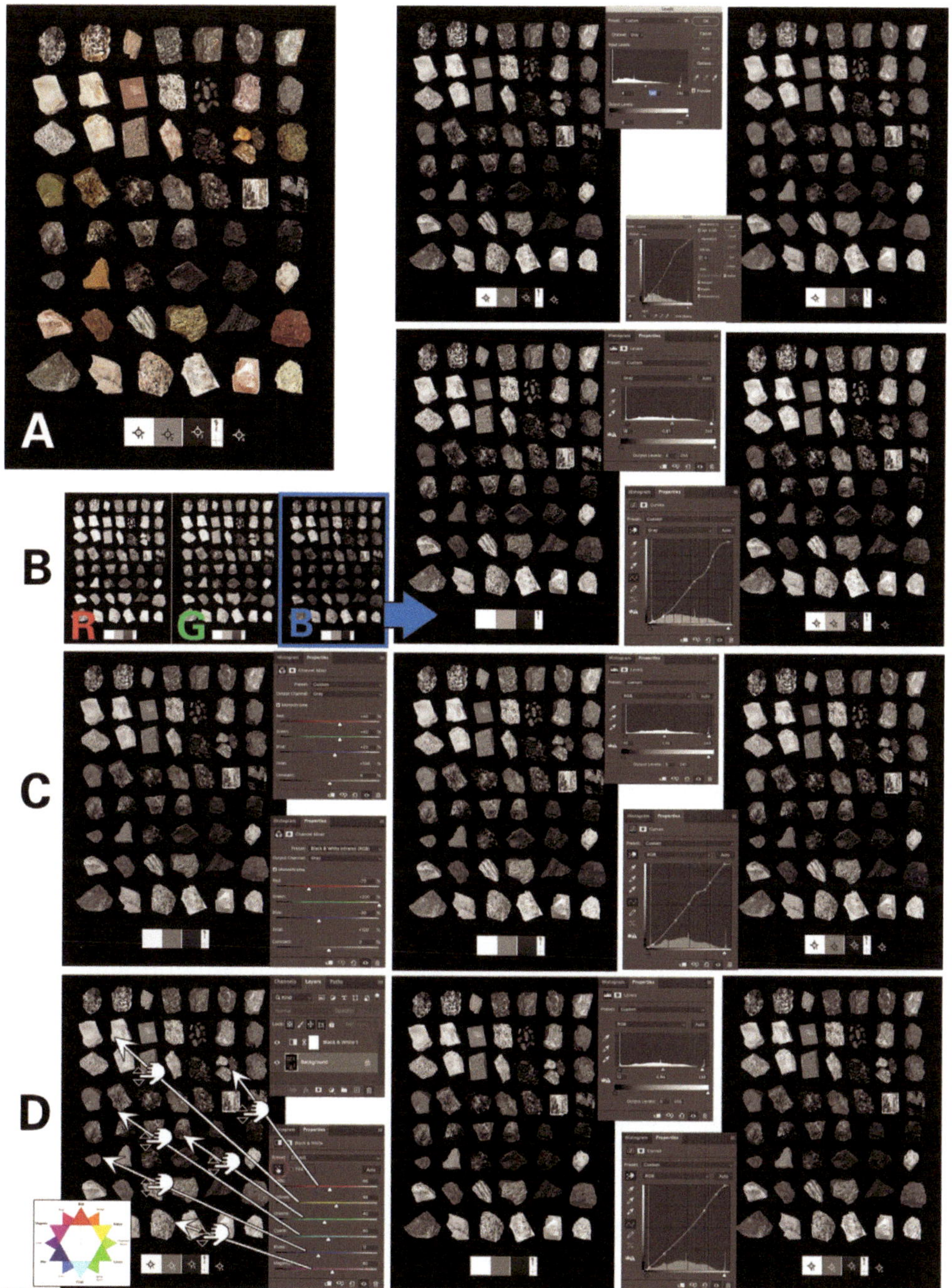

Figure 19.15 This figure demonstrates the outcomes of various methods that can be used to convert an RGB file to a grayscale image. A careful analysis of the results from top to bottom reveal how much more tonal information can be maintained during the conversion using each method. The fastest conversion can occur using the Grayscale Conversion mode and the best result was produced using the Black and White Adjustment Layer. Adobe product screenshot(s) reprinted with permission from Adobe. Image courtesy of Staffan Larsson.

and the black and white image and consider complementary colors when making the adjustments.

4) Finally, convert to black and white. Image >Mode > Grayscale > choose Discard on the popup message menu.
5) Perform the necessary final contrast adjustments. Image > Adjustment > use Levels or Curves.
6) Save As.

Converting RGB Files to Black and White (Grayscale) – Using Black and White Adjustment Layer – Most Advanced

1) Layer > New Adjustment Layer > Black and White.
2) Select the Black and White adjustment layer panel. In an adjustment layer it is possible to go backwards and change settings throughout the procedure.
3) The image will switch to grayscale but remain an RGB image. In the panel adjust the black and white information in RGB and CMYK colors as desired. This method is very precise. Push the small triangles to until you reach the ideal contrast and tones. Compare the RGB image and the black and white image. Always consider complementary colors when performing the adjustments.
4) Convert finally to black and white. Image > Mode > Grayscale > choose Discard on the popup message menu.
5) If necessary, adjust the final contrast: Image > Adjustment > use Levels or Curves.
6) Save As.

Sharpening

It is possible to sharpen an image in several different ways based on output requirements. In each method, different levels of sharpening and control are achieved. It is suggested to sharpen an image using a separate layer on the top of the native image. In this fashion, the sharpening can be changed if using a different output device or a different material. If sharpening an image using a separate layer, set the layer's blending mode to Luminosity. This will avoid color shifts along the edges. It might be useful to sharpen the image multiple times using small amounts. Over sharpening can lead to the production of artifacts or halos around the edges of structures and may appear similar to diffraction effects evident in photomicrographs from an aperture diaphragm.

Sharpening Using Unsharp Masking

The Unsharp Mask sharpens an image by increasing contrast along the edges in an image. It then increases the contrast of neighboring pixels by the amount that is chosen by the operator. Lighter pixels will get brighter and the darker pixels get darker. In addition, it is important to specify the radius of the action. The greater the radius, the more pixels will be involved and the larger the edge effects. Less is more.

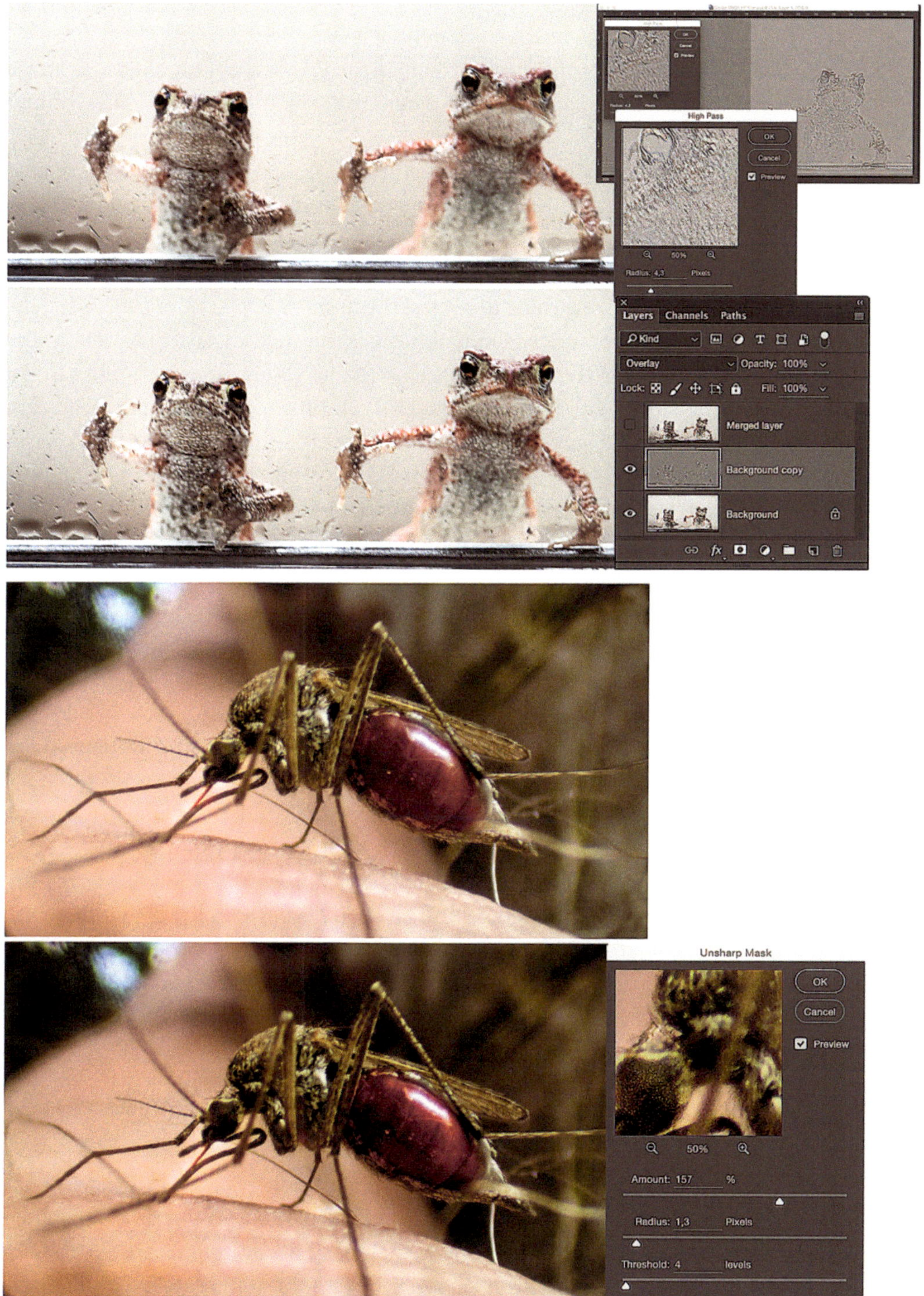

Figure 19.16 This figure shows two outcomes used for image sharpening. At the top is the high band pass method of sharpening and how the dialogue box will appear. Once the filter has been activated; it should be blended using the overlay command. The lower image of the mosquito was sharpened using an unsharp mask filter. Adobe product screenshot(s) reprinted with permission from Adobe. Image courtesy of Staffan Larsson.

Steps for Using the Unsharp Mask Filter

1) Choose Filter > Sharpen > Unsharp Mask. Enable the Preview option. Click the image in the preview window and hold down the mouse to see how the image looks without sharpening.
2) Drag in the preview window to review different parts of an image and click + or – to zoom in or out. Although there is a preview window in the Unsharp Mask dialog box, it's best to move the dialog box so you can preview the effects of the filter in the document window.
3) Drag the radius slider or enter a value to control the number of pixels that will affect the sharpening. The greater the radius value, the wider the edge effect will be. The wider the edge, the more obvious the sharpening. The radius value varies according to the subject matter, the size of the final reproduction, and the selected output. For high-resolution images, a radius value between 1 and 2 is usually recommended. A lower value sharpens only the edge pixels, whereas a higher value sharpens a wider band of pixels. This effect will be less noticeable in print than on-screen, because a 2-pixel radius represents a smaller area in a high-resolution printed image.
4) Drag the amount slider or enter a value to determine how much to increase the contrast of pixels. For high resolution when printing images (300 dpi and large size files), an amount between 150 percent and 200 percent is usually recommended. For low-resolution images (72 dpi and small size files) web, PPT, MS Word etc., an amount between 50 percent and 100 percent is usually recommended.
5) Drag the threshold slider or enter a value assessing how different sharpened pixels differ from surrounding areas before they become edge pixels and sharpened by the filter. For instance, a threshold of 4 affects all pixels that have tonal values that differ by a value of 4 or more from a tonal scale of 0 to 255. So, if adjacent pixels have tonal values of 128 and 129, they will not be affected. This tool helps to control noise or posterization. For images with flesh tones, for example, use an edge mask or try experimenting with threshold values between 2 and 20. The default threshold value (0) sharpens all pixels in the image. To confirm the results of the sharpening put the cursor in the sharpening window and the arrow will shift to a hand. Click the left mouse button up and down and you can evaluate the result before saving.

Sharpening with the High Pass Filter

The High Pass filter retains edge detail in the specified radius where sharp color transitions occur, and it will suppress the information in the rest of the image. A radius of 0.1 pixels works on only edge pixels. The filter removes low-frequency detail from an image and has an effect opposite to the Gaussian Blur filter. In many cases the High Pass filter will give a better result on low contrast images when compared to the Unsharp Mask. It is helpful to apply the High Pass filter on a continuous-tone image before using the threshold command or converting an image to Bitmap Mode. The filter is very useful for extracting line art and large black and white areas from scanned images.

Steps for Using the High Pass Filter

1) Open a file and create a copy of the image file in a new layer.
2) Choose Filter > Other > High Pass. In the layer panel, select and copy the image file and notice the image copy will turn gray. Make sure the Preview option is selected.
3) Click the image in the preview window and hold down the mouse to see how the image looks without the sharpening.
4) You can move around in the preview window to review different parts of the image. It is possible to click + or – to zoom in or out. Although there is a preview window in the High Pass dialog box, it's best to move the dialog box so you can preview the effects of the filter in the document window.
5) Drag the radius slider or enter a value when just enough information of the image breaks through the gray tone. The radius value will vary according to the subject matter, the size of the final reproduction, and the output method. For high-resolution images, a radius value between 1 and 2 is usually recommended. Experience will help guide your work.
6) Next in the layer panel, select the Normal mode from the dropdown menu on the left. Try Overlay, Soft Light, and Hard Light. Each will blend the filter in different amounts and the image will turn from gray with structure to a visible and sharpened picture.

Noise Reduction

When photography transitioned to digital technology, a new image defect accompanied the new technology. The production of noise, a by-product of heat, was created by the sensor. Lack of light, low brightness of light (for example with fluorescence), or a sensor's sensitivity to light are all sources of noise. To minimize noise, photographers can use higher ISO settings on a DSLR camera or use gain or binning with industrial cameras. While these solutions can be helpful, they too can add noise to an image, especially in the regions where there are dark colors or regions with similar tones. Noise will be apparent as grainy specks of texture present across the colored RGB pixels. Noise can occur in chroma (color), which is evidenced as colored artifacts or luminance (grayscale) noise, which will make an image look grainy.

Before starting to process a file and remove noise, zoom in to 100 percent and analyze the image to evaluate the presence and significance of noise. When examining a JPEG file, there may also be JPEG compression artifacts in the image. JPEG compression artifacts will typically be visible as lines or squares of unusual blocks of pixels. Removing JPEG artifacts may or may not benefit from noise reduction methods.

Noise Reduction by Eliminating Luminance Noise (Low Noise): The Despeckle Filter

From the menu bar, choose Filter > Noise > Despeckle. In addition to the possible presence of RGB digital noise, luminance noise will appear as bright and dark patterns of specks. The

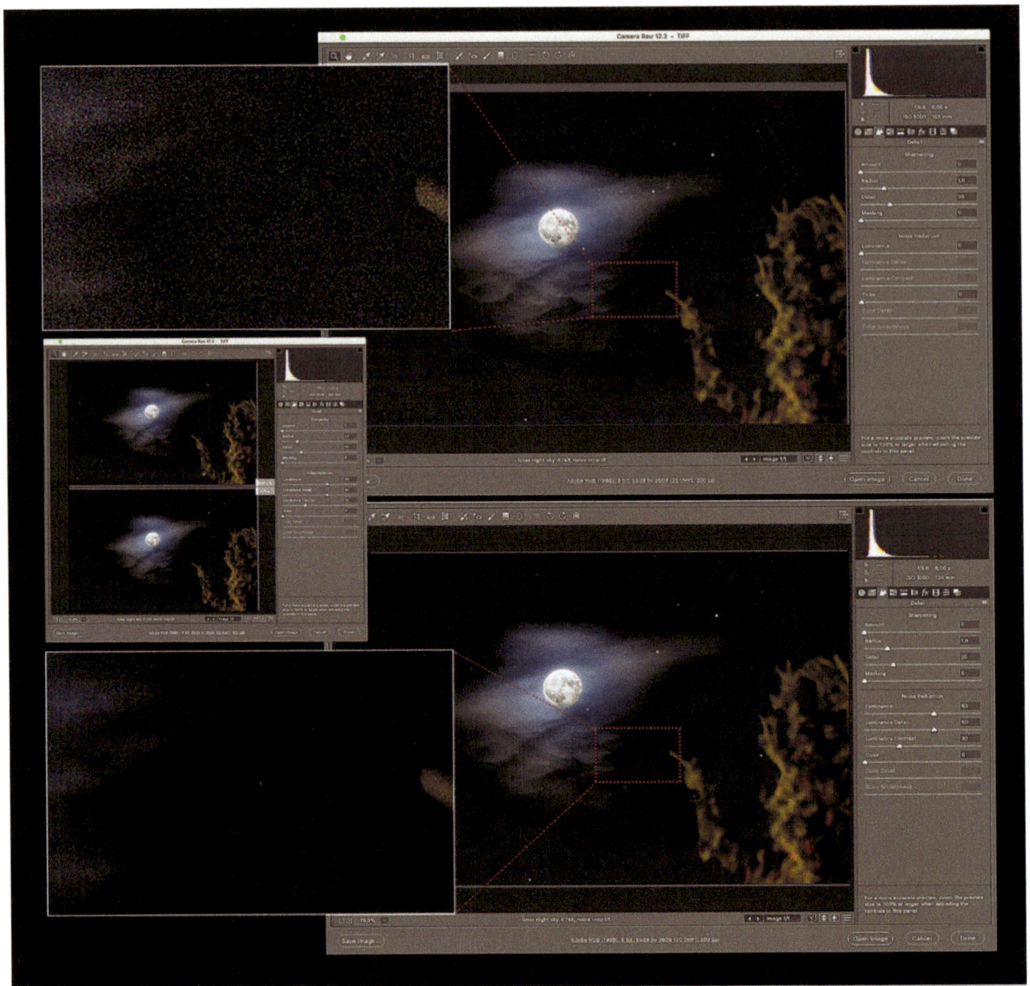

Figure 19.17 This illustration demonstrates how to minimize the presence of noise in a file using the Adobe Camera RAW file convertor. Notice the number pixels containing noise in the top view and then how, when processed using the tools in the RAW file converter tool, the noise can be suppressed. In the RAW file converter window, it is possible to open up smaller windows to compare "before and after" noise reduction. Adobe product screenshot(s) reprinted with permission from Adobe. Image courtesy of Staffan Larsson.

Despeckle filter evaluates edges in a file where significant color change occurs, and it blurs the overall file except for those edges.

Blurring a file removes noise and preserves detail. You can use this filter to remove banding or visual noise that results from scanning magazines or other printed materials. The Despeckle command has no dialog box and no options to choose from so it will not be particularly precise. It can be useful to apply the filter consecutively two or three times in a row.

Noise Reduction Using the Smart Blur Filter

You can also try to Smart Blur if there is more challenging noise. This works for only 8-bit color mode files. Select Filter > Blur > Smart Blur filter, and in the dialogue box enter fractions for both the Radius and Threshold values to control the amount of blurring that will be visible. Try different settings in Quality and Mode to achieve the best result.

Steps for Using the Reduce Noise Filter – Eliminating Luminance Noise and Chroma Noise

The Reduce Noise filter does a very good job of neutralizing the random red, green, and blue pixels while preserving detail in the image. Image parts remain sharp while eliminating RGB noise.

Select Filter > Noise > Reduce Noise. The Reduce Noise dialog box contains a large preview window on the left. The preview option box should be checked. Inside the preview area, you can zoom in and out by clicking plus (+) and minus (−), and navigate around the image using the mouse. 100 percent is recommended to start with. There are various options within this tool:

- Strength and Preserve Details are used to remove luminance noise.
- Reduce Color Noise reduces color noise.
- Sharpen Details – set to 0 percent and do not sharpen using this feature.
- Remove JPEG Artifact option for reducing the appearance of jpeg compression artifacts.

Noise Reduction and Removing Color Noise

Reduce Color Noise is often most obvious and an easy place to start. Use the preview window and zoom in on an area that contains color noise. Check for red, green and blue dots. Set the Reduce Color Noise option to 0 percent, then slowly move the slider towards the right, until the color noise has blended into the image and the noise disappears. Do not move the adjustment too far. Keep an eye on the preview area when judging the result. Click in the preview area and check the result before and after. A value of approximately 60 percent is a good place to start. Release the mouse button to see the final effects of the Reduce Noise filter. The final value will depend entirely on the amount of noise in the original image.

Noise Reduction for Luminance Noise

The next step is to reduce the luminance noise. Compared with color noise (different colored dots) luminance noise contains dots/speckles of varying brightness levels (black, white, and gray). Removing luminance noise is a three-step process.

1) Start by setting the Strength value to 0 percent. Use the same procedure that was used to reduce color noise. Do this until as much of the luminance noise as possible is removed. Click in the preview area and check the result before and after.
2) Next, begin moving the Preserve Details slider towards the right to bring back as much image detail as possible without re-introducing the noise.
3) The last step is to remove as much noise as possible with the Strength slider, then bring back detail with Preserve Details.

Advanced Options for Noise Reduction

If you think too much detail was lost using the Basic options, try the Advanced options. Sometimes one channel contains more noise than the others. Frequently the noise is in the Blue channel. Reviewing the channels will allow you to determine where to concentrate your efforts based on the determination of which is the noisiest channel. It is best to leave the other cleaner image channels alone. Perform the same procedure in each RGB channel if needed. Start by setting the Strength slider to 0. Select Advanced in the top right of the dialog box, then click on the Per Channel tab. Work through one channel at a time using the Channel option and inspect each channel for noise.

Channel options Strength and Preserve Details sliders work in the same way but they affect only the selected channel. Set the sliders to 0. Start moving the Strength slider towards the right to remove as much noise as possible in the channel. Click in the preview area and check the result before and after. When you've removed as much noise as possible, begin adjusting the Preserve Details slider towards the right to bring back some image detail without bringing back too much noise.

Apply different amounts of noise reduction to each channel to achieve the best result. The next step in noise reduction is to switch back over to the main controls in the Overall tab. Fine tune the Strength and Preserve Details sliders, trying to achieve even better results. When you're done with the individual channels, fine tune things with the Overall sliders.

Noise Reduction Removing JPEG Artifacts

Some images will appear as though they have been divided into a pattern of pixels or squares. The visibility of artifacts depends on how much compression was applied when forming the initial file and how many times that file has been compressed. The image size also influences the amount of JPEG artifacts. Large file sizes will have less JPEG artifacts, and small sizes, more. To reduce the appearance of the compression artifacts, select the Remove JPEG Artifact option.

Noise Reduction using the Camera RAW Converter

I recommend you remove/eliminate the noise using the Camera Raw converter. I suggest this because you will have access to all of the image adjustment tools in the Basic settings such as white balance, contrast, saturation etc. The Detail tab includes Noise Reduction and Sharpening. You also have Tone Curve and HSL (Hue, Saturation and Luminance) on a different tab. Note: These steps will work on all versions of Camera Raw.

Steps for Using the Camera RAW Pre-Processor Software Module

Open the file in Photoshop and choose Filter > Camera Raw Filter. You're going to apply a dynamic Smart Filter. To take advantage of this, you'll convert the image to a Smart Object in the following steps:

1) Make a copy of the background layer. Double-click the copy layer to convert it to a floating layer. Give the copy layer a new name and click OK.
2) Right-click in the image window and choose Convert to Smart Object. Choose Filter > Camera Raw Filter.
   a. Using CS6 or earlier, begin by opening the TIFF or JPEG image in Camera Raw. Go to File > Browse in Bridge. When Bridge opens, right-click the image and choose Open in Camera Raw.
   b. As a side note, it is possible to open TIFF or JPEG files directly in Photoshop Camera Raw by changing the settings in Preferences. To do this, go to Photoshop Preferences and select File Handling > Camera Raw Preferences. At the bottom of this window you will find JPEG and TIFF Handling. Choose Automatically Open Supported TIFF. Do the same step with JPEGs. Next time, TIFF and JPEG files open up in Camera Raw.

Camera RAW Image Processing

Main Window Tab: Location One

Image controls for adjustments in exposure, temperature, tones, and color channel etc. Produces image adjustments for optimizing what works best with a particular image.

Detail Tab: Two Triangles Icon

Sharpening and Noise Reduction Sharpening Selection

When a raw file is opened in Camera Raw, you may notice some default settings are already applied to an image, for example sharpening. Sharpening is the opposite of noise reduction. Sharpening enhances the edge contrast and texture detail of file, which also enhances noise that might be present. To improve the noise reduction, decrease the sharpening and noise sliders to zero (0 percent) present in the Detail tab in Camera Raw. Sharpening should always be the last step of an imaging workflow.

Noise Reduction Selection

Image noise includes luminance (grayscale) noise, which makes an image look grainy, and chroma (color) noise, which is visible as colored artifacts in the image.

Note: When performing noise reduction adjustments, first zoom in on the preview image to 100 percent to see the noise reduction previewed.

Noise Reduction adjustments sliders present in the RAW filter convertor detail window. These include:

- Luminance: Reduces luminance noise.
- Luminance Detail: Controls the luminance noise threshold. Higher values preserve more detail but can lead to noisier results. Lower values produce cleaner results but also remove some detail.

- Luminance Contrast: Controls the luminance contrast. Higher values preserve contrast but can produce noisy blotches or mottling. Lower values produce a smooth result but can also exhibit less contrast. Adjusting the Color and Color Detail sliders reduces chroma noise while preserving color detail.
- Color: Reduces color noise.
- Color Detail: Controls the color noise threshold. Higher values protect thin, detailed color edges but can result in color specking. Lower values remove color speckles but can result in color bleeding.

Using the Basic Panel Tools

1) It is useful to adjust the image color, tone, and contrast adjustment at this time and determine what improves an image's visibility. Set the Blacks slider to zero, revealing hidden details in the shadows. Moving forward, the next adjustments become personal preferences using any other Camera Raw adjustments as required. It is practical to make minor edits before further significant edits are performed. This workflow will minimize the enhancing of noise artifacts.
2) Go to the Detail panel and review the Noise Reduction options. Click the Detail tab and zoom in to 100 percent (in Windows, Ctrl + Alt + 0; using a Mac, Command + Alt + 0) to see the enhancements.
3) Luminance noise: First, reduce the luminance noise of an image. Luminance noise will be less distinct than color noise. Be careful not to over eliminate noise. which will create blur on detail. Increase the Luminance value until you see the noise drop out of the image. Evaluate the following settings: Luminance at 40, Luminance Detail at 50, leaving Luminance Contrast to 0 on an image.
4) Reduce Luminance Detail to remove any "pockmarking" that is visible.
5) Increase Luminance Contrast to regain any contrast that was lost when increasing Luminance.
6) Return to Luminance Detail and increase the adjustment until the pockmarking returns. Color noise: The last step is to reduce any color noise in an image. Color noise is generally more noticeable, so the Camera Raw default color noise reduction setting might be around 25 and not 0. A general rule is to match the Color noise slider with the Luminance setting.
7) Before sharpening, switch back to the Basic panel and locate Clarity and Vibrance. Increase the Clarity value to regain edge contrast. The Clarity slider adjusts midtone contrast. A good starting point is Clarity at +20 and Vibrance at +10. The Vibrance slider minimizes clipping when colors approach full saturation: by altering the saturation of lower-saturated colors without affecting the higher ones, more control is maintained.
8) Sharpening should always be the last step of processing.
9) Increase Sharpening to its maximum value.
10) Take Detail down to zero. Set the Amount to 40, increase the Radius to approximately 1.5. Never over sharpen an image. It can overdo things and add noise back into the file. When satisfied, click the OK button and complete additional image processing work in Photoshop.

Pseudo-Coloring Black and White Images

1) Determine the size of the black and white file to be colorized and create a new image document using the RGB color mode at the same resolution and size as the original black and white image. Give the file a name and Save As.
2) In the tool bar menu, select set the foreground color using the two squares near to the bottom of the tool bar. Determine the color for pseudo-coloring using the Color Picker tool. Locate the Paint Bucket tool, select that tool, and hover over the new document and click. The background should be filled with that color.
3) Make the black and white image file active by clicking on it and Select All. Next copy the black and white file. Paste the black and white into the newly colored document. This action will create a new layer.

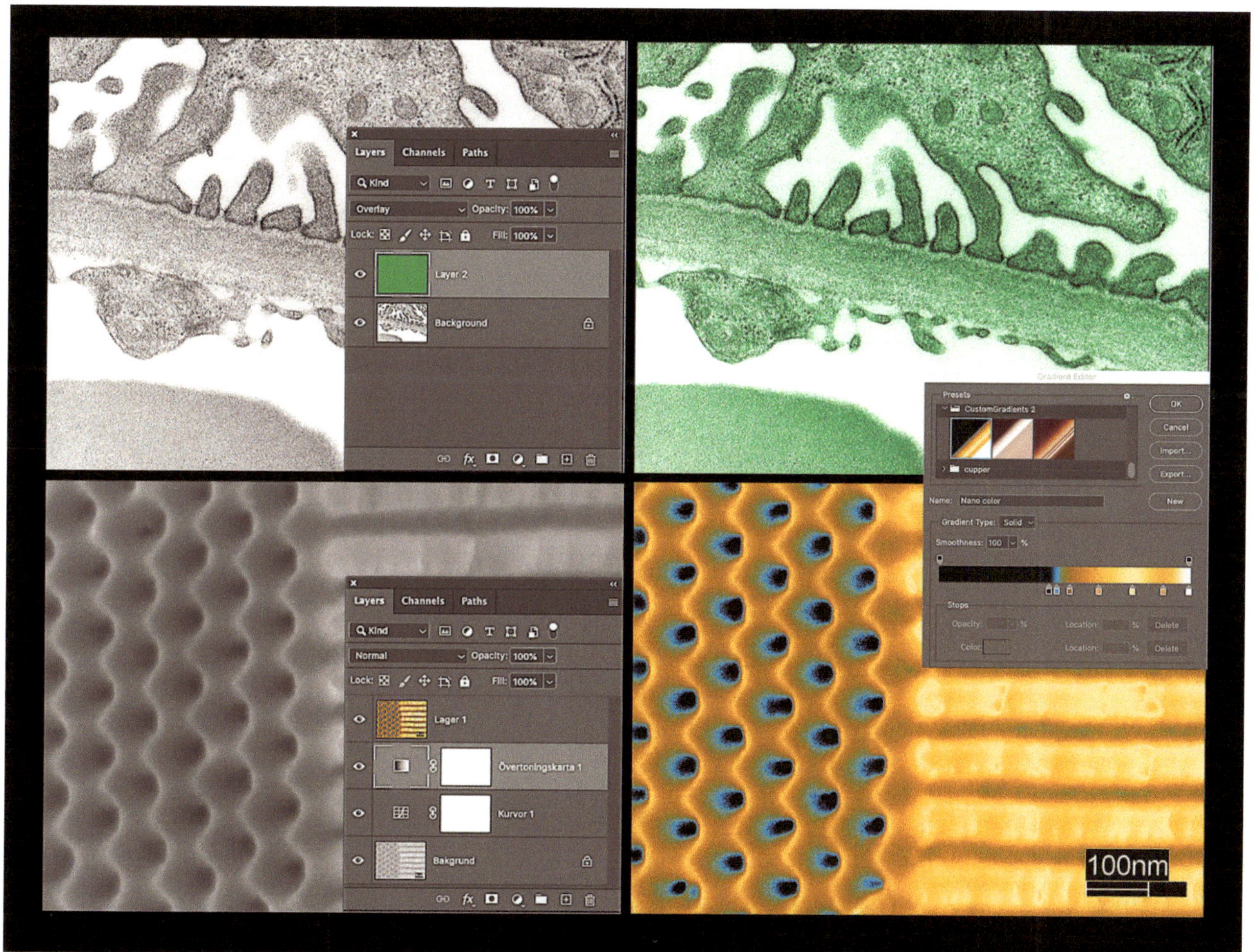

Figure 19.18 This figure shows how grayscale images might appear when pseudo-colored – sometimes called false colorized. On the left, a black and white image is converted to a green image that is more representative of the appearance of this stain. The two layers were blended using overlay. On the right is a more complex colorization using the gradient tool. Adobe product screenshot(s) reprinted with permission from Adobe. Images courtesy of Staffan Larsson.

4) The two layers must have their opacities changed to see the color. In the Layers action window, located near the top is a small dropdown menu with "Normal" displayed. This panel controls the blending of layers. Choose the Multiply option to blend the layers. When satisfied, go to the Layer pull down menu and choose Flatten Image.
5) Adjust the contrast and other image attributes as needed. Complete the image processing by sharpening.

Using the Gradient Adjustment Layer for Colorizing

Click the Gradient Map icon in the Adjustments panel.

1) Choose Layer > New Adjustment Layer > Gradient Map. Click OK in the New Layer dialog box.
2) In the Properties panel, specify the gradient fill you want to use: to choose from a list of gradient fills, click the triangle to the right of the gradient fill, then select the desired gradient fill, and then click in a blank area of the Properties panel to dismiss the list.
3) To customize the gradient, edit the currently-displayed gradient fill. Click the gradient fill, and then modify the existing gradient fill or create a gradient fill in the Gradient Editor.
4) By default, the shadows, midtones, and highlights of the image are mapped respectively to the starting (left) color, midpoint, and ending (right) color of the gradient fill.
5) To choose a color, do one of the following:
   a. double-click the color stop, or click the color swatch in the Stops section of the dialog box. Chose a color, and click OK.
   b. Choose an option from the Color pop-up menu in the Stops section of the dialog box.
   c. Position the pointer over the gradient bar (the pointer turns into the eyedropper), and click to sample a color, or click anywhere in the image to sample a color from the image.
6) To define the ending color, click the right color stop under the gradient bar. Then choose a color.
7) To adjust the location of the starting point or ending point, do one of the following:
   a. Drag the corresponding color stop left or right to the location you want.
   b. Click the corresponding color stop, and enter a value for Location in the Stops section of the dialog box. A value of 0 percent places the point at the far left end of the gradient bar; a value of 100 percent, at the far right end.
8) To adjust the location of the midpoint (where the gradient displays an even mix of the starting and ending colors), drag the diamond below the gradient bar to the left or right, or click the diamond, and enter a value for Location.

Creating Composite Images for Publication

Including images in a journal article remains one of the more expensive elements of publishing. Once an article has been accepted by journal, the acceptance is typically based on allowing only a certain number of images to be included with the article. Additional images can be included, but often the author must pay to include the extras. One way to

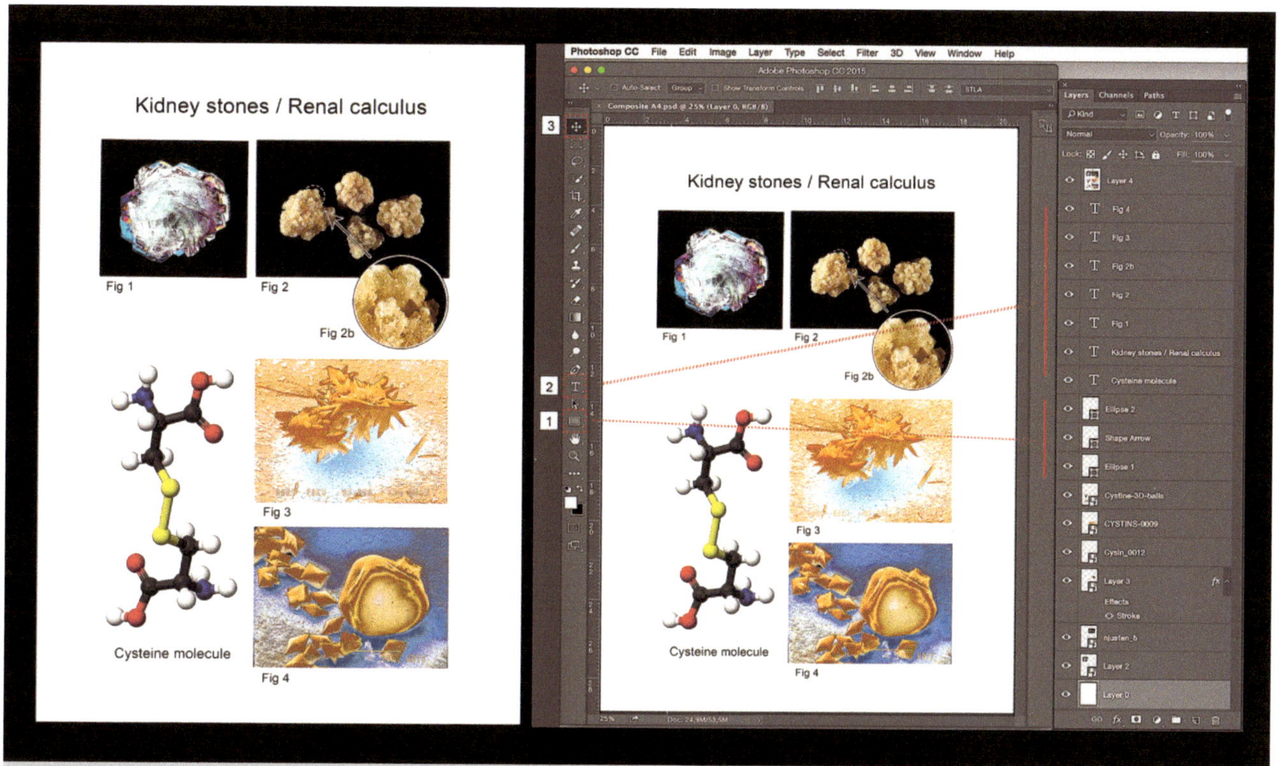

Figure 19.19 This composite reveals the numerous steps and components required to make the graphical image on the left. Highlighted with red lines are the various tools and the layers that indicate where they were applied. Images courtesy of Staffan Larsson. Adobe product screenshot(s) reprinted with permission from Adobe.

resolve the dilemma is to create a composite image that includes multiple images. Making an educational point using a "single image that has multiple frames" is an excellent way to provide readers with a more complete experience with the material for less money.

Composites are relatively easy to assemble in Photoshop or any image-editing program. It is sometimes useful to include a variety of views that provide an overview image and then several close-up photos that reference details for example. Labels, arrows, and frames can also be helpful to complete the composite. You can create a composite using Photoshop in several ways and the following steps are used as an example.

Steps for Making a Composite

1) Before starting to make the composite, it can be helpful to add extras to the workspace and use the toolbar that displays grid, rulers and guides in the display window. To do this go to View > Rulers and, if you want a grid, View > Show > Grid. You can turn it on, or off by selecting Extras located under the View menu. (You will find the setting of the Grid in Photoshop > Preferences > Guides, Grid & Slices, and can save this as a Preset. To work in Guides, use the Moving tool, go to and touch the upper or left rulers, press the left mouse button and drag a line or add more into the working space. To turn on or off, go to View > Show.

2) Next make the file or background that will become the composite's workspace. To do this choose File > New to create a new image. Specify the image size based on the desired print size. For this example, we will create an 8 × 11.5-inch file. To start, go to File > New. In the dialogue box you need to set the new file parameters. Set the width and height to 8 × 11.5-inches. Set the resolution to 300 pixels/inch, and the color mode to 8-bit. Set the background to white. Set the Color Profile to Adobe RGB (1998). It is recommended to show the grid when making a composite for alignment.
3) Next, open and prepare the images you want to insert into the composite. Size, resolution white-balance and contrast should be adjusted and confirmed to have the correct file attributes shared above with the exception of image size. It is important to use unsharpened images. If the desired images for the composite differ in resolution or pixels per inch, they will be displayed in different sizes in the composite image. There are several ways to resize an image and a method you may be familiar with is shared below. Maintain or confirm each file's color space.
4) It is important to have all the images preselected or organized in a folder or some other location. Open each of the files prior to starting to build the composite and determine the file attributes that were shared above. Depending on the number of images that will be included in the composite, assess each image's size. It is imperative to have a plan and frequently it is best to use fewer and bigger images than more and smaller images to make a point. If, for example, four images will be composited, it is easy to see why each file should be 4.25 inches wide and 5.5 inches tall if they are to be the same size. It is also possible to scale images directly into the new document. Something to also consider is whether to include white borders between the images or not. If the files have been resized, they can either be Saved As or the files can be kept open for use in the next step. If you're going to resize the files in the new document, see the method below. Add each new image to the composite image as a separate layer. This process will allow you to re-position and adjust each image individually throughout the compositing process. When all of the images have been assembled into the final image, you can resize the composite file, or move the layers (the added images). If the Layers palette does not appear as a window, it can be made visible by selecting Window > Layers
5) You can add the images to the composite in several ways. It can be practical to limit the number of images used in a composite. It is important to keep in mind that the more photos that are included, the more difficult it will be to see the image detail because the images will become smaller images in the preview window. To add the images, choose File > Open, and open the images to be added to the composite. The composite background image and all of the images should be open at the same time.
6) The first image that will be added to the composite needs to be copied and moved into the new image. To do this, go to Select > All; then Edit > Copy; and then Edit > Paste into the new file. The copied file will create a layer in the composite file.
7) Another way to move a file into a new file can be accomplished by dragging that individual file into the composite file workspace in the following way. Select the Move tool from the toolbar. Click and hold the cursor anywhere in the image that is to be brought into the composite and holding the mouse down, drag the first image into the composite image window. Once located over that space, release the mouse button. The image should be displayed in the composite image window as a new layer. Bring the remaining images into the composite at this time.

8) It is very practical to label the layers to help with future edits of the file if needed. To label a layer, Double-click the word Layer 1 in the Layers palette, and rename each layer as needed to track your layers.
9) To all add images one at a time. Go to one of the open image files and, using the Marquee tool, Select All (Select > All), and copy (Edit > Copy). Click into the new document and Edit > Paste. This will bring each photo into the new document in its own layer.
10) If the image resolutions of the photos that were imported to the composite need to be adjusted, it is possible to resize them in the composite in the following way. First select the image layer that will be resized. Choose Select > All. Then go the Edit menu and select Free Transform. Once made active, using the left corner node of the selected file and, while pressing Shift, adjust the scale of the image until it sized correctly to be used in the new document composite. When finished, double-click the image. It remains possible to resize the images prior to importing them in the composite as well. If you convert the image layer to a Smart Object, an option located in the Layer menu, it will maintain the integrity of the image quality as you size up and down.
11) Close the original image windows so that only the composite image file remains open. Your Layers palette now contains a list including a Background layer and the new layers that may have been renamed. Select a layer and reposition it as needed using the Move tool. It is possible to rotate images as well in the following fashion. In the Layers palette, click once on the layer that you want to adjust, to target the layer. Choose Edit > Free Transform. Be certain that the correct layer is targeted before you transform a layer in the Layers palette.
12) Once the image is selected, under the Edit menu, select Transform. Note the anchor points on all four corners and sides (eight in total) around the edges of the bounding box of the layer. You can resize the layer by dragging the anchor points and then click inside the bounding box and drag the layer to be repositioned. If a layer displays larger than the collage image, drag the layer in any direction until you can see a corner of the image. Use the anchor point on the visible corner to transform the image.
13) Rotate the layer by positioning your cursor just outside the bounding box, clicking, and dragging. You can rotate the layer when your cursor changes to a curved double-headed arrow. In the option bar at the top of the screen, click the tick to accept the transformation, or click the circle with the slash through it to cancel the transformation. The layers can be reordered if needed. To change the order of the layers or the way the layers overlap, select a layer and drag the layer in the Layers palette above or below the other layers as needed.
14) When you have added all images that you want to include, it's time to make the fine adjustments for placement of the individual components. It can be helpful to have borders between images to give the images some space. The amount of white space is governed by personal preferences. To move images, select the image layer to be moved and using the arrow keys, move the image in the direction that is wanted. Using the arrow keys to allow the spaces to be carefully managed across all of the photos.
15) You can align up vertical and horizontal positions using the guides. When the Move tool is selected the align control panel appears at the top. To select the image you want to align, hold down the command key (Mac) or ctrl (Windows) and select the layers with images to line up to the left row and then the images to the right row. Do the same procedure for the upper row and lower row.

The Type Tool

Before using the Type tool there are a few features to become aware of. The Type tool menu will display most of the tools and can be seen by clicking the Type tool in the Tool panel. In the workspace window text options will be shown in the header.

- Horizontal and Vertical Text display
- Font size: there are number of presets of font sizes, but you can also type any number size in here.
- Anti-aliasing options
- Alignment of the text
- Text adjustments pop out. More extended options There are also extended characters and a Paragraphs panel. Go to Window > Character Styles and Paragraph Styles.

Go to the Tool panel and choose T as in Type tool. There are two options. There is a Horizontal Type tool or a Vertical Type tool. If you start using the Horizontal Type tool you can always revert to vertical text. Go to context menu on top of the software window and click the switcher to the left or go to Type > Orientation.

This will be visible as a text layer. There are two ways of adding text to the file. The first and most common way is to use the Point text tool. Start to click on the Text Tool in the Tool palette, click back on your image and start typing. The second way is to use Paragraph Text.

The Point Text Tool

Start typing letters and your text will have an underline and a cursor will show where you are inserting the next letter. When you keep on typing, the text will continue along in one long single line. To get text on a new line, press Enter.

The Paragraph Text Tool

Select the Text tool and drag it out and make a rectangle. Once the box is formed begin typing the text. The text will be constrained by the size of the box. Once the typing is completed, you can grab the text box handles (the little boxes on the middles and corners) and define the size of the area in which you want the text to appear. This can be important for large blocks of text. The Paragraph Text tool will provide more flexibility because of the ability to change the size of the text box.

Once the text has been created, the Move tool can be used to position the block of text where desired. A paragraph text block is easy to center on a file. To center the text, grab one of the handles and drag it to the right edge of your page. Drag the left box to the left edge of your page. Go up into the context menu at the top. Click the Center Text icon. The text will now be centered horizontally on your document.

Editing Text and Transforming Letters

It is possible to change the spacing of the individual letters, change the height of the letters, or the width of letters. It is also possible to change a block of uppercase text to lowercase, or vice versa. To make these changes, activate the Text layer. Double click the T icon and the text box is open for editing, such as changing type, size, colors, subscripts, switch between all caps and small caps, style, and much more.

Rotating the Text Block

Sometimes is might be desired to rotate a text block. To rotate text, use Move tool. Select the text layer and click. Grab a handle one of the corners and rotate. You can also use Edit > Free Transform, and you can also scale up or down.

Shapes

Basic Shapes

Creating circles, rectangles, borders, backing boards, arrows, and frames can provide more impact to a composite. Photoshop offers basic shape tools for working with images. An easy way is to create rectangles and circles. First create a new empty layer. Go to Marquee tool. Select a rectangle or a circle. Drag a rectangle to a suitable spot. Notice the marching ants surrounding the marquee. You can adjust the form later using Free Transform.

Go to Edit > Stroke. In the Stroke window you have choices to fill in. Width: fill in the Stroke width in pixels. It might be useful to make a few tests first. Color: click in the small square and the Color Picker opens up. Drag the small handle and refine the color scale and choice. Stop at the color you want. Go to the shade window and move the small circle with your mouse. Stop when you can see the color you want in the small right window. Press OK. Location: click Outside and click OK. An unfilled rectangle with a border in your chosen color will be created. If you want at wider stroke, go back and add a larger pixel size. If you want to fill it with another color, keep the marching ants, then go to Edit > Fill. In the Fill window, go to Contents and click. You can chose between default black, gray, and white. At the upper part you find Color. To choose the fill color, do the same as you did to color the Stroke. To scale, rotate, or transform the shape use the Move tool. Select the layer and click. Grab a handle on one of the corners and rotate. You can do the same action with Edit > Free Transform.

Custom Shapes

Photoshop features some basic shape tools that can be used when working on photographs or artwork. Rectangles, rectangles with rounded corners, circles and ovals, multisided polygons, straight lines and arrows, and a number of other custom shapes are available to use by clicking and dragging controls.

1) To use a Shape tool, select the appropriate Shape tool or a Pen tool from the Toolbox. If the Pen tool is selected for use with a special pre-designed shape, be certain that the specific shape is selected from the menu in the options toolbar. Select the desired shape from the Options toolbar, and then click and drag to create the shape in the document. It is easy to change the size and appearance of the shape, too. If you know where you want the center of the shape to be located, position the pointer at that location and

draw the shape by clicking and holding the mouse while moving outward from the center and pressing the Alt (Windows) key or Option (Mac) key. The path of the drag using a diagonal direction towards a corner or edge of the frame will affect the shape of the shape. Once satisfied with the size and shape, stop dragging the edge of the shape. It will take a few practices to get the feel of the tool.

2) You can maintain the proportion of a Shape by pressing the Shift key (both Mac and Windows) while you are dragging the cursor when making the shape. This action will maintain the width to height ratio. With the Shift key held, the Rectangle tool creates squares; the Ellipse tool creates circles; the Polygon tool creates proportional polygons; the Line tool creates horizontal, vertical, or diagonal lines.
3) In the Option toolbar or in Properties window, double-click the Shape layer or go to Window > Properties you find the options. To choose the color of the shape, click the color swatches for Fill and Stroke in the options toolbar, and then choose a color from the Color Picker. Once the color has been selected, and next to the right of the color swatches, the stroke width option, and width and height for more exact sizes of the shape can be located.
4) When using custom shapes, it is possible to press the Shift key, which ensures that the shape will maintain a specific height-to-width ratio. More about Custom Shape can be read in step 15.
5) The Option (Mac) or the Alt (Windows) key will create an object that is centered on the point where a mouse is clicked. Without using the Option/Alt key, the object will be created in whichever direction you drag. Pressing Shift and Option/Alt key together helps you create a proportionally constrained object, centered on the point at which you click.
6) To constrain a rectangle or rounded rectangle to a square, to constrain an ellipse to a circle, or to constrain the line angle to a multiple of 45 degrees, hold down Shift when using the tool.
7) If a shape tool is selected and you click rather than drag the tool, you will open a small dialog box that allows you to enter the exact dimensions of the new shape. Click the OK button once the shape has been created to the lower right of (or centered on) the point where you clicked. The dialog box is visible in the figure.
8) When you are dragging a shape, keep the mouse button down and press the spacebar. You can then drag to reposition the shape while you are creating it. Continue to keep the mouse held down, release the spacebar, and finish dragging the object.
9) In the Options toolbar various properties become visible when switching shape tools. The Options bar can allow characteristics of the various shapes to be managed. For example, with the Rounded Rectangle tool active, you can choose the radius of the rounded corners. The Polygon tool offers a simple field in which you choose the number of sides for the shape. When you're using the Line tool, choose the thickness (weight) of the line in the Options bar. It is possible to add arrows to lines by clicking the button to the left of the Weight field in the Options bar.
10) It is possible to change the layer content. With a shape layer selected in the Layers panel, select any shape tool and change the shape's attributes in the Options bar and Properties. You can change or remove both the fill and stroke. If you need to adjust the line for example, use the Direct Selecting tool (inverted arrow) and click

on the line and the Options bar will open. The shape options will now be visible in the Options bar.

11) You can create your own designed shapes with the Pen tool making work paths. Choosing one of the three options in the menu on the left in the Options bar, you can elect to create shapes, work paths (temporary paths used to make selections or masks), or add pixels in the selected shape to your currently active layer. This is done by editing the vector path. You can use the Direct Selection tool to change the course of the path, customizing the appearance of the shape.
12) It is easy to find the Shape layers in the Layers panel. The default name of the Layer is the name of the Shape (Circle 1, Rectangle 2, etc.). You can change the layer name by double-clicking it in the Layers panel if it is useful to do so. In the Layers panel you can see the Shape layer thumbnail includes the shape badge in the lower-right corner. When a Shape layer is selected in the Layers panel, that shape's path is visible in the Paths panel.
13) To scale, rotate, or transform the shape, use Direct Selection tool, Move tool, or select the layer and click. Grab a handle on one of the corners and rotate. You can accomplish the same action by using Edit > Free Transform.
14) The Custom Shape tool has additional features that are different from the basic shape tools. The Custom Shape is located as the last choice in the Shape sub-menu. With the Custom Shape tool selected, go to the right of the Options bar to open sample shapes in Custom Shape picker. The Custom Shape picker offers a number of ready-to-use shapes. In the latest of Photoshop version 2020, there are fewer ready-to-use-shapes, but they are available as free downloads on the Internet. You can find many different shapes. When you click the Gear icon in the upper-right corner, you can choose from a variety of other sets of shapes listed at the bottom of the Custom Shape picker menu.

The Custom Shape Option bar is similar to the basic shapes. You can find and click on the Fill field and the Stroke field to open the options and customize the shape. When you select Custom Shapes from the menu (or use the Load Shapes command to add a set of shapes not in the menu), you are asked whether you want to add the new shapes to the current content of the Custom Shape picker (Append), replace the current shapes (OK), or not add the shapes after all (Cancel). If you select All from the Custom Shape picker menu and click OK, all shapes will be loaded.

Preparing Files for Publication

There are a number of things that can be done when preparing files for printing, which can improve outcomes and meet expectations. Soft proofing or making a virtual image proof is one way to predict how an image will appear in specific conditions when displayed on a computer screen. This simulation provides a visual display that emulates how the image will look when printed on a specific printing press or when using a specific inkjet printer. "Hard proofs" are generally physical prints made from a calibrated inkjet printer and are not the same as soft proofs, which feature electronic display only.

Sometimes when an article is published, authors receive a surprise when, upon opening the journal for the first time, they see the colors of their images on the printed pages can appear different to what was expected. Typically, the images do not display the same color saturation or contrast when compared to the images that were prepared on the computer, and that were observed on the monitor. This difference in this outcome often is the result of a monitor's color space being RGB (Red, Green and Blue), which has a much wider color space (gamut) when compared to a CYMK (Cyan, Yellow, Magenta and "Key" Black) color space used for printing.

Converting RGB to CMYK

When the RGB colors are converted into CMYK colors, many highly colored images will lose saturation and contrast. The expression "what you see is what you get" is not always truthful. Because images can be important evidence used to support a research outcome, using soft proofing can assist scientist photographers in the optimizing of a file prior to submission for reproduction.

It is impossible to print a complete black (0) or white (255) in grayscale because of limitations of printing techniques. For this reason, when preparing files, initially it is useful to begin file evaluation by checking the image's Info panel when adjusting contrast in Curves or Levels. To produce a printable image, it is imperative to keep each of the RGB channels between 5 in the black and 240 in the white. Once established, these numbers can be translated into typical ink limits (max separation in black) for journals and magazines. A coated paper should be set to a range of 300 to 320 percent and newsprint, or uncoated paper to a range of 240 to 260 percent. More has been shared about this recommendation throughout the chapter.

Evaluating an Image's Black Point

Black is an important element in an image and ensuring it is properly displayed or reproduced is vital. To set a black point, first set a measuring point using the Color Sample tool located in blackest part of an image, or in the shadows. Go to Info panel and review the sum of the CMYK result. If it's more than 300 percent try to bring down the total (C X% +M X% +Y X% +K X% = X%) using the Levels or Curves.

There are a numerous standards that have been created to assist in achieving good printing outcomes. One of these standards is the FOGRA39, which is a characterization data set, developed by the German graphic technology research organization, FOGRA, in 2006. This standard was registered in the ICC Characterization Data Registry. It represents the colors used in a typical print made by commercial offset litho press, or onto gloss or matte coated papers that have a maximum ink coverage for the separation of 300 percent.

A color press uses a combination of four printing plates that represents each color featuring Cyan, Magenta, Yellow, and Black (CMYK). The color reproduction is built up by the CMYK pattern or screen dots that are laid down using different sizes. The spaces with very small clusters of mixed CMYK dots will display the white highlight and when the color dots are larger, they represent variations of colors depending on the CMYK mix. As the dot size grows, it continues into the blacks or shadows of an image.

A good quality desktop monitor will assist you in preparing for better output files. Calibration is required. A monitor's default is generally sRGB, which is smaller than Adobe RGB (1998) but can be adequate for basic image output for video display. For print, it is important to set a monitor's color display to Adobe RGB 1998. Today, it is possible to have monitors with a large color space such as Adobe RGB (1998) but frequently these monitors will be expensive and are mostly used by dedicated high-end professional users.

Monitor Settings

An ordinary LCD monitor typically will have a default color temperature of 6500K and only a few will have the capability to make adjustments. If a display with a 6500K color temperature is used for image processing, the results will lead to incorrect colors in the publication. A more standard color temperature for color reproduction is 5000K–5500K. Calibrating the display will help to ensure that colors are represented more accurately when anticipating the reproduction gamut. Be sure when buying a new monitor that at least brightness, color temperature, gamma, and color gamut can be adjusted. A basic monitor calibration was shared previously in the chapter.

Monitor Profiles

Another trick can be when launching Photoshop for the first time, set the standard color space in Preferences. This will remain important for future work and will allow you to have control of the color space for input and output settings from the beginning. To do this, go to the Edit menu and select Color Settings. You can choose from a variety of pre- sets or you configure your own and save it. Depending if you work in the U.S. or Europe there may be slightly different settings. The most common standard settings used with RGB both in U.S.A. and Europe is Adobe RGB (1998). When using CMYK and in the U.S., U.S. Web Coated (SWOP) V2 is the common default and in Europe, Coated FOGRA39. They are also called ICC profiles. There are many specific profiles available. As an example, select "Europe Prepress 3" pre-setting in the settings at the top to the left. (U.S.A. usually uses the similar pre-set North America Prepress 2.) Then set RGB working color space to Adobe RGB (1998) and the CMYK working color space U.S. Web Coated (SWOP) V2). The "Europe Prepress 3" pre-set sets your RGB working color space to Adobe RGB (1998) and the CMYK working color space to Coated FOGRA39. In older versions of Photoshop, it is called Coated FOGRA27 but can be replaced with Coated FOGRA39 through manual selection. You can find this profile easily on the Internet if it is not installed on the system. sRGB with its smaller color space is recommended for Web pages and in PowerPoint and MS Word use. When you are preparing to use an inkjet printer Generic ICC Profiles have been automatically installed when the appropriate printer software has been installed. Select the profile that conforms to the paper of choice.

If you use Adobe Bridge to manage your Color Settings, it will also synchronize and matched the settings in all Adobe CC/CS applications. Find the setting in Bridge: Edit > Color Settings, then select and choose your color space. Continue to the final settings in the Photoshop Color Settings. Select Preserve Embedded Profiles in Color Management

Policies as a recommended suggestion, which allows you to open a file with an embedded ICC-profile. If you are given an image with an sRGB color space (rather than Adobe RGB) leave as sRGB until it is converted to CMYK. If you need to increase the saturation of color, converting the image to Adobe RGB 1998 will be required. When editing images in Photoshop, stay in RGB mode for as long as possible! It is always best to stay in RGB mode when color-correcting an image. When your image has been color corrected, save a master copy in the RGB mode. If needed, you can later modify the file once the maximum color information has been processed. The last step in the workflow should be the conversion to a CMYK profile.

Proofing

It has been stated that an image can change a lot in its color when converted from the primary RGB image to the printed CMYK image. To avoid having to make color adjustments after the image has been converted to CMYK, be sure to proof and make sure that your proof settings are correct. This can minimize this outcome.

To soft proof a file, Go to View > Proof Setup > Working CMYK or select another profile if you have a preference. Working on CMYK files will be similar to your Photoshop settings. There are a lot more choices of profiles available that are designed for print. There are ICC profiles that are made for specific inkjet printers when used with different papers. In the option Custom (Customize Proof Condition) you can customize and design your own settings. Find your ICC profile in the window Device to Simulate. In Rendering Intent chose Relative Colorimetric (for images). If the file will be printed on simpler paper quality, or includes illustrations, use Perceptual and select Black Point Compensation. There can be two useful options for use when selecting proof settings that are found under Custom. Simulate Paper Color and Simulate Black Ink options will reduce the contrast range of your file and can assist in creating a proof view that will be most similar to ink on paper. If desired, you can save the settings for future applications. Saved settings will appear in the Custom Proof Condition drop down menu at the top. To revisit this, go to View > Proof Setup, you will see the custom saved name at the bottom of the list and the image will be displayed using these Proof Colors.

Proofing Colors

To use proofing colors, go to select View > Proof Colors. Making this selection will simulate the result of converting an RGB file to a CMYK Profile. The feature is active when there is a check mark visible next to Proof Colors in the menu. When you are finished with the evaluation, it is best to turn it off. When turning proof colors off, the visible contrast range of the monitor will be extended. This is helpful when working on image detail. Using this tool will allow Photoshop to simulate on the screen how an image will look when printed. The proofed image may reveal a wide range from a small to a significant change, with some differences to be expected.

Compare the Image Before and After

To compare a result before and after proofing can be accomplished by duplicating the window showing the two images in a side-by-side comparison. To do this, click Window > Arrange > New Window displaying the desired file. Photoshop can display a duplicate window of the same file, but this is not a duplicate file, just a duplicate window. Arranged the windows so you can see them easily together. One image will be displayed normally, and the other image will reveal the Soft Proofed since the Proof Colors are active. You may see that the images are similar but there may also be some important differences. Some of the bright, saturated colors often will have shifted and the blue regions will have lost saturation and can become gray and magenta.

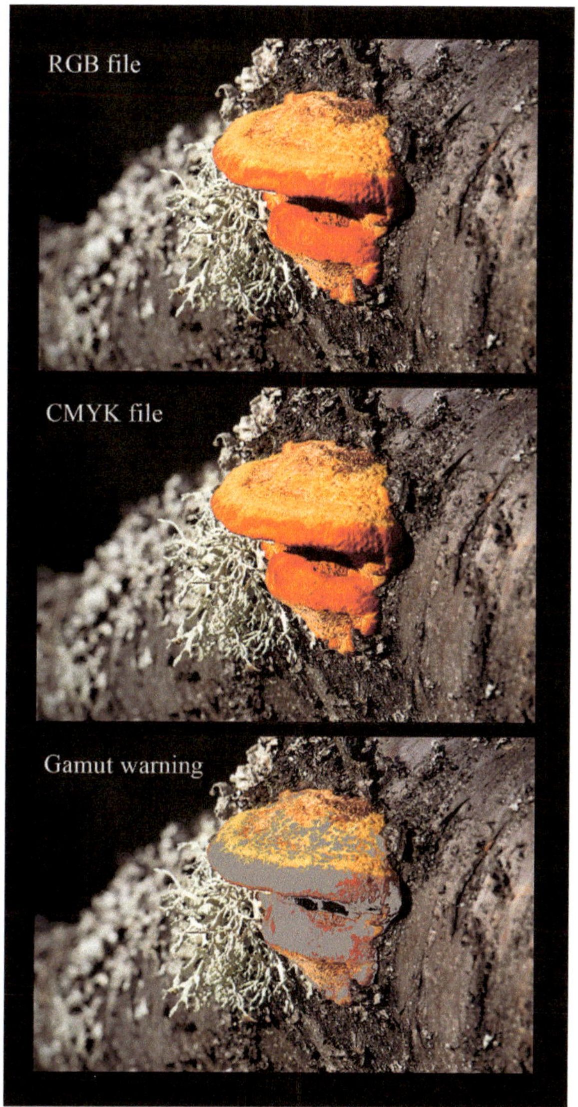

Figure 19.20 This illustration shares how RGB and CMYK files will appear when prepared for soft proof viewing. The gamut warning will be displayed as a gray shadow. The bottom image shows the area that will not reproduce in detail in print. Images courtesy of Staffan Larsson.

Gamut Warnings

Sometimes colors are difficult to display or print. This may be because the gamut is too large, and this can be predicted using the gamut warning. The definition of gamut is a range of colors and tones that can be captured and are available for a particular device such as a monitor or printer. A monitor, which displays RGB signals will typically have a greater color gamut than a printer, which uses CMYK inks and is limited by an ICC Profile. When the color in an image is "out of gamut," it cannot be properly converted for display or output to the selected device. To evaluate gamut warning, Select View > Gamut Warning. Using this function, you will be able see colors that are not able to be reproduced by the chosen device, and they will be displayed as gray. Remember, the device being simulated is configured under View > Proof Setup. Photoshop will reference the selected ICC Profile in this setup in determining the gamut of the output device.

Gray areas will be created in the gamut warning and will indicate all the colors that the printer will not be able to print. Typically, these will be colors that are saturated (rich/strong). The technical term for this situation is out-of-gamut. When the gamut (capability) of the printer is not large enough to handle these strong colors, Photoshop will provide a warning of this condition and hence the name gamut warning. Un-checking this feature (click on View > Gamut Warning again) removes the gray warning tint and the normal presentation of colors appears again. If Photoshop is displaying with Proof Colors (View > Proof Colors), you can see exactly how they will print.

Adjusting Out-of-Gamut Colors

There is a feature in Photoshop used to forewarn operators of issues that may arise when an image is outside of the print gamut, and it can be used to help with troubleshooting these problems. The subject in the image might have some colors that are simply too saturated for reproduction in print. It is the role of the ICC Profile to manage these colors and so you may choose to leave things as they are and go ahead with printing anyway. You may decide to make some manual tweaks to the file using the appropriate tools if you feel you can improve

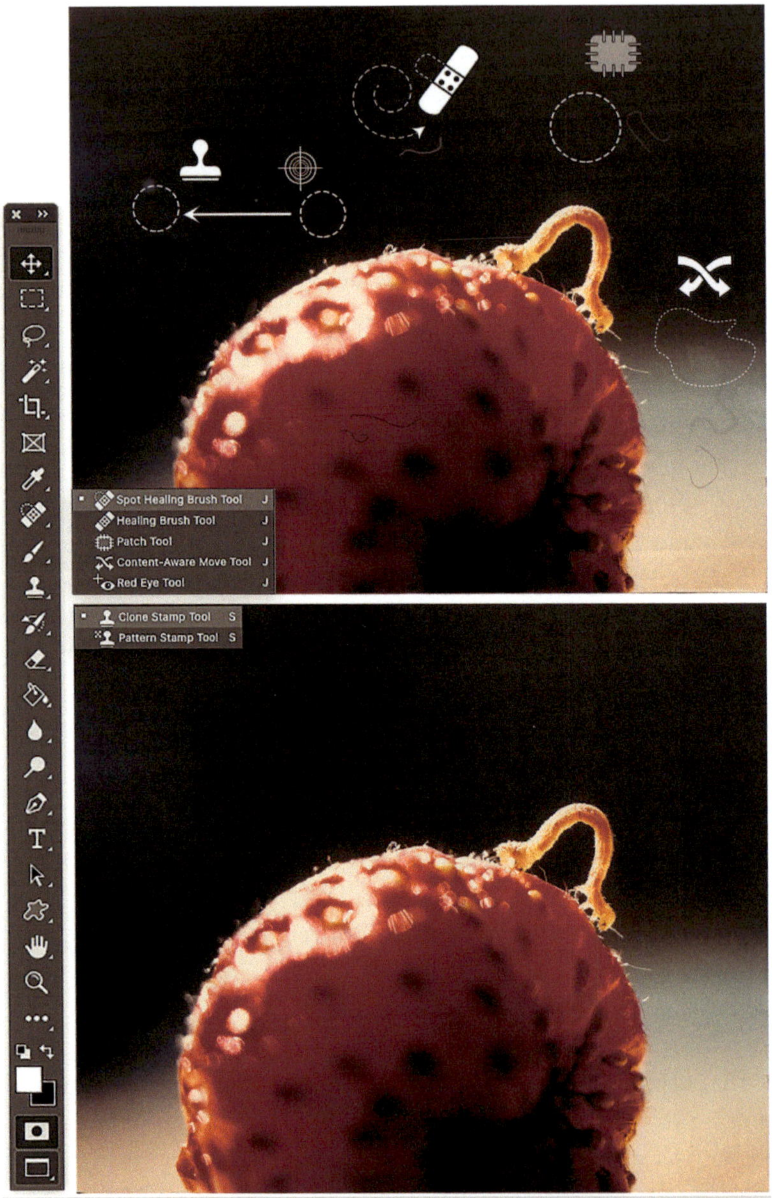

Figure 19.21 This composite image shows the before and after retouching of dirt produced from scanning film. Adobe product screenshot(s) reprinted with permission from Adobe. Image courtesy of Staffan Larsson.

the image. By turning on and off the Gamut Warning you can watch the gray areas disappear or reappear while making adjustments. Adjustment layers for Hue and Saturation, Curves and Levels can be flexible and should be your first choices. Selective Colors, Channel Mixer, and the addition of an Alfa Channel Mask can be used to cover sensitive regions for enhancing or decreasing color and these can be used as an alternative. Small gamut warnings can be ignored if some areas have turned gray, but they don't have a strong presence in the image.

Once a print has been made compare it with the image displayed on the monitor. As your eyes adjust to changes in lighting conditions, hold the print next to the monitor and evaluate for a good match. It can be important to evaluate the print under the right white light conditions. Humans will perceive light in the 5,000K–5,500K color temperature range as white. A light bulb with 5,000 K, CRI of 90 is recommending for a daylight-equivalent print-viewing light source.

Image Retouching after Scanning a Film Image

Sometimes dirt and debris finds its way into digital files. Scanning film pictures often creates this problem. There are two tools recommended for retouching dirt from a file: the Clone tool and the Healing Brush tool. Figure 19.21 shows screen captures of dust being removed from a file. To use these tools, pixels that will be used for replacing the dirt must be identified using the Selection tool. Care must be taken as to where the replacement pixels are selected from.

Suggested Reading

Adobe Photoshop Classroom in a Book (2020 release), by Andrew Faulkner and Conrad Chavez. Adobe Press, 2019.

GIMP 2.6 for Photographers. Image Editing with Open Source Software, by Klaus Goelker. Rocky Nook, 2011.

Image Processing with Image J, by José María Mateos Perez and Javier Pascau. Packt Publishing, 2013.

Index